MW00717155

Organizational Simulation

Organizational Simulation

Edited by

William B. Rouse

Kenneth R. Boff

A JOHN WILEY & SONS, INC., PUBLICATION

Published by John Wiley & Sons, Inc., Hoboken, New Jersey.
Published simultaneously in Canada.

Library of Congress Cataloging-in-Publication Data is available.

ISBN 0-471-68163-6

Printed in the United States of America.

10 9 8 7 6 5 4 3 2 1

TABLE OF CONTENTS

FOREWORD

Modeling and simulation are not new. They have a long history from before modern advances in mathematics, computers and displays. People have been dealing with creative visualization, wishful thinking, counterfactual thinking, and "what iffing" for centuries. However, we have not had efficient and effective means to develop immersive environments with different organizational approaches, chains of command, and best practices. We currently develop a vision and a policy that looks good on paper; we implement it, experience the consequences from the new policy, and then make corrective changes over time. ***Organizational Simulation*** provides an enabling toolkit for people to view, analyze, and try to understand a current organization through interactive simulation, model the changes to an organization as a result of design and policy changes, and ascertain in this synthetic environment what effects, both intended and unintended, are likely to result from these changes.

The Department of Defense (DoD) has long used modeling and simulation technologies to develop interactive "worlds" and games to meet a variety of goals and objectives. These objectives include training and education of personnel; preparing and refining war plans; evaluating new weapon systems during and after research and development; analyzing military strategies and doctrine and tactics; and determining manpower and logistic requirements. The many broad applications of simulation technologies are constantly used to represent quantitative or qualitative states that depict something that exists or could exist in the real world. Given such use, the Defense Modeling and Simulation Office wanted to go beyond current employment of technologies and practices to examine individual behaviors of large numbers of people in different roles as they influence and are influenced by organizational dynamics, processes, and consequences. The project was termed Organizational Simulation or OrgSim. The goals of OrgSim are to help the DoD and the federal government identify, represent, and understand emergent organizational phenomena found in training, organizational structure, management, and policy. In its most mature form, OrgSim will offer new opportunities to study organizational dynamics before one invests in creating a structure to meet the intended function or goal of an organization.

This goal is quite significant as we view the changes in military and civil force structure occurring after September 11, 2001. As we go forward in the new millennium, the world has become increasingly unpredictable in its response to organizational change whether domestically or abroad. In the past, the United States military/acquisition organizational structure was geared to defeat a Soviet peer competitor. The Soviet structure, tactics, techniques, and procedures, along with the advances in technology, moved at a predictable pace and in an expected direction. Our doctrine, military, and organizational structure could change sufficiently quickly to adapt to the slow and often orderly changes to the Soviet doctrine and capabilities. Our military industrial complex was well adapted to fight a large military nation state that had its own attendant slow changing bureaucratic structure needed to recruit, train, and equip its personnel. However, in the future, our forces will most likely be dealing with smaller groups and organizations of insurgents and terrorists who are much less predictable, harder to identify and track, and much more agile in the development of their counters to what we believe is our technological advantage.

The United States military will be dealing with combatants who will not be easily recognizable, who will not follow standard military practices, and who will draw our forces into unfamiliar urban environments. This will reduce our military's overwhelming technological and mobility advantages in narrow streets among a population where the enemy can hide in plain sight within their urban organization. A major approach to defeat this enemy is to win the "hearts and minds" of the population within that environment by first being aware of them as an organization of people and cultures, and secondly accepting the unrestricted threats and how our and their organizations and cultures might adapt. The population can provide the human intelligence needed to distinguish friend from foe. But we still need the ability to model and simulate the relevant organizations whether for combat or for stabilization/reconstruction goals. This includes the full spectrum of military and political missions. Beyond the combat phase, it will take the coordination and cooperation of organizations such as the State Department, intelligence agencies, humanitarian organizations, charities, contractors and other nongovernmental agencies to gain the population's trust and cooperation. This is an enormous organizational understanding and learning problem and challenge that requires our focus and determination to solve it. We need to ensure that we have (a) more behavioral and social data, as well as the means to make relevant use of it, and (b) basic research and theory development in areas such as decision-making, situation awareness, learning, and multi-organizational/multi-cultural modeling.

In the two decades since the Goldwater-Nichols Act of 1986, the military has been struggling to organize and work as joint, combined armed forces. It has been challenging because each military service has its own traditions, ways of doing things, and organizational structure. Getting multiple agencies with entirely different processes, traditions, and structures to work together effectively will be even more challenging. Modeling and simulation of future possible organizational structures and relationships are desperately needed.

In their organization and editing of ***Organizational Simulation***, Bill Rouse and Ken Boff have brought together in one source an extraordinary collection of thought leaders who review and extend the present state of knowledge on the enabling technologies, methods, and tools for organizational simulation. Collectively, this volume offers new multi-disciplinary insights into potential roles of OrgSim, knowledge of individual and collective behaviors, and alternative approaches to modeling these behaviors. A variety of fascinating simulations and games are highlighted in this book, clearly illustrating the capabilities and limitations of what is already possible. We expect that this volume will be an invaluable resource for students, scholars, researchers, developers, and policy makers in the modeling and simulation community.

Paul Chatelier
Michael Lilienthal

PREFACE

ORGANIZATIONAL SIMULATION

From Modeling and Simulation to Games and Entertainment

Edited By

WILLIAM B. ROUSE AND KENNETH R. BOFF

This book took over two decades to come together. It began when we studied the design of complex systems and observed many aircraft designers from several companies wrestling with difficult design issues. It was nurtured by working with executives and senior managers in a wide range of private and public sector enterprises. We were often asked something like, "Wouldn't it be nice if you could drive the future before you wrote the check?"

That possibility was not unreasonable if you were designing an airplane, but what if you were designing an airplane company? What if you were designing an air force? What about a university? The overarching question is how to experience and evaluate substantial organizational change before committing to investing in making it happen.

Along the way, we were motivated by Apple's Knowledge Navigator in 1987 and Sun's Starfire in 1994. The capabilities presented in these technology visions were compelling. However, the technology in those times was not up to the task of supporting the organizational simulation (OrgSim) we had envisioned. What could be done would be very time consuming, very expensive, and not very compelling.

More recently, with the Internet, online games, and picture-like computer-animated movies, our appetites for OrgSim have been whetted. We are now convinced that many pieces of the puzzle are available, despite the fact that the picture on the puzzle box is not yet clear. It is also obvious that there are now many more disciplines, and hence more talent, intrigued by the idea of OrgSim.

We resurfaced our vision in a white paper in early 2003. The Defense Modeling and Simulation Office was intrigued by this vision and committed support for conducting a ***Workshop on Organizational Simulation*** in December 2003 and partial support for subsequently compiling this book. We are most grateful for DMSO's support, without which the workshop and this book would not have been possible.

We are also indebted to the participants in the workshop and the authors of the chapters in this book, many of who were involved in both activities. We feel quite fortunate to be able to draw upon the "best and brightest" from modeling and simulation, gaming, and entertainment, as well as leading thinkers in behavioral science and computing.

The promise of organizational simulation is immense. OrgSim can enhance what we do now, e.g., in design and training, and enable new ways of working. Strategies can be deployed and evaluated in OrgSim so that designers and investors really can drive the future before they buy it. Similarly, people can be trained to operate in a future that does not yet exist. There are also obvious entertainment opportunities. Perhaps the synthetic characters that act in OrgSim can become actual workers, staffing call centers and providing expert advice in a variety of domains.

We are convinced that this range of capabilities will eventually be available. It will happen slowly, over decades, if the natural evolution of the various disciplines involved proceeds as usual. Or, it could happen much faster – less than ten years – if research investments of sufficient magnitude are targeted at a well thought-out portfolio of challenge problems. This is, of course, what we advocate.

This book is the principal product of a project sponsored by Captain Michael Lilienthal of the Defense Modeling and Simulation Office. Mike has been an unwavering supporter of the OrgSim vision, providing financial and conceptual assistance to move this vision to reality. Of equal criticality to the success of OrgSim, in all phases of the project, has been the energetic intellectual engagement of Paul Chatelier, a senior technical advisor with the Potomac Institute for Policy Studies.

The authors are also grateful to Veridian (now General Dynamics) for their highly effective support in the planning and administration of the OrgSim Workshop. Scott Blevins was our onsite IT/computer professional who orchestrated and installed a highly effective wireless intranet that facilitated collaboration among workshop participants. Dr Tom Hughes, a cognitive engineer, helped facilitate the workshop process and made important contributions to documentation of the proceedings. Our sincere appreciation also goes to Jodi Nix and Renee Blanford who played heroic roles in the planning, coordinating and on-the spot troubleshooting that were vital to achieving the goals of the workshop.

Finally, we are very pleased to acknowledge the assistance of Kristi Kirkland of Georgia Tech who served as managing editor in bringing together and integrating all of the elements of this book.

William B. Rouse
Tennenbaum Institute
Georgia Institute of Technology
Atlanta, Georgia

Kenneth R. Boff
Human Effectiveness Directorate
Air Force Research Laboratory
Dayton, Ohio

October 2004

CONTRIBUTORS

Dee H. Andrews is Senior Scientist (ST) with the Human Effectiveness Directorate of the Air Force Research Laboratory. He is a member of the Science and Technology Professional Corps of the U.S. Air Force. Prior to his current position Andrews was the Technical Director of the Warfighter Training Research Division of the Air Force Research Laboratory. In that position he directed the scientific and technical program of the Division. Previously he worked as a research psychologist for both the Army Research Institute for Behavioral and Social Sciences, and the Naval Training Systems Center. His Ph.D., granted in 1980, is in Instructional Systems from Florida State University, He is a Fellow in the American Psychological Association, the Human Factors and Ergonomics Society, and Royal Aeronautical Society of the United Kingdom.

Dr. Herbert H. Bell is the Technical Advisor for the Warfighter Readinesss Research Division of the Air Force Research Laboratory. In that position he coordinates the scientific and technical program of the Division. Previously he was a Senior Scientist within the Division, where he also served as a Branch Chief. Prior to joining the Air Force Research Laboratory, he was a senior ergonomist with Eastman Kodak where he performed a variety of human factors functions. He is a member of IEEE, the Human Factors and Ergonomics Society, and the Psychonomic Society. He received his Ph.D. in experimental psychology from Vanderbilt University.

Aaron Bobick is the Professor and Chair of the IC Division within the College of Computing and the Director of the GVU Center at Georgia Institute of Technology, Atlanta, Georgia. He is also affiliated with the Computational Perception Laboratory (CPL) at Georgia Tech. His research interests are in computer vision and artificial intelligence and he specifically concentrated on modeling and recognition human activities, primarily from video. He earned his BS, MS, and PhD from MIT and he was also on the faculty at the MIT Media Lab from 1990 through 1998, before he joined the faculty at Georgia Tech. He was also on the research staff at SRI from 1987-90.

Kenneth R. Boff, Ph.D., serves as Chief Scientist of the Human Effectiveness Directorate, Air Force Research Laboratory, Wright-Patterson Air Force Base, Ohio. In this position, he has responsibility for the technical direction and quality of a broad multi-disciplinary R&D portfolio encompassing human-engineering of complex systems, training, safety, biotechnology, toxicology, and deployment logistics. He is best known for his work on understanding and remediating problems in the transition of research to applications in the design and acquisition of complex human-systems. Holder of a patent for rapid communication display technology, Boff has authored numerous articles, book chapters and technical papers, and is co-editor of "System Design" (1987), senior editor of the two-volume "Handbook of Perception and Human Performance" (1986), and the four-volume "Engineering Data Compendium: Human Perception and Performance" (1988). Boff actively consults and provides technical liaison with government agencies, international working groups, universities and professional societies. He is founder and technical director of the Department of Defense Human System Information Analysis Center, and founding member and former Chair of the DoD Reliance Human-Systems Interface Technology Panel. Currently, he is serving part-time on the faculty of the Georgia Institute of Technology, School of Industrial and Systems Engineering as an Edinfield Executive in Residence. Until recently, he was the US National Voting Member and Chair for the NATO RTO human factors technology area. In 2003, he received the NATO Scientific Achievement Award. Boff is a Fellow of the Human Factors & Ergonomics Society and the International Ergonomics Association.

Jeff Bradshaw, Ph.D., is a Senior Research Scientist at the Institute for Human and Machine Cognition where he leads the research group developing the KAoS policy and domain services framework. Formerly, he has led research groups at The Boeing Company and the Fred Hutchinson Cancer Research Center. In research sponsored by DARPA, NASA, ONR, and ARL, he is investigating principles of human-robotic teamwork, human-agent interaction, and trust and security for semantic web and semantic grid services. Dr. Bradshaw received his Ph.D. in cognitive science from the University of Washington in 1996. He has been a Fulbright Senior Scholar at the European Institute for Cognitive Sciences and Engineering (EURISCO) in Toulouse, France, is a member and former chair of the NASA Ames RIACS Science Council, and is former chair of ACM SIGART. Jeff serves on the editorial board of the Journal of Autonomous Agents and Multi-Agent Systems, the International Journal of Human-Computer Studies, the Web Semantics Journal, and the Web Intelligence Journal. Among other publications, he edited the books Knowledge Acquisition as a Modeling Activity (with Ken Ford, Wiley, 1993), Software Agents (AAAI Press/MIT Press, 1997) and the forthcoming Handbook of Agent Technology.

C. Shawn Burke, Ph.D. is a Research Associate at the University of Central Florida, Institute for Simulation and Training. Her primary research interests include teams, team leadership, team training and measurement, and team adaptability. Most recently, her research has focused on understanding, measuring, and training for team adaptability. Within this line of research she is currently investigating the impact of stress, leadership, and multi-cultural teams. She has presented at numerous peer-reviewed conferences, has published in several scientific journals and books on the topics of teams and team training, and serves as an ad-hoc reviewer for the *Human Factors* journal and *Quality Safety in Health Care*. She holds a Ph.D. in Industrial/ Organizational Psychology from George Mason University.

Kathleen M. Carley is Professor of Computation, Organizations and Society at the Institute for Software Research International at Carnegie Mellon University. She is also Director of the Center for Computational Analysis of Social and Organizational Systems. Her research combines cognitive science, social networks and computer science. Her specific research areas are computational social and organization theory, group, organizational and social adaptation and evolution, dynamic network analysis, computational text analysis, and the impact of telecommunication technologies and policy on communication, information diffusion, disease contagion and response within and among groups particularly in disaster or crisis situations. Her models meld multi-agent technology with network dynamics and empirical data. Three of the large-scale multi-agent network models she and the CASOS group have developed are: BioWar – a city, scale model of weaponized biological attacks; OrgAhead – a model of strategic and natural organizational adaptation; Construct – a model of the co-evolution of social networks, knowledge networks, personal/organizational identity and capability under diverse cultural and technological configurations; and ORA a statistical tool kit for network and meta-matrix data. She is the current president of NAACSOS – the North American Association for Computational Social and Organizational Science, founding co-editor with Al. Wallace of the journal Computational Organization Theory and has co-edited several books in the area including Computational Organization Theory, Simulating Organizations, and Dynamic Network Analysis.

Paul R. Chatelier, CAPT USN (retired) is an aviation psychologist and research fellow at the Potomac Institute for Policy Studies in Arlington, Virginia where he is responsible for their human factors and training science and technology projects. Prior to this position his many responsibilities included working for the Defense Modeling and Simulation Office, the White House Office of Science and Technology Policy, the Defense Advanced Projects Agency, and the DoD Education Activity. He has co-authored a number of books on human factors and training and has served as a human factors and training specialist on numerous Defense, Navy, and Federal science boards and task forces and technical institute advisory boards. He continues to use his international experience and knowledge as a technical expert and analyst on NATO's human factors, training, and bio-medical programs by serving as a delegate to the NATO Human Factors and Medical Panel. More details on publications and activities can be found in the Marquis' Who's Who series.

Stephen E. Cross is a Vice President of the Georgia Institute of Technology, the Director of the Georgia Tech Research Institute, and a Professor in the School of Industrial and Systems Engineering. Prior to his current position, he was the Director and CEO of the Software Engineering Institute at Carnegie Mellon University. He received a Ph.D. from the University of Illinois at Urbana-Champaign in 1983. He is a Fellow of the Institute of Electrical and Electronics Engineers (IEEE) and a member of the Air Force Scientific Advisory Board.

Frederick J. Diedrich is Leader of the Cognitive Systems Group at Aptima, Inc. His primary areas of research focus on training, decision-making, perceptual-motor influences on cognition, and safety. Dr. Diedrich is leader of the experimental task of the Adaptive Architectures for Command and Control Program.

Elliot E. Entin is a Senior Scientist at Aptima, Inc. Dr. Entin's skills are applied to the design, conduct, and analysis of experiments. He is a member of Aptima's Adaptive Architectures for Command and Control project and leads the Team Adaptive Performance System project.

Irfan Essa is an Associate Professor in the College of Computing, and Adjunct Professor in the School of Electrical and Computer Engineering, Georgia Institute of Technology, Atlanta, Georgia. At Georgia Tech, he is affiliated with the Graphics, Visualization and Usability Center. He has founded the Computational Perception Laboratory (CPL) at Georgia Tech, that aims to explore and develop the next generation of intelligent machines, interfaces, and environments that can perceive, recognize, anticipate, and interact with humans. He is also a founding member of the Aware Home Research Initiative at Georgia Tech. He helped establish a new undergraduate degree of Computational Media at Georgia Tech and is a faculty member associated with the Human Centric Computing (HCC) PhD degree. Irfan earned his SM (1990) and PhD (1994) from the MIT Media Laboratory, where he also worked as a Research Scientist (1994-1996) before joining the Georgia Tech faculty. He earned is BS (1988) from Illinois Institute of Technology. He has received the prestigious awards of NSF CAREER Investigator, Imlay Fellowship, Edenfield Fellowship, and the College of Computing Research, Teaching, and Dean's Awards. His research interests are in computer graphics, vision, animation, and he is specifically interested in building intelligent and interactive machines that can interact with us in a natural manner.

Paul J. Feltovich, Ph.D., is a Research Scientist at the Institute for Human and Machine Cognition, Pensacola, FL. He has conducted research and published on topics such as expert-novice differences in cognitive skills, conceptual understanding for complex knowledge, and novel means of instruction in complex and ill-structured knowledge domains. More recently (with Jeffrey Bradshaw) he has been studying regulatory and coordination systems for mixed human-agent teams. Dr. Feltovich received a Ph.D. in educational psychology from the University of Minnesota in 1981. He was also a post-doctoral fellow in cognitive psychology at the Learning, Research, and Development Center, University of Pittsburgh, from 1978 to 1982. Before joining IHMC in 2001, he served as Professor in the Department of Medical Education, and Director of the Cognitive Science Division at Southern Illinois University School of Medicine, Springfield, IL. He is co-editor (with Ken Ford and Robert Hoffman) of *Expertise in Context: Human and Machine* (AAAI/MIT) and (with Ken Forbus) *Smart Machines in Education* (AAAI/MIT).

Scott D. Fouse is President and Chief Executive Officer of ISX Corporation. His experience ranges from the design and development of custom integrated circuits to the design of intelligent system architectures for advanced military applications. Since 1985, Mr. Fouse has been working closely with DARPA and also operational military groups to develop advanced concepts for intelligent systems in a wide variety of military domains. His current project is the Command Post of the Future, which is being used to provide command and control of Army forces in Baghdad. He is also a member of the Air Force Scientific Advisory Board. He received a M.S. degree in Electrical Engineering from the University of Southern California and a B.S. degree in Physics from the University of Central Florida.

Dr. Richard Fujimoto is a professor in the College of Computing at the Georgia Institute of Technology. He received the Ph.D. and M.S. degrees from the University of California (Berkeley) in 1980 and 1983 (Computer Science and Electrical Engineering) and B.S. degrees from the University of Illinois (Urbana) in 1977 and 1978 (Computer Science and Computer Engineering). He has been an active researcher in the parallel and distributed simulation community since 1985, and has published numerous technical papers on this subject. His publications include a textbook and several award-winning papers. He has led the development of parallel/distributed simulation software systems including the Georgia Tech Time Warp (GTW) simulation executive and the Federated Simulation Development Kit (FDK), both of which have been distributed worldwide. He has given several tutorials on parallel and distributed simulation at leading conferences. He led the definition of the time management services for the DoD High Level Architecture (HLA) effort that has been designated as the standard reference architecture for modeling and simulation in the U.S. Department of Defense. Fujimoto is Co-Editor-in-Chief of *Simulation: Transactions of the Society for Modeling and Simulation International*, and served as an area editor for *ACM Transactions on Modeling and Computer Simulation*. He has served as chair of the steering committee for the Workshop on Parallel and Distributed Simulation (PADS) as well as the conference committee for several other simulation conferences.

Ryland (Ryan) C. Gaskins III is a Senior Research Scientist at the Virginia Modeling Analysis and Simulation Center at Old Dominion University. He received his Ph.D. in Human Factors and M.S. in Industrial Organizational Psychology from George Mason University. Since joining ODU, he has taught both graduate and undergraduate courses in Human Factors, Industrial/Organizational Psychology, Research Methods, Personnel Psychology, Organizational Psychology, Ethics and Introductory Psychology. Dr. Gaskins' research interests are in the area of training and simulation, virtual environments, macroergonomics, persuasive computing, human abilities and task characteristics measurement for selection and placement. His favorite past time is cruising and exploring the Chesapeake Bay onboard his sailboat, Panacea. Dr. Gaskins is originally from Kilmarnock, Virginia and currently resides in Virginia Beach.

Joseph W. Guthrie Jr. is a first year PhD student enrolled in the Applied Experimental and Human Factors Psychology program at the University of Central Florida. He earned a B.S. in Psychology from the University of Tennessee at Chattanooga in 2001 and a M.S. in Industrial/Organizational Psychology from the University of Tennessee at Chattanooga in 2003. He is currently working as a graduate research assistant at the Institute for Simulation and Training. His research interests include usability, the use of simulation to train distributed teams and team performance under stress.

J.C. Herz is a researcher and designer who specializes in multiplayer interaction design, and systems that leverage the intrinsic characteristics of networked communication. Clients include multinational corporations, non-profit organizations, and the Defense Department. J.C. is a fellow at USC's Annenberg school, and is a member of the National Research Council's Committee on Creativity and Information Technology. She is the author of two books, Surfing on the Internet (Little, Brown 1994) and Joystick Nation: How Videogames Ate Our Quarters, Won Our Hearts, and Rewired Our Minds (Little, Brown 1997).

Eva Hudlicka is a Principal Scientist and President of Psychometrix Associates, Blacksburg, VA. Her research interests include cognitive modeling, affective computing, computational psychology, decision support system design, and human-computer interaction. She received her BS in Biochemistry from Virginia Tech, her MS in Computer Science from The Ohio State University, and her PhD in Computer Science from the University of Massachusetts-Amherst. Prior to founding Psychometrix Associates in 1995, Dr. Hudlicka was a Senior Scientist at Bolt Beranek & Newman, Cambridge, MA.

Helen Altman Klein is Professor of Psychology and Member of the Graduate Faculty in the Human Factors Program at Wright State University. She works in a cognitive engineering framework to understand complex natural systems. Her research in cultural differences in cognition has included the domains of civil aviation, multinational peacekeeping, command force modeling, and psychological operations. She also works with the task demands of medical self-management.

Gary Klein, Ph.D., is Chief Scientist of Klein Associates, Inc., a company he founded in 1978 to develop stronger models and methods for studying and improving the way people make decisions in natural settings. In 1985, Dr. Klein and his colleagues developed the Recognition-Primed Decision (RPD) model to explain how people can make effective decisions under time pressure and uncertainty. The research has been extended to training individuals and teams, and to applying a decision-centered design approach to increase the impact of information technology. In addition, Dr. Klein and his colleagues have developed a set of cognitive task analysis methods to study decision-making and other cognitive functions in field settings. Dr. Klein received his Ph.D. in experimental psychology from the University of Pittsburgh in 1969. He was an Assistant Professor of Psychology at Oakland University (1970-1974) and worked as a research psychologist for the U.S. Air Force (1974-1978). He is the author of "Sources of Power: How People Make Decisions" (1998, MIT Press) and "The Power of Intuition" (2004, Doubleday Currency).

James H. Korris is the Creative Director of the Institute for Creative Technologies, University of Southern California. Korris began his work with the Institute's creation in August 1999 as one of the first employees. Korris undertook his assignment at ICT as a natural outgrowth of his work as Executive Director of USC's Entertainment Technology Center. A sponsored, organized research unit of the University of Southern California's School of Cinema Television, the Center is committed to fostering the development of technology related to the production and distribution of entertainment content. Korris came to the Center after a career in studio production, producing and writing. Since May, 1998, he has been producing via korris.film, and writing on assignment. Recent projects include The Killing Yard, starring Alan Alda, a fact-based film centered on the trials following the Attica prison uprising, for Showtime and Paramount. He is a member of the writers' branch of the Academy of Television Arts and Sciences, the Writers Guild of America, the Writers Guild of Canada and the Society of Motion Picture and Television Engineers. He held several creative executive positions at Universal Television and served as a staff producer for Ron Howard's Imagine Films before joining longtime colleague Robert Harris in independent production at Harris & Company. Korris' undergraduate degree in Economics is from Yale University; he was awarded an M.B.A. with distinction at the Harvard Business School. He lives with his wife Stephanie and daughter Amanda in Los Angeles.

Dr. Alexander H. Levis is the Chief Scientist of the Air Force, on leave from George Mason University, Fairfax, VA where he is University Professor of Electrical, Computer, and Systems Engineering and where he heads the System Architectures Laboratory of the C3I Center. He was educated at MIT where he received the BS (1965), MS (1965), ME (1967), and Sc.D. (1968) degrees in Mechanical Engineering with control systems as his area of specialization. He also attended Ripon College where he received the AB degree (1963) in Mathematics and Physics. Dr. Levis is a Fellow of the Institute of Electrical and Electronic Engineers (IEEE) and past president of the IEEE Control Systems Society; a Fellow of the American Association for the Advancement of Science (AAAS) and of INCOSE; an Associate Fellow of the American Institute of Aeronautics and Astronautics (AIAA); and a member of AFCEA. He has received twice the Exceptional Civilian Service medal from the Air Force (1994, 2001) for contributions as a member of the Air Force Scientific Advisory Board and the Third Millennium medal from IEEE. He has taught at the Polytechnic Institute of Brooklyn (1968-1973), headed the Systems Research Dept. at Systems Control, Inc. in Palo Alto, CA (1973-1979), was a senior research scientist at the Laboratory for Information and Decision Systems at MIT (1979-1990), and moved to George Mason University in 1990 where he headed twice the Systems Engineering department. For the last fifteen years, his areas of research have been organization architecture design and evaluation, adaptive architectures for command and control, and the development of decision support systems.

CAPT. Michael G. Lilienthal, MSC, USN is the Special Assistant to Deputy Undersecretary of Defense (Science & Technology). He is working Science and Technology issues concerning Joint Urban Operation requirements within OSD. He graduated from University of Notre Dame where he also received his doctorate in Experimental Psychology specializing in Psychophysical Scaling and Measurement. He is a Fellow of the Aerospace Medical Association and the Aerospace Human Factors Association. He is a certified Human Factors Professional and a member of the Navy Acquisition Workforce. CAPT Lilienthal has been on activity duty for 26 years in the Navy serving in a variety of assignments in Science & Technology, Test & Evaluation, Acquisition, Planning and Policy.

Richard D. Lindheim is the Executive Director of the Institute for Creative Technologies, University of Southern California. With nearly four decades of television experience, Lindheim brings a programming perspective to his work at the Institute. Among his key responsibilities for ICT, he supervises virtual reality training projects developed for the U.S. Army, and oversees outreach to USC's participating schools as well as the entertainment, computer game and computer technology industries at large. Lindheim most recently spent seven years as executive vice president of Paramount Television Group, where he launched the Digital Entertainment Division. Previously he graduated through the ranks of Universal and MCA, beginning as a producer for Universal Television in 1979 and progressing through four posts with MCA Television Group/Universal Studios. They include vice president, current programming, senior vice president series programming, executive vice president, creative affairs and, finally, executive vice president, program strategy. During his tenure with Universal, Lindheim also served as a producer on several television series, and co-created the popular Universal/CBS program "The Equalizer" (1985-89). Transitioning to MCA/Universal from NBC, Lindheim served as vice president, dramatic programs following a nine-year post as vice president, program research for the network. Prior to work in audience studies, he began his entertainment career in 1962 as an administrative assistant in the story department of CBS. Lindheim earned a B.S. in electronic engineering from the University of Redlands, with postgraduate studies in telecommunications and engineering at the University of Southern California. A member of the Writers Guild of America, Lindheim is the author of several publications including two textbooks on television, PrimeTime: Network Television Programming and Inside Television Producing.

R. Bowen Loftin is the Executive Director of the Virginia Modeling, Analysis and Simulation Center, Professor of Computer Science, and Professor Electrical and Computer Engineering at Old Dominion University. He holds a B.S. in Physics from Texas A&M University and an M.A. and a Ph.D. in Physics from Rice University. Since 1983, Dr. Loftin, his students, and coworkers have been exploring the application of advanced software technologies, such as artificial intelligence and interactive, three-dimensional computer graphics, to the development of training and visualization systems. He is a frequent consultant to both industry and government in the area of advanced training technologies and scientific/engineering data visualization. Dr. Loftin serves on advisory committees and panels sponsored by numerous government and professional organizations. Dr. Loftin's recognition includes the University of Houston-Downtown Awards for Excellence in Teaching and Service, the American Association of Artificial Intelligence Award for an innovative application of artificial intelligence, NASA's Space Act Award, the NASA Public Service Medal, and the 1995 NASA Invention of the Year Award. He is the author or co-author of more than one hundred technical publications.

Jean MacMillan is the Chief Scientist at Aptima, Inc. Dr. MacMillan's research has spanned a broad range of topics in human-machine interaction and user-centered system design, including command center design, team decision making, and adaptive instructional design.

Leon McGinnis is Gwaltney Professor of Manufacturing Systems at Georgia Tech, where he also serves as Director of the Product Lifecycle Management Center of Excellence, and Associate Director of the Manufacturing Research Center. Professor McGinnis teaches and leads research in the area of discrete event logistics systems, including warehouse and logistics system design, material handling systems design, high-fidelity simulation methodology, and system performance assessment and benchmarking tools. He is a Fellow of the Institute of Industrial Engineering and a recipient of the Reed-Apple Award by the Material Handling Education Foundation.

Anna McHugh is a Research Associate at Klein Associates, joining the staff in August 2002. Her research interests include uncovering the nature of expertise and skilled performance in teams and individuals. Mrs. McHugh is particularly interested in the study of cultural differences in cognition and how an understanding of such variations can be applied to the development of tools and interventions to enhance multinational collaboration. She is currently leading a project funded by AFRL to expand a model for describing the cultural variances in cognition (Cultural Lens Model), with a focus on understanding Middle Eastern culture. In addition to this work, Mrs. McHugh is leading a project funded by ARL to understand how members of different national cultures view teamwork and function in collaborative settings. She recently completed a project sponsored by OSD to develop tools for assessing higher order cognitive processes and functions (Macrocognition) of small unit Army leaders. The focus of the project was to uncover the cognitive requirements of squad, platoon, and company leaders and to develop means of measuring changes in those cognitive skills as a result of training interventions. Mrs. McHugh also recently completed leadership on a project sponsored by The Centers for Medicare & Medicaid Services (CMS). She and other members of the research team applied Cognitive Task Analysis methodology to uncover aspects of surveyor expertise. She will soon begin to help develop training for new surveyors and preceptors, using the information uncovered. Mrs. McHugh holds a M.S. in Industrial-Organizational and Human Factors Psychology from Wright State University, Dayton, OH, and a B.A. in Psychology and Spanish from Denison University, Granville, OH.

Frederic (Rick) D. McKenzie is an Assistant Professor of Electrical and Computer Engineering at Old Dominion University (ODU). Prior to joining ODU, Dr. McKenzie held a senior scientist position at Science Applications International Corporation (SAIC), serving as Principal Investigator for several research and development projects. While at SAIC he was a Team Lead on a simulation system that encompassed the training requirements of all military services and joint operations. He has had several years of research and development experience in the software and artificial intelligence fields, including object oriented design in C++, LISP, and knowledge-based systems. Dr. McKenzie has also had two years teaching experience in artificial intelligence, software languages, and data structures. Both his Masters and Ph.D. work were in artificial intelligence, focusing on knowledge representation and model-based diagnostic reasoning. He received a Ph.D. in Computer Engineering from the University of Central Florida in 1994.

Janet H. Murray is Professor and Director of the Graduate Program in Digital Media and Information Design in the School of Literature, Communication, and Culture at Georgia Tech. She holds a Ph.D. from Harvard University in English Literature and led projects in humanities computing at MIT before coming to Georgia Tech. Her interactive video projects and her book, *Hamlet on the Holodeck: The Future of Narrative in Cyberspace* (1997) are internationally known. She is currently working on a textbook, *Inventing the Medium* for MIT Press, and a NEH-funded Digital Critical Edition of *Casablanca*.

Mikel D. Petty is Chief Scientist of the Virginia Modeling, Analysis and Simulation Center and Research Professor of Engineering Management and Systems Engineering at Old Dominion University. He received a Ph.D. from the University of Central Florida (UCF) in 1997, a M.S. from UCF in 1988, and a B.S. from the California State University, Sacramento, in 1980, all in Computer Science. Dr. Petty has worked in modeling and simulation since 1990 in the areas of simulation interoperability, computer generated forces, multi-resolution simulation, and applications of theory to simulation. During that time has published over 110 research papers and has been awarded over 35 research contracts. He has served on a National Research Council committee on modeling and simulation and is currently an editor of the journals *SIMULATION: Transactions of the Society for Modeling and Simulation International* and *Journal of Defense Modeling and Simulation*.

Heather A. Priest is a third year PhD student in the University of Central Florida Applied Experimental and Human Factors Psychology program. She received a

BA in Psychology and a MS in Applied Experimental Psychology from Mississippi State University and has worked as a graduate research assistant at the Institute for Simulation & Training since 2001. Ms. Priest's research interests include teams, training, performance under stress, usability and patient safety. In addition, she is a 2003 Interservice/Industry, Training, Simulation and Education Conference (I/ITSEC) scholarship winner and a member of several professional associations, including the American Psychology Association, Usability Professional Association, Computer-Human Interaction and the Human Factors and Ergonomics Society. Ms. Priest is currently an Intern at Aptima Incorporated in Washington DC doing work in multicultural teams, training systems and communication analysis.

Amy R. Pritchett is an associate professor in the School of Industrial and Systems Engineering and a joint associate professor in the School of Aerospace Engineering at the Georgia Institute of Technology. Her research encompasses cockpit design, including advanced decision aids; procedure design as a mechanism to define and test the operation of complex, multiagent systems such as air traffic control systems; and simulation of complex systems to assess changes in emergent system behavior in response to implementation of new information technology. Dr. Pritchett is the editor of Simulation: Transactions of the Society for Modeling and Simulation for the area 'air transportation and aviation'; associate editor of the AIAA Journal of Aerospace Computing, Information, and Communication; technical program chair for the aerospace technical group of the Human Factors and Ergonomics Society; and co-chair of the 2004 International Conference in Human-Computer Interaction in Aerospace (HCI-Aero). Dr. Pritchett serves on the National Research Council's Aeronautics and Space Engineering Board

William B. Rouse is the H. Milton and Carolyn J. Stewart Chair of the School of Industrial and Systems Engineering at the Georgia Institute of Technology. He also serves as Executive Director of the university-wide Tennenbaum Institute whose multi-disciplinary portfolio of initiatives focuses on research and education to provide knowledge and skills for enterprise transformation. Rouse has written hundreds of articles and book chapters, and has authored many books, including most recently *Essential Challenges of Strategic Management* (Wiley, 2001) and the award-winning *Don't Jump to Solutions* (Jossey-Bass, 1998). He is co-editor of the best-selling *Handbook of Systems Engineering and Management* (Wiley, 1999) and edited the eight-volume series *Human/Technology Interaction in Complex Systems* (Elsevier). Rouse is a member of the National Academy of Engineering, as well as a fellow of the Institute of Electrical and Electronics Engineers, the Institute for Operations Research and Management Science, and the Human Factors and Ergonomics Society. He received his B.S. from the University of Rhode Island, and his S.M. and Ph.D. from the Massachusetts Institute of Technology.

Eduardo Salas is Trustee Chair and Professor in the Department of Psychology, University of Central Florida (UCF), where he also holds an appointment as Program Director for the Human Systems Integration Research Department at the Institute for Simulation and Training. In addition, Dr. Salas is the Program Coordinator of the Applied Experimental and Human Factors Psychology doctoral program at UCF. Dr. Salas' research interests include team training, team performance, and organizational safety.

Daniel Serfaty is the founder of Aptima, Inc. His areas of research interest include distributed decision-making, cognitive engineering, and team performance engineering. Over the past twenty years, he has contributed to a better understanding of the structures, competencies, and technologies that combine to optimize complex organizational performance in teams.

Anuj P. Shah is a PhD candidate with the school of Industrial and Systems Engineering at Georgia Institute of Technology. His research focuses on frameworks for analyzing the impact of the technology and process innovation in large-scale socio-technical systems and in enterprises. Prior to starting his PhD he was a scientist with ABB Corporate Research. Over the course of his engineering career, he has worked on several automation systems design and development projects in USA, Europe and India. He has a Masters in Integrated Manufacturing Systems Engineering from North Carolina State University and a Bachelors in Manufacturing Processes and Automation Engineering from Delhi Institute of Technology.

Robert N. Shearer is a Branch Chief in the Warfighter Training Research Division of the Air Force Research Laboratory. He is a Lieutenant Colonel in the U.S. Air Force. Colonel Shearer has held a variety of Air Force jobs including; Intercontinental Ballistic Missile (ICBM) Launch Control Officer, a B-52 Radar-Navigator Bombardier. He has had career broadening assignments in Management Engineering, Organizational Development and served as the Inspector General for the Fighter Wing. Colonel Shearer has had a faculty assignment at the United States Air Force Academy. Colonel Shearer has a Master of Arts Degree in Organizational Communications from Brigham Young University.

Katherine Wilson-Donnelly is a doctoral candidate enrolled in the Applied Experimental and Human Factors Psychology Ph.D. program at the University of Central Florida (UCF) in Orlando, Florida. She earned a B.S. in Aerospace Studies from Embry-Riddle Aeronautical University in 1998 and a M.S. degree in Modeling and Simulation from UCF in 2002. She is also a research assistant at the Institute for Simulation and Training. Her research interests include using simulation and training to improve aviation safety and multicultural team performance.

David Woods, Ph.D. is Professor in the Institute for Ergonomics at the Ohio State University. He has developed and advanced the foundations and practice of

Cognitive Systems Engineering since its origins in the aftermath of the Three Mile Island accident in nuclear power. He has studied team work between people and automation, including automation surprises, through studies in anesthesiology and aviation. He has studied cooperative work systems in space mission operations and he has designed new concepts for aiding cognitive work such as visual momentum and applied them in aviation, space operations, nuclear power, and critical care medicine. Multimedia overviews of his research are available at url: http://csel.eng.ohio-state.edu/woods/. Safety in complex systems is a constant theme in his work; see his monographs: "Behind Human Error" and "A Tale of Two Stories: Contrasting Views of Patient Safety." He was one of the founding board members of the National Patient Safety Foundation, Associate Director of the MidWest Center for Inquiry on Patient Safety of the Veterans Health Administration, and advisor to the Columbia Accident Investigation Board. Dr. Woods received his Ph.D. from Purdue University in 1979 and has been President of the Human Factors and Ergonomic Society. He is a Fellow of that society and a Fellow of the American Psychological Society and the American Psychological Association. He has shared the Ely Award for best paper in the journal Human Factors (1994), a Laurels Award from Aviation Week and Space Technology (1995) for research on the human factors of highly automated cockpits, the Jack Kraft Innovators Award from the Human Factors and Ergonomics Society (2002), and five patents for computerized decision aids.

Greg Zacharias is a Senior Principal Scientist and founder of Charles River Analytics, Cambridge, MA, supporting research efforts in human behavior modeling and agent-based decision support systems. Dr. Zacharias has been a member of the National Research Council (NRC) Committee on Human Factors since 1995, and served on the NRC panel on Modeling Human Behavior and Command Decision Making. He is a member of the USAF Scientific Advisory Board (SAB), the DoD Human Systems Technology Area Review and Assessment (TARA) Panel, and chairs the USAF Human System Wing Advisory Group for Brooks City-Base.

Michael Zyda is the Director of the USC GamePipe Laboratory, located at the Information Sciences Institute, Marina del Rey, California, and the Associate Director of the USC Integrated Media Systems Center. From Fall 2000 to Fall 2004, he was the Director of The MOVES Institute, located at the Naval Postgraduate School, Monterey, California. He is also a Professor in the Department of Computer Science at NPS, and is on extended leave to form the USC GamePipe Laboratory. From 1986 until the founding of the MOVES Institute, he was the Director of the NPSNET Research Group. Professor Zyda's research interests include computer graphics, large-scale, networked 3D virtual environments, agent-based simulation, modeling human and organizational behavior, interactive computer-generated story, modeling and simulation, and interactive games. He is a pioneer in the fields of computer graphics, networked virtual environments, modeling and simulation, and defense/entertainment collaboration. He holds a lifetime appointment as a National Associate of the

National Academies, an appointment made by the Council of the National Academy of Sciences in November 2003, awarded in recognition of "extraordinary service" to the National Academies. He served as the principal investigator and development director of the America's Army PC game funded by the Assistant Secretary of the Army for Manpower and Reserve Affairs. He took America's Army from conception to three million plus registered players and hence, transformed Army recruiting.

CHAPTER ONE

INTRODUCTION AND OVERVIEW

WILLIAM B. ROUSE AND KENNETH R. BOFF

Computer-based simulation has long been used to project the behavior of systems too complex for analytical calculation. Simulation has also been used for many decades to enable human visualization and learning about complex tasks such as aircraft piloting and process plant control. The versatility and cost-effectiveness of training simulators are widely recognized.

Other phenomena for which simulation offers great promise include a variety of types of organizational changes. Organizations are complex social-technical systems. Organizational simulation involves computational representations of people, their behaviors and interactions in response to the environment and each other. The goal may be to predict alternative outcomes in response to changing conditions, or perhaps simply illustrate organizational phenomena. The range of organizations of interest includes loosely coupled groups such as crowds; tightly coupled groups such as teams, units, and forces; and broader groups such as populations.

A key technical challenge for an organization simulation lies in the establishing an effective mix of constructive and human-in-the-loop simulations. Hence; possible organizational simulations include:

- All human actors role-playing in a real environment (e.g. management training exercises)
- All human actors playing within a constructive environment (e.g. man-in-the-loop flight simulation)
- Mixed human in the loop with constructive actors (e.g., ***The Sims***)
- Fully constructive simulations where there are no human actors

To the extent that these simulations include constructive representations of human actors their utility will largely depend on the degree and to which the constructive simulation incorporates appropriate emulation of behavioral and social processes (Pew & Mavor, 1998).

There are numerous domains of application of organizational simulation. For example, large-scale efforts to re-engineer processes and transform private and public sector enterprises often require huge investments in information technology and training to achieve success. Investments of hundreds of millions of dollars are not uncommon. Senior managers would like to experience – via immersive visualization and interaction -- such massive changes before they make decisions to proceed (Rouse, 2004).

As another illustration, coalition operations across national and cultural boundaries can be enabled and enhanced by capabilities for visualizing and experiencing alternative organizational designs. Rapid design and deployment of responses to novel mission requirements such as terrorist events and natural disasters can also be enhanced by such capabilities. Organizational simulation can enable iterative evaluation – by experience – of "pickup" organizations. This could allow early identification of cross-jurisdictional issues, for instance. It could also support evaluation of alternative resolutions of such conflicts via immersive "what if" capabilities.

The maturation of this technology will prove invaluable as a visualization test bed for understanding, designing and evaluating complex organizational architectures for a variety of industrial and military applications in operations, logistics and manufacturing. Users will be able to acquire perspectives on individual and emergent behaviors and performance and their implications for design, acquisition, doctrine and, in general, organizational transformation.

Organizational simulation can enable teams to rapidly discover, experience, and train in new contexts while simultaneously synthesizing the operational doctrine, as well as training and aiding systems, needed to support the team. This allows for evaluation and enhancement of system and organizational concepts prior to committing to their full development and deployment.

Recent world events and rapidly evolving modeling and simulation technologies are precipitating a converging sense of urgency and opportunity that can enable disruptive innovations (Christensen, 1997). Consequently, the goal of this book is to bring together thought leaders from modeling and simulation, behavioral and social sciences, computing, artificial intelligence, gaming, and entertainment. Their assignment was to assess the state of the art, recommend best practices, and formulate recommendations for an R&D agenda whose pursuit will enable needed organizational simulation research, development, and applications.

SCOPE OF ORGANIZATIONAL SIMULATION

It is easy to envision and articulate wonderful, even fanciful, applications of organizational simulation in a wide variety of domains. However, one quickly encounters questions such as:

- How would the organizational simulation actually function to realize these visions?
- What enabling knowledge and technologies would be needed to provide this functionality?
- How mature are the necessary knowledge bases, enabling technologies, and abilities to create the desired functionality?

This chapter provides a framework for answering these questions. The other chapters provide answers to these questions in the context of this framework.

Architecture of Organizational Simulation

The architecture of the organizational simulation shown in Figure 1 embodies key elements needed to enable fully functional capabilities. The representation in this figure is not intended to suggest a software architecture, per se. Instead, it portrays a conceptual model of relationships among several layers of functionality of an organizational simulation.

Since we are emphasizing immersive simulations, it is useful to begin at the top of Figure 1. Users interact with the immersive environment by viewing large-screen displays (e.g., whole walls and rooms), listening and verbalizing to other real and synthetic characters, and gesturing (e.g., walking and pointing). In light of the likely complexity of this environment, users are assisted via training, aiding, and guidance in general. Otherwise, they might get lost in the simulation.

Users' interactions with the simulation happen in a context provided by the organizational story that is playing out. This story might involve command and control (C4ISR) during a military operation, or possibly design and evaluation of "pickup" supply chains in response to emergencies. To an extent, the context is represented in the "script," except that characters – real and synthetic – are free to say and do what they want in this context.

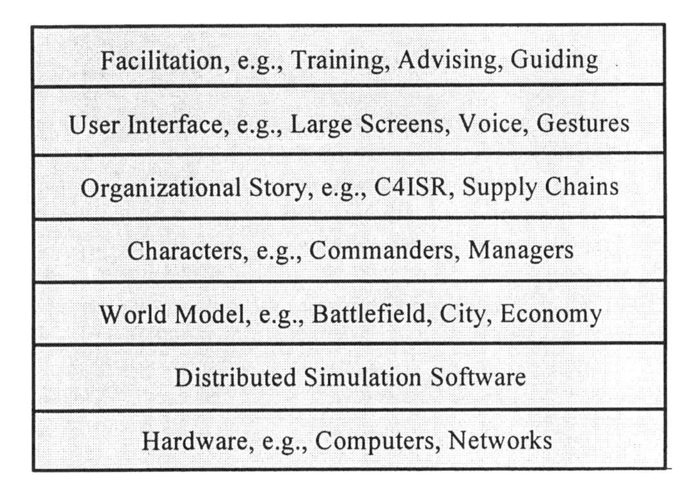

FIGURE 1. Architecture of Organizational Simulation

Characters, e.g., military commanders, supply chain executives, news reporters, and bystanders, populate the organizational story. Some of these characters, particularly the users of the simulation, are usually real humans. However, most of the characters are typically synthetic, perhaps created via intelligent agent technology.

The organizational story plays out and the characters interact in a dynamic world, e.g., a battlefield or city, where actions affect the states of the world, and these states, of course, influence characters' actions. The dynamics of the world, as well as external forces and unseen actors, also affect the states of the world. Typically, the world continues to evolve whether or not users do anything. Often, users actions are intended to bring about desirable world states.

All of the layers discussed thus far are software intensive. Certainly, there is hardware such as visual and audio displays, control devices and, of course, computers, routers, and wires, which enable creating the "virtual reality" of an organizational simulation (Durlach & Mavor, 1995). However, the story, characters, and world are primarily manifested in software, which may be distributed among many computers that are potentially networked across many locations.

This book addresses all layers of this architecture with emphasis on key interactions among layers. Some of the chapters pursue one layer in depth while others report on a wealth of efforts to create, evaluate, and deploy organizational simulations. The latter efforts inevitably address many, if not all, of the layers of the architecture.

Perspectives on Organizational Simulation

The authors of the chapters represent a wide range of perspectives on organizational simulation. Many authors have strong roots in the modeling and simulation community, often from engineering and computing. They build upon a rich legacy of mathematics, models, methods, and tools.

The models and simulations developed by this community are almost always in need of knowledge for representing underlying phenomena and data for estimating parameters with these representations. For organizational simulation, much of this knowledge comes from the behavioral and social sciences. Several chapters address what is fundamentally known about behavioral and social phenomena of potential importance to organizational simulation.

Many immersive simulations are likely to require interactive characters and organizations. Thus, we need to be able to simulate people so that users can interact with them. This has thus far been addressed by employing methods of artificial intelligence, informed by the behavioral and social sciences. There appears to be a consensus that the results of these efforts have been rudimentary to date, although visible progress continues.

It can be argued that the above perspectives tend to be bottom up, building up from what is known, using leading-edge methods and technical tools. An alternative perspective begins with the desired experiences to be created by organizational simulations. This perspective comes from the entertainment and gaming communities (NRC, 1997; King & Borland, 2003), as well as research in online fiction (Murray, 1997). Several chapters represent this perspective.

The differences between bottom-up model building and outside-in experience design create an interesting tension. The model builders often attempt to incorporate as much detail as possible, e.g., creating high levels of visual fidelity because it is technically feasible. Those who design experiences try to incorporate as little detail as possible while still achieving the desired experiences for users.

This tension between wanting to know everything and knowing just enough permeates this book. This is a very creative tension because the two perspectives are complementary. Bottom-up knowledge can inform and enable outside-in aspirations, leading to increasing organizational simulation capabilities.

WORKSHOP ON ORGANIZATIONAL SIMULATION

To gain a better understanding of the various perspectives on organizational simulation, as well as assess the state of knowledge in these areas, a ***Workshop on Organizational Simulation*** was held at the Biltmore Resort in Clearwater, Florida in December 2003. The full workshop report is provided by Boff and Rouse (2004). In this section, we summarize the workshop process and overall findings.

Roughly 50 experts participated from the areas of modeling and simulation, behavioral and social sciences, computing, artificial intelligence, gaming, and entertainment. These experts also included many people with extensive experience in business and military operations.

Participants were asked to assess the organizational simulation capabilities needed to address five challenge scenarios that are elaborated below[1]:

- Enterprise Systems
- NASA Columbia
- Command and Control – C2
- Domestic Crisis – G8
- Joint Urban Operations – JUO

People were assigned to working groups – one per challenge problem – and asked to envision the important phenomena to be incorporated within organizational simulations targeted at these scenarios, as well as the nature of human interaction with these phenomena. They were then asked to consider the functionality needed to create such simulations and the enabling technologies required to provide this functionality. Participants' assessments are summarized following the discussion of the challenge scenarios.

---

[1] Several of these scenarios are discussed in more depth in Rouse's later chapter, "Strategic Thinking Via Organizational Simulation."

Enterprise Systems

The context of the first challenge scenario was the adoption of an enterprise-wide financial management system and the replacement of independent legacy systems in operating units. The value of using this case as a challenge problem is that it provides some vivid examples of how organizational factors can, and often do, have significant implications for the successful introduction of new technologies and their associated processes. Some of the challenges involved in this situation were duplication of efforts with substantial hidden financial costs, skepticism and reluctance to trust and learn a new system, and difficulty in dovetailing the new system into existing accounting processes. The organizational issues involved with this challenge included understanding the functioning of the unit practices, how the awkwardness of existing business practices can be amplified by a new system, and an appreciation of how business pressures, lack of trust, and competition undermine the commitment to challenge.

With these issues in mind one can begin to outline the concepts and phenomena that need to be simulated for Enterprise Systems. These would include, but not be limited to, addressing financial questions and problems, identifying deficiencies, and anticipating the impact of integrated information on the business process. To make all this work, interaction is needed, face-to-face interactions with people who have questions and problems, and with well-informed management team members who have the latest financial information. One could then measure the success of an organizational simulation in this area by seeing the decreased frequency of new questions and problems as well as the number of business process simplifications identified and implemented.

NASA Columbia

Organizational factors that contributed to the Columbia disaster ranged from safety issues to miscommunication, linking organizational factors to mission safety, and trading off multiple kinds of risks in decision-making. The concepts to be simulated included the safety culture, latent organizational effects that impact mission safety, and how decision-making is affected by risk taking. The types of interaction needed to mature the simulation included crew/ground interaction when a mission is in flight, one person serving multiple roles, and how shifts are handed over. Measures of success for an organizational simulation included meeting mission and safety goals, the speed and accuracy of communication and decision-making, and the workload, fatigue and stress burdened on the mission staff.

Command and Control – C2

The context of this challenge was the implementation of an integrated command and control (C2) system, including replacement of a platform-centric orientation with network-centric. The nature of the challenge was to determine how integrated C2 both eliminates the need to think about information in terms of platforms and enables near instantaneous communication among all echelon levels. The organizational issues included understanding the new roles and responsibilities of platform owners and understanding the nature and implications of echelon-skipping communications. This challenge also considered simulation of the impacts of doctrinal changes on C2 practices and outcomes. The types of interaction needed for command and control included the interactions of command teams to be distributed among the echelons and the use of collaborative tools to create one-on-one interactions among echelons. Amongst the measures of the success of organizational simulations are the speed and accuracy of situation assessment, action planning, decision-making, and communications.

Domestic Crisis - G8

The context of this challenge dealt with information technology to support communications for security for multinational meetings hosted in the U.S. The nature of the challenge dealt with planning, training, operations, and logistics during a domestic crisis associated with a G8 economic summit. The group was to discuss who decides what actions to take, who responds with those actions, and later, in lessons learned, who assesses the usefulness and practicality of those actions. The challenge included adjustment of a response plan as the crisis unfolds and the tracking of information for reporting and assessment, among other things. The types of interaction needed included communication among organizational boundaries to replan when constraints emerge. The measures of success were to create a team experience equivalent of handling ten unprecedented crises in one day and a drastic reduction in "what do I do now" radio calls.

Joint Urban Operations – JUO

The issues of interest included the fact that many JUO situations have some degree of novelty for the personnel involved, that most of the information available in the immediate environment is not relevant to accomplishing the needed goals, and that there is often a limited availability of relevant information.

Urban warfare is a relatively new concept for the U.S. military. It requires becoming aware of such things as military operations in the midst of everyday domestic affairs, knowing who's in charge, and anticipating and identifying adversaries' tactics amongst the day-to-day activities of the city. The needs for a JUO organizational simulation include multiple models to show you, your adversary, and your environment as well as plan and skill development. The success of such an organizational simulation can be measured by the speed and accuracy of situation assessment, action planning and decision-making, and minimizing civilian collateral damage.

FUNCTIONALITY AND TECHNOLOGY

Subsequent to the workshop, the considerable content generated by the working groups and plenary sessions was organized, edited, and analyzed to glean a composite set of functionality and technology needs. The somewhat simplistic criterion used to screen the large number of recommendations was that anything recommended by two or more groups merited further consideration.

The resulting 27 functions and technologies are shown in Table 1.

- 11 (41%) recommendations focus on providing the theoretical basis and knowledge needed for simulating characters and their interactions
- 8 (30%) recommendations focus on the required functional nature of simulated characters and their interactions
- 6 (22%) recommendations focus on design and simulation of the environment and means for interacting with it
- 2 (7%) recommendations focus on tools for creating, tailoring, and calibrating organizational simulations

One could interpret this to primarily be a call for increased basic research on the behavioral and social phenomena of importance to the success of organizational simulation as envisioned for the five challenge problems. To an extent, this interpretation is correct.

However, discussions at the workshop also placed great emphasis on pursuing research that would yield computational representations of these phenomena. Great emphasis was also placed on developing tools needed for designing and evaluation such computational representations.

Thus, for example, research yielding empirical results showing that leadership affects organizational change is not sufficient. To create organizational simulations, we need to know how specific leadership behaviors affect particular organizational behaviors. Further, we need to able to create intelligent agents that can exhibit both of these classes of behaviors in reacting to each other.

An overarching conclusion, therefore, is that the research issues in Table 1 should be pursued in the context of creating, evaluating, and deploying organizational simulations. This will greatly increase the value of the knowledge created by this research. It will be very clear how to put this knowledge to work enabling organizational simulations that will, in turn, provide great value in addressing the types of challenge problems discussed here.

OVERVIEW OF BOOK

The remainder of this chapter provides an overview of the subsequent chapters in the book. Table 2 provides an overview of the relationships of these chapters to the issues and needs summarized in Table 1. The cells in black in Table 2 indicate primary relationships; the cells in grey indicate secondary relationships. It is clear that the chapters in this book provide excellent sources for addressing the issues and needs identified by the experts at the workshop.

Functionality		Category
Characters that	exhibit emotions, e.g., wariness or hostility	behavior
Characters that	have personalities	behavior
Characters that	interact with each other	behavior
Characters that	learn from experience & training	behavior
Characters that	respond appropriately	behavior
Characters that	are believable	characteristics
Characters that	belong to formal organizations	relationships
Characters that	serve as coaches	relationships
Immersion that enables	ground-level experience, e.g., by actors	experience
Simulation of	external environment, e.g., regulations	environment
Simulation of	information systems & flows	information flow
Tools for	composing characters & other entities	design
Tools for	tailoring & calibration of OrgSim	evaluation
Technology		**Category**
Theory of	believability of characters	individual traits and/or competencies
Theory of	cognition	individual traits and/or competencies
Theory of	decision making	individual traits and/or competencies
Theory of	emotions	individual traits and/or competencies
Theory of	identify, role & affect	individual traits and/or competencies
Theory of	learning & adaptation	individual traits and/or competencies
Theory of	discourse	interactions among entities
Theory of	human-computer interaction	interactions among entities
Theory of	organizational behavior & learning	interactions among entities
Theory of	organizations, groups & social networks	interactions among entities
Theory of	design	science & engineering
Theory of	information, diffusion & capture	science & engineering
Theory of	modeling & simulation	science & engineering
Knowledge of	character behavior models	modeling

TABLE 1. Functions and Technologies Recommended By Experts at Organizational Simulation Workshop

Chap	Overall Topic	Theoretical basis and knowledge needed for simulating characters and their interactions	Required functional nature of simulated characters and their interactions	Design and simulation of the environment and means for interacting with it	Tools for creating, tailoring, and calibrating organizational simulations
1	Introduction & Overview	Background, Workshop & Framework			
2	OrgSim for Strategic Thinking				
3	OrgSim for Unprecedented Systems				
4	OrgSim for Learning Organizations				
5	Modeling Individuals				
6	Coordinating Joint Activity				
7	Modeling Team Performance				
8	National Differences in Teams				
9	Organizational Performance				
10	Challenges of Modeling				
11	Narrative-Based Models				
12	Agent-Based Models				
13	Petri Net Models				
14	Meta-Matrix Model				
15	Artificial Intelligence Models				
16	Simulating Humans				
17	Simulating Crowds				
18	Immersive Simulations				
19	Creating a We-Based Game				
20	Distributed Simulations				
21	Innovation in Game Communities				

TABLE 2. Mapping of Book Chapters to OrgSim Needs

The chapters in this book are organized in four sections:

- Introduction: This chapter as well as chapters that motivate the needs for organizational simulation to support strategic thinking, design of unprecedented systems, and organizational learning, including the functionality and technology required to enable this support.
- Behaviors: The state of knowledge of individual, group, and team behaviors and performance, how performance can best be supported, how performance is affected by national differences, and how organizational performance can best be measured.
- Modeling: Approaches to modeling and simulating people, groups, teams, and organizations, as well as narrative contexts and organizational environments within which these entities act, drawing from a rich set of modeling methods and tools.
- Simulations & Games: Illustrations of a wide range of fielded simulations, games and entertainment, including the methods and tools employed for designing, developing, deploying, and evaluating these systems, as well as the social implications for the associated communities that have emerged.

Introduction

The four chapters in this section begin with this chapter, "Introduction and Overview," the purpose of which is to set the stage and provide an overall framework for the book. The other three chapters motivate the need for organizational simulation to support strategic thinking, system design, and organizational learning:

- Rouse discusses OrgSim as a means to enhance strategic thinking, using illustrations drawn from business and military strategy issues. The functional and technological implications for OrgSim capabilities are discussed.
- Cross discusses OrgSim for supporting design of unprecedented systems, with illustrations drawn from business and military. He outlines use of OrgSim capabilities for developing operational and training concepts in parallel with system development.
- Andrews, Bell and Shearer discuss OrgSim as a means for supporting organizational learning, with illustrations drawn from military organizations. They review a learning maturity model and argue for the potential use of OrgSim capabilities to increase learning maturity.

Behaviors

The five chapters in this section address human behaviors, both individually and collectively. They provide impressive reviews of the knowledge base in these areas and the implications of this knowledge for organizational simulation:

- Hudlicka and Zacharias discuss requirements and approaches for modeling individuals within organizational simulations, with illustrations of military applications of these approaches.
- Gary Klein, Feltovich, Bradshaw, and Woods discuss the nature of joint activity and its coordination. They introduce the notion of "common ground," elaborate this notion with a wide range of illustrations, and propose how this construct should be embodied in organizational simulations.
- Salas, Guthrie, Wilson-Donnelly, Priest, and Burke provide a rich review of the nature of team performance and alternative approaches to modeling performance for use in organizational simulation.
- Helen Altman Klein and McHugh discuss national differences in teamwork in terms of alternative frameworks for considering cultural characteristics and their impacts on how team members interact and work together, as well as how such cultural phenomena might be embodied in organizational simulations.
- McMillan, Diedrich, Entin, and Serfaty discuss measurement of organizational performance with particular emphasis on empirical testing of the impacts of various organizational characteristics, with illustrations drawn from military organizations.

Modeling

This section includes six chapters that focus on using knowledge of behaviors to create computational models that can underlie organizational simulations. The modeling approaches are targeted at the middle levels of the architecture in Figure 1 – organizational story, characters, and world model – and illustrated in a wide range of domains:

- McGinnis discusses technical challenges of modeling and simulation, using examples of manufacturing and supply chain management to illustrate the difficulties of formally representing complex organizational systems.
- Murray discusses narrative-based models for creating the stories or plots of organizational simulations, with illustrations drawn from popular games and entertainment.

- Shah and Pritchett discuss agent-based models for computational representation of the "actors" within organizational simulations, with illustrations drawn from aviation and air traffic management.
- Levis discusses executable models of decision making organizations, with emphasis on Petri Net models, including an illustration of modeling culturally diverse command and control systems.
- Carley discusses the Meta-Model approach to representing and simulating organizational entities and relationships among these entities, as well as a range of software tools for employing this approach.
- Cross and Fouse discuss artificial intelligence models with emphasis on search, knowledge representation, planning, and intelligent agents. Of particular emphasis is the application of the state of the art in these areas to creating organizational simulations.

Simulations & Games

The six chapters in this section discuss and illustrate experiences in creating fielded simulations and games. Consequently, they address, at varying levels of depth, all the layers of the architecture in Figure 1. These chapters provide insights into what is required to develop, deploy, and support organizational simulations:

- Essa and Bobick discuss simulating humans in terms of modeling faces, bodies, limbs, etc. for the purpose of emulating talking, smiling, gesturing, walking, and so on. They review a range of methods and tools for these aspects of modeling, including their use in contemporary entertainment and games.
- Loftin, Petty, Gaskins, and McKenzie discuss simulation of crowds, including an overall analysis of the requirements associated with such simulations. They discuss the range of crowd behaviors of interest and illustrate various levels of simulation fidelity for military applications.
- Lindheim and Korris discuss immersive environments, drawing upon experiences in the entertainment industry and the initiatives at the Institute for Creative Technologies. Their illustrations clearly argue for designing experiences relative to viewers' perceptions, rather than the "bottom-up" modeling approach typical of engineering.
- Zyda, Mayberry, McCree, and Davis discuss the creation and evolution of a hit online game, ***America's Army***. They chronicle the successive releases of this game, including the motivations for a long series of upgrades. Fixes of problems were, of course, one motivation. More interesting, however, were emergent behaviors of game players that provided unexpected insights.

- Fujimoto discusses distributed simulation with emphasis on use of the High Level Architecture for design and operation of simulations that run on multiple distributed processors. Time management is of particular importance in such simulations to assure that events happen in logical orders in all views of the simulation.

- Herz discusses the social nature of online gaming and its implications for fostering innovation. Online communities are emerging in association with these games, with both practical and social attributes. Practically, these communities are sources of valuable R&D. Socially, various non-game oriented social relationships and support systems have become central for many gamers.

CONCLUSIONS

This book motivates why organizational simulation capabilities are needed, reviews the state of knowledge relative to the types of behaviors that must be simulated, describes approaches to transforming this knowledge into computational models, and illustrates a range of fielded simulations and games.

It is clear that we can and do develop, deploy, and support organizational simulations. Most of these simulations are rudimentary relative to the range of human behaviors that we would like to simulate. While there is a wealth of knowledge of these behaviors, the capabilities of tools for transforming this knowledge into computational models are fairly limited. Consequently, creation and maintenance of organizational simulations can be quite labor intensive.

Tools are the key to substantially advancing the state of the art in organizational simulation. We need much more powerful tools that enable efficient harvesting of our knowledge of human behaviors. Similarly, we also need for human behavior researchers to express their findings in more harvestable form. Perhaps the best way to dovetail these two areas would be research programs that brought the various disciplines together to address challenge problems.

The promise of organizational simulation is immense. OrgSim can enhance what we do now, e.g., in design and training, and enable new ways of working. Strategies can be deployed and evaluated in OrgSim so that designers and investors can "drive the future before they buy it." Similarly, people can be trained to operate in a future that does not yet exist. There are also obvious entertainment opportunities. Perhaps the synthetic characters that act in OrgSim can become actual workers, staffing call centers and providing expert advice in a variety of domains.

We are convinced that this range of capabilities will eventually be available. It will happen slowly, over decades, if the natural evolution of the various disciplines involved proceeds as usual. Or, it could happen much faster – less than ten years – if research investments of sufficient magnitude are targeted at a well thought-out portfolio of challenge problems. This is, of course, what we advocate.

REFERENCES

Boff, K.R., & Rouse, W.B. (2004). Organizational Simulation: Workshop Report. Wright-Patterson Air Force Base: Air Force Research Laboratory, Human Effectiveness Directorate, April.

Durlach, N.I., & Mavor, A.S., (Eds.). (1995). Virtual Reality: Scientific and Technological Challenges. Washington, DC: National Academy Press.

King, B. & Borland, J. (2003). Dungeons and Dreamers: The Rise of Computer Game Culture from Geek to Chic. Emeryville, CA: McGraw-Hill/Osborne.

Murray, J.H. (1997). Hamlet on the Holodeck: The Future of Narrative in Cyberspace. Cambridge, MA: MIT Press.

National Research Council. (1997). Modeling and Simulation: Linking Entertainment and Defense. Washington, DC: National Academy Press.

Pew, R.W., & Mavor. A.S., (1998). Modeling Human & Organizational Behavior: Application to military Simulations. Washington, DC: National Academy Press.

Rouse, W.B. (2004). "Enterprises (As) Systems." Proceedings of IFAC/IFORS/IFIP/IEA Symposium on Analysis, Design, and Evaluation of Human-Machine Systems, September.

CHAPTER TWO

STRATEGIC THINKING VIA ORGANIZATIONAL SIMULATION

WILLIAM B. ROUSE

ABSTRACT

Organizational simulation has the potential to support strategic thinking in new ways. Traditional analytical strategic analyses can be supplemented with experiential perceptions gained by acting in the future and assessing solution characteristics before investing in their actual development and deployment. This chapter considers the issues driving this need. These issues are elaborated in terms of strategy questions to be addressed by organizational simulation. Three scenarios are used to provide the basis for identification of critical functionality to be embodied in organizational simulations, as well the technologies that can enable these functions. These technologies are considered in terms of several levels of sophistication. Critical research priorities are discussed.

INTRODUCTION

Analyses of alternative courses of action can be very important, especially when "big bets" are involved. However, such analyses are inevitably limiting. Once the spreadsheets show that "the numbers work," decision makers' overarching concern is the uncertainties and risks associated with various elements of the analysis. Of course, analytical models can also incorporate these uncertainties.

More often, however, I have found that hesitancy is due to needs to experience a course of action before committing to it. Executives and senior managers – in both private and public sectors – would like to know how a course of action will "feel" before finally committing. Put simply, they would like to test drive the future before buying it.

A central difficulty with meeting this need is the difficulty of efficiently designing, evaluating, and deploying possible futures. Truly experiencing these futures is likely to require immersive, interactive portrayals. This situation can be further complicated by needs to do all this rapidly, e.g., before deploying a military force within 24 hours of recognizing the need for this force.

Organizational Simulation (OrgSim) can provide a means for experiencing the future, acting in this future, and having the future react to you. How could OrgSim enable this? As might be expected, the answer depends almost totally on the strategy questions being asked. The questions of interest in this chapter include:

- How can a new strategy best be deployed?
- What are the organizational implications of a new strategy?
- How will novel situations be addressed with this strategy?
- What are the design implications of this strategy?
- What are the work implications of a new organization?
- How well will the organization perform in the environment?

Note that these questions are more concerned with evaluating and deploying a new strategy rather than developing the strategy. However, as later examples illustrate, using OrgSim for evaluation and deployment of strategies is likely to result in significant changes of these strategies.

The OrgSim functionality needed to address these questions is explored using three scenarios where the strategy of interest involves deploying a new capability and/or way of doing business:

- Enterprise Integration (EI): Implementation of an enterprise-wide financial management system to support planning, reporting and decision making by organizational units as well as the overall enterprise.
- Integrated Command & Control (IC2): Integration of a wide range of information sources into a single source for supporting military command decision making and communications.
- Joint Urban Operations (JUO): Support of planning and training for joint operations in U.S. cities, perhaps in response to natural disasters or terrorist actions, involving federal, state, and municipal agencies.

Each of these scenarios is characterized in terms of the following attributes:

- Context of Challenge
- Nature of Challenge
- Organizational Issues
- Phenomena to Simulate
- Types of Interaction Needed
- Measures of OrgSim Success

Characterization of each scenario in this manner provides the means for identifying functionality needed by OrgSim. Alternative technologies for enabling this functionality are briefly reviewed.

STRATEGY QUESTIONS

Planners and decision makers have numerous strategic questions that they would like OrgSim capabilities to help address. In this section, these types of questions are discussed with a goal of illustrating the ways that OrgSim can support strategic thinking. Subsequent sections place these types of questions in the context of specific scenarios.

How can a new strategy best be deployed?

New strategies involving capabilities such as enterprise-wide information systems and integrated command and control have often focused on acquiring and installing these capabilities, hoping for the promised benefits. Unfortunately, many billions of dollars have been invested to show that this simple strategy often fails. In the private sector, many Enterprise Resource Planning projects have been terminated after huge investments (Economist, 1999). In the public sector, immense investments in air traffic management systems, to name just one area, have not yielded successful enhancements (NRC, 2003).

These failures have, in part, been due to software problems. It often is not software per se, but the difficulties in developing capabilities to satisfy requirements when the user cannot articulate the requirements for the capabilities as well as the way in which these capabilities will be used. There is also the fact that clever users, once presented with new capabilities, will find new uses and hence new requirements for these capabilities.

However, much more salient than software issues have been the people and organization problems associated with deploying these systems. Business process misalignment, inadequate training, and inflated expectations have contributed to immense implementation problems. Such issues need to be addressed before they are encountered during ongoing enterprise operations.

OrgSim could foster strategic thinking about these issues by enabling deployment of a large-scale system in a virtual enterprise configured to match the enterprise targeted for eventual real deployment. This would enable learning what processes have to be redesigned and evaluating potential new designs. It would also enable stakeholders (e.g., managers, employees, and customers) to experience the system prior to its deployment.

What are the organizational implications of a new strategy?

It can be difficult to fully anticipate the organizational implications of new strategies. People often discover new ways of working as they experiment with a new capability. For example, the tendency for commanders to communicate directly with the field, skipping intermediate echelons, emerged from exercises with C2 systems that made this possible, although this phenomenon was not intended.

OrgSim could provide a "laboratory" for exploration of such possibilities. It is essential that this laboratory allows for emergent phenomena – behaviors and consequences that were not foreseen in the development of the particular instantiation of OrgSim. Thus, users of OrgSim cannot be limited to following scripts. Ideally, the synthetic characters should also not be scripted.

Emergent phenomena are more likely to result when users can learn from the simulated world and devise new approached to pursuing goals. Better yet, if the synthetic characters also learn from users' choices and other behaviors. In this way, organizational implications will be discovered rather than just being limited to demonstrations of foreseen behaviors.

How will novel situations be addressed with this strategy?

Planners and decision makers would like to know how new strategies and associated capabilities fare with novel situations, not just those for which they were originally designed. They would like to gain these insights prior to actually encountering novel operational situations. This requires that an OrgSim be easily adaptable to a wide range of scenarios, both in terms of authoring scenarios and "playing" them.

Such novelty can be accomplished in a variety of ways. The environment can change, e.g., a different city, the adversary can change, e.g., an earthquake rather than terrorists, and the goals can change, e.g., humanitarian relief vs. combat operations. Users' responses to novel situations may be for the purpose of gaining experience with the specific novelty as well as gleaning general principles for dealing with novelty. Of course, they also may discover that the new strategy and associated capabilities do not fare well in novel situations.

What are the design implications of this strategy?

The knowledge gained from addressing the first three questions can be very useful for designing how to best employ the new strategy and capabilities from an organizational perspective. For example, tradeoffs between centralization and decentralization may change with new capabilities, especially for novel situations. It might be the case, for instance, that purely decentralized operations hinder an overall recognition of the true nature of novel situations.

Thus, OrgSim could enable assessing doctrinal implications of new capabilities and designing doctrine accordingly. The ability to experiment with capabilities before they really exist will tend to reduce development risks, acquisition costs, and time until deployment. Of course, the risks of doctrinal mismatch can also be reduced in this manner.

What are the work implications of a new organization?

The nature of an organization influences what work is done and how it is done. Priorities among tasks, work methods, and information flows may change when new capabilities are deployed, especially if these capabilities affect the whole enterprise. OrgSim could be used to discover, design, and evaluate work implications.

Typically, work is assumed to continue as before – the EI and IC2 scenarios illustrate this. However, as people gain experience with the new capabilities, perhaps encountering novel situations not anticipated in the design of these capabilities, they will inevitably experiment with new ways of doing things. OrgSim can enable this experimentation. In some cases, these experiments will show that new capabilities should be redesigned before being deployed.

How well will the organization perform in the environment?

Of course, the "bottom line" for most planners and decision makers is how well the organization performs in terms of revenues, profits, lives saved, or battles won. Comparing results in OrgSim to current organizational performance could provide such an assessment. Another possibility is to compare the status quo and new capabilities within OrgSim. Emulating the status quo with OrgSim is likely to provide insights into the basis for current organizational performance.

The insights gained from such evaluations are quite frequently much more important than the numbers. Understanding how new strategies and capabilities will affect the organization, how these impacts will affect work, and the consequences of these changes usually provides deeper insights into the organization and the people who work in it. Typically, this leads to new ideas for improved strategies, capabilities, organizations, and work.

Summary

The overarching questions discussed in this section illustrate how an OrgSim could affect strategic thinking. Experiencing major changes prior to their deployment enables, at the very least, better informed deployment and, quite possibly, redesign of changes before being deployed. The result is better solutions, faster, and cheaper.

This section has explored the potential benefits of OrgSim in general. The next three sections illustrate these benefits in the specific contexts of EI, IC2 and JUO, respectively. Analysis of these scenarios provides the basis for outlining the functionality needed for an OrgSim to support strategic thinking.

ENTERPRISE INTEGRATION

The private sector and, more recently, the public sector have been focused on creating lean, efficient business processes for financial and personnel management, online purchasing, customer services, etc. The means to these ends have been characterized as Enterprise Resource Planning (ERP), Customer Relationship Management (CRM), Supply Chain Management (SCM), Sales Force Automation (SFA), and so on. e-Business is often seen as key to successful ERP, CRM, SCM, and SFA.

This enterprise integration scenario focuses on implementation of an enterprise-wide financial management system. The concern is with how OrgSim capabilities could enable management to anticipate and resolve substantial implementation and adoption issues.

Context of Challenge

The context of interest is an enterprise-wide information system for financial management. Management is concerned with the substantial investment required to acquire this system, its implementation and integration across the enterprise, as well as acceptance and competent use by employees at several levels of the enterprise. Of particular note, adoption of an enterprise-wide financial management system involves replacing a wide range of independent systems currently in use in operating units.

Nature of Challenge

The current independent systems operate well, if not efficiently. However, they are not interoperable. Hence, one cannot "role up" financial reports across operating units. Overall, there is considerable duplication of efforts with substantial hidden costs, e.g., double and triple entries of information. Management expects significant difficulties in dovetailing the new system with existing financial and accounting processes. Initially at least, there is likely to be substantial skepticism and reluctance to trust and learn the new system despite progress in resolving integration difficulties. Consequently, they expect significant competition between adherents of the new and old systems.

Organizational Issues

Management would like to better understand the organization-wide implications of current disparate unit practices, including the basis and functioning of unit practices. They would also like to assess the extent to which the awkwardness of current business processes may be amplified by the new system. They would also like to gain an appreciation of how business pressures, lack of trust, and competition may undermine commitment to change.

Phenomena to Simulate

They would like to simulate people-people interactions in the process of addressing financial questions and problems using both the existing and new systems. This includes addressing likely questions, e.g., handling encumbrances, as well as resolving likely problems, e.g., proposal preparation for new customers. They would also like to simulate the people-people interactions enabled by integrated information and business processes. This includes the impacts of almost real-time financial reporting and enterprise-wide access to reports. They would like these simulations to support identifying deficiencies and designing new business processes. Finally, they would like to simulate people's social response to increased information sharing and loss of control of information.

Types of Interaction Needed

The types of interactions needed for these simulations include:

- One-on-one, face-to-face interactions with people who have questions and problems – with assistance from an "Obi Wan"
- Management team meetings with well-informed team members who have up-to-date financial information
- Veiled hostility by users of legacy systems towards those associated with the new information system

Measures of OrgSim Success

Management will deem OrgSim a success if its use, both for analysis and training, results in decreasing use of legacy systems and, consequently, a decreasing frequency of multiple entries and an increasing coverage of rollup. They also seek a decreasing frequency of new questions and problems. Another key measure is the number of business process simplifications identified and implemented. Finally, they would like estimates of the cost savings due to all of the above.

COMMAND & CONTROL

The Integrated Command & Control (IC2) scenario focuses on integration of a wide range of information sources into a single source for supporting military command decision making (AF-CIO, 2002). The essence of the idea is to enable combat commanders to "publish and subscribe" to a rich set of information sources without having to be concerned with particular sources or destinations. More specifically, commanders can "subscribe" to what they want to know without having to know whether this information came from sensor platforms such as AWACS, JSTARS, Rivet Joint, or any of the large number of available platforms. Similarly, they can "publish" information without having to specifically send this information to those who need it. This scenario builds upon studies of this concept by the Air Force Scientific Advisory Board (McCarthy, et al., 1998, 1999), as well as a more general assessment of developments in command and control (Rouse & Boff, 2001).

Context of Challenge

The context of the IC2 scenario is an integrated information system to support command decision making. Senior military decision makers are concerned with the organizational implications of deploying such a system. In particular, they are interested in the implications of both integrating information sources and adoption of the collaboration technologies often associated with state-of-the-art information systems. These concerns also include needs to understand implications for operational doctrine and the consequences of replacing the enterprise's traditional platform-centric orientation with an intuitively appealing network-centric orientation.

Nature of Challenge

Commanders are familiar with accessing information platforms, i.e., AWACS, JSTARS, Rivet Joint, etc. They also are steeped in hierarchical communications via echelons. However, integrated C2 eliminates the need to think about information in terms of platforms. Further, integrated C2 enables near-instantaneous communication among all echelon levels. It is expected that this will change the nature of command decision making, but there is no consensus on what changes should be made or might emerge (Rouse & Boff, 2001).

Organizational Issues

Senior defense decision makers would like to understand access and use of integrated C2 without reference to platforms. This includes understanding new roles and responsibilities of platform owners. They would like to assess the nature and implications of the "echelon-skipping" communications likely to emerge via information integration and collaboration technology. Overall, they would like to understand the doctrinal implications of integrated C2.

Phenomena to Simulate

They would like simulations that enable command teams to experience the functionality of integrated C2. This includes assessing the responses of echelons to echelon-skipping communications. They would like to project the impacts of doctrinal changes on C2 practices and outcomes. The effects of loosening constraints on informational flow are of particular concern. They would use OrgSim capabilities to perform "What If?" analyses of doctrine.

Types of Interaction Needed

The types of interactions needed for these simulations include:

- One-on-one interactions across echelons via collaboration tool
- Distributed command team interactions across echelons
- Framing and evaluating doctrinal alternatives and "What If?" scenarios

Measures of OrgSim Success

Senior military decision makers will deem OrgSim a success if its use, both for analysis and training, results in C2 systems with significantly increased speed and accuracy of situation assessment, action planning, decision making, and communications. They are, of course, most interested in enhanced mission performance due to these increases.

JOINT URBAN OPERATIONS

Experiences in Somalia and Iraq have clearly illustrated the complications of conducting military operations in cities (Glenn, 2001; Medby & Glenn, 2002). The "terrain" is very complicated and the resident population is often "in the way." There are also the difficulties of collaborating across military services in conducting these operations. The Joint Urban Operations (JUO) scenario elevates the complexity further by assuming that these operations occur in U.S. cities, perhaps in response to natural disasters or terrorist actions. This greatly complicates the notion of "joint."

Context of Challenge

The context of the JUO scenario is urban operations in a U.S. city involving a mix of military and public agencies, e.g., fire and police departments. There are very high risks of collateral damage to the non-combatant urban population (Loftin et al., 2005). The operations require rapid formation of cross-jurisdictional command

and control capabilities. Succinctly, this scenario involves military operations in the midst of daily domestic affairs

Nature of Challenge

A key element of this scenario involves knowing who's in charge – who can decide and authorize decision making. This involves identifying and negotiating governance conflicts and gaps, i.e., issues for which there are overlapping responsibilities and issues for which no organization perceives that it is responsible. There are, of course, the immense complications and risks of conducting military operations in urban business, shopping, and living areas. A significant element of this is the need to anticipate and identify adversaries' tactics and actions amidst predominantly day-to-day activities

Organizational Issues

From an operations perspective, a key issue involves understanding the current and emerging situation amidst largely irrelevant information. This requires, in turn, understanding the nature of day-to-day, domestic activities – what's "normal." Then, of course, there is a need to understand potential adversaries' tactics, actions, and likely attempts at concealment. Responding to such adversaries requires addressing the nature and implications of governance conflicts and gaps.

Phenomena to Simulate

Planners and decision makers need tools to visualize and represent information flows due to normal urban activities. Also needed are means to visualize cues of adversaries' likely activities and the information flows associated with these activities. There is also a need to simulate civilian responses to urban conflict in terms of behaviors, attitudes, etc (Loftin et al., 2005). Finally, an OrgSim will need an immersive street-level simulation of all of the above.

Types of Interaction Needed

The types of interactions needed for these simulations include:

- Street-level immersion as well as higher-level views of normal urban activities and adversaries' likely activities
- Military operations amidst a largely civilian population with mostly irrelevant activities and associated information

- One-on-one interactions between individuals and groups across public and military agencies (Loftin et al., 2005)
- Distributed command team interactions across agencies in terms of information flows, including impacts of governance misalignments

Measures of OrgSim Success

Similarly to the IC2 scenario, Org Sim success will be judged in terms of increased speed and accuracy of situation assessment, action planning, decision making, and communications. Analysis and training enabled by OrgSim should also decrease the time until the conflict situation is under control and population risks are mitigated. Also important will be minimization of collateral damage to civilian personnel and assets.

FUNCTIONAL REQUIREMENTS

The scenarios have illustrated the potential benefits of using OrgSim to support strategic thinking in three different contexts. The remainder of this chapter focuses on the functional implications for an OrgSim that provides these benefits. Consideration of this functionality provides a basis for considering technologies needed to realize these functions.

Table 1 summarizes the types of interactions portrayed in the three scenarios in terms of several aspects of OrgSim:

- Modes of Interaction
- Views & Information Flows
- Tasks & Experiences
- User Support

The cells marked with an X indicate the scenarios where this type of interaction is noted. It is important to note, however, that virtually all of these types of interaction apply to all three scenarios.

Modes of Interaction

These aspects of interaction emphasize human-human dealings where some of the humans may be synthetic characters – synthespians. These interactions take place one-on-one or in groups or teams. Synthespians can be well-informed, friendly or possible hostile. They learn from interacting with users and respond appropriately. Consequently, they can articulate a line of reasoning and argue with users. They may even hate users and behave accordingly.

Types of Interaction Needed	Scenario		
	EI	IC2	JUO
Modes of Interaction			
One-on-one face-to-face interactions	X		
One-on-one interactions across echelons		X	
Face-to-face team meetings	X		
Distributed team meetings & collaboration		X	X
Interactions with hostile stakeholders	X		
Interactions with well-informed stakeholders	X		
Views & Information Flows			
"Street-level" immersions			X
High-level "God's eye" views			X
Operations amidst irrelevant activities			X
Operations amidst irrelevant information			X
Information flows across people & echelons	X	X	X
Tasks & Experiences			
Planning, design, & evaluation		X	
Executing & monitoring	X	X	X
Impacts of governance conflicts & gaps			X
User Support			
Assistance from "Obi Wan"	X		

TABLE 1. Types of Interaction in OrgSim

Views & Information Flows

These aspects are concerned with both street-level perspectives, where users can walk around in OrgSim, and higher-level views and overviews where users can see everything at once. In many realistic worlds, many of the activities and most of the information have little to do with the users' goals. Nevertheless, these activities and information are coherent, making sense in their own context. One of users' key tasks is sifting through all these activities and information to determine what is relevant to the tasks at hand. For all of this to work, information has to flow realistically among people and across organizational echelons.

Tasks & Experiences

These aspects of interaction relate to what users do in OrgSim. In all three scenarios, users devise courses of action and execute them. These courses of action might be project proposals (EI), air tasking orders (IC2), or escape plans (JUO). OrgSim users also experience things happening to them, e.g., discovering a governance conflict between the National Guard and local police. OrgSim is, of course, oriented towards providing experiences of organizational phenomena.

User Support

Support includes a range of mechanisms for training and aiding users. Of course, OrgSim in itself can be viewed as a training mechanism. To the extent that the "What If?" capabilities of OrgSim are used to support investment decisions, it is also an aiding mechanism. Beyond OrgSim itself, there are likely to be needs to aid users in the use of OrgSim. This reflects the frequent need to provide aids for users of sophisticated aids (Rouse, 1991). Such aids need to embody knowledge about the use of the more sophisticated aiding.

Beyond "Obi-Wan" knowing how best to use OrgSim, it can also be useful to have aids that have deep understanding of the context of the simulation. In this way, the user can focus on the novelty of the adversaries' tactics, for example, rather than contextual details, e.g., where to find the stairs. In such situations, aiding serves a staff function that provides decision makers key information when needed.

Methods & Tools

Various methods and tools are needed to enable the types of interactions just discussed. There must be prototyping tools that enable creating experiences quickly. Methods of immersion are needed to foster perceptions that OrgSim experiences are "real." Adaptation mechanisms are needs to support learning and more proactive modifying of experiences. These methods and tools require means of measurement for evaluation, feedback, and adaptation.

ENABLING TECHNOLOGIES

The last section outlined, in a broad sense, the functionality required of an OrgSim to support the types of strategic thinking portrayed in the scenarios. This section addresses the technologies needed to enable this functionality. To do this, we first must consider what is meant by "technology."

The National Research Council (1997) reported on a study of the linkages between defense and entertainment technologies. This study categorized technology needs in terms of five classes:

- Computer-Generated Characters: Adaptability, Individual Behaviors, Human Representations, Aggregations/Disaggregations & Spectator Roles
- Technologies for Immersion: Image Generation, Tracking, Perambulation & Virtual Presence
- Networked Simulation: Higher-Bandwidth Networks, Multicast and Area-of-Interest Managers & Latency Reduction
- Standards for Interoperability: Virtual Reality Transfer Protocol, Architectures for Interoperability & Interoperability Standards
- Tools for Creating Simulated Environments: Database Generation and Manipulation, Compositing & Interactive Tools

There is little doubt that these classes of technologies are important to the types of OrgSim depicted in this chapter. However, the essence of the capabilities portrayed seems to be lost in this classification. An alternative classification scheme, admittedly tailored to the needs outlined in this chapter, is quite simple: visualization, interaction, animation, and intelligence. Infrastructure could be added as a category, but this does not add unique requirements.

Table 2 relates functional requirements to enabling technologies classified using these four categories. The entries in Table 2 provide a rich panorama of the types of technology capabilities needed to provide the range of functionality suggested by the three scenarios. Creating these capabilities will truly present significant research challenges.

The technology for enabling believable character actions and speech is fairly well developed as evidenced by many state-of-the-art computer games. While it is unlikely that someone would confuse a synthetic character with a live character, the context and pace of these games often prompts suspension of disbelief. Natural language dialog with characters will require extending the state of the art. This capability exists for well-structure dialogs, but does yet adapt well to free-

Functions	Visualization	Interaction	Animation	Intelligence
Modes of Interaction	Believable Character Actions & Speech	Natural Language Dialog with Characters	Ongoing Substantive Discourse with Characters	Characters Have Rich Knowledge Bases
Views & Information Flows	High Fidelity Portrayals of Flows & Views at Multiple Levels	Views Change in Response to Actions & Changing Information	Dynamic Phenomena Drive Information Flows & Views	Characters' "Off-Stage" Interactions Affect Flows & Views
Tasks & Experiences	Tasks & Consequences Portrayed Realistically	Characters' Interactions Affect Tasks & Consequences	Nature of Tasks & Consequences Evolve in Time	Characters Affect Nature of Tasks & Consequences
User Support	Obi-Wan Looks & Acts Reasonably	Obi-Wan Responds Appropriately	Obi-Wan Proactively Coaches	Obi-Wan Has Deep Knowledge
Methods & Tools	Rapid Prototyping & Evaluation Tools			

TABLE 2. Functions vs. Technologies

form dialog. For example, characters need to be smart enough to dismiss a question or comment as not relevant. Capabilities for ongoing discourse will require much more. For instance, synthetic characters will need to be able to recall yesterday's conversation and employ it to contextualize a current question or comment. The ultimate will be characters that actually understand domains and share this knowledge with real characters or users. Such characters will be able to provide advice and mentor other characters, both synthetic and real.

Some games and simulations do fairly well in terms of providing high-fidelity portrayals of information flows and views of the environment at multiple levels. You can both survey the worlds at a high level and zoom in to gain a more street-level perspective. Views are affected by information flows in that actions by both synthetic and real characters affect what is seen.

However, in most games the reactions of the world to what users do are quite limited. Most of the world remains fairly static until one moves to the next view. It would be more sophisticated if dynamic models of the company, economy, and world drove information flows and, consequently, views at all levels. In this way, views would change whether or not users acted.

Ultimately, OrgSim needs worlds where characters perform much of their roles "off stage." Meetings happen and battles are fought out of sight, but nonetheless affect how information flows and the content of views at all levels. This is essential because the world that OrgSim helps one understand does not all happen in one place, on one screen.

A central issue in any simulation is what users do. There might be strategies to be developed, puzzles to be solved, or battles to be fought. As a baseline, one needs tasks and consequences to be portrayed realistically. The next level would be characters whose interactions affect what is done (tasks) and what happens (consequences). To the extent that this is scripted, there should be a rich set of branches and possibilities.

A richer dynamic environment would allow the nature of tasks and consequences to evolve in time. The set of things that need to be done and the impacts of doing them would depend on what has gone on before. Experiences would not repeat. Instead, one experience would build on another. What needs to be done would depend on what has been done.

The richest environment would allow the world to be defined by the synthetic and real characters. Their choices and actions, as well as the consequences of these choices and actions, will determine what needs to be done next. In this way, the "community" of synthetic and real characters will define the agenda and priorities. While this may prompt elements of standard scripts, the play of the game will be anything but standard.

Obi-Wan is the user's guide, mentor, and coach. This support focuses on both helping in the use of OrgSim and providing domain-specific advice and guidance in the context of the simulation. At the very least, Obi-Wan should look and act reasonably. The advice and guidance should make sense at the point in the play when it is provided.

An Obi-Wan that responds appropriately to queries, perhaps via natural language voice input, would provide more natural and powerful support. This might be manifested in either a "talking to yourself" mode or as teammate or colleague in the play of simulation – just as the original Obi-Wan was manifested in the movies. Quite often, users anthropomorphize such support and this adds to the experience.

At a higher level of sophistication, Obi-Wan could proactively coach users, both in the use of OrgSim and domain-specific activities. This is more sophisticated in that the support will frequently be intrusive, stopping the play to explain and tutor. To be fully useful, Obi-Wan will need to understand when and how to intervene; when the need is aiding and when the need is training (Rouse, 1991).

Perhaps the highest level Obi-Wan has deep knowledge of the fundamental values and principles of importance in the simulated world. As in the movies, Obi-Wan tutors and mentors the OrgSim initiate in these mysteries. Considering the nature of the scenarios discussed earlier, this type of support is definitely a "tall order."

The last row of Table 2 is rather different from the other rows in this tabulation. The other rows focus on levels of sophistication of functionality and, by implication, a wealth of needs for understanding and knowledge. The last row is concerned with translating this understanding and knowledge into working instances of OrgSim. As indicated, the need is also to be able to make this translation relatively quickly. Quite a bit is known about crafting "virtual realities," but it is difficult to do this quickly (Durlach & Mavor, 1995).

Considering the breadth of implications of Table 2, one might ask whether enough is known to support the creation of the needed tools (Pew & Mavor, 1998). In light of the panorama of knowledge discussed in the other chapters in this book, it seems reasonable to conclude that much is known already. However, tools that enable manifesting this knowledge in synthetic characters with some depth, for instance, are quite limited. Doing this quickly, perhaps due to needs to address a novel situation, are very limited.

These limits are due, in part, to the gulf between those discovering fundamental knowledge of social behaviors, for example, and those creating and using tools to craft simulated worlds. Knowledge that X affects Y is useful, but what is really needed is knowledge of how X specifically affects Y. This would enable creating tools for crafting specific instances of this phenomenon. Unfortunately, the mismatch between the paradigms of research and practice can be rather significant.

CONCLUSIONS

In this chapter, the focus has been on the ways in which OrgSim capabilities can support and enhance strategic thinking. The central idea concerns the value of being able to experience solutions before committing to them, whether these solutions are enterprise information systems, new command and control concepts, or potential operational doctrine for new missions. These types of solutions tend to involve "big bets" and planners and decision makers would like both analytical and experiential insights before committing the enormous resources typically required.

The chapter began by outlining the types of questions asked by these planners and decision makers. To consider how best to answer these types of questions, we used three scenarios from private and public sector enterprises. This provided insights into the functionality needed for OrgSim to support answering these strategic questions. This functional analysis then mapped into consideration of technology needs.

There is much that we need to know – and research we need to pursue -- to realize the functionality outlined. However, as many of the other chapters in this book ably demonstrate, much is known already, across many key disciplines ranging from behavioral and social science to computing and engineering. Thus, while we do not know everything, we know a lot.

The greatest difficulty involves translating scientific knowledge into computational representations for specific domains. Such representations are needed to support creation of the experiential environments that embody OrgSim capabilities. We particularly need tools to create these environments, especially tools that enable doing this relatively quickly. A focus on tools would also help to drive both the knowledge sought and the form of the knowledge should take.

This means that we need to move beyond research whose sole purpose is tabulation of what affects what, e.g., incentives affect motivation, and seek specific relationships that are computationally useful. An example of a more specific representation of knowledge is:

- Knowledge of the bases of new product plans
- Increases organizational commitment to execution of these plans and
- Leads to broader sharing of plans that
- Tends to result in external knowledge of plans and
- A changed competitive environment
- More so in domains with relatively short product life cycles and
- Less so in domains with longer product life cycles
- Where research plans tend to be more open
- Due to sources of funding for long-term research.

The former simpler statement can only be incorporated into a simulation heuristically. The latter statement provides clear guidance for a rule-based representation of the phenomena for one specific type of incentive, i.e., knowledge sharing. Much more attention should be paid to creating forms of knowledge that can be more easily operationalized and provide the bases for OrgSim capabilities.

This chapter opened with a characterization of decision makers' needs to gain a "feel" for the impacts of major investments before committing to them. The executives and senior managers articulating these desires also indicated another desire. "Once we experience the future, " one executive commented, "and decide we like it, it would be great to be able to flip a switch and immediately have it."

This desire suggests an intriguing possibility:

- Why can't a compelling OrgSim environment become the actual world?
- Why can't synthetic characters become actual workers?
- Why can't the simulated organization become the actual organization?

Of course, this would mean that OrgSim would not only be a means for supporting strategic thinking, but also that strategic thinking about OrgSim would be a means to new types of organizations and enterprises. This possibility is indeed compelling.

REFERENCES

AF-CIO. (2002). Air Force Information Strategy. Washington, DC: The Pentagon, August.

Christensen, C.M. (1997). The Innovator's Dilemma: When New Technologies Cause Great Firms to Fail. Boston, MA: Harvard Business School Press.

Durlach, N.I., & Mavor, A.S., (Eds.). (1995). Virtual Reality: Scientific and Technological Challenges. Washington, DC: National Academy Press.

Economist. (1999). ERP: RIP? The Economist, June 24, www.economist.com/surveys /PrinterFriendly.cfm?Story_ID=322811.

Glenn, R.W., (Ed.). (2001). Capital Preservation: Preparing for Urban Operations in the Twenty-First Century. Santa Monica, CA: Rand.

Loftin, R. B., Petty, M.D. Gaskins, R.C. & McKenzie, F.D. (2005). Modeling Crowd Behavior For Military Simulation Applications. In W.B. Rouse & K.R. Boff, Eds., Organizational Simulation: From Modeling and Simulation to Games and Entertainment. New York: Wiley.

McCarthy, J., et al. (1998). Information Management to Support the Warrior. Washington, DC: Air Force Scientific Advisory Board.

McCarthy, J., et al. (1999). Building the Joint Battlespace Infosphere. Washington, DC: Air Force Scientific Advisory Board.

Medby, J.J., & Glenn, R. W. (2002). Street Smart: Intelligence Preparation of the Battlefield for Urban Operations. Santa Monica, CA: Rand.

National Research Council. (1997). Modeling and Simulation: Linking Entertainment and Defense. Washington, DC: National Academy Press.

National Research Council. (2003). Securing the Future of U.S. Air Transportation: A System in Peril. Washington, DC: National Academy Press.

Pew, R.W., & Mavor. A.S., (1998). Modeling Human & Organizational Behavior: Application to military Simulations. Washington, DC: National Academy Press.

Rouse, W.B. (1991). Design for Success: A Human-Centered Approach to Designing Successful Products and Systems. New York: Wiley.

Rouse, W.B., & Boff, K.R. (2001). Impacts of next-generation concepts of military operations on human effectiveness. Information · Knowledge · Systems *Management*, 2 (4), 347-357.

CHAPTER 3

USING ORGANIZATIONAL SIMULATION TO DEVELOP UNPRECEDENTED SYSTEMS

STEPHEN E. CROSS

ABSTRACT

A user-centered development process based on organizational simulation is proposed as a means to mitigate the risks commonly experienced in developing and adopting software-based systems with unprecedented characteristics. A system is unprecedented if the requirements for the system are inconsistent or not well understood; the system architecture (both hardware and software) is not known or is inadequate to achieve the requirements (both functional and nonfunctional), or (3) the system's acquisition and development teams have not worked together to develop a similar previous system. The risks of building and fielding such systems on a predictable schedule with a predictable cost are very high. Not only are these development risks high, but the risks associated with successfully inserting the system into the intended user organization are high because of culture change issues that often inhibit acceptance of new technology and systems. The user-centered development process explored in this chapter supports the concurrent discovery of requirements, the development and analysis of system architecture, and the facilitated change of culture to aid adoption of the new system.

INTRODUCTION

Consider the following. A senior executive of a large enterprise, perhaps the CIO or CFO, envisions improved financial management through use of a common information technology (IT) system. Each business unit within the corporation currently uses individual and independently developed IT systems that adhere to each unit's own business processes. A "one size fits all" solution has never been tried before in this organization. A contractor is hired to develop the common system. The contractor meets many times with user groups throughout the corporation. The development time is extended several times because of requirements added from these meetings. There are delays and budget overruns, but a system is finally delivered. However, it is not greeted with enthusiasm. The business units resist the new system for a variety of reasons. Some resist because the system does not fully meet their needs. Indeed, the contractor had sought to field a system that met the common functionality requirements in each unit with plans to later upgrade the system to meet additional needs. Some units don't adopt the new system because to do so requires these units to change their business processes. Even though they agree such improvements would lead to greater efficiency and more accurate budgets, the cultures are resistive to change. As a result, the vision to migrate to a common IT system is a dismal failure. Such scenarios are an all-too-often occurrence (Flowers, 1996).

Why? Often the requirements are not fully understood or there are different interpretations of requirements between the users and the developers. Management and development approaches to address these issues are well known (Royce, 1998). But there are two issues that are not well supported by development methodologies or management approaches. The first occurs when an entirely new kind of system, an unprecedented system (Beam, 1989), is being developed for which no one has the requisite experience to define requirements. In this case, requirements are discovered during the system development, suggesting the need for a methodology to support co-evolution of requirements and the system. A second related issue occurs when an organization's processes have to change in order for the organization to adopt a new software system. In this case, cultural change should be managed concurrently with system development.

What if requirements discovery, system development, and culture change could be supported concurrently by organizational simulation? That is, what if users could play a much more active role in the definition, development, and adoption of new software systems?[1] What if a team of developers, users, and their managers could co-evolve these systems while discovering better ways to work through experiencing what it would be like to work that way? That is, what if systems could be developed concurrently with the discovery of more effective processes enabled by those systems? What if this could be done in a way to break through cultural barriers by motivating the need to change? What if all of this could be done much faster than the "specify-delay-surprise" cycle of current IT system development?[2]

In this chapter, we explore how organization simulation techniques could be used to develop unprecedented systems. From this we determine five requirements that organization simulation must be able to support. The approach described is being pursued as an initial research project at the newly established Tennenbaum Institute at Georgia Tech, a multi-disciplinary initiative focused on enterprise transformation. A central tenet is enabling teams to discover and experience new ways of work by concurrently developing a systems architecture while developing

[1] By new, we mean IT systems with new or enhanced functionality. Often such systems are the result of the integration of legacy systems

[2] A survey of 1,500 software projects conducted by the Standish Group [www.standishgroup.com] found that 31% of projects are canceled before they are completed, while another 53% are late and, on average, over budget by almost 90 percent. Only 16% are completed on time and within budget. Of those projects that do eventually deliver, the average final product contains only 61% of the originally specified features.

the experience necessary to learn how to use the new system even if work processes have to change as a result. Using organizational simulation concepts, we describe a methodology to allow users to experiment in their domain to discover better ways of doing work and to experience that work during system development. Experiencing the new system may also prove to be a proactive way to support culture change. We believe this approach will also prove useful in changing established work cultures that in turn will lead to faster migration-to and acceptance-of new software systems. The measure of effectiveness will be the ability to significantly reduce the cycle time from the start of requirements development until a system is actually used for its intended purpose.

In the following sections, we first illustrate the approach through two scenarios, then examine the characteristics of high performance teams and successful approaches to organization change, and finally examine a five part research approach to formulating a new methodology for the development of unprecedented systems.

ILLUSTRATIONS OF UNPRECEDENTED SYSTEMS

The 1987 Air Force Scientific Advisory Board study described by Beam (1989) defined a precedented system as "those for which (1) the requirements are consistent and well understood, (2) the system architecture (both hardware and software) is known and adequate for the requirements, and (3) the acquisition and development teams have worked together to develop a similar previous system." Hence an unprecedented system is one where one or more of these three characteristics are not satisfied. The report goes on to examine the risks associated with each of the three characteristics and proposes development methodologies that are best suited for each risk. Clearly, the financial system mentioned in the previous section is unprecedented because the acquisition team (in this case representatives of the end-users and their management) and the development team have not worked on a system such as this before. For this type of 'unprecedentedness,' the waterfall or block upgrade development approach would be appropriate depending on magnitude of requirements and architecture risks. These characteristics of precedented systems are still true, but the complexity, size, and nature of today's IT systems suggest some added characteristics.

We are now at the threshold of a new kind of unprecedented system called the system-of systems or net-centric systems. Such systems involve many legacy or new systems, each capable of a stand alone system function in their own right, that are integrated into a "super system."' Examples in the DoD include the Future Combat Systems[3] and FORCENET[4]. Examples in the commercial sector are typified by integrated supply chain management systems that involve a major

---

[3] www.darpa.mil/tto/PROGRAMS/fcs.html
[4] www.usni.org/Proceedings/Articles03/PROmayo02.htm

corporation and its suppliers. These kind of systems are unprecedented because of the reasons outlined in the 1987 report and also because (1) no one has ever even experienced what it would be like to use such a system, hence it is impossible to state the requirements, and (2) the very suggestion of such a system requires radical rethinking about how work is done in the organization and this requires a commitment to change work process and culture. Two examples are presented to illustrate the need for an approach to concurrent requirements discovery, process invention, culture change, and system development using organizational simulation. The first case is based on the author's experience during Operation Desert Storm in 1990. The second example is a case study in the future tense, that is an envisioned capability.

Story 1 – DART

In 1990, the author was a program manager at the Defense Advanced Research Projects Agency in charge of research in knowledge-based planning and scheduling (Cross, et al., 1994). In August, he was tasked to help the United States Transportation Command (TRANSCOM) create a semi-automated capability to support the analysis of operational plans to ensure adequate transportation and logistics support. This occurred during the initial planning for Operation Desert Storm (ODS). At this time the Central Command, under General Swartzkopf, was generating a new plan every day, sometimes twice a day. But it took over four days for TRANSCOM to respond with an analysis using the existing Joint Operations Planning and Execution System (JOPES) of the World Wide Military Command and Control System (WWMCCS). General Schwartzkopf was not shy is stating his displeasure in the lack of timely analytical support. The task assigned was to develop a new system within ten weeks that would decrease the analysis time from four days to less than four hours. Technically this was not difficult because of the existence of commercial client server solutions (rather than the existing mainframe) in the same domain used by the airline industry and the expectation to explore new approaches that is part of DARPA's mandate. But early in the project, on the second day, it was obvious that the user community and their support contractors had strong reservations about adopting new technical solutions with which they were unfamiliar and which might require a change in they way they worked. In fact, no one on that second day could envision how transportation and logistics analysis could be accomplished in less than four hours.

To make a long story short, a system called the Dynamic Analysis Replanning Tool (DART) was delivered in seven weeks. DART enabled military planners and analysts to respond to General Schwarzkopf and his staff in less than 30 minutes. This included the time necessary for a trained military planner or analysts to learn the new system. This system is still being used today in command centers around the world. How did this happen? We spent a great deal of time building trust in the team of users, developers, and managers. This was done through a game we invented called "a day in the life of the transportation planner." At anytime during this game, anyone could ask, "What would you do if General Schwarzkopf demanded an answer this instant? He's on the phone and he wants to talk to you NOW!" Despite this happening about once a day to the senior officer we supported at TRANSCOM, this was rather fun and provided a venue to explore new ways to solve the problem at hand.

We started with a representation of the current process flow and got the users to envision improvements that would lead to great efficiencies. As trust among the team members gelled, new innovative ideas were proposed. One such idea was to produce a visualization of the operations plan and analysis in the form of a Gantt chart and allow spreadsheet like manipulation of that representation. It turned out that such a representation was exactly the mental model military folks had of such plans, though the medium in which they were required to work was transaction oriented based on tabular data displays. Another idea was to put the new plan representation on a teleconference so that planners and analysts from TRANSCOM and CENTCOM did not need to co-locate to resolve plan constraints. This simple idea was fundamental to removing time inefficiencies from the analysis process. In short, the users working in trust with the developers invented powerful new ways to conduct work. Through the game played, they began to experience what it would be like to work that way. This directly helped change the culture to be more accepting of new technology and new work process.

Simple role playing games, like "a day in the life of the transportation planner," provided an organizational simulation capability that proved helpful for creating an accelerated systems development methodology in which requirements were discovered concurrently with inventing new processes and changing the culture so that the new system was readily adopted and used. The example discussed in the next section is more visionary and illustrates additional requirements for organizational simulation.

A current DARPA program, Command Post of the Future, uses an enhanced development process from that used in the DART project. The "double helix" process[5] seeks to co-evolve scenario-based descriptions of possible CONOPS (Concepts of Operation), for example the user's view of requirements and desired functionality, with the system. People act out the scenarios, using the current instantiation of system functionality, to better understand and to experience the CONOPS. Based on these experiences, the CONOPS is refined and a better understanding is obtained of the functionality that needs to be addressed in the system. While this work represents the state of the art in terms of user-developer collaboration for rapid system builds, one can imagine more complex scenarios with both disruptive changes in technology and operational concepts.

---

[5] www.isx.com/projects/cpof.php

Story 2 -- Robotics-Agents-People - or RAP – Teams

A RAP team is an envisioned future team of robots, agents, and people. Consider the following scenario based on the August 2003 hotel bombing in Indonesia. Jakarta, Indonesia is the site for the upcoming ASEAN-US business conference. The meeting is expected to include high profile government finance and trade representatives as well as the CEOs of major international corporations. After years of relative stability, the leaders of fundamentalist sects are once again dividing the predominantly Muslim population of Indonesia. Throughout the turmoil, the largely pro-west government attempts to sustain normal international relations. From prison, cleric Abu Bakar Ba'asyir issues a decree that western cultural influences are to be shunned, including modern commerce, which is viewed as undue pro-western influence. A US joint command center team is mobilized and deployed to Southeast Asia with the task of assisting the Indonesian government in stabilizing the local unrest and thwarting any possible terrorist actions. Specifically, the US team is asked to support the Indonesian government on two tasks: (1) enhance public awareness and strengthen public participation in the fight against terrorism and (2) enhance early detection through increased anticipation and proactive intelligence[6]. Interacting with US, coalition and local security forces, the command center is confronted with a volatile and deteriorating situation to which they are forced to respond. They have never worked together before, though each member is skilled in his/her specific job. There are also a few new team members including a White House observer and a virtual team member, an intelligent system that can provide interactive advice on Indonesia and translate English into passable Bahasa Indonesian, the national language of the Republic of Indonesia. There are also operating constraints on the US team that represent an unprecedented situation. The rules of engagement direct the team to operate in shadow mode in support of the Indonesians. All of their actions are to be in direct support of their Indonesian counterparts without telegraphing any US involvement (e.g., such that any actions taken are to appear as consistent with Indonesian doctrine and culture). At the same time, the US team is to be ready to provide an instantaneous transfer of situational assessment and intelligence support to an off-shore Joint Task Force that is prepared to conduct a rescue operation, should the need arise.

This scenario raises some interesting questions. Without an organizational simulation capability, how would the team develop high performance team experience in the short time it has to plan and travel to the crisis location? How would it integrate and tailor its IT systems to support teamwork, some of which will likely include new processes invented for the unprecedented tasks? What if

---

[6] Though the role of the US in the scenario is fiction, the objectives are realistic. Along with enhancing physical security, these objectives were assigned by the Indonesian president to the Coordinating Minister for Cultural Affairs shortly after the hotel bombing and were described in a speech by the Indonesian ambassador to the United States. See www.embassyofindonesia.org/

our assembled team had the task to take a current hostage rescue plan (and the information gathering and other intelligence functions associated with that) and modify it into an executable plan to provide support?[7] Assume the team can quickly practice on the existing hostage rescue scenario. What improvements do they see they need to make in their own working relationships? What can they learn about Indonesian culture and doctrine (i.e., what might an Indonesian have done differently)? How can the team quickly modify the scenario to the new situation and practice it? How should the team roles and responsibilities change? What team cultural issues have to be resolved in order to operate in the background (e.g., the US team leader, General Halftrack, as he is only an invisible support to an Indonesian minister)? How can the team quickly make its systems interoperate with the Indonesians, in a disguised way and in a way that protects classified information? What new processes might have to be invented 'on the fly' to support these constraints? Suppose the team has access to a virtual team member - the new Department of State automated intelligent system for Indonesia (an interactive oracle one can query with questions about appropriate phrases, cultural norms, etc. – i.e., an Indonesian Knowledge Navigator[8]). How might an automated team member, one with which no human member of the team has ever interacted, become a trusted and useful member of the team?

The story illustrates the need to simultaneously support requirements discovery, system development, and culture change. It suggests that users in the field could tailor and extend the system based on processes they discover during training. It suggests the positive role of experiencing new ways of work while creating the system and the expertise to use the system. Before commenting on the technical requirements of an organizational simulation capability to support developing of such systems, it is necessary to consider the social dimensions of teams and culture change that such a capability must also support.

ROLE OF TEAMS AND CULTURE CHANGE

Teams need a crystal clear understanding of their purpose, accepted leadership, clear roles and responsibilities, effective processes, and excellent communication (MacMillan, 2001). We examine each one briefly.

Purpose. This may seem obvious but it is often ambiguous. Macmillan describes the five attributes of a "clear purpose" to be clarity, relevant, significant, achievable, and urgent. In the Jakatra scenario, the purpose may seem clear, but each team member may have slightly different interpretations and none of the team

[7] Embassies maintain and frequently update such plans, any new team would first look there as a place to start.

[8] See weblog.infoworld.com/udell/2003/10/23.html for more on the Knowledge Navigator, a classic visionary video.

members have ever supported a mission exactly like the one they are now called upon to support. Rehearsal is an obvious way to communicate purpose.

Leadership: This is also often vague and sometime contested even in military organizations. In a joint or coalition environment, the leader may not be the military officer of the highest rank or the civilian with most impressive sounding title. The leader may change based on task at hand.

Roles and Responsibilities (R&R). High performance teams are characterized by each team member knowing what to do and when to do it, as well as having trust in the other team members to do the same. This requires trust in capabilities, trust that those capabilities will be implemented as required, and anticipation of needs of other team members that can be met by another team member. In the DART story, the team of military planners and analysts invented new roles based on the envisioned system.

Effective Processes: A process is a description of the steps taken to accomplish some end goal. The team's initial task is to define its processes and then to master them. In the Jakarta scenario, new processes will have to be invented on the fly. The team will have to master them very quickly. A requirement of the DART system was to integrate training into the system so that a trained military planner or analyst could learn to use the new system within 30 minutes. As roles and responsibilities changed, so did workflow and work processes. The integrated training was instrumental in creating experience, and even enthusiasm, for the new ways of work.

Excellent Communication: Efficient processes connote excellent communication. For example, there would be no radio calls of the form "what do I do now?" or "why haven't you sent me the form yet?" Instead verbal communication will only be used when it is the most efficient and effective way to transit actionable information required by another team member. In the Jakatra story, creating and learning these communication patterns are critical to rapid decision making.

Because of the critical role of teams in future unprecedented systems – both teams of acquirers and developers and the teams of humans that will use the new system to do work – we state a key principle: design the training first. System development should start with a simulation of the system. This simulation should be used to aid discovery of requirements for the actual system, to invent new more powerful processes, and to gain team experience in use of the system as an approach for changing the culture into one that will accept the system. The software substrate of the simulation, namely its architecture, should be used as a basis for the systems architecture.

Attention to culture change issues is a critical concern since many good systems experience a kind of "organization foreign body rejection phenomena." That is, people refuse to use the new system because to do so requires a fundamental change in how they work. Kotter and Cohen (2002) have written extensively on culture change. They advocate an eight-step process to facilitate culture change.

1. Establish a sense of urgency
2. Create a guiding coalition
3. Develop a vision and strategy
4. Communicate the changed vision
5. Empower broad-base action
6. Generate short-term wins
7. Consolidating gains and producing more change
8. Anchor new approaches in culture.

The DART story illustrates each of these steps. Because there was a crisis defined by both a well-understood and urgent military need, it was easy creating a sense of urgency and establishes a guiding coalition of users and developers. But there was resistance to change nonetheless. One key element of success of this project was not for the author to speak at meetings, but to have the senior general at TRANSCOM plus the most respected military users speaks about their visions and how the new system was helping realize that vision. Another key to success was the weekly release of new functionality and a clear communication of how that functionality addressed the vision. Measures of success assigned by the senior general at the beginning of the project were used to demonstrate progress.

The Jakarta story illustrates slightly different needs. Though there is still a sense of urgency, all of the other issues of culture change need to be addressed. It seems fundamental that a key to supporting the culture change is the direct involvement of the actual end-users in the development and their use of the evolving system to support the creation of their team experience. An unanswered question we hope to explore in our research is if by building systems by the above mentioned principle, one can change the culture in more proactive and positive ways rather than through the recognition of a crisis and the possible fear that that create. That is, can culture change become fun and productive?

REQUIREMENTS FOR ORGANIZATIONAL SIMULATION

As previously discussed, an approach to unprecedented systems must address concurrent requirements discovery, system development, and culture change. In this section we discuss five requirements: architecture, process, scenarios, agents, and measurement.

The key ideas in modern software engineering are architecture focused development and component prototyping. An architecture basis supports engineering analysis of systems attributes such as scalability, evolvability, survivability, and usability (Bass et al., 2003). Prototyping has long been used as a method to aid expression of requirements since it is often difficult for end users to express these requirements in English. Hence prototypes provide an effective means to elicit requirements and create a shared understanding of them between developers and users.

Architecture

As software development involves a team (e.g., an architect, developers, coders, testers, and users), a commitment to defined and repeatable processes has been important for ensuring projects produce desired result on time and on budget. The following example illustrates these concepts. In the DoD, it is common to talk in terms of operational architecture or view, technical architecture or view, and systems architecture or view[9] (DoD, 2003). These are defined as follows.

Operational View (OV): A description (often graphical) of the operational elements, assigned tasks, and information flows required to accomplish or support the war fighting function. It defines the type of information, the frequency of exchange, and what tasks are supported by these information exchanges.

Technical View (TV): A minimal set of rules governing the arrangement, interaction, and interdependence of the parts or elements whose purpose is to ensure that a conformant system satisfies a specified set of requirements. The technical view identifies the services, interfaces, standards, and their relationships. It provides the technical guidelines for implementation of systems upon which engineering specifications are based, common building blocks are built, and product lines are developed.

Systems View (SV): A description, including graphics, of the systems and interconnections providing for or supporting war-fighting functions. The SV defines the physical connection, location, and identification of the key nodes, circuits, networks, war fighting platforms, etc., and specifies system and

[9] Different terms, but the same intent is used in other literature, e.g., (Royce, 1998).

component performance parameters. It is constructed to satisfy Operational View requirements per standards defined in the Technical View. The SV shows how multiple systems within a subject area link and interoperate, and may describe the internal construction or operations of particular systems within the architecture.

The OV describes how the system will be used from the users' perspective and is related to the Concept of Operations or CONOPS. For the software or system engineer, the OV will be described in use cases and may be expressed in a formal language such as the UML (Unified Modeling Language) (Booch, et al. 1999). The TV describes the building codes with which systems will be constructed. The SV is the actual architecture of the system to be fielded. It satisfies the needs as defined in the OV and adheres to the constraints imposed by the TV.

Technologies now exist for XML compliant modeling of architectures. Tools and methods exist for evaluating the architecture from the perspective of scalability, real-time performance, survivability, and other nonfunctional attributes. For example, the SEI's Architecture Tradeoff Analysis Method (ATAM)[10] is a commonly utilized analysis technique for architecture assessment. Some work has been done on executable and model-based architectures, which enable early execution in a simulation environment.

A new approach in the software engineering community, the Model Driven Architecture (MDA), attempts to take advantage of the adoption of software architecture approaches. MDA is designed to document the organization's business, business rules, and structure of its information infrastructure such that this expression of the business need for software can be used to guide IT projects (Brutzman & Tolk, 2003). This prevents smaller IT projects from having to "re-imagine" the business case for the software and having to "re-define" reusable business rules. One of the ideas for MDA is to document the business rules using UML-based use cases during the design and implementation of the individual software projects. The power of the MDA approach is that the people that know the business will develop and maintain the business model; while the software engineers will develop and maintain the design and documentation of the software, also using the UML. MDA and UML provide the maintainable enterprise documentation of the "as-is" system and an expressive, implementable statement of the "to be" system. Levis has pushed these concepts further in creating executable architectures based on a model of the domain of the system (Handley & Levis, 2003).

Modern software engineering exploits the benefits of prototyping to understand user requirements and an early commitment to architecture to ensure adherence to desired nonfunctional properties such as interoperability and scalability. Figure 1 illustrates these concepts. These capabilities however do not support rapid realization of new systems in contexts of unprecedented problems.

---

[10] Described at www.sei.cmu.edu/ata/ata_init.html.

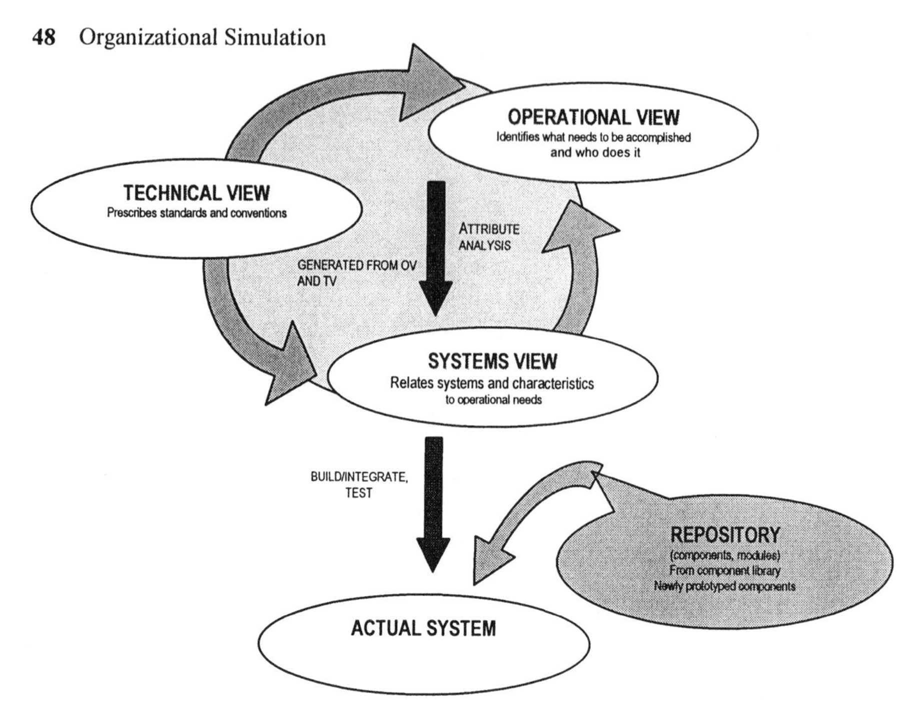

FIGURE 1. Graphical Overview of Modern Software Engineering

Such problems need to be addressed by teams of people and their automated support systems. In order to support the creation of the high performance team experience, such teams need to discover ways to work effectively and to train rapidly while concurrently evolving and developing an integrated, more capable IT system. The capabilities of rapid prototyping and architecture based development, especially pulling on newer concepts of model-based architecture and executable architecture set the stage though for a new, more powerful development methodology that will make possible better team processes.

Process

One popular approach to the definition and use of software process has been the Software Engineering Institute's Capability Maturity Model® (CMM®). The CMM is a process model. Process models represent best processes used by high performance organizations, in this case high performance software engineering organizations. A process is set of activities, methods, and practices that people employ to develop and maintain software and associated products (e.g., project

---

® Capability Maturity Model and CMM is registered in the U.S. Patent and Trademark Office by Carnegie Mellon University.

plans, design documents, test cases, and documentation). A CMM may contain process definitions for project management, engineering management, organization training and assessment, measurement and analysis, and continuous improvement. A review of software processes and software process improvement is at (Cross, 2002).

It is interesting there has been little written about the commonality between the work of the software process improvement community and the workflow and business process re-engineering communities. The latter communities have been evolving their capabilities so that they are now closely aligned with the modern precepts of software engineering. Abstractly, one can view this work as progressing through three stages and poised to make the leap into a fourth stage directly related to organizational simulation concepts. In their recent book, Smith and Fingar (2003) describe three phases of business process improvement over the past 80 years.

Phase 1: Based on the management theory of Fredrick Taylor in the early 20th century, process was implicit in manual work processes. These processes were not automated. An example is an assembly line. An example today might be a software development process such as quality assurance. A software development organization might have an approach for testing, such as regression testing on code delivered from a development organization to a testing organization. The process followed by the testers is written down in a notebook and perhaps testing literature to tell them what they should do at each step of the testing process.

Phase 2: In the past two decades many organizations have sought to reengineer their business processes. This has involved a one-time manual redefinition of a process and compilation of it into an Enterprise Resource Planning (ERP) or other off-the-shelf tool. The tool might be enhanced with a document-centered workflow management system. But the process is implicit and static. An example from software development might be quality assurance process improvement where peer reviews are added at two phases of software development - on the software design and on the software source code. Effectively this process improvement is an example of moving quality assurance earlier in to the development cycle (e.g., code reviews, design reviews). In practice, such an improvement has often led to a 6x increase in software quality as measured by the number of defects (or bugs) found during compiled source cost test. But the key issue here is that the new and improved process is captured and encoded in a static form, often in a process notebook or a web based library. The process user must read and comprehend the process and commit to its use in order to realize its benefits.

Phase 3: In the third wave, the design goal is the support for change. Explicit process representation is separate from the application. The ability to change is valued as much, if not more, than the creation of the product the process is focused upon. A current ongoing example in business is the GE Digitization Initiative.

Here the company is seeking to transform its enterprise through agile customization of its processes and the processes of its suppliers in order to meet any competitor along any measure of success. We might call this the "teams' present experience."

The goal is to have automated systems respond in synchronization with teams of people in ways that support how those people want to work right now. Such is at the forefront of software development right now. There have been many attempts to build automated process engines and automated tool support to guide software development teams. And there has been extensive debate in the software community about the best kind of processes to use in a given situation. For example, when the software development requires ultra high reliability, as in a medical instrumentation system, extensive regression testing is appropriate. When a software system has a high dependency on human-computer interaction, rapid prototyping and user-centered testing approaches are more appropriate. A process toolkit that could respond to the demands of the needs of the software testing team in the context of the software under test might prove highly valuable.

We believe that organization simulation concepts will enable a new phase.

Phase 4: Our work points to a new fourth phase which we might also call the "teams' future experience." We know from many other domains that the team that trains together, that builds trust and mutual respect for each other's capability to perform a given task, performs better. An experimental framework that allows teams to practice their work in the context of a scenario and then change workflows and processes on the fly to experience different ways of working would conceivably produce useful team experience more rapidly. Such an environment would allow teams to focus on developing the five essential traits of high performance as previously discussed (clear purpose, accepted roles, accepted leadership, effective processes, and excellent communication). Acquiring these team traits in minimal time will also require an experimental framework capable of supporting the change in culture. It is this "Phase 4" vision that suggests the need for a "to be" approach for developing next generation IT systems such as those motivated by the previously described scenario. By combining the iterative nature of architecture development, with organization modeling like what is done with software teams or the work of Carley (2003) with crisis action organizations, we have a powerful basis for organizational simulation.

Scenarios

Scenarios are user-defined portrayals of desired functionality in the context of real problem solving. They are stories. We envision a richer form of scenarios that allow exploration of multiple story lines and that allow narrative form expression of the stories with automated translation into XML compliant language suitable for

UML-type use case representation[11]. Authoring new scenarios based on interactive narrative technology will facilitate the construction of highly complex simulations. Authoring environments and interactive applications that combine narrative structures, interactivity, and the design of game worlds will maximize replay as well as comparison of outcomes across multiple parameterized instantiations. Domain analysis will be used to extract narrative elements (character types, situations, events) from the organizational domain; these narrative elements are incorporated into the authoring system that end-users will use to explore and describe the architecture's operational view. Analyzing the results of a complex team simulation to determine where process breakdowns resulted in poor performance may be difficult due to the complexity of simulation traces. By analyzing such traces and turning them into a story, narrative abstraction highlights the important chains of cause and effect leading to a breakdown. By presenting these causal chains in a narrative form (focusing on the conflicts between team members and emotional reactions to these conflicts); the lessons of a given simulation run are made more compelling and memorable. Narrative abstraction makes use of AI models of narrative structure to take a simulation trace and display it to the user as a narrative, perhaps visually representing the progress of the simulation as a sequence of images or as interactions between real-time, 3D characters.

Agents

Mateas (2002) has written extensively on stories and automated agents. His ABL (A Behavior Language) is a real-time agent architecture for complex graphical agents capable of pursuing both individual and team goals. The real-time, graphical agents can be used to depict an organizational simulation at the level of individual people. ABL is a reactive planning language; an ABL agent consists of a library of behaviors with reactive annotations. ABL has a number of mechanisms for authoring behaviors that continuously react to the contents of its working memory, and thus to sensed changes in the world. The details of sensors depend on the specific world and agent body. To support rapid development of agents for simulations, we are developing an authoring framework for assembling agents for organizational simulations out of a library of preexisting agent types and common behaviors.

Measurement

One of the many challenges will be to understand the characteristics of the organization well enough to predict and forecast its behavior. One key subset of the data needed is a measure of the interactions between members of the many

---

[11] This works builds on the research of Professor Janet Murray, which she discusses elsewhere in this book.

small teams that accomplish the work of the organization. Collection of this small team data provides the parameterization needed for organizational simulation. Once such simulations have been used to affect organizational change, the same infrastructure used to collect team data can then be used to monitor the effects of the changes to the organization at the important small team level.

Putting it all together

We envision a new process for requirements discovery, culture change, and system evolution which is an advance over the current "as is" process shown in Figure 1. Our vision of the "to be" state is depicted in Figure 2. We add a user-centered and user-controlled outer loop to enhance the current practice of modern software engineering. In this outer loop, rapid realizations of end systems (initially possible stubbed systems with roles played by humans and then incrementally enhanced prototypes) are embedded into an experimental framework that supports real play in the context of the intended problem-solving environment. Instrumentation collects data, some quantitative (such as the number of communications that take place), and some qualitative (such as how effective the team members felt the problem solving was). Team members can change the story line through narrative descriptions of scenarios. These tools are linked to use case tools such that automated transformation of a narrative form results in a modified use case. We believe such an environment will allow users to quickly state and then improve their processes and iterate on representations of use case and architecture that will concurrently transform the system into a usable end product.

SUMMARY

We have described the role organizational simulation can play in developing unprecedented systems. By unprecedented, we mean systems that support teams of people, possibly net-centric or system of systems, which have never before been built. Hence, no one has experience in using or building such systems. This means that requirements have to be discovered by a team of people while they invent new work processes. Given the high failure rate of introducing new software intensive systems into organizations, it is argued that organizational simulation should be used early and throughout system development. One key principle is that system design and development should start with simulation and training. The architectures created for that purpose should also be used to develop and analyze the system architecture. In parallel with the system development, exercises and games should be played to develop trust and enthusiasm for the new system. By this means, the teams learns to work together in new situations, learns to trust and use the system, invents and advocates the use of improved processes, and by such changes the culture for the better.

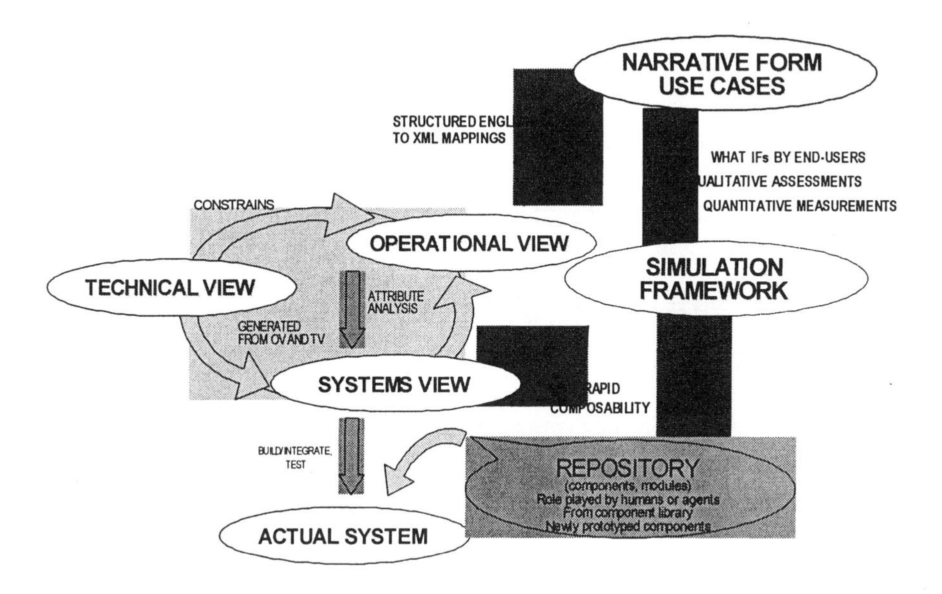

FIGURE 2. Evolving an Unprecedented System Using An Organizational Simulation Capability

REFERENCES

Bass, L., Clements, P., and Kazman, R. (2003). Software Architecture in Practice, 2nd Edition, Boston: Addison Wesley.

Beam, W. (1989). Adapting Software Development Policies to Modern Technology, Washington: National Academy Press.

Booch, G., Rumbaugh, J., and Jacobson, I. (1999). Unified Modeling Language User Guide, Reading MA: Addison-Wesley.

Brutzman, D. & Tolk, A. (2003). JSB Composability and Web Services Interoperability via Extensible Modeling & Simulation Framework (XMSF), Model Driven Architecture (MDA), Component Repositories, and Web-based Visualization, Washington D.C: US Air Force, JSB Program Office.

Carley, K. (2003). Computational Organizational Science and Organizational Engineering, Simulation Modeling Practice and Theory, 253-269.

Cross, S. (2002). "Reflections on the Process Revolution," IEEE Tutorial on Software Management, 6th Edition, Reifer, D (Ed). Hoboken (NJ): Wiley-IEEE Computer Society Press.

Cross, S, Thompson, E, Fouse, S., & Smith, A. (1994). A Methodology for Intelligent Systems Technology Transition Through Technology Demonstrations, Proceedings of the 7th International Symposium on Artificial Intelligence, 111-118.

DoD Architecture Framework Working Group. (2003). The DoD Architecture Framework Version 1.0 (Final Draft). Washington D.C: Department of Defense.

Flowers, S. (1996). Software Failure: Management Failure, Chichester UK: John Wiley & Sons.

Handley, H. and Levis, A. (2003) Organization Architectures and Mission Requirements: A Model to Determine Congruence, Journal of Systems Engineering, 184-194.

Kotter, J. and Cohen, D. (2002). The Heart of Change, Boston: Harvard Business School Press.

MacMillan, P. (2001). The Performance Factor: Unlocking the Secrets of Teamwork, Nashville: Broadman & Holman Publishers.

Mateas, M. and Stern, A. (2002). A Behavior Language for Story-Based Believable Agents. Working notes of Artificial Intelligence and Interactive Entertainment. Ken Forbus and Magy El-Nasr Seif (Eds.) Menlo Park: AAAI Press.

Royce, W. (1998). Software Project Management: A Unified Framework, Reading (MA): Addison-Wesley.

Smith, H. and Fingar, P. (2003). Business Process Management: The Third Wave. Tampa: Meghan-Kiffer Press.

CHAPTER 4

THE LEARNING ORGANIZATION AND ORGANIZATIONAL SIMULATION[1]

DEE H. ANDREWS, HERBERT H. BELL AND ROBERT N. SHEARER

ABSTRACT

In this chapter, we examine the role Organizational Simulation can play in helping organizations of the twenty-first century become more effective Learning Organizations. We begin by defining the learning organization and providing a brief review of the concepts that underlie learning organizations. We then explore the potential interaction between learning organizations and organizational simulation. Finally, we discuss the challenges, opportunities and recommendations for a way ahead in Organizational Simulation.

THE LEARNING ORGANIZATION CONCEPT – DEFINITION AND CONCEPTUAL FOUNDATIONS

During the 1980's it became clear to many business strategists and organizational development experts that some organizations were more flexible and nimble at gaining competitive advantage and improving their value chains then their competitors. Examination of these organizations indicated that one of the fundamental differences was that more successful organizations appeared more capable of learning from both the positive and negative experiences that they and other organizations experienced. These more successful organizations were described as effective learning organizations. The distinguishing characteristic of a learning organization (LO) is its skill "...at creating, acquiring, and transferring knowledge, and at modifying its behavior to reflect new knowledge and insights." p. 80 (Garvin, 1993).

All organizations are Learning Organizations (LO) to some degree. Regardless of whether they bake bread, field professional sports team, deliver education, deliver social services, or provide national security; every organization creates and uses new knowledge. However, all organizations are not equally effective LOs. As Gavin's LO definition points out, effective LOs are skilled in

[1] The opinions expressed in this chapter are those of the authors and do not necessarily represent the official views or policies of the Department of Defense or the Department of the Air Force.

creating, acquiring and transferring knowledge. In addition, good LOs have the ability to modify their behavior based on the new knowledge and insights. This modification of behavior is not just at the upper ends of the organization; good LOs have determined how to stimulate the entire organization from top to bottom with an understanding of the necessity to make positive changes based upon new knowledge.

As we have considered the purpose and promise of Organizational Simulation (OrgSim), we have been struck with how well that concept and the concept of the LO mesh. In this chapter, we will explore ways in which the LO concept can drive OrgSim developers toward better models and simulations. It is important to note that the OrgSim community still has a long way to go to achieve this goal. However, we believe that developing quality simulations of, and for, organizations can have great usability as organizations strive to become effective LOs. In turn, we believe that developing better LOs can be a major reason for the development and testing of OrgSims.

While the formal LO construct is relatively new, LOs have existed in some form since humans first began to organize themselves into groups. For example, armies that learned lessons more quickly than their foes typically defeated their adversaries. Nations that learned how to grow crops and trade more efficiently became powers in their eras. The current interest in LOs is being driven by changes in the business world that have brought the need for better LOs into focus.

Companies in highly competitive environments have constantly been searching for ways to improve their entire business enterprise. The notion of formalizing the need for constant behavioral change based on new learning has become necessary, as companies have come to truly realize that their employees are their most important resource. Such a catch phrase has been used in the past to make the employees feel that they were valued. However, employees often got the sense that large organizations viewed employees as necessary evils as demonstrated by the huge downsizing actions in the 1980's and 1990's. Some companies however, have come to truly believe that the progress of the organization is fundamentally bound to the capability of all its employees to change their behavior in the workplace. These companies have found that only when every employee was not only allowed, but also actually encouraged to think in new ways, did the company truly progress toward its strategic vision.

Organizations are seeking to improve performance and the bottom line. This frenetic search has resulted in a plethora of initiatives such as TQM (Total Quality Management) and BPR (Business Process Reengineering). Companies are finding that such efforts very often succeed or fail depending on a number of variables such as skills, attitudes and organizational culture. It also appears that many implementations are geared to highly specified processes, defined for anticipated situations. Current interest in Learning Organizations derives from the recognition that these initiatives, by themselves, often do not produce the desired organizational results. Organizations need to:

- Cope with rapid and unexpected changes
- Provide flexibility to cope with dynamically changing situations
- Allow front-line staff to respond with initiative based on customer needs vs. being constrained by business processes established for different circumstances

As Tom Peters put it, "To meet the demands of the fast-changing competitive scene, we must simply learn to love change as much as we have hated it in the past." p. 55-56 (Peters, 1978).

Organizational Simulation

At the same time that the LO construct gained credibility; the modern notion of an OrgSim construct became feasible. Not only have the computational models required for calculating the many permutations and combinations of a modern organization become commonplace, but also advances in programming techniques have made it possible to start to simulate the significant complexities of modern organizations. Perhaps even more importantly, the research and practice communities in a variety of disciplines (e.g., psychology, sociology, economics, anthropology, etc.) started to gather the necessary data to develop valid models of organizations (Pew and Mavor, 1998). We are at just the beginning of this era. Despite the millennia of human organization, we still do not know very much about how to properly model organizations. However, we are hopeful current activities in this area are just the start.

THE LITERATURE OF THE LEARNING ORGANIZATION

The implication of the concept of the Learning Organization is that organizations can learn. The construct "Organizational Learning" (OL) has been the subject of considerable attention in a broad range of disciplines including: Social Psychology, Organization Theory, and Business Management (de Geus, Morecroft, & Sterman, 2000). All of these disciplines are interested in the question of how organizations learn. There is a great deal of divergence of thought on what OL really means.

Philosophy of an LO

Senge (1990) in his seminal work on the subject defines the Learning Organization as an organization in which individuals cannot *not* learn because learning is a key part of the organization's fabric. Senge argued that the concept of the Learning Organization is very relevant given the increasing complexity and uncertainty of the organizational environment. Indeed the only real and sustainable competitive advantage organizations have is the rate at which organizations can learn faster than the competition. Senge, et al (1999) suggested a number of key questions for senior management as they strive to develop a better LO.

- Is the organization willing to examine and challenge its sacred cows?
- What kinds of structures has the organization designed for this testing?
- When people raise potentially negative information, does the organization shoot the messenger?
- Does the organization show capabilities it didn't have before?
- Does the organization feel as if what it knows is qualitatively different – "value-added" from the data its taken in?
- Is the knowledge accessible to all of the organization's members?

Argyris (1977) and Nevis, DiBella and Gould (1995) also highlight the importance of experiential learning. Argyris characterizes organizational learning as the process of "detection and correction of errors" (p. 116) and observes that organizations learn through individuals acting as agents for the company. He describes how these learning activities would in turn, facilitate or inhibit a group or system of factors that may be called an organizational learning system. A similar emphasis on experience is reflected in Nevis et. al. (1995) definition of organizational learning as "the capacity or processes within an organization to maintain or improve performance based on experience" (p. 73). Nevis argues that learning is a systems-level construct because that learning stays within the organization, even after individuals leave.

Characteristic Activities of a Learning Organization

Garvin (1993) identified a number of activities that are common across effective LOs. First LOs routinely practice systematic problem solving. Such problem solving is guided by a systems level view of the organization and its relationship with its environment. In addition, such problem solving emphasizes the priority of data over assumptions. Once potential solutions had been identified these effective LOs often under take demonstration projects to test the solution and learn from that experience. In addition, these LOs also provide incentives for risk taking and rewarded personnel for productive failures.

Kerka (1995) states that most learning organizations work on the premise that 'learning is valuable, continuous, and most effective when shared and that every experience is an opportunity to learn.' Kerka argues that the following characteristics appear in some form in the more popular conceptions of learning organizations:

- Provide continuous learning opportunities.
- Use learning to reach their goals.
- Link individual performance with organizational performance.
- Foster inquiry and dialogue, making it safe for people to share openly and take risks.
- Embrace creative tension as a source of energy and renewal.
- Are continuously aware of and interact with their environment.

The Five Disciplines

Another way to think about organizational learning is to consider the five disciplines offered by Peter Senge (1994). These are learning disciplines that include lifelong programs of study and practice. And as the name suggests, these are not easily attained or mastered by organizations—they more accurately describe a culture, a way of doing things—almost at the reflex level of human behavior:

- Personal mastery – learning to expand our personal capacity to create results we most desire and creating an organizational environment which encourages all its members to develop themselves toward goals and purposes they choose.
- Mental Models – reflecting upon, continually clarifying, and improving our internal pictures of the world, and seeing how they shape our actions and decisions.
- Shared Vision – building a sense of commitment in a group, by developing shared images of the future we seek to create, and the principles and guiding practices by which we hope to get there.
- Team Learning – transforming conversational and collective thinking skills, so that groups of people can reliably develop intelligence and ability greater than the sum of individual members' talents.
- Systems Thinking – a way of thinking about, and a language for describing and understanding forces and interrelationships that shape the

behavior of systems. This discipline helps us see how to change systems more effectively, and to act more in tune with the larger processes of the natural and economic world.(pp. 6-7)

Organizational learning: Adaptive Learning vs. Generative Learning

All organizations learn. How and why they learn sets apart the good Learning Organizations from the also-rans. Whether an organization merely adapts to new situations or actually generates the thinking required to produce new situations can mean the difference between survival and extinction.

The current view of organizations is based on adaptive learning, which is largely concerned with coping. Senge (1990) notes that increasing adaptiveness is only the first stage, companies need to focus on Generative Learning or "double-loop learning" (Argyris, 1977). Generative learning emphasizes continuous experimentation and feedback in an ongoing examination of the very way organizations go about defining and solving problems. In Senge's (1990) view, Generative Learning is about creating – it requires "systemic thinking," "shared vision," "personal mastery," "team learning," and "creative tension" between the vision and the current reality of Generative Learning, unlike adaptive learning, requires new ways of looking at the world.

To maintain adaptability, organizations need to operate themselves as "experimenting" or "self-designing" organizations and they should maintain themselves in a state of frequent, nearly continuous change in structures, processes, domains, goals. This focus should be maintained even in the face of apparently optimal adaptation

Activities and Processes of Good Learning Organizations

Over the years, observers of good Learning Organizations have defined activities and processes that are hallmarks for their success. These are some of those processes:

- Strategic and Scenario Planning – approaches to planning that go beyond the numbers, encourage challenging assumptions, thinking 'outside of the box'. They also allocate a proportion of resources for experimentation.
- Competitor Analysis – part of a process of continuous monitoring and analysis of all key factors in the external environment, including technology and political factors. A coherent competitor analysis process that gathers information from multiple sources, sifts, analyzes, refines, adds value and redistributes is evidence that the appropriate mechanisms are in place.

- Information and Knowledge Management – using techniques to identify, audit, value (cost/benefit), develop and exploit information as a resource (known as IRM – information resources management); use of collaboration processes and groupware e.g. Lotus Notes, First Class to categorize and share expertise.
- Capability Planning – profiling both qualitatively and quantitatively the competencies of the Organization.
- Team and Organization Development – the use of facilitators to help groups with work, job and organization design and team development – reinforcing values, developing vision, cohesiveness and a climate of stretching goals, sharing and support
- Performance Measurement – finding appropriate measures and indicators of performance; ones that provide a 'balanced scorecard' and encourage investment in learning (see, for example, Measuring Intellectual Capital).
- Reward and Recognition Systems – processes and systems that recognize acquisition of new skills, teamwork as well as individual effort, celebrate successes and accomplishments, and encourages continuous personal development.

EXAMPLES OF ORGANIZATIONS STRIVING TO BE EFFECTIVE LEARNING ORGANIZATIONS

Tracing the key elements of quality learning organizations through existing organizations can be somewhat difficult because key concepts manifest themselves differently in different organizations. However, here we present three examples of military organizations that are striving to improve their LO capabilities.

U.S. Department of Defense and Terrorist Organizations

The war on terrorism has many elements. Chief among them are the U.S. Department of Defense (DoD) and the terrorist organizations now plaguing the world. Al Qaida has proven itself to be a formidable LO. It has been organized into many separate cells which each appear to have a considerable amount of autonomy. It has shown itself quite capable of learning from past successes and failures and modifying its behavior based upon that learning. For example, the dastardly attack of 9/11 came about after numerous attempts to highjack aircraft around the world. Two notorious plots that were foiled was one to hi-jack aircraft and fly them into the Eiffel Tower in Paris, and a plot to hi-jack a half dozen U.S.

jetliners over the Pacific Ocean and crash them. Although those plots were discovered and the attacks never happened, there is evidence that Al Qaida planners learned lessons about gathering intelligence, protecting tactical plans, and penetrating airport security from the failed plots. U.S. anti-terrorism authorities have repeatedly warned that the war on terrorism will take many years to conclude. We believe that is largely because the enemy is a shrewd LO.

The U.S. DoD is working hard to improve as an LO. The Secretary of Defense has made "transformation" of the DoD a major objective for many years to come. Transforming such a large organization (well over two million people) from a cold war institution ready to take on the Soviet Union, to a flexible, responsive organization that can take on shadowy, splintered terrorist cells is a monumental task. Secretary of Defense Donald Rumsfield has said,

> "It's not good enough to be capable of fighting big armies and big navies and big air forces on a slow, ponderous basis. We have to be able to move quickly and have to be agile and have to have a smaller footprint. And we have to be able to deal with the so-called asymmetrical threats, the kind of threats that we're facing with terrorists and terrorist networks. So I think the people in this department understand it and that they're making good progress on it. It's a difficult thing to do with a great big institution like this. If the transformation initiatives under way are as successful as the department believes they will be, I think there will be some success." (p. 1) (Sample, 2003)

Al Qaida has the advantage of being much smaller than the DoD. While there are many examples of large organizations that have made significant progress towards being more effective LOs, it is accurate to say that the smaller an organization is, the better chance it has of being able to implement key LO characteristics more quickly across the organization. Small organizations generally have fewer organizational levels and shorter communication lines so that LO leadership at the top is more quickly and easily communicated to the rest of the organization. However, while Al Qaida has those advantages, since 9/11 it has had its capability to communicate significantly curtailed because the U.S. and its allies have focused tremendous intelligence, military and financial assets on isolated Al Qaida cells.

The Office of the Under Secretary of Defense (2002) has launched a Training Transformation effort to enable the DoD's move towards a better LO. The OSD has stated, "The Department of Defense is transforming. ... Among the principal determinants of that transformation are: a new continuously changing environment, the need for improved and expanded 'jointness,' and the opportunities made possible by advanced technologies." (p. 1) The Chairman of the Joint Chiefs of Staff, General Richard B. Myers said, "To evolve into a decisive superior force, transformation must spread across doctrine, organizations, and training – not just materiel solutions." (p.2)

Efforts like Training Transformation are examples of the DoD's intent to become a more effective LO. Strides in transformation are already obvious, but it is clear the DoD leadership is intent on accelerating the pace of change. Key lessons were learned from the wars in Afghanistan and Iraq, and time will tell if the organizational behavioral change required by those lessons is accomplished successfully enough to defeat a tough foe in Al Qaida.

Valid OrgSim has been, and will be, of tremendous utility as DoD both seeks to transform itself and seeks to disrupt and destroy all of the world's Al Qaida cells. U.S. and allied intelligence agencies will use valid models of Al Qaida to model and forecast Al Qaida organizations and its planning and operational capabilities. This capability will allow the U.S. and allies to make accurate predictions about future Al Qaida targets and modes of operation. However, building valid OrgSim models of Al Qaida has proven to be difficult. The terrorist organization obviously operates as deeply in the shadows as possible, and its cultural elements are often difficult for western agencies to understand fully. Its leadership and members evidence a fanatical adherence to its diabolical mission, perhaps to a degree difficult for westerners to understand. This means that not only will OrgSim developers have to model accurately Al Qaida's organization and operational capability, but also this fanatical devotion to the organization's mission. In addition, OrgSim developers will have to provide quality models of how western intelligence and military organizations will react to the Al Qaida LO model. This indeed is a tall challenge.

U.S Navy's Integrated Learning Environment

The U.S. Navy has embarked on a major effort to improve its capability as an LO. It became to clear to Navy leadership that in order to make the transformation called for by senior leaders in the government, major changes would have to be made in how the Navy prepared its personnel. While the effort is happening across the entire Navy, the major task of re-organizing and re-vectoring the organizations that prepare Navy personnel has fallen to the Chief Learning Officer of the Navy. The Navy's CLO is the Commander of the Navy Education and Training Command (NETC), a Vice Admiral.

The Chief of Naval Operations laid out five key priorities for the Navy that can only be accomplished through a robust LO. (Hollis, 2004).

- Manpower – How will the Navy best compete for personnel with the private sector? How will it recruit and retain the best people?
- Current Readiness – How will the Navy insure that all personnel are properly trained?

- Future Readiness - How can the Navy get the right information to the right people at the right time?
- Quality of Service – How will the Navy improve the professional and personal growth of each sailor and Navy employee?
- Alignment – How will the quality of life of Navy personnel be improved?

To accomplish the five priorities described above, the Navy has developed a five-vector model for their personnel that define routes for professional development, personal development, leadership, certificates and qualifications.

Navy learners are given responsibility, along with their leaders, to define their learning needs along each of the five dimensions. The personnel then map out a learning path, and the Navy helps them in meeting their learning needs as they progress through their careers. To assist Navy personnel and their leaders, the Navy has developed a Learning Portal and a Learning Management System. Rosenberg (2001) defines a learning portal as "a web-based, single point of access that serves as a gateway to a variety of e-learning resources on the Web. Using a knowledge management approach, a learning portal can access and distribute e-learning information, programs, and other capabilities to employees" (p. 155). He also said, "While portals provide gateways to learning resources, learning management systems provide the functionality. A learning management system uses Internet technologies to manage the interaction between users and learning resources." (p. 161).

Saundra Drummer, director of the learning strategies division at the Naval Education and Training Command, said, "The object of the whole process is to create a dynamic learning environment, with the idea of increasing the learners' control and responsibility for their own learning. The idea is to reduce the gap between the learning environment and the performance environment," (p.1). (Hollis, 2004).

While the Navy's effort to create an effective LO is still in its early stages, the Navy is committed to the concept of more productive organizational learning. This goal of making the learning environment and the performance environment more closely related is at the heart of the LO construct. OrgSim can be of great assistance in meeting this goal as it provides decision makers with robust and valid models of how the Navy's LO function is operating. Where are the strong points and weak points? Of significant benefit will be OrgSim models that can show accurately the interactions of personnel and the training system as sailors progress through their careers. OrgSim can help Navy leaders define optimal career-long "horizontal" training pipelines that forecast how sailors can best move from initial skill levels to expert levels, not just in their Navy specialties but in their complete development. OrgSim can also be of help in defining optimal "vertical" training systems along each stop in a sailor's progression. For example, OrgSim's training models and data can help to define optimal mixes of training method and media in learning venues (e.g., formal courses, self-study, mentoring).

POTENTIAL BENEFITS TO A LEARNING ORGANIZATION FROM USE OF ORGANIZATIONAL SIMULATION

OrgSim provides the opportunity to model the heterogeneous behavior of different individuals each having unique information, decision rules, and goals. This heterogeneous micro behavior gives raise to different macro behaviors within the organization. To properly model organizational learning a modeler needs to have multiple scenarios characterized by different combinations of variables and variable levels to discover the patterns that create unique clusters of outcomes. The modeler then needs to combine models of individual agents with models of the internal policies, procedures, and information technologies used by various organizations in order to predict how organizations will evolve their informal networks and how those networks will impact performance. There is a need to provide means to allow decision makers to conduct "what if" exercises to examine the implications of alternative organizational designs. Given various combinations of internal structure and external environmental forces (e.g., economy, job market) organizations could examine the impacts of redesigning internal processes, divesting corporate enterprises, or restructuring the workforce.

Applications of OrgSim to the LO

We believe that valid OrgSim can be of great value to organizations as they attempt to become better LOs. This section describes some ways that we believe OrgSim can advance the LO construct in organizations seeking to be more effective at intelligent behavioral change.

There have not been many empirical studies done to examine the effectiveness of the LO construct. Such a finding is not surprising because measuring the progress of an entire organization is very difficult. Here are two examples of how the developers and users of the two concepts can benefit from associating them.

Complex organizations can be structured in a myriad of ways. These organizational structures can range from simplified "flat" structures with few managers and managerial layers, to complex "hierarchical" structures with many managers and managerial layers. The general trend over the last twenty years has been to "flatten" organizations whenever possible. The goal is to eliminate middle management layers thus forming a stronger connection between top managers and organizational members who develop and deliver goods and services.

We have seen this trend to some degree in the military as the Goldwater-Nichols Act in Congress sought to strengthen the "joint" nature of the Department of Defense and make jointness super ordinate to the individual services (e.g., Army, Air Force, Navy and Marines). We still do not know very much about the best way to structure organizations to make them better LOs. Arguments could be made that flatter organizations are better LOs because it is easier for senior managers to more immediately communicate to their subordinates the necessity of behavioral change and to guide that change. Flat organizations allow subordinates to get the knowledge to senior managers quicker and with fewer "gate keepers". On the other hand, one might make the argument that more organizational layers make for better LOs because they allow more managers more time to think about new knowledge and lay the course for behavioral change within the organization.

Valid simulations of organizations would allow the managers and organizational development experts to play "what if" games with the structure of the organization. Which alternatives facilitate organizational learning? Such a purpose for OrgSim would create a significant challenge because it would require OrgSim developers to understand, model and simulate complex organizational interactions that forecast behavioral outcomes based on new organizational knowledge.

Over the years there have been a number of theories of organization. Managers and researchers have concerned themselves with framing organizations under different theories primarily because they inform the way that an organization is defined and structured. Muchinsky (2003) describes the three main organizational theory schools. The Classical theory, formed in the first part of the 20th century, contained four main principles:

- Functional principle – work should be divided into units that perform similar functions.
- Scalar principle – refers to the chain of command that grows with levels added to the organization.
- Line/staff principle – line functions have the primary responsibility for meeting the major goals of the organization, while staff functions support the line's activities.
- Span-of-control principle – refers to the number of subordinates a manager is responsible for supervising.

The Neoclassic theory was born in the 1920's with the Hawthorne studies and became popular in the 1960's. As the name implies, the Neoclassical theory honored many of the principles of the Classical theory but sought to update and make them more human using behavioral research results. For example, the neoclassicists believed that functional principle of division of labor could cause employees to feel isolated and perceive their work to be undervalued.

The third major school of thought in organizational theory grew out of the biological sciences. Systems theory presented the model of interlocking parts. "A complex organization is a social system; the various discrete segments and functions in it do not behave as isolated elements. All parts affect all other parts. Every action has repercussions throughout the organization, because all of its elements are linked." (Scott, Mitchell & Birnbaum, 1981, p.44).

Is there a most correct organizational theory? Should new theories be formed to better explain today's existing organizations? Perhaps most importantly for the objectives of this chapter, which theory will result in the most effective LO for a particular organization? Researchers interested in these questions can benefit from valid OrgSim models because they will allow the researchers to quickly develop theoretical constructs, try them out on working models of all types of organizations, and come to data-based conclusions about their potential for LO purposes. A major challenge in using OrgSim models for LO improvement is the necessity to come to grips with the significant verification and validation issues involved in testing these models. Managers could benefit from quality OrgSim models because the models could allow those managers to apply abstract organizational theories to LO problems and determine which theories might help structure the most effective LO.

Process Modeling

One might look for gross measures such as profitability increases for a for-profit company, or victories in war for a defense organization, but there are so many factors that effect such measures that it is very challenging to know if such increases are due to changes wrought by adherence to LO principles. We believe that it is essential to measure process as well as product if one is to make rational decisions about the effectiveness of an organization's learning endeavors. Valid OrgSim models can be of significant help with this process measurement question. It can allow managers and researchers to watch the measurable changes within an organization as learning efforts are undertaken in the OrgSim model. Since organizational change in the real world can be a very time consuming activity, often taking place over years, it is difficult to isolate exactly which organizational elements are most important for the learning change. Valid OrgSim can allow observers to gain valuable insights to the change process in a fairly short time. For this to occur, OrgSim should take into account the following organizational culture elements.

Organizational Culture Elements in Organizational Simulation

To understand these process elements in an organization it is useful to define those elements that make up the organizational culture. Carrell, Jennings and Heavirn

(1997) tell us that, "Organizational culture can be more precisely defined as the underlying values, beliefs, and principles that serve as a foundation for an organization's management system as well as the set of management practices and behaviors that both exemplify and reinforce those basic principles." (p. 569).

Carrell, et. al. (1997), offer an alternative definition of organizational culture as, "a set of key characteristics that describe the essence of an organization." (p. 569). They list a set of these key characteristics:

- Values. The dominant values espoused by an organization.
- The philosophy that guides an organization's policies towards its employees
- Norms of behavior that evolve in working groups
- Politics. The rules of the game for getting along in the organization.
- The climate of work, which is conveyed by the physical layout and the way people, interact.
- Behaviors of people when they interact such as the language and demeanor: the social interaction.

At the present, as described in other chapters of this book, OrgSim has a difficult time in validly modeling most of the elements of culture defined by Carrell, et. al. How well can OrgSim currently model accurately Carrell, et. al.'s. cultural elements? Our estimate is as follows:

- Values – reasonably well
- Policy philosophy – not very well
- Norms – not very well
- Politics – not very well
- Climate – reasonably well
- Interaction Behavior – not very well

It is fair to say that OrgSim is still in its relative infancy when it comes to being predictive about any of the cultural elements. However, we believe that as the value of OrgSim becomes clearer to research funding agencies, continual progress will be made in this area.

Once OrgSim can validly model and predict organizational outcomes in organizations, our understanding of the Learning Organization will greatly improve. The ability of an organization to learn from its past successes and failures, and the past successes and failures of other organizations, is wholly dependent on the organization's people. Since those people reside in the organizational culture, an ability to model the culture will help us predict how quickly the organization's people can ascend a learning curve.

The Learning Management Maturity Model

At the beginning of this chapter we opined that there is a continuum of LO effectiveness. Every organization, by its very existence, is an LO, but some organizations are further along the continuum than others when it comes to their learning effectiveness. Moore (2002) describes a path by which the learning capabilities of organizations mature. He calls his model the "Learning Management Maturity Model (LM3)". "Learning management is the process of planning, directing, and accounting for the resources, participants and outcomes of learning activity within an organization." (p. 1). The continuum consists of five stages of maturity. Figure 1 shows the title of each stage and briefly describes these five stages.

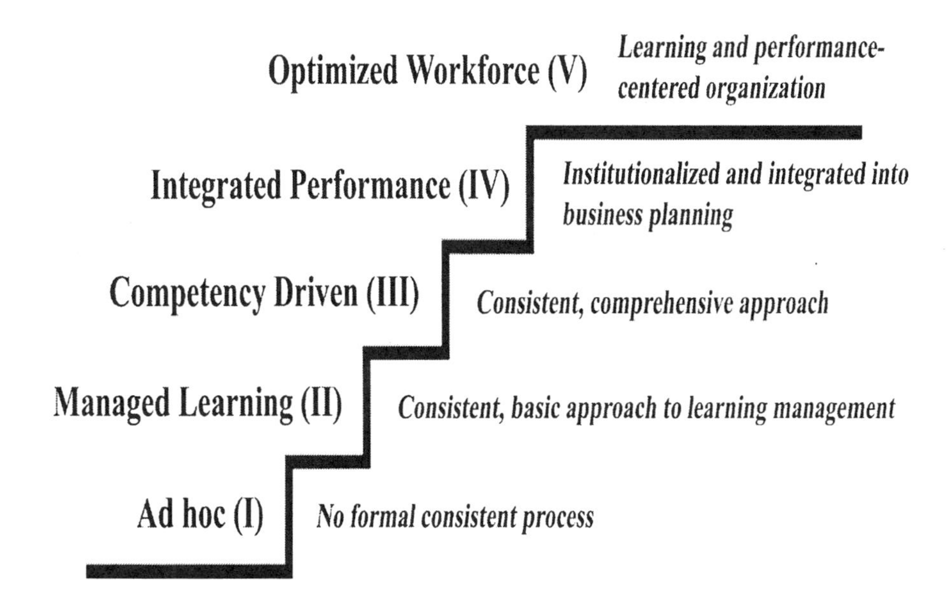

FIGURE 1. The Learning Maturity Management Model (LM3)
(Used with permission: Copyright © 2002-2004 THINQ Learning Solutions, Inc.)

About the model Moore says,

> "A maturity model is a framework for characterizing the evolution of a system from a less effective state to a highly ordered and highly effective state. The Learning Management Maturity Model is such a framework for learning management. It judges the maturity of an organization's learning management processes and identifies the key practices required to increase the maturity of these processes. LM3 further serves as a means for modeling, defining, and measuring the maturity of the processes used by learning professionals and the organizations they support. The term *maturity* implies that organizations must grow their capabilities over time in order to produce repeatable success in learning management." (p. 1)

Moore considers the following to be key characteristics of each learning stage:

Stage 1 – Ad Hoc

- Many incomplete, informal approaches for managing learning
- Unpredictable learning outcomes
- Little organizational support for learning and development
- Mix of manual and automated systems
- Departmental Learning Management Systems in place but not consistent or connected

Stage 2 – Managed Learning

- Enterprise Learning Management Systems in place (centralized or interconnected)
- Common terminology and nomenclature
- Common, repeatable processes
- Simple, standards-based integration with learning content
- Use of common tools and techniques for key learning management processes
- Learning plans in place for basic jobs and activities of the workforce

Stage 3 – Competency-Driven

- Senior management support for learning management
- Competency models in place to build a proficient workforce
- Competency-based learning and skill assessments in use to align learning against developmental needs of the organization
- Learning modalities (live, self-paced) blended for increased efficiencies
- Consistent use of tools and techniques for learning management processes
- Learning Content Management System providing central repository for content development and delivery
- Universal, assumed interoperability between content and Learning Management System
- Prescriptive and self-initiated learning occurring
- Collaboration and mentoring reinforcing knowledge transfer
- 360-degree feedback instruments in place for skill and performance management

Stage 4 – Integrated Performance

- Active senior management support for integration of business planning and learning/performance initiatives
- Business goals aligned with work activities
- Interactive dialog among work force a consistent part of activity management and coaching
- Developmental planning occurring in relation to accomplishment of work activities
- Learning and knowledge management systems beginning to integrate with enterprise portal technologies

- Results of learning and work performance measured and correlated against key business performance metrics

Stage 5 – Optimized Workforce

- Flexible, learning and performance-centric organization structure

- Improvements to the learning and performance management environment actively encouraged

- Daily work and learning activities directly linked to business priorities

- Shared understanding and appreciation throughout workforce of the factors and influences affecting the business

- "Connected" organization sharing knowledge and content assets with key components of its extended enterprise

- Just-in-time blended learning and performance support tools increasing skill proficiency and organizational performance

Moore's experience after working with many organizations of varying types is that most organizations are at Stage II (Managed Learning) on the continuum. He has very seldom found organizations at Stage III (Competency driven), and he has not yet found any organizations at Stage IV (Integrated Performance) or Stage V (Optimized Workforce).

It is not surprising that Moore hasn't found organizations operating above Stage III. Operating at the upper stages of his model is complex and time consuming, and requires a significant expenditure of organizational resources. However, more importantly, it requires a solid commitment from the entire organization, especially the leadership to move beyond Stage II.

We believe that OrgSim can be used to help organizations move into the higher stages of the LM3 model. By allowing organizational managers to try out in simulated fashion organizational responses to various alternatives to move into the higher stages, OrgSim can provide needed impetus to LM3 improvements. Some examples will help to explain the point:

- An organization wishes to move to Stage II. OrgSim could help the organization test out the Enterprise Learning Management System that it has put in place. Various sets of OrgSim data could be run through the Enterprise System to see if the results provide the kind of organizational learning path desired by management.

- An organization wishes to move to Stage III. OrgSim could be used to help determine if the "competency models (are) in place to build a proficient workforce." Various models of the workforce at different competency levels could be constructed using OrgSim. Then, different organizational results could be simulated using different competency models.
- An organization wishes to move to Stage IV. OrgSim could be used to model the results of organizational learning, and then measure and correlate those results against key business performance metrics. It is difficult to model such complex measurements and correlations presently.
- An organization wishes to move to Stage V. OrgSim can be of great help as the organization strives to determine if there is a "shared understanding and appreciation throughout the workforce of the factors and influences affecting the business". Modeling the understanding and appreciation of one employee about these matters will be difficult enough. However, modeling a shared understanding throughout the entire workforce will be a major accomplishment. Management could use significant help from OrgSim in reaching a conclusion about how their entire organization thinks about learning.

Sample "What If" Questions

This section provides some sample "what if" questions that those interested in improving learning organizations might consider posing and answering via OrgSim. As mentioned earlier, modern organizations are usually so complex that it would take a long time to answer these questions if all researchers and managers had to examine were actual organizations. OrgSim will provide a mechanism by which these questions can be addressed in a timely and economically fashion. Needless to say, time and cost savings will not be very important if the OrgSim models that are used to answer these "What if" questions are not based on valid organizational principles and data. We assume those assumptions will be met as OrgSim advances. Sample "what if" questions that OrgSim could help answer:

- "What if the organization changed its hiring practices and hired employees who had either higher or lower entry level skills? How would such changed policies affect the organization's capability to learn?" (A corollary question is, "How do our current hiring practices affect our capability to move the LO forward?")

- "What if the organization increases or decreases the amount of training of all types that we provide for current employees? Will such a change affect our capability to improve as a LO?"
- "What if the organization improves the climate for decision making so that more employees are involved in more decisions (e.g., production, quality control, and marketing)? Will the extra time required for such a climate change produce better organizational learning? Will the majority of employees feel empowered for such a change? Will they be interested in more decision making power?"
- "What if the organization improves its capability to capture and archive 'lessons learned' from current operations? Will it be able to wisely make use of such data? Will it be worth the resources expended? Will the organization be a better learning organization with such data?"
- Greenberg and Baron (1997) tell us that, "Organizational culture is a cognitive framework consisting of attitudes, values, behavioral norms, and expectations shared by organization members. Once established these beliefs, expectancies, and values tend to be relatively stable and exert strong influences on organizations and those working in them." (p. 471). Given Greenberg and Baron's definition of organizational culture one might ask, "What will be required to change an organization's beliefs, expectancies, and values if it is determined that the culture is not a LO?"

CONCLUSION

Challenges and Opportunities

Since the industrial revolution, leaders, workers and researchers in the industrialized world have attempted to determine how best to "fit" the worker into work organizations. Early in the twentieth century, as the assembly line became a major part of organizational functioning in many industries, researchers and others began to worry that the worker, as a mere cog in the machine, was being de-humanized. In the later half of the twentieth century a variety of attempts were made by various organizations to bring worker's talent, skills and creativity into the production process. Organizations began to realize that competitive advantage lay at least as much in their members' capability to improve the organization as in the organization's other capital assets. In fact, in the 1980's, as the information age dawned, organizations began to recognize that perhaps their only competitive advantage lay in their people. This concept lay at the heart of the Total Quality Management movement. Now, instead of trying to determine how best to "fit" the worker into the organization, the LO movement sees it as vital that the organization is built around the unique capabilities and creativity of the worker.

Once an organization has determined to become a more effective LO, the challenge lies in determining how to realize that goal. No organization should assume that the journey is inexpensive. It is certain that significant expenditures will be necessary to provide the organizational learning capabilities required. Funds will have to be committed to providing the learning opportunities necessary. Time for leader and employee learning will have to be allocated. Even if an organization is successful at merging the learning function and the operational function, such a transformation will bring costs for training and for tools such as Internet portals and learning management systems. We believe OrgSim can be of great use to managers and consultants as they determine how best to organize, learn, and invest to become a better LO. OrgSim will provide viable opportunities for managers and consultants to play "what if" games to determine necessary LO inputs and outputs. Return on investment analyses should be soberly approached as an organization examines it's LO options. OrgSim models and data can be of immense help in this enterprise.

Another challenge and opportunity for OrgSim is to help organizations determine how their LO compares to that of competitors. Valid OrgSim models, not just of their own organization but of rivals as well, will allow organizational leaders to determine if they need to pay more attention to their LO efforts. As organizations become more attuned to the role that a quality LO can play in the organization's long term strategy, it can make tactical LO adjustments relative to its competitors. Clearly, our ability to structure and run OrgSim models is at a fairly low level of sophistication when it comes to modeling rival organizations, but we see tremendous payoff as these capabilities are developed.

Recommendations

This chapter has discussed the intersection of two powerful constructs, OrgSim and the Learning Organization. We have attempted to demonstrate here how both constructs can benefit from their association by those who would use them. Significant challenges remain before the optimal intersection can be achieved. However, as the constructs mature, the means to actualize them evolve, and our understanding of how best to use them grows, we can foresee huge leaps in our ability to make better LOs a reality.

We have the following recommendations for both the OrgSim and the LO communities.

- The LO community (or at least the organizational researchers who study the phenomenon) should produce analytical and, especially empirical research concerning the effectiveness of LOs. In order to produce OrgSim models that will be of most benefit to LOs, OrgSim developers need to know much more about what makes an organization an effective LO. Research results might examine various return on investment metrics for LOs. Our examination of the LO literature for this chapter did not reveal much analytical or empirical data or information about LO effectiveness.

- OrgSim developers should structure OrgSim models that capture the culture that has arisen in organizations that are examples of good LOs. Our belief is that a strong commitment to the principles of effective Learning Organizations fundamentally changes the corporate culture. Presently, there are only a few examples of effective LOs. We believe their cultures are different than the cultures found in organizations who have not made the LO commitment. It will be important for OrgSim developers to examine the handful of good LOs to build accurate corporate culture models.
- Along the same lines as the previous recommendation, we recommend that the OrgSim community should carefully examine existing LO literature. Only by understanding the current conceptions and research (limited though it is) about effective LOs will OrgSim developers and researchers be able to capture the LO fundamentals. This examination will in turn allow OrgSim models to be fashioned in the most optimal manner to aid LO decision makers.
- Finally, we believe it is vital that the OrgSim community examine the role of Learning Officers in organizations to determine how they function, what their needs are, and how OrgSim can best help them. The LO literature continually reinforces the need for a specific individual with a committed LO organization. That person is the focal point and impetus for making sure the organization has properly planned for and resourced the LO commitment. Our experience is that the chances for successful commitment to Organizational Learning are closely bound to the skill of the Chief Learning Officer. OrgSim should capture the CLOs unique role.

REFERENCES

Argyris, C., & Schön, D. (1978) Organizational learning: A theory of action perspective, Reading, Mass: Addison Wesley.

Carrell, M.R., Jennings, D.F. and Hevrin, C. (1997). Fundamentals of Organizational Behavior. Upper Saddle River, N.J.; Prentice Hall.

de Geus, A. P., Morecroft, J. D. W., & Sterman, J. D. (Eds.) (2000). Modeling for learning organizations. Washington DC: American Psychological Association.

Dewey, J. (1938). Experience and education. New York: Simon and Schuster.

Forrester, J. (1961). Industrial dynamics. Cambridge, Ma: MIT Press.

Garvin, D. (1993) Building a Learning Organization. Harvard Business Review, Vol. 71, No. 4

Garvin, D. A. (2000). Learning in action. A guide to putting the learning organization to work. Boston, Mass.: Harvard Business School Press.

Greenberg, J. and Baron, R.A. (1997). Behavior in organizations (6th Ed.). Upper Saddle River, N.J.: Prentice Hall.

Hollis, E. (2004). U.S. Navy: Smooth Sailing for Education, Chief Learning Officer. February 2004.

Kerka, S. (1995). The learning organization: Myths and realities. Retrieved on 24 July, 2004 from http://www.infed.org/biblio/learning-organization.htm.

Moore, C. (2002). The Learning management maturity model. Thinq, Inc. Retrieved from on 24 July, 2004 from http://www.thinq.com/pages /white_papers.htm.

Muchinsky, P.M. (2003). Psychology applied to work. (7th Ed.) Belmont, CA: Wadsworth/Thomson Learning.

Nevis, E. C., DiBella, A. J. and Gould, J. M. Understanding organizations as learning Systems. Sloan Management Review, Winter 1995, Vol. 36, No. 2.

Office of the Under Secretary of Defense for Personnel and Readiness. (March, 2002) Strategic plan for transforming DOD training. Washington D.C.: Pentagon.

Pedler, M., Burgoyne, J. and Boydell, T. (1991). The learning company. A strategy for sustainable development. London: McGraw-Hill.

Peters, Tom (1987) Thriving on chaos. New York, NY: Harper & Row Publishers, Inc.

Pew, R. W. and Mavor, A.S. (Eds.) (1998). Modeling human and organizational behavior. Washington, D.C.: National Academy Press.

Rosenberg, M.J. (2001). E-Learning: Strategies for delivery knowledge in the digital age. New York: McGraw-Hill.

Sample D. (2003). Military transformation tough, important and progressing, Secretary Says. American Forces Information Services News Articles. Retrieved on 24 July, 2004 from http://www.defenselink.mil/news/Aug2003 /n08212003_200308214.html.

Schein, E.H., (1992). Organizational culture and leadership. San Francisco: Jossey-Bass Publishers.

Scott, W.G., Mitchell, T.R., & Birnbaum, P.H. (1981). Organization theory: A structural and behavioral analysis. Homewood, IL.: Richard D. Irwin.

Senge, P. M. (1990). The Fifth Discipline. The art and practice of the learning organization. London: Random House.

Senge, P.M., Kleiner, A., Roberts, C. Ross, R. and Smith, B. (1994). The fifth discipline fieldbook: Strategies and tools for building a learning organization. New York: Doubleday/Currency.

Senge, P., Kleiner, A., Roberts, C., Ross, R., Roth, G. and Smith, B. (1999). The Dance of change: The challenges of sustaining momentum in learning organizations. New York: Doubleday/Currency.

Further reading

Argyris, C. and Schon, D.A., (1978). Organizational learning: A theory of action perspective. Reading, Mass.: Addison-Wesley Publishing Company.

Deming, E. (1993). The new economics for industry, government, education. Cambridge, Ma: Massachusetts Institute of Technology.

Dodgson, M. (1993). Organizational learning: A review of some literatures. Organizational studies. Vol. 14, No. 3 (375-394).

Easterby-Smith, M., Burgoyne, J. and Araujo, L. (eds.) (1999). Organizational learning and the learning organization. London: Sage.

Schön, D. A. (1973). Beyond the stable state: Public and private learning in a changing society. Harmondsworth, Middlesex, U.K.: Penguin.

Senge, P., Cambron-McCabe, N. Lucas, T., Smith, B., Dutton, J. and Kleiner, A. (2000). Schools that learn: A fifth discipline fieldbook for educators, parents, and everyone who cares about education. New York: Doubleday/Currency

Watkins, K. and Marsick, V. (eds.) (1993). Sculpting the learning organization. Lessons in the art and science of systematic change. San Fransisco: Jossey-Bass.

CHAPTER 5

REQUIREMENTS AND APPROACHES FOR MODELING INDIVIDUALS WITHIN ORGANIZATIONAL SIMULATIONS

EVA HUDLICKA AND GREG ZACHARIAS

ABSTRACT

The past decade has witnessed a growth of interest in modeling and simulation of individuals and organizations. From a research perspective, these simulations are important because they enable explorations of both individual and organizational decision-making mechanisms. From an applied perspective, these simulations aid organizational design, improve the realism and effectiveness of training and assessment systems, help in the assessment of enterprise workflow, and aid in behavior prediction. The human behavior modeling community has traditionally been divided into those addressing *individual behavior models*, and those addressing *organizational and team models*. And yet it is clear that these extremes do not reflect the complex reality of the mutually-constraining interactions between an individual and his/her organizational environment. In this chapter we argue that realistic models of organizations may require not only models of individual decision-makers, but also explicit models of a variety of individual differences influencing their decision-making and behavior, such as cognitive styles, personality traits, and affective states. Following a brief overview of relevant research in individual differences, and cognitive architectures, we outline the knowledge, representational, and inferencing requirements for two alternative approaches to modeling the individual within an organizational simulation: one focusing on cognitive architectures and the other centered on profile-based social network models. We illustrate each approach with concrete examples from existing prototypes, and discuss how these two approaches could be integrated within organizational simulations. We conclude with a discussion of some of the critical challenges in modeling individuals within organizational simulations.

INTRODUCTION

The human behavior modeling community has traditionally been divided into researchers and practitioners addressing *individual behavior models*, and those addressing *organizational and team models*. And yet it is clear that these extremes do not reflect the complex reality of the mutually-constraining interactions between an individual and the organization environment within which s/he operates. We cannot effectively model the individual if we ignore the organizational constraints within which s/he operates, nor can we effectively model an organization if we abstract away the behavioral idiosyncrasies of the individuals who make up that organization.

Several recent efforts have attempted to bridge this gap by introducing more complex individual models into the traditional social network approaches to organizational modeling, by exploring multi-agent simulations, and by introducing organizational components into individual simulations. In this chapter we continue this trend and suggest that effective organizational models must consider the effects of the individual on an organization, and that models of organizations may therefore require not only complex models of the constituent individuals, but also take into account the range of individual differences within and across these individuals, and the ensuing behavior variations.

The purpose of this chapter is three-fold. *First*, we make the case for embedding individual models within organizational simulations, and for developing guidelines when such models may be needed for a particular application. *Second*, we argue that for these models to be sufficiently realistic, they must represent the types of behavioral variations and individual idiosyncrasies that characterize real human behavior. *Third*, we provide examples of two approaches to modeling the individual within an organizational context, and we discuss the knowledge, data, representational, and inferencing requirements of each approach. These three themes are briefly outlined below, and elaborated in the remainder of this chapter.

Organizations Are Composed of Individuals

The extent to which individual behavior is constrained and facilitated by the environment, including the social and organizational environment, and the extent to which the individual can shape that environment, is one of the great unanswered questions regarding the human condition. It underscores broad historical questions such as the role of individual politicians and military leaders in the various political and military achievements (and disasters) throughout history: Do individuals make history or does history make the individual actors? We know intuitively that both cases are true, but we lack sufficient understanding of the complex interactions to define the conditions that facilitate the predominance of one or the other set of influences and constraints.

On a smaller scale, we can ask questions such as "Could a different set of individuals in the USS Vincennes chain of command have avoided the shooting down of a civilian airliner?" "Would an alternative organizational culture within NASA have prevented the space shuttle Columbia tragedy?" Answering such questions definitively is, of course, impossible since we can't reset the clock of history. But with simulations of sufficiently high fidelity, we *can* explore broad trends in individual/organizational behavior and potential scenario outcomes.

However, when attempting to simulate the individual within an organization we face a Catch 22 problem. On the one hand, we cannot construct empirically-justified models until we have more data. On the other hand, the sheer complexity of the studies required to obtain such data frequently makes them infeasible, without the aid of computational models to help guide and focus the critical questions of empirical interest.

For these reasons, a number of researchers have recently highlighted the importance of computational models for organizational science research, centered on the inclusion of "realistic" models of the individual decision-makers operating within his/her organizational context. (e.g., Carley & Hill, 2001; Saunders-Newton, 2003; Handley and Levis, 2001). Two key questions naturally arise from this approach:

- When is it necessary to include individual decision-maker models within an organizational simulation? What are the specific characteristics of the organizational context, the task, and the individual that require their explicit inclusion?
- What are the criteria for selecting their level of resolution and fidelity, to assure adequate level of realism and validity of the associated organizational simulations?

To answer these questions we must first understand the conditions (from the organizational, the task, and the individual decision-maker perspectives) that facilitate or inhibit the influence of the individual decision-maker's behavior on his/her organizational context. This will then allow us to determine whether individual models need to be explicitly included in a particular organizational simulation, and determine their level of resolution and fidelity.

In particular, we need to understand which individual characteristics are more or less likely to affect organizational behavior and performance and, therefore, are more or less required for inclusion in high-fidelity simulation representations. We also need to understand which organizational characteristics amplify (or attenuate) the behaviors of different individuals, across a range of tasks, and across a range of individuals. When constructing simulation models, we must on the one hand avoid the possibility of *abstracting away* existing individual variations, when such variations play a critical role in organizational decision-making and outcomes, and on the other hand avoid the unnecessary difficulty of introducing individual models and behavior variations when they are not necessary for the simulation objectives.

To address these issues we must develop effective means of modeling the individual within an organizational simulation, along with an ability to model a range of individual variability and organizational "states," to explore the space of interactions that facilitates or hinders both individual performance and organizational objectives.

Individual Behavior is Not "Nominal"

In a recent review of personality effects by Revelle (1995), there are sections entitled "All people are the same", "Some people are the same", and "No person is the same". In other words, while there are similarities in individual responsiveness, behavior also varies across and within individuals: Different people respond differently to an identical set of stimuli, and the same individual may react differently to the same set of stimuli at different times, depending on a variety of factors, including his/her affective state, level of fatigue, and many other individual differences, task and environmental factors. Individual behavior is influenced by variability in a range of internal and external factors, and by the interactions among them.

Currently, the majority of existing simulations focus on modeling *nominal* behavior, resulting in simulations that can be highly predictable across distinct situations and uniform across individuals, types of groups and organizations, and cultures. While easier to implement, such simulations are clearly unrealistic. A number of authors in this volume address the necessity for realism in assuring effectiveness of organizational simulations. Rouse, for example, highlights the utility of experiential perceptions gained by acting in the future (Rouse, 2005). But to be effective, such simulated "experiences" must reflect the individual's idiosyncrasies, which necessitates a high-degree of realism in modeling the individual decision maker.

Depending on the context and objectives, realism in modeling and simulation takes different forms. Several papers in this volume address what we term realism of expression. For example, Essa and Bobick discuss the elements necessary to assure "interaction plausibility" (Essa & Bobick, 2005). Zyda and Lindheim discuss the importance of VR and immersiveness (Zyda et al., 2005; Lindheim & Korris, 2005). An equally critical aspect of realism concerns the models' ability to generate behavioral choices that appear realistic. This aspect of realism also takes different forms, with one of the more critical ones being the ability of the simulation model to generate the types of variations in the synthetic characters' behavior that is characteristic of human behavior. This includes also the cultural phenomena highlighted in several chapters in this volume (Klein et al., 2005; Levis, 2005). Individual models thus require an ability to represent a range of behavioral variability, and corresponding variability in the perceptual and decision-making processes that precede the selection of particular behavior.

The central point we address in this chapter is the necessity of, and approaches to, modeling a variety of interacting individual differences, including cultural differences, in human behavior models of individual decision-makers. We provide an overview of a broad range of individual differences and their effects, and include specific examples to illustrate how some of these factors can be represented in individual models.

Approaches to Modeling Individuals within Organizations

What are the modeling alternatives available to the researcher or practitioner who wishes to explicitly model the individual within an organizational context? In this chapter, we discuss two possible approaches, each representing a distinct methodology for individual modeling. One alternative, the *cognitive or agent architecture,* focuses on emulating the actual human decision-making processes, in terms of the structures and processes that mediate the variety of processing mechanisms required for action selection, including perception, decision-making, planning, and goal selection. The second alternative, a *profile-based approach*, is less concerned with the "how" of the decision-making, than with the "what," that is, the eventual behavioral outcome. Whereas the cognitive architecture approach aims to emulate the human reasoning processes, the second alternative aims to reason about the human, possibly by simulating the reasoning of an expert observer attempting to predict the individual's behavior from available data.

Each of the approaches can be embedded within an organizational context in a distinct manner. The *cognitive architecture* approach requires that selected individuals within an organization be modeled by distinct instances of the architecture. These are then linked to the representation of the organization as a whole via a set of well-defined links, which capture the mutual influences between the individual and his/her organizational context, in terms of the specific communication or controlling signals. Figure 1 shows a high-level schematic of this approach.

In the *profile-based social network* approach, the individual's group or organizational context is captured in terms of a number of psychosocial relationships that are explicitly represented within this profile. These relationships then allow an explicit representation of the individual's social network, and support inferences about the mutual influences between the individual and his/her social (organizational) environment. Social network models are commonly used in organizational simulations (e.g., Carley, 2005). However, typically the individuals within these networks, usually represented as distinct nodes, are highly abstracted and do not capture the complexity and variability of individual decision-making. Embedding a full individual profile within a social network enables the resulting organizational simulation to take into account the broad range of individual characteristics that determine his/her decision-making, and the effects of these decisions on the organization as a whole. Figure 2 illustrates this approach.

These two approaches have overlapping but distinct knowledge, data, representational, and inferencing requirements. The choice of the approach for a particular organizational simulation depends on the needs and objectives of the simulation, the type of data and knowledge available, and on any implementation constraints. These distinct requirements are described below, followed by examples of each approach, and an outline of how each approach could be integrated within an organizational simulation.

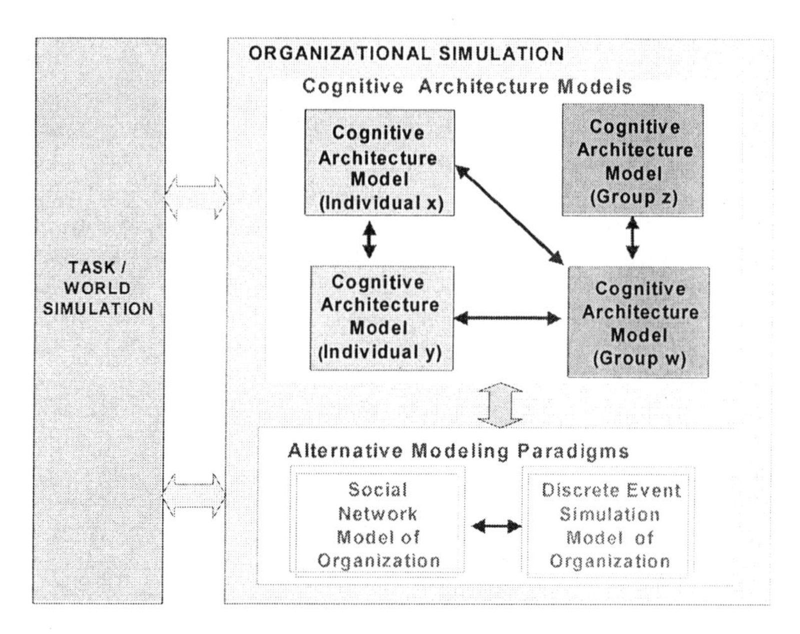

FIGURE 1. High-Level Schematic Illustrating the Cognitive Architecture Approach to Modeling an Individual within Organizational Simulations. Arrows linking modeled entities (individual, groups) represent communication links.

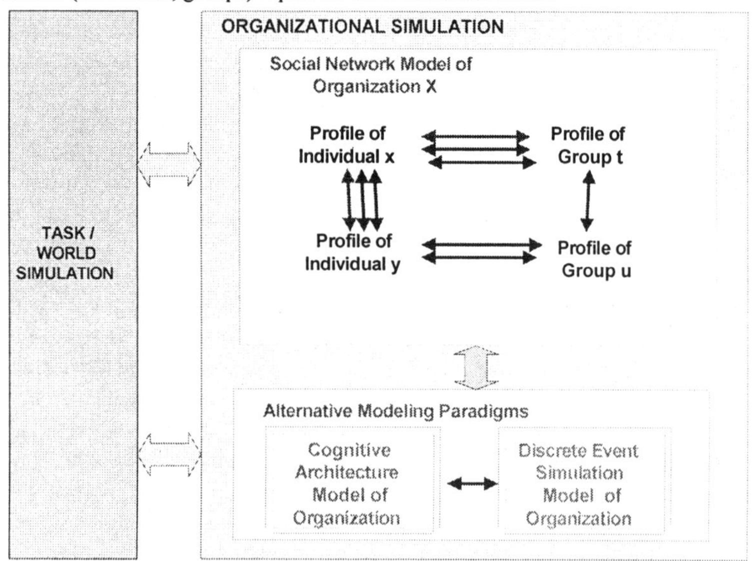

FIGURE 2. High-Level Schematic Illustrating the Profile-Based Social Network Approach to Modeling an Individual within Organizational Simulations. Arrows linking distinct "profiles" represent psychosocial relationships among individuals and groups.

RELATED RESEARCH AND BACKGROUND INFORMATION

This section provides a brief overview of existing research in the areas most relevant to modeling individual behavior within organizational simulations. We do not discuss organizational simulations per se, since several chapters in this volume address this area (e.g., chapters by Carley, 2005; Andrews et al., 2005; and Cross, 2005). Instead, we focus on providing background information on the effects of a variety of individual differences and determinants of behavior, and on the existing individual human behavior models.

Effects of Individual Differences on Decision-Making and Behavior

Individual behavior is determined by range of internal and external factors, and by the interactions among them. Depending on the breadth of focus, these have variously been termed *behavior determinants* (Hudlicka et al., 2002; 2004), *behavior moderators* (Pew & Mavor, 1998), and *individual differences* (Revelle, 1995), and include a variety of static and dynamic factors, most notably basic cognitive abilities and skills, personality traits, and affective states. These factors, along with the decision-maker's individual history, in turn determine the individual's internal mental dynamic context consisting of activated beliefs, expectations, attitudes and goals, which eventually lead to the selection of a particular observable behavior. Variability among these factors then causes the types of behavior variability observed in humans (but generally not represented in models). It should also be noted that the recently emphasized cultural differences are ultimately expressed in terms of particular values of these individual factors. Thus to adequately understand and simulate the effects of cultural differences we must first understand how they map onto the various individual differences which then directly influence individual behavior.

For expository purposes, it is helpful to categorize these factors along several dimensions, including the following:

- Source of variability: cognitive, personality, or affective
- Degree of permanence: static vs. dynamic
- Level of abstraction: fundamental (e.g., working memory capacity, the intelligence "g" factor), or composite (e.g., values, beliefs, attitudes)

Below, we provide a brief overview of several key categories of these factors, beginning with the fundamental categories of cognitive, personality, and affective factors, and concluding with the complex factors that comprise the decision-maker's internal dynamic context and cultural factors.

Cognitive Factors

Cognitive individual differences influence a variety of perceptual and inferential processes mediating all stages of the decision-making process: perception, situation assessment, problem solving and planning, learning, and action selection. These factors can be divided into three subcategories, reflecting their level of abstraction and task/skill specificity: *general ability factors, stylistic factors, and specific ability factors* (Hudlicka, 2002a; 2003a). We briefly describe each category below.

The *general ability factors* have been studied most extensively, both at the aggregate level of general intelligence (the "g" factor), and, more recently, at the level of elementary cognitive tasks (e.g., attention speed and accuracy, memory retrieval speed, working memory capacity, etc.) (Revelle & Born, 1999). The general ability factors reflect the limiting or baseline conditions of the fundamental information processing capabilities of the cognitive apparatus. The most extensively studied generic cognitive factors relate to attention and memory, and a number of empirical studies document the effects of a variety of affective states and personality traits on these processes (e.g., capacity, accuracy, speed, and vigilance of attention; capacity, speed, accuracy of working memory; and retrieval and encoding processes associated with long term memory). The influence of the general cognitive abilities is the greatest in unfamiliar situations and during the early stages of skill acquisition and training (Ackerman et al., 1989), correlates with other broad criteria (e.g., success in college) at 0.3 to 0.5 levels, and correlates with specific skill cognitive factors (i.e., a person who scores high on a "g" test is more likely to score high on specific skill test, all else being equal) (Revelle & Born, 1999).

Stylistic factors reflect consistent biases or preferences in information processing (e.g., preference for visual vs. textual vs. oral information delivery; preference for diverse sources of information vs. single, trusted source, simple vs. complex information, analytic vs. intuitive inferencing, etc.). These factors include both domain-independent factors (e.g., preference for visual vs. text format; preference for abstract vs. concrete information), and highly domain-specific and task-dependent factors (e.g., preference for task vs. 'people' data, etc.). While some empirical data about these factors are available (see Hudlicka, 2004b for a literature review), they have not been studied as extensively as the general cognitive factors, nor have they been deconstructed to a level of resolution that maps them onto the corresponding cognitive architecture structures and processes.

Finally, the subgroup of *specific ability factors* refers to specific complex reasoning and problem-solving skills, whether domain independent (e.g., what-if and abductive reasoning, meta-cognitive abilities), or domain specific (e.g., economic data interpretation, adversary psychological assessment, ability to maintain subordinate morale, etc.). Individuals differ in specific skill levels largely due to training and experience, but also as a function of personality and temperament, due to general cognitive abilities, and due to the current affective state. For example, experts' performance is typically fast and accurate, anxious experts' performance is slightly slower to maintain accuracy, whereas novices tend to be slow and inaccurate.

Personality Factors

Personality traits and temperament have been extensively researched by all subdisciplines within psychology, and a number of trait sets have been defined, via distinct methods (e.g., factor-analyses of self-report data, biologically-inspired theories, clinical observations). Currently, the most widely accepted trait set within academic psychology is the Five Factor Model (Big 5) (Costa & McCrae, 1992), consisting of (openness, conscientiousness, agreeableness, extraversion, neuroticism (OCEAN)). A number of other n-factor models have been proposed, ranging from Eysenck's "Giant 3" (neuroticism, extraversion, and psychoticism) (Eysenck, 1991) to Cattell's 12 factor model (Cattell, 1971). More biologically-oriented psychologists speak of systems mediating approach (BAS), avoidance (BIS), and fight or flight (FFS) (Gray, 1990). Clinicians speak of distinct personality styles (e.g., avoidant, passive-aggressive, narcissistic, etc.). Political and social psychologists consider high-level traits relevant to leadership and interpersonal decision-making style (e.g., authoritarianism, dogmatism, ease of persuasion, locus of control, need for power, need for affiliation, etc.) (Holsti, 1977; Shaw & Post, 1999).

Trait effects can be both *structurally-oriented* and *functionally-oriented. Structurally-oriented* effects influence the types of schemas stored in long-term memory (LTM) (e.g., higher proportion of self- and threat-oriented memory schemas associated with low emotional stability (Matthews & Deary, 1998)), and preferential processing pathways among functional components of the cognitive architecture. F*unctionally-oriented* effects influence the dynamic characteristics of states such as onset triggers, ramp up and decay rates, and maximum intensities (Eid, 2001) (e.g., low emotional stability correlates with lower trigger thresholds, steeper growth, slower decay and higher intensity for fear and anxiety states, as well as more generalized expressions of anxiety (Matthews and Deary, 1998)).

These effects are in contrast to states, discussed below, which tend to produce transient changes that influence the dynamic characteristics of a particular cognitive or perceptual process (e.g., attention and WM capacity, speed, and accuracy)). For example, low emotional stability (also referred to as neuroticism) is correlated with a predominance of negative, threat- and self-related schemas in long-term memory, as well as schemas pertaining to affect (Matthews et al., 2000b). It is also correlated with attentional and situation assessment preference for self-related and affective stimuli, and a bias towards negative appraisal (Matthews et al., 2000b). Low emotional stability is also associated with increased likelihood of negative affect and punishment-avoiding behavior, while high extraversion is associated with increased likelihood of positive affect and reward-seeking behavior (Matthews & Deary, 1998).

Affective Factors

Psychologists have studied emotions extensively. Over the past two decades, emotion research has witnessed resurgence, in large part due to advances in neuroscience methods and availability of computational methods for emotion recognition, expression, and modeling. Emotions have a broad range of effects on all aspects of perceptual and cognitive processing, and recent research has demonstrated that appropriate affective functioning is necessary for decision-making (Damasio, 1994). Suggestions have been made to refer to multiple types of rationality, to provide terminology that accurately reflects the role of emotion in decision-making, and to counteract the still prevalent belief in the outdated "rational" vs. "emotional" dichotomy. For example, Lisetti and Gmytrasiewicz (2002) proposed the term $rationality_1$ to refer to rational decision-making requiring emotion, and $rationality_2$ to refer to formal logical reasoning, which may or may not be used during a particular decision-making sequence.

Emotional states can be categorized by their *duration* (transient *emotions* vs. long-lasting *moods*); by their *degree of differentiation and behavioral specificity* (*affective states* such as like/dislike or approach/avoid reactions vs. more specific *emotions* such as joy, pride, sadness, fear, shame, anger, etc.); by the *degree of cognitive involvement* and consequent *individual behavioral variations* possible (*basic emotions*[1] such as fear, sadness, happiness, anger typically show fewer variations than *complex emotions* such as pride, shame, jealousy, guilt); and by a number of additional factors such as triggers, manifestations, and degree of voluntary control. Emotions can also be categorized in terms of the underlying dimensions, such as positive and negative affect (Watson and Clark, 1992), energetic and tense arousal (Thayer, 1996), hedonic tone, energy and tension (Matthews et al., 1990), and valence and arousal (Watson & Tellegen, 1985; Russell, 1979) For a brief overview of these issues see Hudlicka (2003c). For extensive discussions see Ekman and Davidson (1994), Lewis and Haviland (1993), Davidson et al., (2003), and Forgas (2000, 2001).

The specific effects on attention and cognition of a number of affective states have been studied extensively (e.g., anxiety and fear, anger and frustration, positive and negative affect, etc.). Emotional influences exist both at the "lower" levels of processing, causing changes in attentional and working memory processing, and contributing to selective enhancement or inhibition of particular cues or memory schemas (e.g., attention orientation during an acute fear episode, increased working memory capacity correlated with positive affect), and at

---

[1] The notion of basic emotions remains somewhat controversial.

"higher" processing levels, involving goals, situation assessments, expectations, and self schemas (e.g., complex feedback relationships between affective state and self-schemas (Matthews et al., 2000a)). Examples of specific findings are shown in Table 1.

Anxiety and Attention (Williams et al., 1997; Mineka & Sutton, 1992).

Narrowing of attentional focus

Predisposing towards detection of threatening stimuli

Affective state and Memory (Bower, 1981; Blaney 1986)

Mood-congruent memory phenomenon – positive or negative affective state induces recall of similarly valenced material

Obsessiveness and Performance (Persons & Foa, 1984)

Delayed decision-making

Reduced ability to recall recent activities

Reduced confidence in ability to distinguish among actual and imagined actions and events

Narrow conceptual categories

Affect and Judgment & Perception (Isen, 1993; Williams et al., 1997)

Depression lowers estimates of degree of control

Anxiety predisposes towards interpretation of ambiguous stimuli as threatening

Positive affect promotes heuristic processing (Clore, 1994)

Positive affect increases estimates of degree of control (Isen, 1993)

TABLE 1. Effect of Emotions on Cognition: Examples of Empirical Findings

Dynamic Mental Constructs

We outlined three categories of the fundamental factors influencing decision-making: cognitive abilities, personality traits, and emotional states. These fall within the broad category of individual differences and have been the focus of much research. Here we focus on higher-level, composite mental constructs mediating decision-making and action selection, such as situation assessments and beliefs, values and goals, expectations and attitudes; that is, on constructs that comprise an individual's dynamic mental context within which decisions are made and behavior selected.

In computational terms, some of these are best thought of as collections of activated memory schemas (e.g., the activated LTM schemas representing the current situation assessment or currently most active goals). Others are more heterogeneous functional entities, which cannot be classified in terms of a single schema but reflect rather a complex constellation of structures, processes, and particular preferences and biases. Examples of the latter category are attitudes, which consist of motivational, cognitive, affective and behavioral components (Aizen & Fishbein, in press).

Much recent research has focused on the role of these constructs in decision-making and behavior. For example, the area of naturalistic decision making and recognition-primed decision-making highlights the importance of situation assessment in decision-making, and the associated constructs of situations and expectations (Klein, 1997; Endsley, 2000). Recent attitude research focuses on the critical role played by attitudes. In particular, on the importance of attitudes towards *specific behaviors,* as opposed to general *attitudes towards people or objects*, in influencing individual behavior (Aizen & Fishbein, in press).

The mental constructs which play these key roles in determining behavioral choices are activated by a collection of internal and external factors, including the individual differences discussed above, and represent more or less stable configurations of mental schemas, which are more or less readily activated under specific circumstances. While much progress has been made in developing, refining and evaluating theories of situation assessment and decision-making (e.g., Klein, 1997; Endsley, 2000) and some studies focus on individual differences (Gugerty & Tirre, 2000) and stress (e.g., Orasanu, 1998; Cannon-Bowers & Salas, 1998), little attention has been paid to explicitly incorporating affective and personality influences within these theories at a level that would enable computational modeling. Several recent exceptions to this are the work of Matthews, (e.g., Matthews et al., 2000a), who attempts to identify the mechanisms of trait and state effects, and Hudlicka, who developed a methodology and architecture for modeling the effects of individual differences on recognition-primed decision-making (Hudlicka, 1997; 2002a; 2003a; Hudlicka & Pfautz, 2002).

Cultural Factors

Last, but not least, we turn to cultural factors as additional behavior determinants. The influence of culture on behavior has received much attention lately, and is of interest to human behavior modelers for a variety of reasons, ranging from basic research questions to applied objectives in the business community, government and the military. The applications of interest include improved understanding of business negotiations in multi-cultural business settings; diplomacy and peace negotiations in government; and multi-national task forces, peacekeeping operations, and asymmetric warfare in the military. There is an extensive literature in psychology and cultural anthropology discussing characteristics of cultures, research methods, and the effects of culture on decision-making and social interactions (e.g., Matsumoto, 2001; Renshon & Duckitt, 2000; Triandis & Suh, 2002; Cooper & Denner, 1998; Codevilla, 1997; Hofstede, 1991).

In spite of the vast literature addressing cultural issues, there is relative paucity of attributes defined at a sufficient level of specificity to enable computational modeling and inferencing; that is, cultural characteristics which could be operationalized to enable computational models of the effects of culture on individual (and organizational) decision-making, culture-based profiling, and, more importantly, likely to yield useful behavior predictions for particular individuals, groups, and organizations.

Perhaps the most prominent set of "cultural" factors is that identified by Hofstede (1991), consisting of the following: Power Distribution, Individualism-Collectivism, Femininity-Masculinity, Uncertainty Avoidance, and Short vs. Long-Term Orientation. This list was recently augmented by Klein and colleagues (2002), and termed the "cultural lens" model. Klein's set of factors augments the Hofstede factors to include hypothetical-concrete reasoning, casual attribution, and contrasting synthesizing.

There are several problems with the existing cultural factors that limit their utility and applicability to computational modeling. Specifically:

- The *factors are identified at very abstract levels*, which have little predictive value for specific behaviors selected by particular individuals or groups, operating within particular situational and environmental constraints. For example, how will knowing that a particular individual comes from a culture with low power distribution help us predict whether s/he likely to participate in a violent demonstration to be held tomorrow?

- The *factors are insufficiently operationalized* to serve as a basis for computational modeling and behavior prediction. In other words, it is difficult to translate factors such as power distribution, femininity/masculinity, or dialectical thinking into specific artificial intelligence inferencing formalisms and knowledge.

- The *factors lack adequate breadth* to capture all of the relevant and necessary cultural variables. In spite of the existing literature, the actual number of specific cultural factors identified is surprisingly small. In addition to the two problems above, the identified factors thus do not begin to capture the complexity of cultural influences, nor do they adequately address the problems of multiple and possibly conflicting cultural influences, or specific individual factors that may amplify (or inhibit) a particular cultural influence. Recent attempts to address these issues, and to develop a practical cultural-profiling approach, have explored a notion introduced by Karabaich (1996) that proposes to consider each group to which an individual belongs as representing a distinct culture; that is, the assumption that *every group creates its own culture* (Hudlicka, 2003d). The term culture in this approach is thus not limited to nations or ethnic groups, as has generally been the case in the literature (Matsumoto, 2001; Hofstede, 1991), but is instead broadened to include any group that influences the individual's behavior. This approach is motivated by the observation that national and ethnic groups are in fact not as diagnostic with respect to behavior prediction as are smaller groups to which an individual belongs (e.g., student group, social group, political group, family, etc.).

In closing, it should also be noted that any cultural influence ultimately functions at the individual level, and must therefore be translated to one or more of the individual behavior determinants outlined above. Thorough understandings of the critical behavior determinants, and a determination of the mappings of the cultural factors onto these behavior determinants, are therefore critical to the effective modeling of cultural differences.

Modeling the Individual

Computational cognitive models have a long history in artificial intelligence and cognitive science, going back to the seminal work of Newell on the general problem solver, which led to the Soar architecture (Newell, 1990), and Selfridge's Pandemonium in the 50's (Selfridge, 1959) which laid the foundation for the now-popular blackboard systems and blackboard cognitive architectures (e.g., COGNET (Zachary, 1992)). Since then, researchers have explored models of various perceptual, cognitive and motor processes, in isolation or within integrated architectures, using a variety of methods. These range from analytic closed-form models to a wide variety of simulation-based models, including task-based, symbolic, connectionist, and hybrid models.

Recently, the integrated-architecture approach, which aims to model end-to-end information processing required for intelligent adaptive behavior, has become the most prominent method, and forms the basis for intelligent behavior in synthetic agents and robots. Associated developments in virtual environments and robotics have further motivated the development of intelligent agents, capable of functioning in simulated or real environments. Below we provide a brief overview of representative cognitive architectures used to model individuals. We will not address task-based modeling environments such as MicroSaint (Lee, 2004) or IMPRINT (Mitchell, 2000), or models focused on a single component or function, such as EPIC (Myers & Kieras, 1997). This emphasis is not meant to imply that these approaches are irrelevant to organizational modeling, but rather reflects our belief that the most promising initial approach is via cognitive architectures, which are well suited for representing individual behavior, including the effects of individual differences, and may also be suitable for representing the behavior of a group or organizational component.

The cognitive architecture approach builds on the ideas of Newell and his unified theory of cognition (Newell, 1990), which suggest that cognition cannot be effectively investigated in terms of models of isolated phenomena. Instead, the individual processes and structures must be integrated into comprehensive models, capable of emulating the entire "see-think-do" sequence of human information processing, *and* be able to exhibit adaptive behavior. This view has led to the emergence of a variety of architecture-based models, which share several key characteristics:

- Implemented primarily in terms of symbolic representational and inferencing mechanisms, as opposed to connectionist, dynamical systems or analytical models -- although, some architectures may contain such components.

- Consisting of modules, processes and structures intended to emulate structural and functional components of human cognition (e.g., sensory stores, working memory, long-term memory, attention, multi-tasking, perception, situation assessment, decision making, planning, learning, goal management, etc.). This is in contrast to models which abstract performance onto simpler input-output relationships, without the explicit representation of the mediating structures and processes, such as the popular task-based models.

- Functioning in simulated scenarios in a particular domain and attempting to perform an actual simulated task.

Below we provide a brief overview of several representative cognitive architectures, including also recent attempts to model individual differences and emotions, and conclude with a brief discussion of validation issues. For more extensive and detailed reviews of existing cognitive architectures see Pew and Mavor (1998) and Ritter et al. (1999).

Cognitive Architectures

Among the first implemented cognitive architectures were Soar (Symbolic Operator Architecture) (Newell, 1990), and ACT (Adaptive Character of Thought) (Anderson, 1990). These architectures were developed with distinct aims. Soar aims to model learning and intelligent behavior and uses rules and rule chunking as the primary representational and inferencing mechanisms. ACT was initially aimed to model lower-level memory processes, and account for observed empirical data, but was later developed into a full-fledged agent architecture, using a combination of rules and semantic nets. Each of these architectures currently has an extensive user community, primarily in academic settings but more recently also in government, primarily within the military. Soar has been incorporated into a variety of synthetic agents and used in training simulations to model aircraft and helicopter pilots.

Coming from a different tradition, and built for a different purpose, is the Sim_Agent architecture. Sim_Agent was developed by Sloman and colleagues in the 1980's, with the objective to explore the architectural components and processes necessary to exhibit adaptive behavior, including emotions, and address the fundamental questions of what specific architectural features are necessary for different types (and complexity) of cognition, emotion, and behavior (Sloman, 2000; 2003).

Over the past decade, a number of new architectures have been developed, including BBN's OMAR (Operator Model Architecture), focusing on modeling multi-tasking and multiple operators in the air traffic control domain, and using a hierarchical representation of procedures (Deutsch et al., 1993); NASA's MIDAS (Man-Machine Integrated Design and Analysis System), used for human-machine system design, primarily within the commercial aviation cockpit (Corker, 2001; 2001; Laughery et al., 2001); COGNET (COGnition as a Network of Tasks) is used to model multitasking, interface design and adversary models and uses a blackboard architecture (Zachary, 1992); SAMPLE (Situation Awareness Model for Pilot-in-the-Loop Evaluation) focuses on modeling recognition-primed decision making in a variety of settings, including piloting and air traffic control and uses a combination of fuzzy logic, belief nets, and rules (Zacharias et al., 1995; Harper et al., 2000, 2001); and MAMID (Methodology for Analysis and Modeling of Individual Differences) designed to explicitly model the effects of a variety of individual differences on perceptual and decision-making processes (Hudlicka & Billingsley, 1999; Hudlicka, 2002a; 2003a; 2003b). Many other architecture-based models have been developed in academic laboratories, as research vehicles, as components of synthetic virtual agents, or in the context of robotics (see Shah & Pritchett, 2005).

Modeling Individual Differences and Emotions

With increasing awareness of the effects of individual differences on behavior, and increasing need for more realistic simulations, attempts have begun to incorporate individual differences effects within cognitive architectures and agents. Much of this work builds on existing research in psychology and neuroscience addressing emotion and personality processes. Models that focus on modeling individual differences range from individual processes to integrated architectures. The most frequently modeled process has been *cognitive appraisal,* whereby external and internal stimuli (emotion elicitors) are mapped onto a particular emotion. Several alternatives have been hypothesized for these processes in the psychological literature (Ortony, Clore & Collins, 1998; Frijda, 1986; Lazarus, 1991; Scherer, 1993; Smith & Kirby, 2000; Scherer et al., 2001). A number of these models have been implemented, both as stand-alone versions, and integrated within larger agent architectures (e.g., Scherer, 1993; Velasquez, 1997; Canamero, 1998; deRosis et al., 2003; Breazeal, 2003; Bates et al., 1992; Elliot et al., 1999; Andre et al., 2000; Martinho et al., 2000). Other emotion model implementations include models of emotions as goal management mechanisms (Frijda & Swagerman, 1987), models of interactions of emotion and cognition (Araujo, 1993), and effects of emotions on agent belief generation (Marsella & Gratch, 2002). Examples of *integrated architectures* focusing on emotion include most notably the work of Sloman and colleagues (Sloman, 2000), and more recent efforts to explicitly model the effects of a range of interacting individual differences on cognition and behavior (Hudlicka, 2002a; 2003a; 2003b), and efforts to integrate emotion effects in Soar (Jones et al., 2002) and in ACT (Ritter et al., 2002).

State-of-the-Art

Several of the architectures mentioned above have been prototyped, and in some cases evaluated, in simulation-based training and tutoring systems, and used in human-machine system design. Examples of these applications include the following: tutoring (ACT-R), intelligent synthetic agents or robots interacting with autistic children to facilitate social skill acquisition (Dautenhahn et al., 2002; Michaud & Theberge-Turmel, 2002; Blocher & Picard, 2002), synthetic agents populating virtual environments for training (e.g., Marsella & Gratch, 2002), tactical pilot simulations (TacAir-Soar, SAMPLE), pilot error modeling (MIDAS, OMAR), air traffic controller modeling (OMAR, SAMPLE), Aegis air defense models (COGNET), soldier and commander modeling (SAMPLE, MAMID), and adversary decision-making models (ACT-R, COGNET). More recently, there has been increased interest in incorporating cognitive architectures in interactive games, both commercial and military. It should be noted however that at this point no "fielded" system uses models as replacements of humans.

Cognitive modeling is an active research area that is beginning to emerge from the laboratory into a variety of applied settings. Several of the architectures above are available to the public, although in most cases the learning curve tends to be rather steep, particularly for novices to the modeling area. Research is currently under way to facilitate rapid model understanding and model sharing through a development of modeling ontologies (e.g., Napierski et al., 2004).

Challenges and Validation Issues

Much progress has been made in cognitive modeling of individual behavior, and many challenges remain, including the following: incorporation of more complex perceptual processing beyond visual stimuli; including non-verbal communication; modeling more sophisticated planning (the majority of the existing architectures do not model complex planning); modeling learning and skill acquisition; continuing to refine individual differences models; and representation of situational and organizational factors.

A broader challenge is to reduce the development costs of these models and to make them more robust. Both of these issues have recently begun to be addressed by an Office of Naval Research initiative, which also includes the development and sharing of ontologies mentioned above, to enable more rapid model development. Efforts are also under way to systematically evaluate and compare the performance of distinct models (Tenney et al., 2003; Gluck & Pew, 2002).

Perhaps the most important challenge is model validation. To date, there is no validated architecture, either at the individual module level, or at the integrated architecture level as a whole. There has, however, been progress with validating single-process models, for example, a variety of memory experiments, language acquisition, and multi-tasking.

Unfortunately, there is a frequent tendency to confuse verification (the model does what it was programmed to do), with validation (the model corresponds to the actual phenomenon being represented – that is, human decision-making and behavior), with the latter obviously being the more challenging, and meaningful, criterion of success. Current efforts to systematically compare human performance with multiple models, such as the work of Tenney and colleagues (2003), represent a promising trend in this area.

REQUIREMENTS FOR MODELING THE INDIVIDUAL DECISION-MAKER

In this section we outline the knowledge and data necessary for modeling the individual within an organizational context, as well as the associated representational and inferencing requirements. We discuss these requirements in the context of the two approaches described earlier in this chapter: a *cognitive architecture* and a *profile-based social network model*.

The objective of the *cognitive architecture approach* is to emulate the structures and processes used by the human decision-maker. The resulting architecture can then function in a simulation environment to represent the individual decision-maker's behavior for training purposes, and can also help in behavior prediction. In contrast to this, the *profile-based approach* does not require knowledge that allows emulation of the actual decision processes. Instead, these models require knowledge that enables automatic inferencing by a decision-aid (e.g., an expert system) *about* the decision-maker, to derive additional knowledge about the decision-maker's profile, and to predict likely decisions and behavior, within a particular context.

The distinguishing characteristic of the *cognitive architecture approach* is thus the need to identify the internal structures and processes mediating the performance of interest, and to capture these in terms of the architecture components; that is, the *architecture modules*, the *mental constructs* manipulated by these modules, and the *algorithms* comprising processing within these modules. Typically the performance of interest will be a set of concrete tasks within the domain of interest.

In contrast, the *profile-based approach* requires knowledge and data characterizing a decision-maker and predicting his/her behavior, and the knowledge of how to use the available data to derive the information of interest (e.g., particular unknown characteristic or likely behavior).

Ideally, a profile would consist of a small set of factors from which all other characteristics and likely behaviors could then be predicted. Unfortunately, the complexity of human personality and decision-making precludes a characterization in terms of a single set of orthogonal covering dimensions. To obtain a detailed characterization of an individual, and define the behavior determinants, it is therefore necessary to analyze the person from a variety of perspectives and at varying levels of abstraction. This requires the use of multiple sets of profile attributes, whose relative importance for particular behavior predictions may vary, depending on the operational context.

There are thus overlapping but distinct requirements for the types of knowledge necessary to construct these two types of models, for the data required to populate the model structures, and for the data required to support dynamic simulations. For example, the individual profile may include the individual's situations, expectations, and goals. However, while in the cognitive architecture approach these situations, expectations and goals are *derived by the architecture* via the emulated decision-processes, in the profile-based approach they are either *provided directly by the modeler* (as input data), or derived via some inferencing mechanism that makes no attempt to emulate human decision-making, but instead simply captures the observed regularities (e.g., situation A frequently leads actor X to generate behavior B).

These distinctions (and similarities) should be kept in mind as we discuss the specific knowledge and data requirements for each approach below. We distinguish between *knowledge*, which defines the models' structures and processes, and may include long-term memory schemas and the rules that manipulate them vs. *data*, which define the contents of these structures (e.g., individual profile data) and enable the dynamic simulation of the model (e.g., dynamic data representing the changing environment, including the social organizational environment, and the evolving task).

Knowledge and Data Requirements

Below we briefly describe the four primary sources of knowledge and data for developing models via a *cognitive architecture approach* and a *profile-based approach*. These sources serve as the basis for the following aspects of the models: (1) theories of decision-making to emulate within the cognitive architecture approach; (2) the long-term memory schemas used in the cognitive architecture approach; (3) the mappings among these schemas that enable the actual transformation of incoming data to selections of particular adaptive behaviors[2] within the architecture approach; (4) the structure and contents of the individual profiles used in the profile-based approach; and (5) data necessary for the dynamic simulation of the evolving context during model execution.

Sources of Knowledge and Data

There are three primary sources of knowledge and data for developing models of individuals, whether based on cognitive architectures or profiles: (1) existing empirical literature; (2) task analysis and knowledge elicitation interviews; and (3) empirical studies collecting specific required data. We briefly discuss each of these below.

Existing Empirical Data. Finding existing empirical data adequate for building a cognitive architecture model, or an individual profile, is challenging. *First*, the majority of the existing data were collected from laboratory experiments, in tightly controlled conditions, and using specific populations (usually university undergraduates). It is not always clear how well such data generalize to other settings and other populations. *Second*, the particular modeling objectives may not coincide with the specific conditions and hypotheses for which the data were collected, thus, again, possibly compromising the degree of applicability. This is particularly relevant for cognitive architecture models, where detailed data regarding the internal structures and processes are required, but rarely available.

Third, the representational and inferencing requirements of computational models currently exceed our ability to collect corresponding empirical data. In other words, we have the means to construct computational models that represent a variety of internal mental constructs (e.g., goals, expectations) and processes (e.g., goal selection, expectation generation) for which it would be difficult to

---

[2] Some of these mappings can be represented declaratively in the architecture LTM, in terms of rules, beliefs nets, or some other representational formalism, while others may be encoded in terms of the procedures and algorithms within the architecture modules.

unequivocally identify corresponding human data. The existing objective empirical data are, of necessity, obtained from externalized measures of the sensory stimuli presented to the human decision maker, and from his/her corresponding motor responses. These data thus reflect input/output mappings and focus on the periphery of the sensory-motor apparatus (attention, perception, and quantifiable motor responses). Inferring the variety of intermediate, internal structures and processes that mediate the transformation from sensory cues to motor responses, even with the use of "introspective" knowledge elicitation techniques (see below), is a very difficult and inexact process.

Nevertheless, the picture is not utterly bleak. For example, with respect to modeling individual differences, there are many studies that indicate the existence of reasonably robust patterns of responsiveness, for example, effects of particular traits and states on attention, working memory, and perception, as outlined above. Empirical methods exist that may be able to capture some of the intermediate structures and processes, for example, a variety of verbal protocol studies used to assess situation awareness, and possibly the associated mental constructs (i.e., situations, expectations) (Endsley, 2000). It may also be possible that some of the recently popular neuroimaging methods may eventually provide useful data, although currently many psychologists remain skeptical of their immediate utility for modeling integrated architectures.

Similarly, many correlations have been identified among particular individual characteristics (e.g., high extraversion, low neuroticism, and trust), and between particular characteristics and specific behaviors (e.g., trait-anxiety and checking behaviors). Such correlations can then serve as a basis for defining individual profiles, as well as for defining the knowledge used to derive additional information from the available profile data; that is, deriving new characteristics from existing characteristics, or deriving likelihood of behaviors from existing individual characteristics.

Task Analysis and Knowledge Elicitation. Task analysis and knowledge elicitation methods represent a critical component of both cognitive modeling and individual profile development. These methods may be used in the context of a formal empirical study, with the aim of producing statistically significant results or, more typically, they are used with a small number of subjects to provide idiographic "expert data." Task analysis methods have traditionally been used in human factors and human-system design, to provide data for work design, task allocation, and user interface design (e.g., Schraagen et al., 2000). Knowledge elicitation techniques have been used in expert system design to obtain data from which expert rule or case bases can be constructed (e.g., Cooke, 1994; Geiwitz et al., 1988).

The aim of these methods is to capture the structures and data, and to some extent the processes, that mediate decision making and skilled task performance. While subject to many of the caveats above regarding accessibility of internal mental structures and processes, a number of these methods are increasingly being used in rigorous empirical studies, to help characterize mental models, knowledge manipulation rules, and, more recently, to provide data for the development of computational cognitive models and architectures.

A subcategory of knowledge elicitation techniques of particular interest to modelers are the *indirect elicitation techniques*, which aim to capture the internal structures and rules mediating skilled reasoning that may not be readily articulated in response to direct questions. A number of these methods have been developed, falling into two broad categories: (1) methods derived from multivariate statistical analysis techniques such as multidimensional scaling (Schiffman et al., 1981), and hierarchical clustering (Johnson, 1967); and (2) methods derived from symbolic questionnaire and interview methods (e.g., semantic differential (Osgood et al., 1957) and repertory grid analysis (Kelly, 1955).

As we begin to expand cognitive architectures to include affective and personality influences, the existing task analysis techniques which focus on cognitive aspects of processing must be augmented to include affective and personality factors. An example of such an enhancement is the Cognitive Affective Personality Task Analysis method developed by Hudlicka (2001), which uses existing empirical data coupled with structured interviews with experts to generate larger spaces of possible behaviors that capture the variations due to a variety of individual differences, primarily personality traits and affective states.

Empirical Studies Collecting Required Data. As mentioned above, in many, if not most, cases, the specific empirical data required for a particular modeling application do not exist in the literature. Ideally, in these situations, the necessary empirical studies would be conducted focusing on collecting the exact data (e.g., particular types of individuals conducting the task of interest under specific circumstances). Unfortunately, most cognitive model development efforts do not have the time and budgets to also incorporate the types of extensive empirical studies that would be required to collect the necessary data. If such studies are conducted at all, they may be part of a later validation effort, and are typically limited. Fortunately, this is beginning to change as the funding agencies recognize the necessity of conducting tightly coupled computational-empirical efforts and as they increasingly encourage validation experiments. It is hoped that as the modeling community begins to develop more interdisciplinary teams among psychologists, cognitive scientists and AI researchers, such studies will become an integral part of the model development and validation process.

Cognitive Architecture Model Development

Three factors help constrain the large space of possibilities for cognitive architecture definition: (1) the simulation objectives (both research and applied); (2) existing relevant theories; and (3) knowledge and data availability. We briefly discuss each of these below. We later illustrate this approach via a concrete example.

Simulation Objectives. A successful cognitive architecture-based simulation of individual behavior requires a well-defined set of research questions or specific performance requirements, within a narrowly constrained set of tasks. These factors then determine the representational resolution required to adequately capture the cognitive processing of interest; that is, to identify the aspects of the perceptual and decision-making processes that must be represented in greater detail, and those which can be abstracted, as well as the type of data that will be required to support dynamic simulation.

For example, if the focus is on explicit models of low-level perceptual processing, then the architecture may need explicit representations of a variety of sensory memories, in addition to the working and long-term memories, and a hierarchy of perceptual features, to allow the mapping of raw data onto the higher-level task-relevant perceptual features. If, on the other hand, the focus is on situation assessment, then it may be possible to dispense with low-level perceptual processing and focus instead on the explicit representation of a variety of high-level situations (aggregate mental models of the external or internal states and events), the detected stimuli that serve as the data for deriving these situations, the possible sets of future alternatives for each situation, and the mappings among these data and constructs (e.g., data-to-situations; situations-to-expectations).

A focus on the effects of individual differences (traits, states, and cognitive characteristics) requires a corresponding set of representations of these factors, and the structures and processes influenced by them. Thus, for example, architectures aiming to simulate the effects of stress on decision-making are likely to require explicit representations of the affect appraiser processes that generate the stress state, as well as the specific influences of stress on memory, attention, perception, all aspects of decision-making, and action selection. The more complex the construct in question, the more "distributed" its representational requirements will be within the architecture. Thus, for example, a focus on the role of attitudes requires explicit representation of the various components of attitude (cognitive, motivational, affective), in addition to representing the 'subject' of the attitude (e.g., person, event, behavior), as well as any other relevant factors that may influence the attitude application (e.g., personal and social norms, current social context).

Relevant Theories. The architecture components and processes are further constrained by the existing theories regarding the phenomena of interest. Thus, for example, the current theories of recognition-primed decision making (RPD; Klein, 1997) highlight the central role of situation awareness and situation assessment (Endsley, 2000), and suggest a range of specific structures (situations, expectations, long-term memory situation and expectation schemas, working memory storing currently active situations, etc) and processes (generating situations from cues, generating expectations from situations, priming by situations and goals, priming by individual history, etc.). Similarly, current theories of affect appraisal suggest multiple stages and multiple levels of resolution for the automatic and deliberate appraisals, as well as a series of elicitor types (Lazarus, 1991; Smith and Kirby, 2000), which in turn help determine the architectural structures and processes, as well as the data being manipulated by them.

The challenge here is that most existing empirically-supported theories provide only a partial glimpse of the complex cognitive processing required for intelligent, adaptive behavior. In constructing a cognitive architecture it is therefore necessary to collect these existing pieces into a unified whole. This exercise often reveals gaps, inconsistencies and contradictions within the existing theories, thereby stimulating further refinement. It is precisely this observation that motivated Newell's unified theories of cognition that contributed to the emergence of the integrated cognitive architectures discussed earlier.

Data Availability. Ideally, the architectural definition would be driven by the project objective on the one hand, and the theoretical constraints on the other. In reality however, it is also necessary to consider the availability of the data: if we cannot obtain the data to populate the model, then it will be of limited utility. As discussed above, while much empirical data are available about human performance, and some about the internal structures and processes mediating human decision-making, currently the representational resolution of cognitive models exceeds our capability to obtain the corresponding data. For example, we can construct detailed models of goal hierarchies and goal processes, but currently no validated means exist to unequivocally identify the corresponding constructs in long-term memory, or their dynamic use during decision-making. Thus if we wish to maintain some level of correspondence with observable reality, the architecture model resolution must match the range of available or obtainable data.

Profile-Based Social Network Model Development

This approach to modeling the individual within an organizational context combines two existing methods used to characterize individuals and organizations, and predict their behaviors: *individual profiling* and *social network analysis. Individual profiling* consists of collecting relevant information about a particular individual and using this information to: (1) infer the individual's most relevant characteristics, goals and motivations; (2) assess the individual's strengths and vulnerabilities, and, (3) predict the likelihood of particular behaviors. *Social network analysis* consists of constructing a graph representing the key individuals, groups, or organizations with which a particular individual interacts, and the relationships among them. Depending on the level of resolution of this social network, the relationships among these entities can range from simple characterizations of interactions (e.g., X does (or does not) communicate with Y) to complex analyses of the types of interactions (e.g., X trusts Y; Z obtains resources of type A from W, etc.).

Typically, these approaches are used separately. Below we discuss a means of integrating these approaches to model the complexity of individual decision-makers within their broader social and organizational context.

The exact structure of the individual profile depends on several considerations: (1) the *objective of the decision-aid*; (2) the nature of the *application domain*; (3) the *availability of knowledge* to support the desired inferences about the aspects of the decision-maker; and (4) the types of *decisions and behavior of interest*. Below we identify the general categories of behavior determinants (see Table 2) and outline a number of possible specific factors that may need to be included in the profile. The knowledge necessary to construct these profiles comes from the sources outlined above, including psychological empirical studies, but more frequently from a variety of "field" sources where profiling is used, including forensic applications and a variety of law enforcement and military applications (e.g., FBI and CIA profiling methods, Psychological Operations (PSYOP) methods in the Army, etc).

The collection of factors (attributes) used to characterize an individual is referred to as a "profile"; and the specific values assigned to these attributes for a particular individual are referred to as attribute values. The process of assigning these values to a particular individual during the modeling process is referred to as profile instantiation. The result of a profile instantiation is the creation of a particular instance of the generic profile, with the filled-in attribute values representing a specific individual.

Human behavior is determined and influenced by a large number of factors, and a number of approaches can be used to characterize an individual decision-maker and predict his/her behavior. These factors range from such high-level and relatively mundane data as demographics (age, race, marital status, ethnicity, religion, socioeconomic status), through characterizations of the decision-maker's personality and temperament (e.g., extraverted, stable, high risk and anxiety tolerant) and specific individual history (e.g., recent relevant experiences, early formative experiences), to complex constructs such as attitudes, beliefs, values, and goals. If possible, the profile should also include a range of typical behaviors and the preferred behavior for accomplishing desired goals. Depending on the specific application, it may also contain assessments of the decision-maker's strengths and vulnerabilities. Table 2 summarizes the broad categories of these profile factors.

While some of the profile attributes resemble the type of knowledge required for development of a cognitive architecture (e.g., beliefs, goals), the key difference between the profile-based approach and the cognitive architecture approach lies in the manner in which this information is generated. In the cognitive architecture approach, these constructs are generated dynamically during the emulation of the individual's decision-making process. In the profile-based approach, this information is either provided by the modeler, or derived from existing data, and

Demographic Info.	Training & Education
Individual History	Role
Intra / Inter personal Conflicts	Vulnerabilities, Pressure Points
Psychological Factors	Psychosocial Relationships
Attitudes / Beliefs	Situation Assessments
Goals / Goal Personnel	Goals Scripts
Info. Environment	Data Triggering Beliefs

TABLE 2. Categories of Behavior Determinants in Individual Profiles

then used to derive further information about the individual. No attempt is made to emulate the actual human information processing sequence generating these constructs.

In addition to the individual information about the particular person of interest, the profile also needs to capture the decision-maker's social environment. This is accomplished by including information about the decision-maker's relationships to other individuals and groups in his/her environment. These relationships include communications (with whom does the individual communicate, how, about what subject, how often), relationships with family and friends (parents, children, relatives, spouse, colleagues, confidants), groups with which the individual is affiliated (church, professional societies, sports, social groups). For each relationship, whether with an individual or a group, the nature of the relationship should be indicated

There are several categories into which these relationships can be classified, including: power and control (x controls y), affective tone (x likes y, y dislikes z), trust (x distrusts y), resources (x obtains money from y). It is the explicit representation of these relationships that then enables the representation of the individual's psychosocial environment, and the construction of the types of social networks that are typically used in organizational models. However, in contrast to most social networks approaches in organizational modeling, the richness of the individual profile enables a corresponding increase in complexity of the individual model, within the broader social and organizational context.

Depending on the simulation objective, and the nature of the domain, the type and level of detail for the different attributes in the profile will vary. In some cases, simple demographic data may be adequate. For example, actuarial predictions generally use very high-level, simple demographic characteristics to predict longevity and risk of disease. These data are adequate for the types of large-sample and long time-frame applications of interest to insurance companies. In most cases however, more specific information may be necessary to predict likelihood of specific behavior, changes in beliefs and attitudes, emergence of particular strengths or vulnerabilities, or changes in particular relationships. Table 5 in the next section illustrates a sample profile used for a PSYOP decision-aid.

Use of Profile for Behavior Prediction. The discussion above highlights the diverse types of knowledge necessary to construct an individual profile. To use this profile information for behavior prediction, additional knowledge must be collected to enable the derivation of missing profile information and, eventually, the prediction of likely decisions and behavior. This information is based on stable patterns among particular individual characteristics (e.g., characteristic X correlates with characteristic Y; characteristics A and B predict behavior C, etc.).

This knowledge can be either correlational or causal. For example, knowing that a decision-maker has a number of aggressive personality traits indicates that s/he is likely to engage in more aggressive behaviors and may experience higher-than-normal frequency of hostile or aggressive affective states; knowing that the decision-maker has engaged in particular behaviors in the past, or has expressed admiration for certain behaviors, is a good indication that s/he will select such behavior in the future.

As described earlier, this type of knowledge comes from a variety of sources, both empirical research studies and field applications. A variety of long-term psychological assessment studies are relevant here, which attempt to correlate particular profile characteristics with specific beliefs, attitudes, and behaviors.

Of particular interest here are theories about decision-making and behavior selection, since they provide templates indicating which factors are critical in the decision and behavior selection. For example, recent work by several attitude researchers indicates that one of the most critical factors predictive of particular behavior is the individual's attitude towards this behavior (Aizen, 2001; Aizen & Fishbein, in press). This is in contrast to attitudes towards the object, which have limited utility in predicting specific behaviors.

Representational and Inferencing Requirements and Options

Earlier we discussed the knowledge and data required to emulate, or reason about, an individual's decision-making. To be effective in simulating or profiling individual decision-makers, and predicting their behavior, this knowledge must be structured into formats that can be used by some type of automated inferencing methods to derive the information of interest. In the case of cognitive architecture models, these representational structures and automated methods aim to emulate the actual decision-making performed by humans. In the case of the profile-based approach, the automated methods may aim to simulate the reasoning that would be performed by a skilled analyst to predict individual decisions and behavior, or they may simply implement any appropriate method capable of deriving the desired information about the individual from the existing data.

We now briefly outline the requirements and some of the available options for representing the required knowledge, and for the inferencing mechanisms necessary. We discuss the representational and inferencing requirements together, due to their mutually constraining relationships: the selection of a particular representational requirement defines the options for the associated inferencing, and vice versa.

Cognitive Architectures

Three primary representational choices must be made for a cognitive architecture model: (1) the structure and functioning of the overall architecture, that is, the *architecture topology and processing*; (2) the structure and functioning of the individual modules comprising the architecture, that is the *individual module processing algorithms*; and (3) the *structure of the long-term memory* associated with the architecture modules. These representational choices then dictate, to a large extent, the associated inferencing options.

Architecture Topology and Processing. As outlined earlier, a number of options exist for designing the overall architecture. There are certain '*paradigms*' that may be followed, partially represented by existing systems (e.g., Soar's rules, ACT-R's semantic nets and rules, CogNET's blackboard and knowledge sources). A number of architectures use a structure that aims to emulate the decision sequence of the human decision-maker. This is true for the variety of BDI (belief / desire / intention) architectures (Rao & Georgeff, 1995), and the various see-think-do architecture models such as SAMPLE (Harper, 2000; 2001) and MAMID (Hudlicka, 2003a; 2003b).

Within each of these architectures, the overall processing sequence may be sequential (module n executes *after* module n-1), parallel (all modules execute simultaneously and any sequencing requirements are enforced by the availability of the emerging data), or some combination of these two. In some cases a type of partial pseudo-parallelism is also possible, where data from previous cycles may be available for future execution cycles. The choice of a particular architecture topology depends primarily on the requirements of the modeling objectives; that is, if we wish to model processes requiring parallelism, we need to use a parallel processing paradigm. The choice is also influenced by implementation characteristics and constraints, such as real-time operational requirements for training, communications bandwidth limitations associated with distributed simulations, etc. As mentioned above, these design decisions are, at this point, largely in the realm of art rather than science, and more research is required before a series of principles or "cookbook" recommendations can be collected to guide the model design process.

Module Processing Algorithms. Each of the modules within a cognitive architecture model represents a particular aspect or stage of the decision processing sequence. The algorithms within each module implement a selected theory of decision-making the architecture is intended to emulate, and that theory in turn determines how the information available within the module is combined to derive new information. For example, in the case of a situation assessment module, the module's algorithms would select the incoming or existing cues and combine them into higher-level situations reflecting the state of the world, the self, the task or the organizational and social context.

It is within each of these modules that the mutually constraining effects of the representational formalism and the inferencing algorithms are most evident. Typically, the algorithms that implement the processing within each module use a combination of procedural and declarative knowledge representations. In most cognitive architectures, the majority of the domain-specific knowledge is encoded in terms of declarative representations such as rules, semantic nets, propositions, or belief nets. Each of these representations necessitates particular interpretive procedures that perform the actual inferencing; that is, procedures that derive new information from the existing knowledge. These interpretive mechanisms may operate under the control of the module's algorithm, or their invocation may be a function of the available data and architecture's goals, or some combination of the two. Below we briefly discuss some of the alternatives for representing knowledge within a cognitive architecture; that is, the alternative knowledge representation formalisms for constructing the architecture's long-term memory (LTM).

Long-Term Memory Structure. As described above, a number of choices exist for representing the agent's long-term memory (e.g., rules, belief nets, semantic nets, propositions). The specific choice for a particular architecture depends on several factors, including the following: (1) the objective of the particular architecture; (2) available theories guiding a particular choice, and (3) nature of the problem and knowledge available. For example, since both the available data, and the human decision-making processes, are characterized by large degrees of uncertainty, it may be appropriate to use representational and inferencing approaches capable of explicitly representing uncertainty, such as the increasingly popular Bayesian belief net formalisms (Pearl, 1986). In situations where the objectives of the architecture are to emulate specific memory processes, and evaluate particular theories, these requirements may dictate the representational choices, e.g., rules (Newell, 1990) or semantic nets (Anderson, 1990).

Profile-Based Social Network Models

The representational requirements of the individual profile model are generally simpler than those of a cognitive architecture, since no attempt is made to emulate the actual structure of the human decision-making process. Instead, the objective is to use the existing data to derive the information of interest, possibly by simulating a skilled analyst's reasoning process regarding how some *other* human or organization perceives the world, makes decisions, and effects actions. Note the distinction between *emulating* human decision-processes, with an inherent focus on replicating the underlying structures and processes, and *simulating* the reasoning of a skilled analyst, with a focus on generating the same output but disregarding the means through which the output is obtained.

In spite of its relative simplicity compared with the cognitive architecture approach, a modeler using the profile-based approach must still make a number of choices regarding the type of information to include in the profile, the profile structure, and the syntax of the individual factors represented within the profile. As is the case with cognitive architecture approaches, the representational choices are closely tied with the choices of the associated inferencing methods that will be used to derive missing profile data and predict behavior.

Individual Profile Structure. The key requirement for the profile representational formalism is to capture the necessary knowledge in a format that can be used by some type of an automated inferencing method to derive the information of interest. The profile structure must therefore match the needs of the associated inferencing methods. Minimally, the information must be stored in terms of constituent components that make the key factors accessible. Thus "raw" data should not be stored in the form of free text, but rather in terms of highly structured data elements, which make specific factors explicit and readily accessible to the inferencing methods.

For example, consider the representation of the decision-maker's individual history, which could be represented as free text or a set of strings in a profile attribute labeled "Individual history". To use such data effectively, the automated inferencing would require sophisticated parsing or natural language understanding algorithms to identify the critical elements. A better alternative is to represent this information in highly structured templates, which explicitly encode, for example, the events in which the individual participated, the other actors present, the behavior of each actor and the results in terms of changed states of critical individuals or objects. This format then makes these critical pieces of information readily accessible to the associated inferencing methods. Similarly, when representing the decision-maker's attitudes and beliefs, one should make explicit their constituent components, such as the *believer* (e.g., Agent A), the *belief itself* (e.g., is trustworthy), and the *object of the belief* (e.g., Agent B). Tables 3 and 4 from Hudlicka and Zacharias (2002a) show examples of syntactic structures for two profile attributes from the C2WARS (Command and Control Warfare Analysis and Reasoning System) PSYOP decision-aid developed by Charles River Analytics: individual history, and attitudes and beliefs (Hudlicka & Zacharias, 2002b; Hudlicka et al., 2002).

In addition to profile attributes characterizing the individual, a variety of meta-information also needs to be represented; that is, information about the nature of the available data. This includes characteristics such as reliability, certainty, source, recency, staleness, importance, and the like. This information can then be

Event type	{military engagement | military service | foreign travel | foreign alliances | religious affiliations | political conflict | personal conflict | political alliance | personal alliance | mentor | role model | friendship | romantic relationship}
Key Participants	{individual | group}
Affect associated with event	{joy | anger | fear | affection | disgust | dislike}
Outcome	{facilitated | hindered} {<goal><attitude><belief><preference> <state>} {for | against} {individual | group | entity} {high | medium | low}

TABLE 3. Profile Attribute Syntax: Individual History

used to determine the utility of the available data, the best means to combine and select existing data to derive new information, and the certainty and reliability of the newly derived information.

Inferencing Methods. Once the profile knowledge is structured in a suitable manner, the inferencing methods can manipulate it to derive the desired information. In some cases the same syntax may be used by multiple inferencing methods, if the necessary translators are available to convert the profile contents into the appropriate knowledge representation scheme. Thus, for example, the information in the profile outlined above can be translated into data required by rules or belief nets, and processed by the associated inferencing mechanisms. Such translations typically involve relatively simple mechanical transformations of the available data. It should also be noted that the profile as a whole is well suited for case-based reasoning.

<table>
<tr><td>Entity</td><td>{<self> | <individual> | <group>}</td></tr>
<tr><td>Attitude / Beliefs about Personality / Temperament</td><td><psychological parameter>*</td></tr>
<tr><td>Attitude / Beliefs about Goals</td><td><goals>*</td></tr>
<tr><td>Attitude / Beliefs about Goal Scripts</td><td>{<goal> <goal script>+}*</td></tr>
<tr><td>Attitude / Beliefs about Existing Intrapersonal Conflicts</td><td><intrapersonal conflict>*</td></tr>
<tr><td>Attitude / Beliefs about Information Environment</td><td><information environment></td></tr>
<tr><td>Attitude / Belies about Psychosocial Relationships</td><td><psychosocial relationships></td></tr>
<tr><td>Attitude / Beliefs about Vulnerability / Pressure Point</td><td><vulnerability / pressure point>*</td></tr>
</table>

TABLE 4. Profile Attribute Syntax: Attitude / Beliefs

EXAMPLES ILLUSTRATING THE TWO ALTERNATIVE APPROACHES

In this section we present examples of the two alternative approaches to individual modeling discussed above: a *cognitive architecture* and a *profile-based social network model*, highlighting in each case the means and ability of representing individual differences, and the ability to model the influence of those differences on behavioral outcomes.

Modeling the Individual in Terms of a Cognitive Architecture

As indicated earlier, there are a number of possible architectures that could be used to model the individual within an organizational simulation. Given the critical role that individual behavioral variations may play in organizational behavior, it is important to select an architecture that is capable of representing these variations, in a rapid and empirically justified manner.

Below we describe in detail an architecture (and modeling environment) developed specifically to address the modeling of a range of multiple, interacting individual differences: MAMID (Methodology for Analysis and Modeling of Individual Differences), developed by Psychometrix Associates (Hudlicka, 2002a; 2003a; 2003b). MAMID is well suited for modeling individuals within organizations because of its ability to represent the effects of multiple, interacting individual differences in terms of parametric manipulations of the architecture structures and processes.

The initial MAMID prototype demonstrated an ability to model the effects of selected individual differences within an Army peacekeeping demonstration scenario (Hudlicka, 2003a; 2003b). Below we describe the individual differences modeling methodology, outline the key components of the MAMID cognitive architecture, and illustrate its operation and results from an initial evaluation.

MAMID is a generic methodology for modeling a range of multiple, interacting individual differences within symbolic cognitive architectures, via parametric manipulations of the architecture processes and structures. The underlying thesis of the approach is that the combined effects of a broad range of cognitive, affective, and trait individual differences, as well as a variety of cultural and individual history factors, can be modeled by varying the values of these parameters, rather than the architectural components themselves (Hudlicka, 1997; 2002a; 2003a). This then allows a rapid specification of cognitive architecture configurations capable of modeling a wide range of individual stereotypes, represented by distinct individual differences profiles. This thesis is motivated by a number of recent research findings in personality theory, emotion research, and neuroscience emotion research (Williams et al., 1997; Revelle, 1995; Matthews & Deary, 1998; Matthews, 2004; Fellous, 2004).

The architecture parameters control speed of module processing, capacity of working memory associated with each module, and attributes of internal mental constructs such as cues, situations, expectations and goals (e.g., threat level of cues, salience of situations, desirability of goals, etc.), and structure and contents of long-term memories mediating perceptual and cognitive processing. Distinct individual types (e.g., normal, anxious, aggressive) are represented by distinct individual profiles, which are then mapped onto specific configurations of the architecture parameters. The parameters cause "micro" variations in architecture processing, (e.g., number and types of cues processed by the attention module, number and types situations derived by the situation assessment modules; focus on goal A vs. goal B), which then lead to "macro" variations in observable behavior (e.g., high trait-anxious team leader requires more time and resources for a particular operation than a low trait-anxious; high-anxious operator misses a critical cue on operating console due to attentional narrowing, failing to diagnose an electrical malfunction, etc.). Figure 3 illustrates the general relationship between a representative set of individual differences, the architecture parameters, and the architecture itself. Figure 4 provides an expanded view of the MAMID architecture.

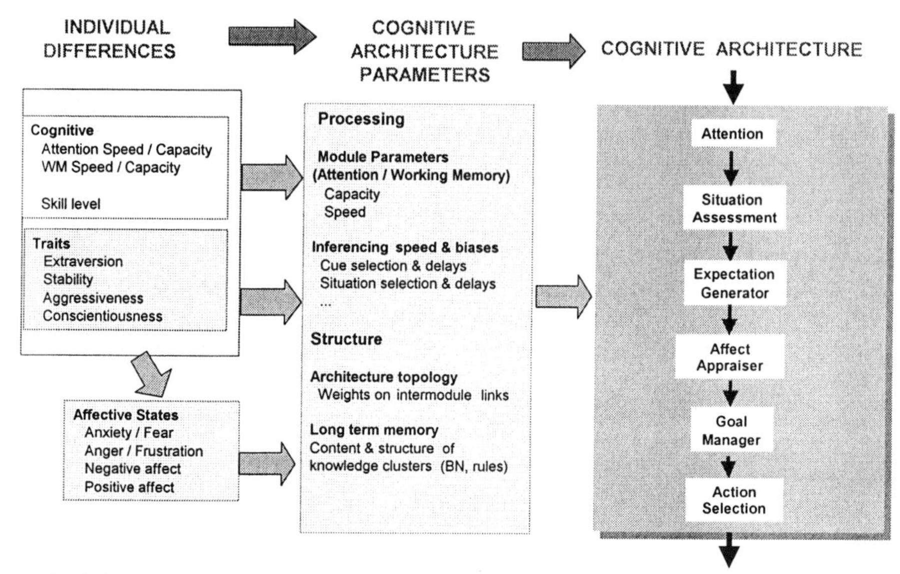

FIGURE 3. Schematic Illustration of MAMID Modeling Methodology

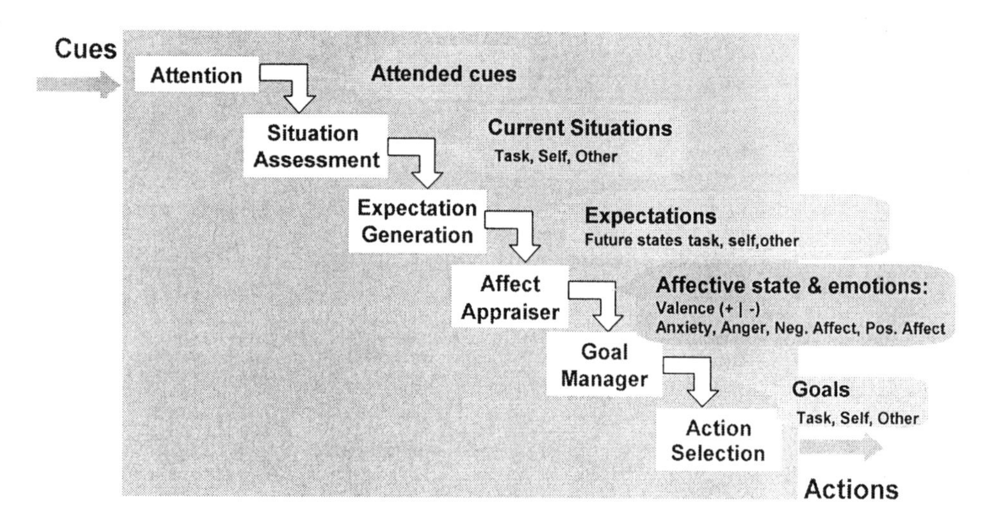

FIGURE 4. Schematic of the MAMID Architecture Showing Distinct Modules and Associated Mental Constructs

The parameterized cognitive architecture implementing this modeling methodology consists of seven modules, which map the incoming stimuli (cues) onto the outgoing behavior (actions), via a series of intermediate internal representational structures (situations, expectations, and goals). To effectively model the effects of affective states (e.g., stress, anxiety, frustration, boredom), the MAMID architecture includes an explicit representation of the affect appraisal process, which derives the agent's current affective state from a combination of external and internal cues. The MAMID architecture modules are as follows:

- *Sensory Pre-processing*, translating the incoming raw data into high-level task-relevant perceptual cues.
- *Attention*, filtering the incoming cues and selecting a subset for further processing.
- *Situation Assessment*, integrating individual cues into an overall situation assessment.
- *Expectation Generation*, projecting the current situation into one or more possible future states.
- *Affect Appraiser,* deriving the affective state from the variety of influencing factors: static (traits, individual history) and dynamic (current affective state, current situation, goal, expectation).
- *Goal Manager,* selecting the most relevant goal for achievement.
- *Action Selection*, selecting the most suitable action for achieving the current goal within the current context.

MAMID was developed in the context of an Army peacekeeping training scenario. An evaluation in this context indicates the feasibility of the approach to model a broad range of individual differences and their effects on individual behavior and mission outcome (Hudlicka, 2003a; 2003b). Figure 5 illustrates in detail the internal processing of two instances of MAMID architecture, representing a "normal" and a "high-anxious" commander encountering a particular problematic situation (hostile crowd) during a peacekeeping mission. The "stickies" in the center column of the figure indicate, for each module, the types of biases resulting from the anxious trait and state. Comparing the nature of the constructs in each module for the normal and anxious commanders then indicates how these biases influence the constructs derived within each module. Figure 6 provides a summary of the distinct behaviors produced by normal, anxious, and aggressive commanders.

Note the differences in the types, frequency, and duration of distinct actions by the different commander types. For example, the anxious commander spends more time in situation assessment and communication, and moves more slowly than the normal and the aggressive commanders.

The results illustrated in Figures 5 and 6 demonstrate the ability of the MAMID methodology and architecture to model distinct agent individual profiles within the architecture parameter space, and to generate distinct commander behaviors and mission outcomes, resulting from distinct individual profiles, for the selected individual differences and within the constrained environment of the demonstration scenario. A key component of the architecture that provides this capability is the architecture's multi-stage affect appraisal module, which dynamically generates distinct emotions in response to distinct external and internal contexts. These emotions, in conjunction with distinct personality traits, then give rise to different behavior in response to the same set of external circumstances. The MAMID testbed environment also serves as a research tool for investigating broader issues in computational cognitive modeling, by supporting the rapid testing of theoretical hypotheses regarding alternative mechanisms of influence of a variety of individual differences factors on perceptual and cognitive processing.

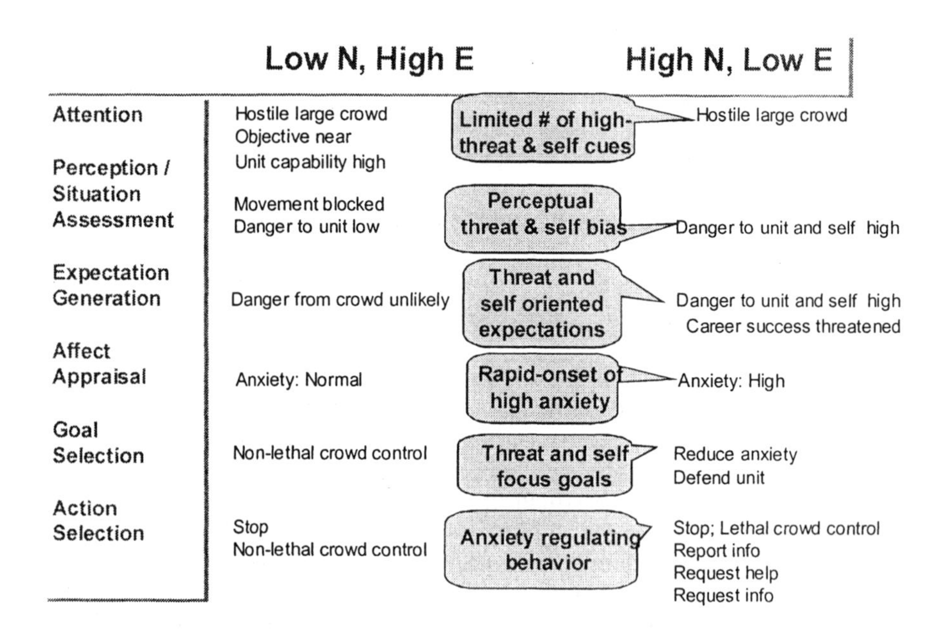

FIGURE 5. MAMID Models of 'Normal' and 'Anxious' Commanders: Summary of Mental Constructs Within Each of the MAMID Modules During the Processing of the "Hostile Crowd' Surprise Situation

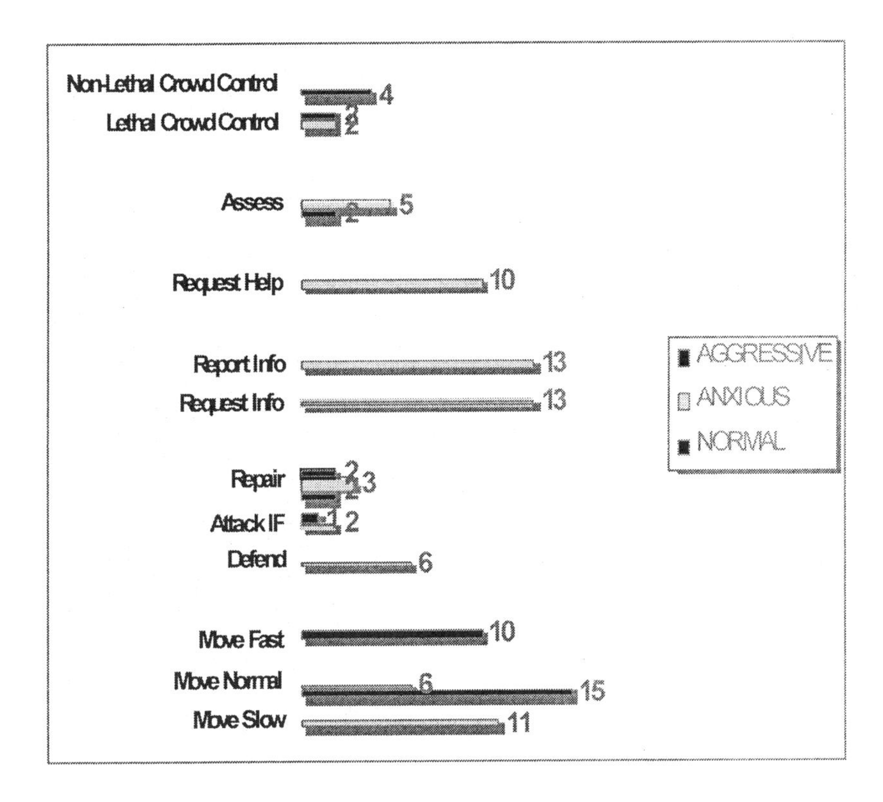

FIGURE 6. Summary of Behavior Generated by the MAMID Models of the 'Normal', 'Anxious', and 'Aggressive' Commanders

Although the initial prototype and the full implementation were developed within the military domain, the modeling methodology and architecture are domain-independent and applicable across a broad variety of domains. MAMID thus represents a promising approach for application to additional domains (e.g., variety of team training settings, neuropsychological assessment, edutainment and gaming, etc.). To increase effectiveness and 'experiential believability', the cognitive architecture should be coupled with an enhanced user interface, ideally using immersive virtual reality environment techniques to enhance the synthetic agent believability, and enhanced interactive visualizations to provide improved, rapid access to the numerous internal structures and processes, such as those shown in Figure 5. Additional enhancements can also be implemented within both the architecture and the testbed environment.

The MAMID architecture parameter space appears sufficiently rich to model additional high-level constructs, including a variety of cultural differences, such as activity orientation and uncertainty management (Klein et al., 2002), impulsiveness, leadership, workload, etc. It would be useful to explore the utility of additional architecture modules (e.g., planning), and alternative processing sequences (e.g., parallel or distributed processing among modules). It is also important to conduct further evaluations of the relative influence on behavior of stable factors (e.g., memory schemas), transient factors (e.g., emotions), and the global constraining factors of the current situational contexts, including the social context, such as the organization or group within which the individual functions. Following the social network example below, we briefly outline augmentations that would be required to integrate the cognitive architecture individual model within an organizational context.

Modeling the Individual in Terms of a Profile Within a Social Network

We now describe an alternative approach to modeling individual behavior, based on the use of profiles. This approach makes no attempt to emulate the human decision-making process, but rather aims to collect as much relevant information about the individual as possible in terms of a profile consisting of the critical behavior determinants, and past behaviors, as outlined earlier. A series of mappings among these determinants then enable the derivation of additional profile information, from existing data, as well as the prediction of likely behaviors. Similarly to the cognitive architecture approach described above, the profile-based approach is able to represent a variety of individual differences that cause variations in individual behavior. By explicitly including information about the individual's social environment (e.g., types of relationships to other individuals and groups), this approach also supports inferencing about individual behavior within an organizational context.

Below we describe a specific profile-based approach developed by Charles River Analytics and implemented within a prototype PSYOP decision-aid. This approach originated with C2WARS and is now being extended under the IODA (Information Operations Decision Aid) program (Hudlicka et al., 2002; Hudlicka et al., 2004). We first describe the individual behavior determinant profile, discuss how its components are used to derive additional individual data and behavior predictions, and provide examples of rules deriving this information.

Individual Behavior Determinant Profile

Below we outline a catalog of behavior determinants we have identified as relevant to PSYOP decision aiding, but which also apply to other domains. These determinants include individual differences and behavior moderators, but go beyond these to include the individual's goals, characteristic beliefs and attitudes. The profile also include the psychosocial and information environment within which the individual operates, represented in terms of the variety of relationships of the individual to his/her social environment. These relationships then define the individual's *social network* and are a critical component of modeling the individual's organizational milieu. The profile attributes are shown in Table 5. A subset of the profile attributes describes the individual's social network. An example of a social network generated from the IODA profile information is shown in Figure 7.

<u>Deriving Profile Values and Predicting Individual Behavior</u>

The profile in Table 5 summarizes the knowledge about the individual, both static and dynamic. To derive additional information from the existing profile data, and to generate behavior predictions, the profile-based approach needs a means of manipulating this knowledge to derive the information of interest. This can be accomplished via a number of means. The approach adopted for the IODA decision-aid described here uses a combination of rules and belief nets, which map the individual profile attributes onto other attributes, eventually allowing the derivation of the likely predicted behavior for the individual.

A variety of mappings may be constructed and encoded in the rules and belief nets, relating different attributes within the profile (e.g., relating traits with attitudes and goals), and eventually resulting in the derivation of the likelihood of particular behaviors of interest (e.g., selection of a particular strategy to achieve a particular goal). These mappings are derived from the types of knowledge described earlier; that is, a combination of knowledge from academic and applied psychological studies (e.g., correlations or particular trait configurations with specific attitudes, beliefs, values and goals), and practical field knowledge (e.g. correlations of particular characteristics with likely behavior).

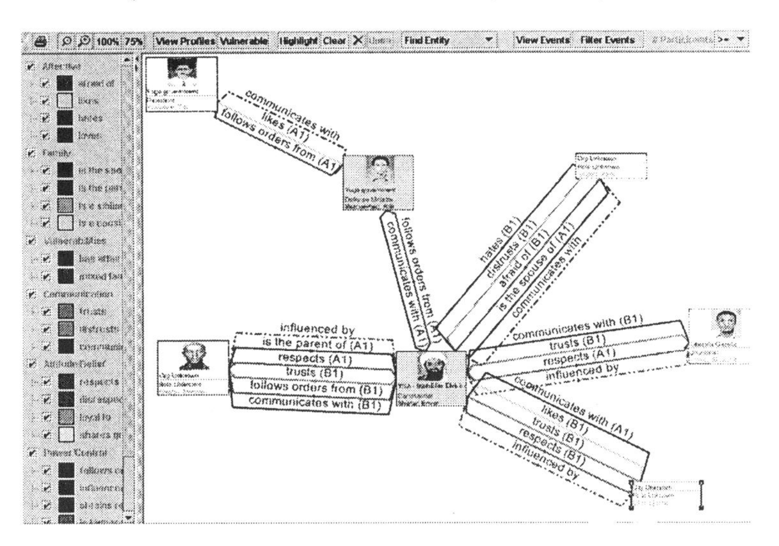

FIGURE 7. Example of a Social Network for a Set of Fictitious Individuals Modeled within the IODA Decision-Aid

Demographic Info.	Training & Education	Attitudes / Beliefs
Age Gender Marital status Children Geo location Socioeconomic status Social affiliation Religion Ethnicity	Institutions attended Location of each institution Duration of attendance Primary mentors Primary peers Primary mentees Performance	Entity Attitude / Belief re: Personality / Temperament Attitude / Beliefs re: Goals Attitude / Beliefs re: Goals Scripts Attitude / Beliefs re: Existing Intrapersonal Conflicts Attitude / Beliefs re: Information Environment Attitude / Beliefs re: Psychosocial Relationships Attitude/ Belief re: Vulnerabilities /PP
Psychological Parameters	**Existing Intrapersonal (Intragroup) Conflicts**	**Vulnerabilities, Pressure Points**
Interpersonal Attitudes Collaboration Style Cognitive Decision Making Style Decision Making Information Requirements Temperament Affect	Role conflicts Loyalty conflicts Goal conflicts Belief / attitude conflicts Personality elements Subgroups / Factions	Sources of fear / pleasure Preferred individual / group Avoided individual / group Need for / need to avoid a specific situation Physical diseases Physical stressors Psychological disorders Preferred / Avoided maneuvers Preferred / Avoided units Preferred / Avoided timing of attack Preferred / Avoided targets
Individual History	**Psychosocial Relationships**	**Information Environment**
Event type Key participants Outcome	Nature of relationship Entities involved Direction, if applicable	Link id Entities linked Role Frequency Format of information Type of information Level of trust Special vulnerabilities of link, if any
Goals / Goal Personnel	**Goal Scripts**	**Beliefs / Situation Assessments**
Goal Entity Entity State / Condition Entity Attitude / Belief Entity Affective State Event to occur wrt entity Goal status Goal importance Goal triggering conditions	Goal Means of goal satisfaction Entities necessary for goal satisfaction Entity belief / attitude necessary for goal satisfaction Entity affective state necessary for goal satisfaction Other goal script preconditions & constraints	Entity Belief Perception triggering belief

TABLE 5. List of Individual Profile Behavior Determinants Deriving Profile Values and Predicting Individual Behavior.

Figure 8 shows a high-level schematic illustrating this process. First, 'raw' data provided from a variety of sources are formatted and stored in the profile. Second, these data are used by rules and belief nets capturing theoretical psychological knowledge and applied operational knowledge, to infer additional data. Finally, all of the available profile data are used to derive the individual's likely behaviors.

The profile-based approach described above was implemented in several prototypes, using fictitious but realistic operational scenarios, and focusing on distinct objectives. The C2WARS prototype implemented a proof-of-concept demonstration, instantiating profiles of specific individuals within a Balkans-like PSYOP training scenario, deriving additional individual information from existing data, and focusing on the identification of specific vulnerabilities (e.g., "Mixed Family Loyalties" shown in Figure 9). The IODA prototype implemented a different training scenario, involving civil-war context in a multi-ethnic environment, and focused on behavior prediction, expanding the profile and inferencing accordingly.

An example of a belief net deriving the vulnerability "Mixed Family Loyalties" is shown in Figure 9. Examples of behavior prediction rules are shown in Table 6 (Hudlicka & Zacharias, 2002c); examples of rules deriving profile attributes are shown in Table 7.

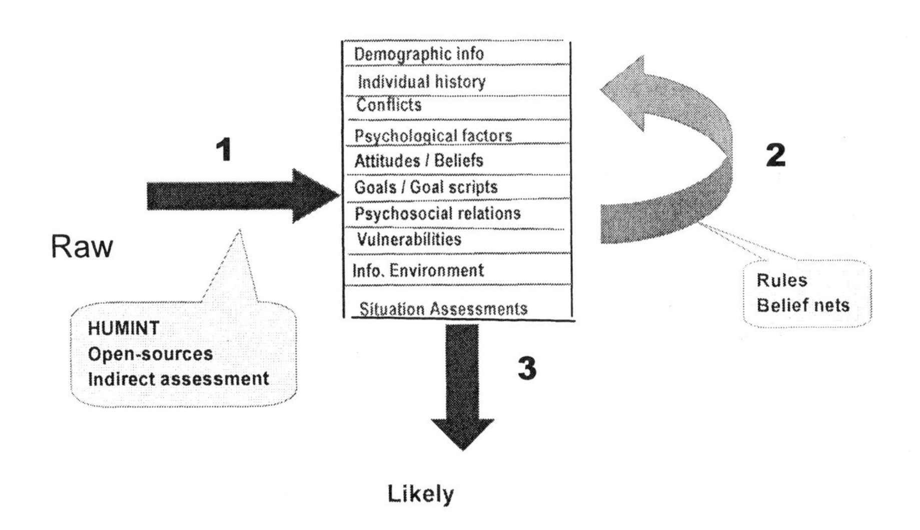

FIGURE 8. High-Level Schematic Illustrating Individual Profile Use

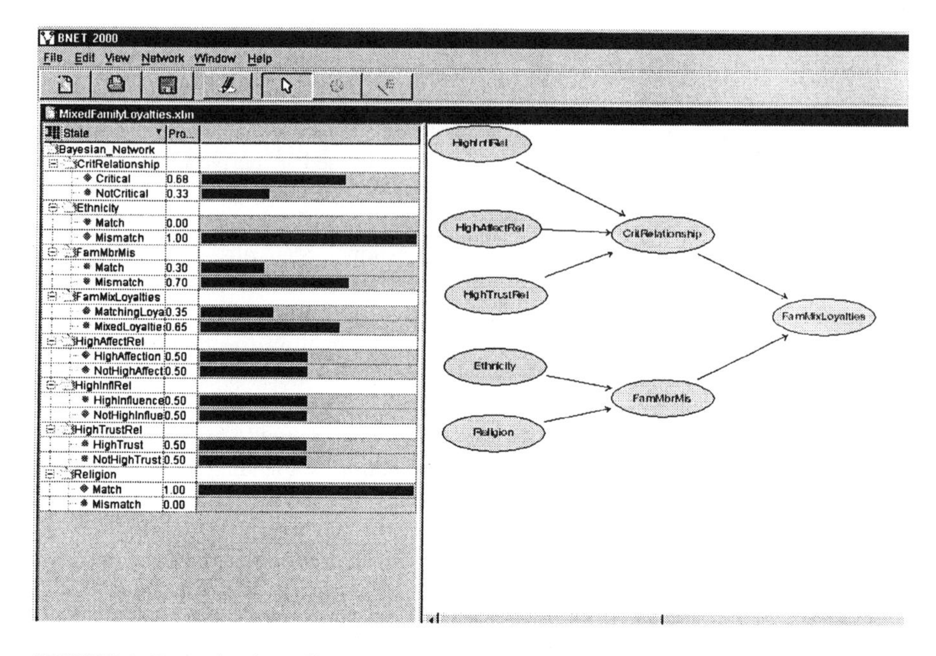

FIGURE 9. Example of a Belief Net Deriving a Particular Individual Vulnerability: "Mixed Family Loyalties"

IF	Person Y's religious extremism = high Person Y's event history includes violence	THEN	Person Y's likelihood of violence = high
IF	Person Y's event history does not include violence	THEN	Person Y's likelihood of violence = low
IF	Person Y's goals include "Peaceful co-existence" Person Y's religious extremism = low	THEN	Person Y's likelihood of violence = low
IF	Person Y's event history includes violence	THEN	Person Y's likelihood of violence = high
IF	Person Y has goal "Peaceful co-existence"	THEN	Person Y's likelihood of violence = Low

TABLE 6. Examples of Behavior Prediction Rules

High preference for face-to-face contact

IF (extraversion=high) AND (agreeableness = high)

THEN

(preference for face-to-face contact = high)

(preference for team collaborative decision-making = high)

(diverse sources of information = high)

High need for collaborative decision making

IF (need for affiliation = high) And (Trust = high) AND (paranoia = low)
THEN (Need for consensus = high) (need for collaborative decision making = high)
(general relationship with subordinates = empowers)

Rigid control over staff

IF (need for agreement from subordinates = low) (abusive to subordinates = high) (dogmatism = high) (authoritarianism = high) (assertiveness = high) (aggressiveness = high)

THEN (rigid control over staff = high)

High need for consensus

IF (task orientation = high) AND (empathy = low) AND (assertiveness = high) AND

THEN (desired level of consensus = low) (desired feedback from staff = analytical)

Information environment

IF (# communicates-with < norm) THEN (add 'restricted-communication-connections)

IF (agreeableness = low) (extraversion = low) (trust = low) (preference-for-face-to-face decision making = low) THEN (restricted-communication-connections)

TABLE 7. Examples of Rules Deriving Profile Attributes from Existing Data

INTEGRATING INDIVIDUAL MODELS WITHIN ORGANIZATIONAL SIMULATIONS

We now briefly discuss how the two approaches described above would be integrated within organizational simulations, and what enhancements would be required. In each case, the primary objective is to capture the set of relationships between the individual and the organization, and the resulting interactions and mutual influences. Exactly what information about the organization is required, how this information is represented, and at what level of abstraction the organization is represented depend on the objectives of the particular simulation. In the context of these considerations, each of the approaches discussed here has a distinct set of knowledge requirements necessary to represent the additional 'organizational' interactions and influences.

Integrating Cognitive Architecture Models within Organizational Simulations

There are several ways in which a cognitive architecture can be integrated within an organizational simulation. A given instance of an architecture can represent an individual within the organization, so that multiple instances can represent multiple interacting individuals within the organization. Under some circumstances a given instance of a cognitive architecture can also be used to represent a group of individuals, a component of the organization, or even the organization as a whole, depending on the level of modeling fidelity required by the simulation objectives.

When using this approach to represent an individual acting within an organization, we must first characterize the types of interactions that occur between the individual and the organization, and then augment the corresponding models accordingly, by adding schemas and knowledge that: (1) represent these interactions; (2) allow the perception and parsing of the "messages" inherent in these interactions; (3) enable modeling of the effects of these messages on the processing in both models; and (4) enable the generation of meaningful responses.

This type of augmentation will typically require additional content of the knowledge structures, both knowledge and data in long-term memories, and the knowledge necessary to use and manipulate these (e.g., rules, belief nets), as well as the generation of simulation-driven dynamic data that provide information about the organizational context. However, incorporating a cognitive architecture within an organizational simulation is not likely to require changes to the architecture itself; that is, the modules, their internal algorithms, and the mental constructs they manipulate.

The exact nature of the additional schemas required, and the dynamic data generated during a simulation, depend entirely on the nature and objectives of the associated organizational simulation, but are likely to include the following:

- Types of information provided by or about the organizational context, in terms of messages sent to, and received from, other members of the organization or other subgroups.

- Explicit representation of other relevant entities within the organization; e.g., individuals, departments, subgroups, the organization as whole, as well as specific resources.

- Some degree of baseline knowledge about these entities and their roles within the organization; e.g., individual x is head of department y; department z has n units of resource r.

- Relevant states and behavioral repertoire of the other active entities within the organization (individual, organizational subgroups, or the organization as a whole); e.g., 'person x knows fact b', 'person z believes fact w', 'department y can perform procedure d', 'president x is pleased with outcome y', 'person q cannot perform action t', 'organization p is lacking resource r'.

This type of knowledge, along with the necessary inferencing procedures for manipulating this knowledge, will then enable the individual model to participate in a simulation of the behavior of the organization as a whole. Thus, for example, rules relating the states of the other entities within the organization (individuals, departments) to likely behavior will enable the individual represented by the cognitive architecture to reason about the likely future behavior of the entities of interest, which may influence his/her own belief state, expectations, affective state, goal activation and/or behavior selection.

The simulation objectives, and the associated modeling fidelity requirements, also determine whether the organization is represented as a single uniform entity, with which the individual model interacts, or whether the organization itself consists of multiple models, representing the organization at varying levels of aggregation, including both individuals and groups. The simulation objectives also determine what aspects of the individual and organization characteristics and behaviors are represented, and thus what types of communication and interaction is possible.

Integrating Profile-Based Social Network Models within Organizational Simulations

By explicitly representing the individual's social relationships and information environment, the behavior determinant profile is already well suited for integration within an organizational simulation context. Depending on the nature and requirements of the overall organizational simulation, integrating an individual profile may only require adding the corresponding content to the existing profile structures; that is, the nature and type of relationships between the individual and his/her organization. Depending on the representational resolution of the organization, this may involve adding multiple relationships to other individuals comprising the organizations, or adding relationships to subgroups within the organization represented as a single model entity (e.g., a single node in a social network may represent a group of individuals or a component of the organization, such as a department).

To support the additional inferencing required, the knowledge base associated with the profile must be augmented. For example, the rules or belief nets must be added to use the additional organizational knowledge and derive from it the necessary data, such as an appropriate response to a particular message arriving from a different component of the organization.

SUMMARY AND CONCLUSIONS

In this chapter, we first motivated the need for including explicit models of individuals within organizational simulations to adequately capture the mutual influences between the individual and his/her organizational context. We highlighted the need to model individual differences, including personality traits and affective states, to assure an appropriate degree of realism. We reviewed relevant research, focusing primarily on the effects of a variety of individual differences on decision-making and behavior, and existing cognitive architecture models capable of modeling individual decision-making and behavior.

We then described in detail the knowledge and data, and representational and inferencing requirements for two alternative approaches to modeling the individual within organizational simulations: a *cognitive architecture* approach, represented by the MAMID architecture, and a *profile-based social network model* approach, represented by the IODA decision-aid. Both approaches are capable of representing a variety of individual differences, and both are well suited for integration into larger organizational simulations. They differ however both in the methods used, and the knowledge and data required. The cognitive architecture approach aims to emulate actual human decision-making, via explicit modeling of the structures and processes that mediate perception, decision-making and behavior selection. The profile-based social network model focuses instead on constructing a profile of critical determinants that influence individual decision-making and behavior. This information provides the basis for deriving additional individual characteristics and eventually behavior predictions. Which of these two approaches is appropriate for a particular organizational simulation depends on a combination of factors, including the objectives of the simulation, and the data and knowledge available about the individuals.

We illustrated each approach with concrete examples of existing prototypes, suggested how each could be integrated within organizational simulations, and what additional knowledge, data and inferencing would be required. Specifically, the cognitive architecture approach can be used in two distinct manners. First, particular individuals of interest would be modeled by distinct instances of cognitive architectures, which would then be integrated into the overall organizational simulation. Additional schemas would be required in the architecture's knowledge-bases to enable perception of relevant organizational data (e.g., messages from other members or subgroups within the organization), to reason about these (e.g., select appropriate goals relating to self, other members of the organization, the organization as a whole, or some combination of these), and to generate appropriate behavior (e.g., either self- or other-oriented behavior, including interactions with the organization). Second, a cognitive architecture can be used to represent a group of individuals or a component of the organization acting in unison. The assumption underlying this approach is that in some circumstances, it is possible to apply the individual decision-making model to a group as a whole.

The profile-based social network model shares the social network component with several existing organizational simulation models. The distinguishing feature of the approach presented here is the degree to which individual differences are represented within the nodes in the social network representing the individual decision-makers. The individual profile consists of a rich collection of behavior determinants, which enable a simulation of a wide variety of individual behavioral variations.

We can thus envision an organizational simulation consisting of a variety of models, representing individual and group decision-making at varying levels of detail, as necessitated by the simulation objectives and enabled by available knowledge and data. At the highest level of resolution, these would include models of individuals, each represented by an instance of a cognitive architecture. At an intermediate level of resolution, these would include instances of cognitive architectures representing groups of individuals or components of the organization (e.g., departments). At the lower levels of resolution, an entire organization could be represented by a single instance of a cognitive architecture. A variety of other options are of course possible, integrating multiple representational formalisms and modeling methods, where cognitive architecture models would be integrated with social network models.

Conclusions

As organizations increase in size and complexity, and as the complexity and pace of the various systems we manipulate exceeds the capacities of the unaided human mind, we are faced with a difficult dilemma. On the one hand, we should structure organizations in such a way that the idiosyncrasies of the individual members do not adversely influence the functioning of the organization as a whole, that is, to somehow neutralize the individual idiosyncrasies. This would then eliminate the need to model the individual within an organizational context, since individual variation would quickly be absorbed within the organizational safety systems and thereby made irrelevant. On the other hand, it is clear that the continued success and survival of organizations depends in large part on the creativity of its individual members, that is, precisely on the individual idiosyncrasies which form the basis of creative solutions to particular challenges the organization encounters, both internal and external.

In order to optimize organizational structure to enhance individual creativity on the one hand, and limit the possibly adverse effects of individual behavior on the other, we must improve our understanding of the complex interactions between the individual within the organizational context. Such improved understanding requires computational modeling approaches, due to the complexity of the phenomenon, which precludes purely non-computational empirical studies. These computational approaches must be able to adequately model individual behavior, particularly the types of individual differences that give rise to idiosyncratic behavior, which can be both beneficial and detrimental to the organization as a whole. Understanding the effects of individual behavioral variations on the organization as a whole is also critical for any type of organizational behavior predictions.

Both of the approaches to modeling the individual described here are well suited to modeling the individual within an organization. However, we have yet to develop adequate criteria to determine which approach is best suited for which application. Systematic evaluations of the various modeling approaches are necessary to identify their benefits and shortcomings, across various contexts. Additional research is necessary to explore these, and other, modeling approaches, in various contexts, to define these criteria, and to contribute to defining a concrete set of guidelines for developing organization simulations. Such guidelines would help determine which method to use when, what level of representational abstraction is most appropriate for a particular application or objective, and how distinct modeling approaches may be combined.

Challenges

A number of challenges must be addressed in human behavior modeling, both at the individual and at the organizational level. We briefly outline four of these below.

Lack of Empirical Data

Construction of computational models, particularly models of agent architectures in field settings (as opposed to simpler, isolated phenomena in laboratories), requires empirical data that are currently not available. Frequently we must rely on extrapolations of existing data, and often we are limited to only qualitative effects. Model-motivated, focused empirical studies are required to provide the necessary data for computational models of human decision-making. Constructing profile-based models also requires large quantities of data, which may not be readily available. Methods are needed to analyze individual behavior to allow for automated or semi-automated construction of individual profiles.

Effort Required to Construct a Model

The labor-intensive nature of human behavior model development is a recognized problem. Developing knowledge-acquisition tools, shared domain ontologies, and interactive GUIs helps somewhat, but does not solve the inherently difficult problem of attempting to emulate results of phylogenetic and ontogenetic development. Cognitive architectures capable of learning are likely to help somewhat, and other methods must be developed to enable rapid model construction for large organizational simulations, requiring a number of individual interacting models.

Model Brittleness and the Necessity for Repeated Fine-Tuning

Both the knowledge bases (agents' LTM) and the model parameters exhibit this problem. It results from a combination of factors, including the use of symbolic representations, and the complexity of both the model and the task. As above, automating the knowledge acquisition process and including learning would help, as would the use of alternative or hybrid representational and inferencing formalisms (e.g., connectionist or spreading activation techniques).

Validation

Computational models are powerful tools in the behavioral science researcher's repertoire. It is possible to construct agent models with distinct personalities, individual history, and affective reactiveness, resulting in observable differences in behavior. It would however be a leap of faith to then claim that such models represent what actually takes place in the mind. It is critical to distinguish between verification ("the simulation does what we programmed it to do") and validation ("the simulated model's structure, processing and output correspond to the natural phenomena we are attempting to model"). All too frequently one hears claims about an agent model that has been validated by demonstrating reasonable behavior within a very narrowly defined simulation context. We must guard against such hubris and focus instead on the development of validation methodologies and criteria that would advance the state of the art, and produce "sharable" models and model components which can be cross-validated by different researchers in different contexts.

Future Work and Research Agenda

Significant progress has been made in individual and organizational modeling in the past decade, and much more remains to be accomplished. Some of the key issues that must be addressed include the following:

- Identification of empirically-justified criteria for determining when an individual model is required within an organization simulation, and when the organizational context is necessary for effective individual modeling; translation of these criteria into systematic procedures guiding the model design.

- Identification of criteria for determining the type of individual model required, that is, its level of resolution, the specific cognitive and affective processes requiring explicit representation in the model, etc.

- Development of a set of guidelines for determining the best alternatives for model development, in terms of the knowledge and data required, and the representational formalism and inferencing strategies necessary

- Definition of the types of mutual interactions and constraints between the individual model and the organizational model.

- Conducting large-scale, coupled computational-empirical studies to generate the necessary empirical data required to design and develop empirically justified simulation models.

- Validation studies evaluating both the model's ability to emulate human decision-making (whether at the individual, organizational, or combined levels), and the benefits and/or drawbacks of applying such models in practice.

- Streamlining model development through the use of standardization of representational formalisms, shared ontologies, plug-and-play model components, and scenario libraries.

REFERENCES

Ackerman, P.L., Kanfer,R., & Cudeck, R. (1989) (Eds.): *Learning and Individual Differences: Abilities, Motivation, and Methodology.* (pp. 297-341). Hillsdale, NJ: LEA.

Ajzen, I. (2001). Nature and Operation of Attitudes. *Annual Review of Psychology*, 55, 27-58.

Ajzen, I. (1991). The theory of planned behavior. *Organizational Behavior and Human Decision Processes,* 50, 179-211.

Aizen, I., & Fishbein, M. (in press). The Influence of Attitudes on Behavior. The influence of attitudes on behavior. In D. Albarracín, B. T. Johnson, & M. P. Zanna (Eds.), *Handbook of attitudes and attitude change: Basic principles.* Mahwah, NJ: Erlbaum.

Anderson, J. (1990). *The adaptive character of thought.* Hillsdale, NJ: LEA. http://act.psy.cmu.edu.

Andre, E., Klesen, M., Gebhard, P., Allen, S. & Rist, T. (2000). Integrating Models of Personality and Emotions in Lifelike Characters. In *Proceedings of IWAI.* Siena, Italy.

Andrews, D.H., Bell, H.H. & Shearer, R.N. (2005). The Learning Organization and Organizational Simulation. In W.B. Rouse & K.R. Boff, Eds., *Organizational Simulation: From Modeling and Simulation to Games and Entertainment.* New York: Wiley.

Bates, J., Loyall, A.B., & Reilly, W.S. (1992). Integrating Reactivity, Goals, and Emotion in a Broad Agent. In *Proceedings of the 14th Meeting of the Cognitive Science Society.*

Blaney, P.H. (1986). Affect and Memory. *Psychological Bulletin*, 99 (2), 229-246.

Blocher, K. & Picard, R.W. (2002). Affective Social Quest: Emotion Recognition Therapy for Autistic Children. In Dautenhahn, K., Bond, A.H., Canamero, L. and Edmonds, B. (2002). *Socially Intelligent Agents: Creating Relationships with Computers and Robots. Dordrecht, The Netherlands: Kluwer Academic Publishers.*

Bower, G.H. (1981). Mood and Memory. *American Psychologist,* 36, 129-148.

Breazeal, C. (2003). Emotion and Sociable Humanoid Robots. *International Journal of Human-Computer Studies*, 59 (1-2), 119-155.

Canamero, D. (1998). Issues in the Design of Emotional Agents. In *Proceedings of Emotional and Intelligent: The Tangled Knot of Cognition.* AAAI Fall Symposium, TR FS-98-03, 49-54. Menlo Park, CA: AAAI Press.

Cannon-Bowers, J. A. & Salas, E. (1998). *Decision Making Under Stress.* Washington, DC: APA.

Carley, K.M. (2005). Organizational Design and Assessment in Cyber-space. In W.B. Rouse & K.R. Boff, Eds., *Organizational Simulation: From Modeling and Simulation to Games and Entertainment.* New York: Wiley.

Carley, K.M. Hill, V. (2001). Structural Change and Learning Within Organizations". In *Dynamics of organizational societies: Models, theories and methods.* Alessandro Lomi, (Ed.). Live Oak,CA: MIT Press/AAAI Press.

Cattell, R.B. (1971). *Abilities: their structure, growth and action.* NY, NY: Houghton Mifflin.

Clore, G.L. (1994). Why emotions require cognition. In *The Nature of Emotion.* (P. Ekman and R.J. Davidson, Eds.). Oxford: Oxford University Press.

Codevilla, A.M. (1997). *The Character of Nations.* NY: Basic Books.

Cooke, N. J. (1994). Varieties of knowledge elicitation techniques. *Intl. Journal of Human-Computer Studies*, 41, 809-849.

Cooper, C.R., & Denner, J. (1998). Theories Linking Culture and Psychology: Universal and Community-Specific Processes. *Annual Review of Psychology*, 49, 559-584.

Corker, K. (2000). Cognitive Models & Control: Human & System Dynamics in Advanced Airspace Operations. In *Cognitive Engineering in the Aviation Domain,* N. Sarter and R. Amalberti (Eds.). Mahwah, NJ: Lawrence Earlbaum Associates.

Corker, K. (2001). Air-ground Integration Dynamics in Exchange of Information for Control. In *Transportation Analysis*, L.. Bianco, P. Dell'Ormo and A. Odani eds.). Heidleberg, Germany: Springer Verlag, http://ccf.arc.nasa.gov/af/aff/midas/MIDAS _home_page.html

Costa, P. T, & McCrae, R. R. (1992). Four ways five factors are basic. *Personality and Individual Differences*, 13, 653-665.

Cross, S.E. (2005). Using Organizational Simulation to Develop Unprecedented Systems. In W.B. Rouse & K.R. Boff, Eds., *Organizational Simulation: From Modeling and Simulation to Games and Entertainment.* New York: Wiley.

Damasio, A.R. (1994). *Descartes' Error: Emotion, Reason, and the Human Brain.* NY: Putnam.

Dautenhahn, K., Bond, A.H., Canamero, L. & Edmonds, B. (2002). *Socially Intelligent Agents: Creating Relationships with Computers and Robots. Dordrecht, The Netherlands: Kluwer Academic Publishers.*

Davidson, R.J., Scherer, K.R., & Goldsmith, H.H. (2003). *Handbook of Affective Sciences.* NY: Oxford University Press.

DeRosis, F., Pelechaud, C., Poggi, I., Carofiglio, V., & De Carolis, B. (2003). From Greta's mind to her face. *International Journal of Human-Computer Studies*, 59 (1-2), 81-118.

Deutsch, S., Adams, M., Abrett, G., Cramer, N., & Feehrer, C. (1993). Research, Development, Training, and Evaluation: Operator Model Architecture (OMAR). Software Functional Specification. AL/HR-TP-1993-0027. WPAFB, OH: Air Force Material Command. http://www.sover.net/~nichael/misc/omar/ifac.html.

Eid, M. 2001. Advanced Statistical Methods for the Study of Appraisal and Emotional Reaction. In *Appraisal Processes in Emotion*. K.R. Scherer, A. Schorr, T. Johnstone (Eds.). NY: Oxford University Press.

Elliot, C., Lester, J., & Rickel, J. (1999). Lifelike pedagogical agents and affective computing: An exploratory synthesis. In M. Wooldridge and M. Veloso, (Eds.). *AI Today.* Lecture Notes in AI. NY: Springer-Verlag.

Ekman, P. & Davidson, R.J. (1994). *The Nature of Emotion.* .Oxford: Oxford University Press.

Endsley, M. R. (2000). Theoretical Underpinnings of Situation Awareness: A Critical Review. In *Situation Awareness Analysis and Measurement*, M.R. Endsley & D.J. Garland, eds. Mahwah: NJ: LEA.

Essa, I, & Bobick, A. (2005). Simulating Humans. In W.B. Rouse & K.R. Boff, Eds., *Organizational Simulation: From Modeling and Simulation to Games and Entertainment.* New York: Wiley.

Eysenck, H. J. (1991). Dimensions of personality: the biosocial approach to personality. In J. Strelau & A. Angleitner (Eds.), *Explorations in temperament: international perspectives on theory and measurement.* London: Plenum.

Eysenck, M.W. (1997). *Anxiety and Cognition: A Unified Theory.* Hove, E. Sussex: Psychology Press.

Fellous, J-M. (2004). From human emotions to robot emotions. In *Proceedings of the AAAI Spring Symposium: Architecture for Modeling Emotion: Cross-Disciplinary Foundations.* AAAI Technical Report SS-04-02. Menlo Park, CA: AAAI Press.

Fishbein, M. (1967). Attitude and the prediction of behavior. In M. Fishbein (Ed.), *Readings in attitude theory and measurement.* New York: Wiley.

Forgas, J.P. (2000). *Feeling and Thinking: The Role of Affect in Social Cognition.* Cambridge, UK: Cambridge University Press.

Forgas, J.P. (2001). Introduction: Affect and Social Cognition. In *Handbook of Affect and Social Cognition.* Mahwah, J.P. Forgas, Ed. NJ: LEA.

Frijda, N.H. & Swagerman, J. (1987). Can Computers Feel? Theory and Design of an Emotional System. *Cognition and Emotion,* 1 (3), 235-257.

Frijda, N.H.(1986). *The Emotions.* Studies in Emotion and Social Interaction. New York: Cambridge Univ. Press.

Geiwitz, J., Klatzky, R.L., & McCloskey, B.P. (1988). Knowledge Acquisition Techniques for Expert Systems: Conceptual and Empirical Comparisons. Santa Barbara, CA: Anacapa Sciences, Inc.

Gluck, K. & Pew, R. (2002). The AMBR Model Comparison Project: Round III – Modeling Category Learning. In *Proceedings of the 24th Annual Conference of the Cognitive Science Society*, George Mason University, VA.

Gray, J. A. (1994). Three Fundamental Emotion Systems. In *The Nature of Emotion.* (P. Ekman and R.J. Davidson, Eds.). Oxford: Oxford University Press.

Gugerty, L.J. & Tirre, W.C. (2000). Individual Differences in Situation Awareness. In *Situation Awareness Analysis and Measurement*, M.R. Endsley and D.J. Garland, eds. Mahwah: NJ: LEA.

Handley, H. & Levis, A. (2001). Incorporating Heterogeneity in Command Center Interactions. *Information, Knowledge, and Systems Management,* 2, 297-309.

Harper, K. A., S. S. Mulgund, et al. (2000). SAMPLE: Situation Awareness Model for Pilot-in-the-Loop Evaluation. In *Proceedings of the 9th Conference on Computer Generated Forces and Behavioral Representation*, Orlando, FL. http://www.cra.com/sample.

Harper, K., N. Ton, et al. (2001). GRADE: Graphical Agent Development Environment for Human Behavior Representation. In *Proceedings10th Conference on Computer Generated Forces and Behavior Representation*, Norfolk, VA.

Hofstede, G. (1991). *Cultures and Organizations: Software of the Mind.* New York: McGraw Hill.

Holsti, O.R.. (1977). The 'operational code' as an approach to the analysis of belief systems. *Final report to the National Science Foundation.* Grant # SC 75-15368. Duke University.

Hudlicka, E. (1997). Modeling Behavior Moderators in Military HBR Models. TR 9716. Lincoln, MA: Psychometrix Associates, Inc. (see also Pew and Mavor, 1998).

Hudlicka, E. (2001). CAPTA: Cognitive Affective Personality Task *Analysis. In Proceedings of the 45th Meeting of the HFES.* October, 2001.

Hudlicka, E. (2002a). This time with feeling: Integrated Model of Trait and State Effects on Cognition and Behavior. *Applied Artificial Intelligence*, 16:1-31. 2002.

Hudlicka, E. (2002b). Bibliography of Technical and Research Literature Resources for C2WARS Development. TR R97310a. Cambridge, MA: Charles River Analytics.

Hudlicka, E. (2003a). *Methodology for Analysis and Modeling of Individual Differences: MAMID - Final Report.* Psychometrix Technical Report 0003. Blacksburg, VA: Psychometrix Associates, Inc.

Hudlicka, E. (2003b). Modeling Effects of Behavior Moderators on Performance: Evaluation of the MAMID Methodology and Architecture, In *Proceedings of BRIMS-12*, Phoenix, AZ, May.

Hudlicka, E. (2003c). To Feel or Not To Feel: The Role of Affect in HCI. *International Journal of Human-Computer Studies*, 59 (1-2), 1-32.

Hudlicka, E. (2003d). Personality and Cultural Factors in Gaming Environments. *Proceedings of the Workshop on Cultural & Personality Factors in Military Gaming.* Alexandria, VA: Defense Modeling and Simulation Office. (See also Psychometrix Associates Report 0306.)

Hudlicka, E. (2004a). Two Sides of Appraisal. In *Proceedings of the AAAI Spring Symposium-Architectures for Modeling Emotion*, TR SS-04-02. Menlo Park, CA: AAAI

Hudlicka, E. (2004b). *Summary of Factors Influencing Decision-Making and Behavior.* Technical Report 0403. Blacksburg, VA: Psychometrix Associates, Inc.

Hudlicka, E. & Billingsley, J. (1999). Representing Behavior Moderators in Military Human Performance Models. In *Proceedings of the 8th CGF Conference.* Orlando, FL.

Hudlicka, E. & Pfautz, J. (2002). Once more with feeling: Augmenting Recognition Primed Decision Making with Affective Factors. In *Proceedings of the 46th meeting of the Human Factors and Ergonomic Society.* October, Baltimore, MD.

Hudlicka, E. & Zacharias, G. (2002a). *C2WARS: Command and Control Warfare Analysis and Reasoning System: Target Profile Specification.* Report R97312. Cambridge, MA: Charles River Analytics, Inc.

Hudlicka, E. & Zacharias, G. (2002b). *C2WARS: Command and Control Warfare Analysis and Reasoning System: Requirements Specification.* Report R97313. Cambridge, MA: Charles River Analytics, Inc.

Hudlicka, E. & Zacharias, G. (2002c). *C2WARS: Command & Control Warfare Analysis and Reasoning System: Target Vulnerability Assessment.* Report R97314. Cambridge, MA: Charles River Analytics.

Hudlicka, E., Zacharias, G., & Schweitzer, J. (2002). Individual and Group Behavior Determinants: Inventory, Inferencing, and Applications. In *Proceedings of the 11th CGF Conference*, Orlando, Fl, 21-28.

Hudlicka, E., Karabaich, B., Pfautz, J., Jones, K., & Zacharias, G. (2004). Predicting Group Behavior from Profiles and Stereotypes. In *Proceedings of the 13th BRIMS Conference,* Arlington, VA.

Isen, A.M. (1993). Positive Affect and Decision Making. In *Handbook of Emotions*. J.M. Haviland and M. Lewis, eds. New York, NY: The Guilford Press.

Johnson, S.C. (1967). Hierarchical Cluster Analysis, *Psychometrika,* 32, No. 3.

Jones, R., Henninger, A. & Chown, E. (2002). Interfacing Emotional Behavior Moderators with Intelligent Synthetic Forces. In *Proceedings of the 11th Conference on Computer Generated Forces and Behavior Repersentation*, Orlando, FL.

Karabaich, B. (1996). *Target Anslysis Concept.* Technical Memo. Leavenworth, KS: Karabaich Strategic Information Systems.

Kelly, G. (1955). *The Psychology of Personal Constructs*. New York: Norton.

Klein, G. A. (1997). The recognition-primed decision (RPD) model: Looking back, looking forward. In C, Zsambok & G. Klein (Eds.), *Naturlistic decision making*. Mahwah, NJ: Erlbaum.

Klein, H. & McHugh, A.P. (2005). National Differences in Teamwork. In W.B. Rouse & K.R. Boff, Eds., *Organizational Simulation: From Modeling and Simulation to Games and Entertainment*. New York: Wiley.

Klein, H.A., Pongonis, A., & Klein, G. (2002). Cultural Barriers to Multinational C2 Decision Making. In *Proceedings C2 Research & Technology Symposium.*

Laughery, R., Archer, S. & Corker, K. (2001). Modeling Human Performance in Complex Systems. *In Handbook of Industrial Engineering,* G. Salvendy (Ed.) NY: Wiley Interscience -Human Performance Modeling.

Lazarus, R.S. (1991). *Emotion and Adaptation.* New York, NY: Oxford University Press.

LeDoux, J.E. (1989). Cognitive-Emotional Interactions in the Brain. *Cognition and Emotion*, 3(4), 267-289.

LeDoux, J.E. (1992). Brain Mechanisms of Emotion and Emotional Learning. *Current Opinions in Neurobiology*, 2(2), 191-197.

Lee, Y. (2004). Software review: Review of the Tools for the Cognitive Task Analysis. *Educational Technology & Society*, 7 (1), 130-139.

Levis, A. (2005). Executable Models of Decision Making Organizations. In W.B. Rouse & K.R. Boff, Eds., *Organizational Simulation: From Modeling and Simulation to Games and Entertainment*. New York: Wiley.

Lewis, M. & Haviland, J.M. (1993). *Handbook of Emotions.* NY: The Guilford Press.

Lindheim, R.D. & Korris, J.H. (2005). Application of Immersive Technology For Next-Generation Simulation. In W.B. Rouse & K.R. Boff, Eds., *Organizational Simulation: From Modeling and Simulation to Games and Entertainment.* New York: Wiley.

Lisetti, C.L. & Gmytrasiewicz, P. (2002). Can a Rational Agent Afford to Be Affectless? *Applied Artificial Intelligence,* 16 (7-8), 577-609.

Marsella, S. & Gratch, J. (2002). A step toward irrationality: Using emotion to change belief. In *Proceedings of the 1st International Joint Conference on Agents and Multiagent Systems.* Bologna, Italy.

Martinho, C., Machado, I. & Paiva, A. (2000). A Cognitive Approach to Affective User Modeling. In *Affective Interactions: Towards a New Generation of Affective Interfaces.* (A. Paiva, ed.). New York: Springer Verlag.

Matsumoto, D. (2001). *The Handbook of Culture and Psychology.* NY: Oxford.

Matthews, G., Derryberry, D., & Siegle, G.J. (2000a) Personality and emotion: Cognitive science perspectives. In S.E. Hampson (Ed.), *Advances in personality psychology* (Vol. 1). London: Routledge.

Matthews, G., Davies, D.R., Westerman, S.J. & Stammers, R.B. (2000b). Human Performances Cognition, Stress, and individual differences. Philadelphia, PA: Taylor and Francis.

Matthews, G., and Deary, I.J. (1998). *Personality Traits.* Cambridge, UK: Cambridge.

Matthews, G. (2004). Designing Personality: Cognitive Architectures and Beyond. In *Proceedings of the AAAI Spring Symposium: Architecture for Modeling Emotion: Cross-Disciplinary Foundations.* AAAI Technical Report SS-04-02. Menlo Park, CA: AAAI Press.

Meyer, D.E. & Kieras, D.E. (1997). A Computational Theory of Executive Cognitive Processes and Multiple Task Performance, Part 1, Basic Mechanisms. *Psychological Review,* 104 (1), 2-65.

Mineka, S. & Sutton, S.K. (1992). Cognitive Biases and the Emotional Disorders. *Psychological Science,* 3(1), 65-69.

Michaud, F. & Theberge-Turmel, C. (2002). *Mobile Robotic Toys and Autism: Observations of Interaction.* In Dautenhahn, K., Bond, A.H., Canamero, L. and Edmonds, B. (2002). *Socially Intelligent Agents: Creating Relationships with Computers and Robots. Dordrecht, The Netherlands: Kluwer Academic Publishers.*

Mitchell, D. (2000). *Mental Workload and ARL Workload Modeling Tools.* Army Research Laboratory Report ARL-TN-161. Aberdeen, MD. Army Research Laboratory.

Napierski, D., Young, A., & Harper, K. (2004) Towards a Common Ontology for Improved Traceability of Human Behavior Models. In *Proceedings of the 13th Conference on Behavior Representation in Modeling and Simulation (BRIMS*), Arlington, VA.

Newell, A. (1990). *Unified theories of cognition.* Cambridge, MA: Harvard University Press. http://www.soartech.com/htmlonly/technology.soar.html

Orasanu, J. (1998). Stress and Naturalistic Decision Making: Strengthening the Weak Links. In R. Flin, E. Salas, (Eds.), *Decision Making Under Stress: Emerging Themes and Applications*. Aldersho, Hants, UK: Avebury

Ortony, A., Clore, G.L., & Collins, A. (1988). *The Cognitive Structure of Emotions*. NY: Cambridge University Press.

Osgood, C. E., Suci, G. J., & Tannenbaum, P. H. (1957). *The measurement of meaning*. Urbana, IL: University of Illinois Press.

Pearl, J. (1986). Fusion, Propogation, and Structuring in Belief Networks. *AI*, 29(3): 241-288.

Persons, J.B. & Foa, E.B. (1984). Processing of fearful and neutral information by obsessive-compulsives. *Behavior Res. Ther.*, 22, 260-265.

Pew, R.W. and Mavor, A.S. (1998). *Representing Human Behavior in Military Simulations*. Washington, DC. National Academy Press.

Rao, A.S. & Georgeff, M.P. (1995). *BDI Agents: From Theory to Practice*. Tech. Rep. 56. Melbourne, Australia: Australian Artificial Intelligence Institute.

Renshon, S.A. & Duckitt, J. (2000). *Political Psychology: Cultural and Crosscultural Foundations*. Washington Square, NY: NY University Press.

Revelle, W. (1995). Personality Processes. Annual Review of Psychology, 46, 295-328.

Revelle, W. & Born, W. (1999). Modeling Personality Differences in Personality and Cognition. Technical Report. Evanston, IL: Northwestern University.

Ritter, F , Avramides, M., & Councill, I. (2002). Validating Changes to a Cognitive Architecture to More Accurately Model the Effects of Two Example Behavior Moderators *Proceedings of 11th CGF Conference*, Orlando, FL.

Ritter, F.R., Shadbolt, N.R., Elliman, D., Young, R., Gobet, F. & Baxter, G.D. (1999). *Techniques for modelling human performance in synthetic environments: A supplementary review*. Technical Report No. 62. ESRC Centre for Research in

Development, Instruction and Training. Nottingham, UK: Department of Psychology, University of Nottingham.

Rouse, W.B. (2005). Strategic Thinking Via Organizational Simulation. In W.B. Rouse & K.R. Boff, Eds., *Organizational Simulation: From Modeling and Simulation to Games and Entertainment*. New York: Wiley.

Russell, J.A. (1979). Affective space is bipolar. *Journal of Personality and Social Psychology*, 37, 345-356.

Salas, E., Guthrie, Jr., J.W., Wilson-Donnelly, K.A., Priest, H.A. & Burke, S. (2005). Modeling Team Performance: The Basic Ingredients and Research Needs. In W.B. Rouse & K.R. Boff, Eds., *Organizational Simulation: From Modeling and Simulation to Games and Entertainment*. New York: Wiley.

Saunders-Newton, D. (2003). Reflections on Credible Computational Models of Society," in *Proceedings on the Workshop on Cultural & Personality Factors in Military Gaming*. Alexandria, VA: Defense Modeling and Simulation Office.

Scherer, I.R., Schorr, A. & Johnstone, T. (2001). *Appraisal Processes in Emotion: Theory, Methods, Research.* Oxford: Oxford. University Press.

Scherer, K. (1993). Studying the Emotion-antecedent appraisal process: The expert system Approach. *Cognition and Emotion*, 7, 325-355.

Schiffman S., Reynolds, L. , & Young, F.W. (1981). Introduction to Multi-Dimensional Scaling: Theory, Methods, and Applications. NY, NY: Academic Press.

Schraagen, J.M., Chipman, S.F. and Shalin, V.L. (2000). *Cognitive Task Analysis.* Mahwah, NJ: LEA.

Selfridge, O. (1959). Pandemonium: A paradigm for learning. *In Symposium on the mechanization of thought processes.* London, UK: HM Stationary Office.

Shah, A.P. & Pritchett, A.R. (2005). Agent-Based Modeling and Simulation of Socio-Technical Systems. In W.B. Rouse & K.R. Boff, Eds., *Organizational Simulation: From Modeling and Simulation to Games and Entertainment.* New York: Wiley.

Shaw, E. & Post, J. (1999). Review of Commander Information and Interaction Profile: Recommendations for Operationalization. Report. Washington, DC: Political Psychology Associates, Ltd.

Sloman, A. (2000). Architectural Requirements for Human-like Agents Both Natural and Artificial. In Human Cognition and Social Agent Technology. K. Dautenhand, Ed. Amsterdam / Philadelphia: John Benjamins

Sloman, A. (2003). How many separately evolved emotional beasties live within us? In Emotions in Humans and Artifacts, R.Trappl, P. Petta, & S. Payr (Eds.). Cambridge, MA:The MIT Press. http://www.cs.bham.ac.uk/~axs/cogaff.html.

Smith, C. & Kirby, L. (2000). Consequences Require Antecedents: Toward a Process Model of Emotion Elicitation. In *Feeling and Thinking: The Role of Affect in Social Cognition.* J.P. Forgas, (Ed.). Cambridge, UK: Cambridge University Press.

Tenney, Y., Diller, D., Pew, R., Godfrey, K., & Deutsch, S. (2003). The AMBR Project: A Case-Study in Human Performance Model Comparison. In *Proceedings of BRIMS-12*, Phoenix, AZ, May.

Triandis, H.C. & Suh, E.M. (2002). Cultural Influences on Personality. *Annual Review of Psychology*, 53, 133-160.

Thayer, R.E. (1996). *The origin of everyday moods.* New York: Oxford University Press.

Velaszuez, J.D. (1998). Modeling Emotion-Based Decision-Making. In *Proceedings of "Emotional and Intelligent: The Tangled Knot of Cognition".* AAAI Fall Symposium Series, TR FS-98-03. Menlo Park, CA: AAAI Press.

Watson, D., & Clark, L.A. (1992). On traits and temperament: General and specific factors of emotional experience and their relation to the five-factor model. *Journal of Personality, 60,* 441-446.

Watson, D. & Tellegen, A. (1985). Toward a consensual structure of mood. *Psychological Bulletin*, 98, 219-235.

Williams, J.M.G., Watts, F.N., MacLeod, C., & Mathews, A. (1997). *Cognitive Psychology and Emotional Disorders.* NY: John Wiley.

Zacharias, G.L., Miao, A.X, Illgen, C., & Yara, J.M. SAMPLE: Situation Awareness Model for Pilot-in-the-Loop Evaluation. (1995). In *Proceedings of the 1st Annual Conference on Situation Awareness in the Tactical Air Environment.* Wright-Patterson AFB, OH: Crew Systesm Ergonomics Information Analysis Center. 1995.

Zachary, W. (1992) COGNET. http://www.chiinc.com/cognethome.shtml.

Zyda, M., Mayberry, A., McCree, J., & Davis, M. (2005). From Viz-Sim to VR to Games. In W.B. Rouse & K.R. Boff, Eds., *Organizational Simulation: From Modeling and Simulation to Games and Entertainment*. New York: Wiley.

CHAPTER 6

COMMON GROUND AND COORDINATION IN JOINT ACTIVITY

GARY KLEIN, PAUL J. FELTOVICH, JEFFREY M. BRADSHAW, AND DAVID D. WOODS

ABSTRACT

Generalizing the concepts of *joint activity* developed by Clark (1996), we describe key aspects of team coordination. Joint activity depends on *interpredictability* of the participants' attitudes and actions. Such interpredictability is based on *common ground*—pertinent knowledge, beliefs, and assumptions that are shared among the involved parties. Joint activity assumes a *basic compact*, which is an agreement (often tacit) to facilitate coordination and prevent its breakdown. One aspect of the Basic Compact is the commitment to some degree of aligning multiple goals. A second aspect is that all parties are expected to bear their portion of the responsibility to establish and sustain common ground and to repair it as needed. We apply our understanding of these features of joint activity to account for issues in the design of automation. Research in software and robotic agents seeks to understand and satisfy requirements for the basic aspects of joint activity. Given the widespread demand for increasing the effectiveness of team play for complex systems that work closely and collaboratively with people, observed shortfalls in these current research efforts are ripe for further exploration and study.

INTRODUCTION

> Attending Anesthesiologist: {re-entering the operating room mid-case} "Nice and tachycardic."
>
> Senior Resident Anesthesiologist: "Yeah, well, better than nice and bradycardic."

To most of us, this exchange is a mystery. We don't have a clue about what the two anesthesiologists were saying to each other. To the anesthesiologists, this highly coded conversation carried a great deal of meaning. They relied on their common ground to clearly communicate even when using very abbreviated comments, and their comments helped them maintain their common ground.

The two anesthesiologists were working together on a neurosurgery case to clip a cerebral aneurysm. The patient had been anesthetized, but the surgeons had not yet exposed the aneurysm. The anesthesiologists had little to do during a relatively long maintenance phase, so the attending anesthesiologist used this slow period to check on another case.

Then the anomaly occurred—a fall in heart rate that is termed a bradycardia. The resident anesthesiologist quickly detected the disturbance and acted to correct it. He also paged the attending anesthesiologist. The resident's intervention resulted in a temporary overswing—the patient's heart rate became too rapid, rather than too slow. In this surgery, an increased heart rate is better than a decreased heart rate.

In their discussion above, the attending anesthesiologist, who had looked at the monitors as he entered the operating room, used three words to describe his understanding of the way the patient's status had changed, inviting some explanation. The resident, in seven words, corrected the attending anesthesiologist's understanding, enabling them to work together to diagnose the reason for the bradycardia.

Performance depends on coordination, as cognitive work is distributed among different team members. If we want to improve teamwork, we need to better understand the nature of coordination and its requirements. That is the purpose of this chapter, to build on Clark's (1996) in-depth analysis of the coordination needed to carry on a conversation in order to explore issues of coordination more generally.

Clark (1996) has described how the simple transaction of two people talking to each other places a surprising number of demands on coordination. Sometimes we have conversations with people we have just met, sometimes with people we have known all our lives. We have to follow what the other person is saying while preparing to insert our own thoughts and reactions. We have to manage the turn taking in order to start speaking at an appropriate time. These handoffs in a conversation are difficult to coordinate in order to sustain a smooth flow of comments that are at least nominally linked to each other—unlike long, awkward silences or tangles where everyone tries to talk at once. Conversations are in large part undirected—the route taken depends on the topics raised and the reactions made, rather than the following of a rigorous script or the guidance of an external leader. In many ways, a conversation is a microcosm of coordination.

Clark's observations are valuable for understanding joint activity in general. Though previous accounts of team coordination (e.g., Klein, 2001; Malone & Crowston, 1994; Zalesny, Salas, & Prince, 1995) have identified features of effective coordination, Clark's description of joint activity during conversations seems to provide a much stronger basis for understanding team coordination. This chapter extends Clark's work on conversations to a whole range of other coordinated activities.

However, this much being said about the usefulness of Clark's work in clarifying the nature of joint activity, there is a caution that should be raised. As

Spiro, Feltovich, Coulson, and Anderson (1989) have shown, any single model or any single analogy can be limiting and misleading as an account of a complex phenomenon. Thus, the dynamics of a conversation will capture some aspects of coordination, such as taking turns and developing collaborative products, but miss other aspects such as following a plan or taking direction from a leader. The analogy of a conversation misses situations marked by the need for precise synchronization. It misses the aspect of planned coordination involving a dedicated team, such as a relay team, that relies on arduous training and preparation to carry out tight sequences of actions.

In preparing this chapter, we studied the nature of joint activity in three (non-conversational) domains that require coordination. Relay races demand very tight coordination and careful planning (Klein, 2001). Driving in traffic requires coordination among strangers who will likely never meet, and who must be counted on to follow official as well as informal rules (Feltovich, Bradshaw, Jeffers, Suri, & Uszok, in press). Coaching high school football teams involves understanding and adhering to the directives of leaders (Flach, personal communication, December 16, 2003). Throughout this chapter we use examples from these domains to elaborate on or contrast with the activity of having a conversation.

We considered several definitions of team coordination. Olson, Malone, and Smith (2001) offer the following definition: "Coordination is managing dependencies between activities." This definition includes only one of the processes involved in teamwork and organizational dynamics. One of its omissions has to do with the process of resolving issues of conflicting and interacting goals.

The definition of coordination offered by Zalesny et al. (1995) emphasizes temporal dependencies oriented around a common goal. They conceptualize coordination as "the complementary temporal sequencing (or synchronicity) of behaviors among team members in the accomplishment of their goal." (p. 102)

Klein (2001) has stated, "Coordination is the attempt by multiple entities to act in concert in order to achieve a common goal by carrying out a script they all understand." (p. 70) This definition anchors the concept of coordination to a script. This emphasis is understandable, because coordination requires some partially shared scaffold around which to structure and synchronize activity. However, in examples, such as driving on a highway or engaging in a conversation, the shared scaffold is minimal. There need be no overriding common goal for a conversation, and there is almost surely none for highway drivers (except self-protection). There is no overriding script in either case, other than routines and conventions (e.g., of speech) used by a society. The participants are not working together to carry out a plan. The participants are not a formal team. They are coordinating in order to have productive or safe transactions.

The Zalesny et al. (1995) and Klein (2001) definitions address the issue of goals. However, the concept of working toward a common goal fails to capture the interdependencies of how multiple goals interact and conflict (Clancey, 2004; Cook & Woods, 1994). It misses cases where the parties do not have any common goal except for the goal of working cooperatively in order to achieve their individual goals. Coordinating parties have to relax their own dominant goals and develop a joint agreement to work together—to adjust and communicate with each other as they all benefit from participating in the joint activity despite the costs of coordination.

None of these definitions does a good job of covering all of the examples we considered. We will provide our own definition at the end of this chapter, after we have had a chance to explore the nature of joint activity further.

The concepts of joint activity and team coordination are obviously very closely related. Entering into a joint activity requires the participants to coordinate because at least some of their actions affect the goal-directed activities of others.

The next three sections examine joint activity in some detail: the *criteria* for joint activity, the *requirements* for carrying out joint activities, and the *choreography* of joint activity (see Figure 1). The criteria for joint activity are that the parties intend to work together, and that their work is interdependent, rather than performed in parallel without need for interaction. If these criteria are to be satisfied, the parties to the joint activity have to fulfill requirements such as making their actions predictable to each other, sustaining common ground (the concept of common ground will be elaborated later in the chapter), and letting themselves be directed by the actions of the other parties. The form for achieving these requirements—the choreography of the joint activity—is a series of phases that are guided by various signals and coordination devices, in an attempt to reduce the coordination costs of the interaction.

CRITERIA FOR JOINT ACTIVITY

A joint activity is an extended set of behaviors that are carried out by an ensemble of people who are coordinating with each other (Clark, 1996, p. 3). The length of time of a joint activity can vary greatly. For example, the exchange between a customer and a cashier in a convenience store during a purchase may be on the order of a minute or two. Performance of a duet may take an hour or so. Two researchers writing a paper in collaboration may take months, or even years.

We have identified two primary criteria for a joint activity: the parties have to *intend* to work together, and their work has to be *interdependent*.

Intention to Generate a Multi-Party Product

"It's not cooperation if either you do it all or I do it all" (Woods, 2002). To be a truly joint activity, the activity should be aimed at producing something that is a genuine joint project, different from what any one person could do working alone.

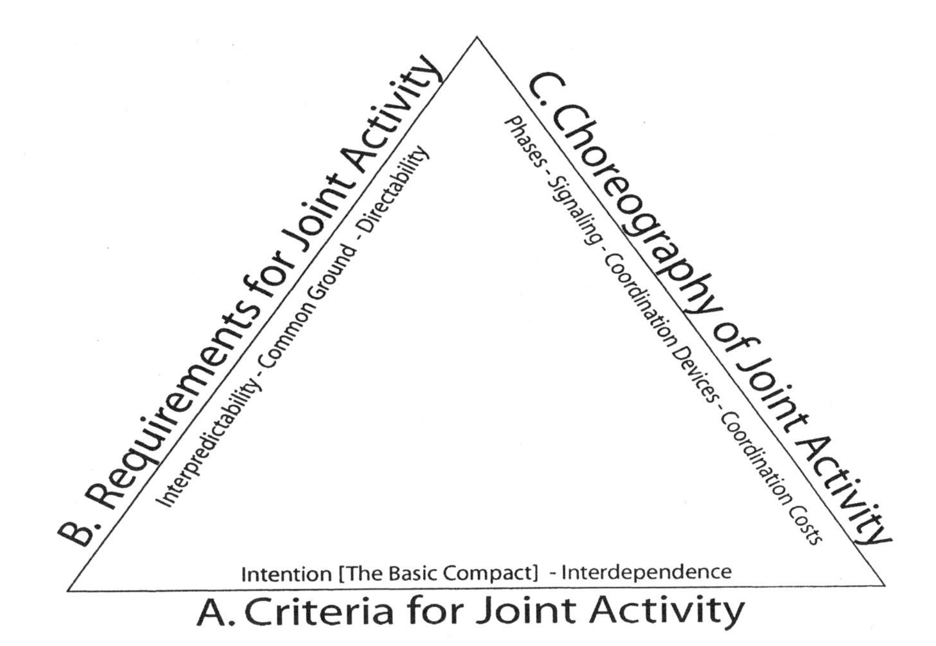

FIGURE 1. Description of Joint Activity

People engage in joint activity for many reasons: because of necessity (neither party, alone, has the required skills or resources), enrichment (while each party could accomplish the task, they believe that adding complementary points of view will create a richer product), coercion (the boss assigns a group to carry out an assignment), efficiency (the parties working together can do the job faster or with fewer resources), resilience (the different perspectives and knowledge broaden the exploration of possibilities and cross check to detect and recover from errors), or even collegiality (the team members enjoy working together).

We propose that joint activity requires a "Basic Compact" that constitutes a level of commitment for all parties to support the process of coordination. The Basic Compact is an agreement (usually tacit) to participate in the joint activity and to carry out the required coordination responsibilities. Members of a relay team enter into a Basic Compact by virtue of their being on the team; people who are angrily arguing with each other are committed to a Basic Compact as long as they want the argument to continue.

One aspect of the Basic Compact is the commitment to some degree of goal alignment—typically this entails one or more participants relaxing some shorter-term local goals in order to permit more global and long-term goals to be addressed. These longer-term goals might be shared goals (e.g., a relay team) or individual goals (e.g., drivers wanting to ensure safe journeys). A second aspect of the Basic Compact is a commitment to try to detect and correct any loss of common ground that might disrupt the joint activity.

We do not view the Basic Compact as a once-and-for-all prerequisite to be satisfied, but rather as a continuously reinforced or renewed agreement. Part of achieving coordination is investing in those things that promote the compact as well as being sensitive to and counteracting those factors that could degrade it.

The Basic Compact can vary in strength. Physically or culturally distributed teams such as international peacekeeping missions may have difficulty maintaining the Basic Compact (Klein, H.A. and McHugh, 2005). When that happens, the amount of coordination that is possible is reduced. Some teams will deliberately simplify their plans because they realize they are not capable of complex interactions.

All parties have to be reasonably confident that they and the others will carry out their responsibilities in the Basic Compact. In addition to repairing common ground, these responsibilities include such elements as acknowledging the receipt of signals, transmitting some construal of the meaning of the signal back to the sender, and indicating preparation for consequent acts.

The Basic Compact is also a commitment to ensure a reasonable level of interpredictability. Moreover, the Basic Compact requires that if one party intends to drop out of the joint activity, he or she must inform the other parties.

The Basic Compact is illustrated by an example from Cohen and Levesque (1991)—arranging a convoy of two cars—as a prototype for establishing common ground in a distributed team. A critical requirement for making the convoy work is that all team members understand their obligations to each other and sign onto the Basic Compact.[1]

Example 1. A Convoy as a Distributed Team.

There are two agents in this example: Alice and Bob. Bob wants to go home after a party, but doesn't know the way. He does know that Alice knows the way and that her destination is not far from his home. So all Bob has to do is to follow Alice. But that may not be enough. Bob cannot follow Alice if she drives too quickly through traffic, runs red lights, and so forth. Therefore, it is best if Alice knows Bob is going to be following her until he gets within familiar range of his home. If they use an intermediary to organize their convoy, they may be uncertain about what the intermediary arranged. So Bob and Alice need a direct affirmation that they are both signing on to drive in a convoy.

Bob and Alice may fail to affirm their commitment to drive in a convoy. Thus, if an intermediary mentions to Alice that Bob will be following her, but Bob isn't sure if Alice has agreed, then the door is open for all kinds of confusions. For example, if Bob sees Alice going down lots of side streets, he may infer that Alice

[1] We have simplified the example described by Cohen and Levesque (1991).

doesn't really know her way and abandon the convoy, leaving Alice to worry about where she lost Bob, and circling back to try to find him. And if Bob makes a momentary wrong turn, Alice might conclude Bob now knows his way so she can abandon the convoy and leave him to find his own way.

While she is in the convoy, Alice has to drive differently, adapting her own goals (e.g., not letting her attention wander too much, slowing her speed, refusing to maneuver around knots in the traffic, refusing to take a shorter but more complicated route). The compact signals her willingness to adapt what would be her ways to achieve her goals in order to participate in the joint activity and the goals that it supports.

If one party abandons his or her intention, he or she must make this known to the rest of the team. All sorts of events can occur to make an agent believe a commitment is no longer reasonable. But the Basic Compact requires parties to communicate to each other if they plan to abandon the joint activity. If Bob and Alice had talked to each other before the party ended and agreed to ride in a convoy, then their agreement—their Basic Compact—would require each of them to signal to the other before heading off alone. The convoy example shows how difficult it is to arrange a Basic Compact by proxy.

Interdependence of the Actions of the Parties

In joint activity what party "A" does must depend in some significant way on what party "B" does and vice versa (Clark 1996, p. 18). Two musicians playing the same piece in different rooms are not involved in the joint activity of performing a duet. If two researchers decide to have one author write the first half of the paper and the other the last, with the final product merely assembled together at the end, then this is parallel—not joint—activity. A group of Army units that set out on related missions which fit within some sort of synchronization schedule but have no further interaction, are not engaged in a joint activity because their efforts have no mutual influence. Joint activity emphasizes how the activities of the parties *interweave* and *interact.*

REQUIREMENTS FOR EFFECTIVE COORDINATION IN JOINT ACTIVITY

In reviewing different forms of coordination, we have identified three primary requirements that cut across domains: the team members have to be *interpredictable*, they have to have sufficient *common ground*, and they have to be able to *redirect* each other.

Interpredictability

Coordination depends on the ability to predict the actions of other parties with a reasonable degree of accuracy. It follows that each party also has a responsibility to make his or her actions sufficiently predictable to enable effective coordination.

Predictability includes accurate estimates of many features of the situation—for example, the time needed by all of the participants to complete their actions, the skill needed, and the difficulty of the action.

Shared scripts aid interpredictability because they allow participants in joint activities to form expectations about how and when others will behave, thus enabling absence of activity to be seen as meaningful (e.g., inasmuch as the other is not doing X, then the element Y must not be present because the doctrine or script would lead someone who sees Y to do X).

Interpredictability is greatly increased when the partners can take on the perspective of the others. The heuristic of assuming that others will see the events the same way we see them, and that they will react the way we would, is reasonable in many settings. But it is not as powerful as being able to decenter and imagine the events from the other person's point of view.

Common Ground

Perhaps the most important basis for interpredictability is common ground (Clark & Brennan, 1991), which refers to the *pertinent* mutual knowledge, mutual beliefs, and mutual assumptions that support interdependent actions in some joint activity. Common ground permits people to use abbreviated forms of communication and still be reasonably confident that potentially ambiguous messages and signals will be understood. Short of being able to rely on common ground to interpret such communications, every vague or ambiguous referent would have to be unpacked, at great cost to the parties in the transaction. For example, in a relay race, as runner A approaches runner B and announces "stick," this one word has a clear meaning to runner B and requires no breath-wasting elaboration. Runner A doesn't have to tell runner B, "I am getting close to you and am just now reaching out the baton, so why don't you extend your right hand back to let me place the baton in your palm."

The Basic Compact includes an expectation that the parties will repair faulty knowledge, beliefs, and assumptions when these are detected. Common ground is not a state of having the same knowledge, data, and goals. Rather, common ground refers to a *process* of communicating, testing, updating, tailoring, and repairing mutual understandings (cf. Brennan, 1998). Moreover, the *degree of quality* of common ground demanded by the parties can vary due to the particulars of people, circumstances, and their current objectives. Two parties in friendly chit-chat may be satisfied by simple head-nods or sustained attention to confirm that they are understanding each other well enough; two people working out a contractual agreement may demand and work toward—through more extensive mutual testing, tailoring, and repair—evidence for more precise mutual understanding (see Cahn & Brennan, 1999, p. 2 "grounding criteria"). In general, common ground is what makes joint activity and coordination work (although each joint action, in turn, serves to change common ground).

Common ground can be characterized in terms of three basic categories (Clark, 1996): *initial common ground, public events so far,* and the *current state of the activity.*

Initial common ground includes all the pertinent knowledge and prior history the parties bring to the joint activity. It involves not only their shared general knowledge of the world, but also all the conventions they know that are associated with their particular joint task. For an example of such conventions, think of the general procedure to be followed by a surgical team in conducting a particular kind of surgical procedure in a particular hospital setting (different hospitals can have slightly different procedures, as can particular surgical teams). It also includes what parties know about each other prior to engagement—for example, the others' background and training, habits, and ways of working.

Public events so far includes knowledge of the event history—the activity the participants have engaged in together up to the present point in the joint activity. For instance, it is important to know that options may have been lost irretrievably due to time or resource dependence (e.g., in surgery, tissue may have been cut or excised and cannot be uncut or put back). Once everyone is informed about the pertinent aspects of previous history, the participants' ongoing work together provides them with further information about each other. Also, as the course of the joint activity unfolds, it establishes precedents regarding how certain situations have been handled, how the parties have negotiated the meaning of things and events, who has been established as leader or follower, and the like. Precedents established during the process of carrying out the joint activity are important coordination devices. As we bring new people into an evolving situation we have to work out a range of issues about how they will come up to speed in the absence of their participation in the public events so far.

The *current state of the activity* also provides cues to enable prediction of subsequent actions and the formulation of appropriate forms of coordination. The physical "scene" provided by the current state serves as a kind of accumulated record of past activity, and often makes salient what is most critical in the scene for further operation. It must be remembered, however, that parties can also be fooled by such situational framing into falsely thinking they are viewing the same thing and interpreting it the same way (Koschmann, LeBaron, Goodwin, & Feltovich, 2001). In addition to the state of the literal work environment at a particular point in the activity, "props" of various kinds are often available to portray the current state—for example, the pieces placed on the chessboard. Sarter, Woods, and Billings (1997) and Christoffersen and Woods (2002) have discussed the importance of "observability" for effective coordination, and we see the various forms of observability as bases for solidifying common ground.

In reviewing various forms of team coordination, we found that some of the most important types of mutual knowledge, beliefs, and assumptions are about:

- Roles and functions of each participant;
- Routines that the team is capable of executing;

- Skills and competencies of each participant;
- Goals of the participants, including their commitment to the success of the team activity; and
- "Stance" of each participant (e.g., his or her perception of time pressure, level of fatigue, and competing priorities).

Although common ground depends greatly on these kinds of shared knowledge, we obviously do not mean to imply that all parties in joint activity need to know and think identically. Parties in joint activity represent different points-of-view by way of their particular roles or duties in the activity, or because they represent distinct backgrounds, training, or skills. It is likely that understanding and aligning these different perspectives can at times require added effort, for instance, in the need for more "repair" in the process of communication. But such diversity of perspective may actually improve performance by requiring the participants to negotiate and reconcile different pertinent understandings (e.g., Feltovich et al., 1996; Spiro, Feltovich, Coulson, and Anderson, 1989).

We can differentiate several activities in which teams often engage in order to support common ground. These are:

- Structuring the *preparations* in order to establish initial calibration of content, and to establish routines for use during execution.
- *Sustaining* common ground by inserting various clarifications and reminders, whether just to be sure of something or to give team members a chance to challenge assumptions.
- *Updating* others about changes that occurred outside their view or when they were otherwise engaged.
- *Monitoring* the other team members to gauge whether common ground is being seriously compromised and is breaking down.
- Detecting *anomalies* signaling a potential loss of common ground.
- Repairing the loss of common ground

In relay races, baton passes are tricky and are often the key to success or failure. The exchange must happen within a limited stretch of the track. The current runner and the new (handoff) runner try to make the exchange with both of them going at as near full speed as possible. Hence the waiting runner starts to accelerate before the baton pass so that he or she can be in full motion at the time of the handoff. Ragged hand offs waste precious fractions of a second. But the precise timing of the handoff cannot be directed by the coach of the team. A coach's attempts to tell the runners exactly when to make the exchange would add confusion. The runners themselves must signal to each other when and where to pass the baton.

Common ground can vary in quality over time and in different situations but can never be perfect. That is one of the functions of the sustaining and monitoring functions—to catch the discrepancies before they become serious. Common ground enables a team to coordinate in order to achieve tasks, but it is not a work output in itself.

Directability

Christoffersen and Woods (2002) have identified "directability" as an important aspect of coordination, because it is so central to the resilience of a team. We also see it as central to the interdependence of actions. If the way that one partner performs a task has *no* effect on the other partner, then they are stove-piped, perhaps working in the same space but not coordinating with each other. Sometimes, one partner runs into difficulty, and the other partner notices this and adapts. At other times, one partner will find some way to signal to the other to alter course, and to modify the routine they are carrying out. Directability refers to deliberate attempts to modify the actions of the other partners as conditions and priorities change. Taking direction and giving direction are another whole aspect of the interplay in coordination when there are levels of management and control (Shattuck & Woods, 2000).

THE CHOREOGRAPHY OF JOINT ACTIVITY

Each small phase of coordination is a joint action, and the overall composite of these is the joint activity (Clark, 1996, pp. 3, 30-35, 125). For example, a couple on a dance floor is engaged in a joint activity (the entire dance) and in joint actions (each sequence within the dance). Two waltz dancers must recognize in an ongoing manner the changes of body pressure and posture of the other that indicate the direction of the next movement and they must respond appropriately. Failure to recognize the appropriate cues and, just as importantly, failure to respond appropriately in the necessary time frame (in this case nearly instantaneously) results in at best bad dancing, and at worst something one might not characterize as "dancing" at all.

The choreography of a joint activity centers on the *phases* of the activity. The choreography is also influenced by the opportunities the parties have to *signal* to each other and to use *coordination devices. Coordination costs* refer to the burden on joint action participants that is due to choreographing their efforts.

Three-Part Phases

According to Clark (1996), what actually gets coordinated is a phase. A phase is a joint action with an entry, a body of action, and an exit (although there may also be other embedded joint actions within any joint action). Coordination is accomplished one phase at a time in a joint activity. Therefore, the most basic representation of the choreography of joint action is the entry, actions, and exits for each phase, with coordination encompassing each of these. Often, the phases do not have any official demarcations—the parties themselves determine which part of the phase they are in, based on the work they have accomplished together.

The exiting of a phase can be difficult to coordinate—each person performing an action needs evidence that he or she has succeeded in performing it (e.g., when you press an elevator button, you want to see a light go on). Similarly, joint closure is needed for coordinated actions. For example, during a conversation the passive team members have a role in signaling to generate the belief in the one talking that the action was understood. This signaling can be done through head nods, grimaces, paraphrases, and so forth.

Thus, communication proceeds on two tracks: the official business can be thought of as the task work (Salas & Cannon-Bowers, 2001), while the conformations, corrections, and completions can be thought of as the teamwork or the communication overhead. This overhead constitutes the coordination costs of the interaction. Note that the distinction can sometimes blur—for example, what is the task work versus the teamwork in dancing a waltz?

Both conversations and relay races can be described in terms of joint action phases. In a relay race, one phase is the handoff of the baton. That is when the teammates have to coordinate with each other. In a conversation, one phase is the handoff of turn taking, as one person yields the floor to the other. That is when the participants engage in joint action to effect a smooth transition.

The overall structure of a joint activity (e.g., "writing a collaborative research article") is one of embedded sets of a series of phases throughout many possible layers of joint activity. The "writing of the paper" is itself a joint activity, which will have within it sub-joint activities, e.g., "creating the outline," which will themselves have embedded joint activities, e.g., "collecting the major resources to be used," and so on, all the way down. Synchronizing entry and exit points of the many embedded phases involved in complex joint activity is a major challenge.

The phases of a joint activity can be more or less "scripted," that is, more or less prescribed in how the activity is to be conducted. Precise scripting may be mandated or may emerge because of law, policy, norms of practice, and the like (Clark, 1996, p. 30). An example of tight scripting is the handling of evidence in a crime lab. A somewhat less scripted example is the carrying out of a particular kind of surgical procedure by a surgical team. Even less scripted is the work of a design team working to produce a new company product or logo.

The entire joint activity can be more or less scripted—as can be each of the phases. In addition to regulatory coordination mechanisms that are officially mandated, various macro guides can serve to coordinate the course of entire joint activities. Examples are plans for an activity worked out in advance by the participants, and the prior extensive outline worked out by authors involved in writing a joint manuscript. Motivations for standardizing procedures include the desire to prevent interactions where some earlier move impedes some necessary later move, and also better anticipation of necessary coordination points (e.g., Shalin, Geddes, Bertram, Szczepkowski, & DuBois, 1997). However, no matter the extensive nature or degree of foresight incorporated into any prior "plan," there will always be a need for adjustment, further specification, or even drastic revision (Clancey, 1997).

We observe that conversations are not as scripted or planned as relay races. The direction that conversations take is continually being reshaped, depending on what has been said earlier, what has just been said, and the motivations of the parties for what they want to accomplish through the conversation (e.g., to get to the bottom of a deep issue, or, at a social gathering, to get rid of an unwanted conversational partner).

Signaling

The choreography of joint activity depends on the way the participants signal to each other about transitions within and between phases. The participants may also signal such things as their intentions, the difficulties they are having, and their desires to redirect the way they are performing the task. The importance of signaling can be seen in the way drivers coordinate with each other on the highway.

Feltovich et al. (in press) have identified the task of driving in traffic as a challenge of coordination. One of the distinctive properties of driving in traffic is that it depends on coordination among people who have never met each other and in no sense form a team. Yet if their actions are not skillfully coordinated, the result can be chaotic and even fatal. We rely on the coordination of strangers as we travel at high speeds on crowded expressways, and when we make split-second decisions about when to make a left turn against oncoming traffic. When we look away from the road if no immediate hazard is apparent, we trust the driver in front of us not to unexpectedly come to a halt when we aren't watching.

Driving in traffic captures a new item of coordination: the drivers cannot rely on personally established precedent for coordination. Hence, there is a great deal of local control. Except in extraordinary circumstances, nobody is directing traffic, telling the drivers what to do. Yet, interpredictability is crucial if hazards of the road are to be avoided.

Driving in traffic depends heavily on signaling (see also Norman, 1992). The individual drivers are constantly providing cues for coordination, the most salient being signals to indicate intention to turn or to change lanes, or (through the brake lights) the act of slowing down. If drivers see a potential problem ahead, they may pump their brakes in advance of slowing down in order to alert the car behind them and reduce the chance of being rear-ended.

Driving in traffic depends on a Basic Compact that affirms that everyone will follow the rules. However, there may be defensible motives for drivers to escape from the Basic Compact, as when they are responding to an emergency by rushing someone to the nearest hospital. At such times, drivers may turn on their emergency blinkers to signal to other drivers that their actions are no longer as predictable and that they temporarily consider themselves not bound by certain aspects of the Basic Compact. Automobiles are all equipped with emergency blinkers to permit drivers to signal this kind of state during unusual events.

Signaling only works if the other participants notice the signals. Clark (1996) explains that coordinated conversation depends on the way the participants manage attention—their own attention and the attention of others. Generally speaking, we can only give full conscious attention to one thing at a time. We can redirect attention very quickly, sometimes too quickly. Our efforts to attend to a particular strand of a situation are easily subverted by distractions. Thus, part of the coordination effort in a conversation is to coordinate attention, helping the participants gain, hold, and relinquish attention. The coordination of attention is done via such things as gazes, pointing, and instructions—and by making repairs when coordination goes wrong. Appropriate direction of attention is such a challenge that people often expect interference, disruptions, and interruptions. Then they try to work around these when they occur.

Thus, part of the choreography of joint action is to direct the attention of other participants to signals, or to use signals to direct their attention to relevant cues in the situation (Moore & Dunham, 1995; Woods, 1995). If one partner believes a coordination phase has been completed and wants to exit that phase, it may be necessary to direct the other partner's attention to the cue that marks the phase completion.

Signaling carries a responsibility to judge the interruptability of the other partners, as well as to understand what signaling modality would be the most appropriate for the situation. Careless signaling can degrade performance. Dismukes, Young, and Sumwalt (1998) have described how interruptions have contributed to a number of aviation accidents as when, for example, air traffic controllers request information from pilots who are going through a flight checklist, leading to the omission of critical steps. The very use of a checklist is a coordination device intended to protect pilots from factors such as interruptions because of how the checklist can structure activities in ways that are visible to others.

Coordination Devices

The choreography of joint activity is shaped by the use of coordination devices (Clark, 1996, pp. 64-66). These devices include highly diverse mechanisms of signaling. Many of these signals, coupled with common ground, serve to increase interpredictability among the parties. Examples of typical coordination devices are agreement, convention, precedent, and situational salience.

> **Agreement:** Coordinating parties can explicitly communicate their intentions and work out elements of coordination. This category includes, in addition to language, diverse other forms of signaling that have shared meaning for the participants, including signs, gestures, and displays.
>
> **Convention:** Often prescriptions of various types and degrees of authority apply to how parties interact. These can range from rules and regulations to less formal codes of appropriate conduct. These less formal codes include norms of practice in a particular professional community as well as established practices in a workplace. Coordination by convention depends on structures outside of a particular episode of joint activity.
>
> **Precedent:** Coordination by precedent is like coordination by convention, except that it applies to norms and expectations developed within the ongoing experience of the joint activity. As the process unfolds, decisions are made about the naming and interpretation of things, standards of acceptable behavior and quality (e.g., what is established by this particular surgical team, during the course of a surgical procedure, as the standard for adequate cauterization of a vessel), who on the team tends to take the lead, and so forth. As these arise and develop during the course of the activity, they tend to be adopted as devices (or norms) of coordination for the remainder of the activity.
>
> **Salience:** Salience has to do with how the ongoing work arranges the workspace so that next move becomes apparent within the many moves that could conceivably be chosen. During surgery, for example, exposure of a certain element of anatomy in the course of pursuing a particular surgical goal can make it clear to all parties involved what to do next. Coordination by salience is produced by the very conduct of the

> joint activity itself. It requires little overt communication and is likely to be the predominant mode of coordination among long-standing, highly practiced teams.

Let us look at an example, one in which the authors, having recently communicated only via e-mail regarding their paper, decide that this has been insufficient for resolving some issues and that they need to talk on the phone (see also Brennan, 1998 for a complementary account of a set of e-mail exchanges). They try to arrange this phone conversation by e-mail:

> **F:** Hi, K, our e-mails have left a lot of questions and confusion in my mind. I really think we need to talk if we are going to get some of this settled.
>
> **K:** Yes, F, I've been feeling the same way. I'm free all tomorrow afternoon. How's that for you?
>
> **F:** Any time between two and three is okay for me, how about two o'clock?
>
> **K:** Let's book two o'clock! Want me to call you or do you want to call?
>
> **F:** I'll do it. Catch you at two tomorrow. By the way, what is your cell number in case I don't reach you in the office?

There are several coordination devices involved in this exchange. There is a convention that one responds to the overtures of friends and active collaborating colleagues with a relatively prompt return communication. There is precedent at work: such exchanges on the paper have been held from the location of the authors' offices in the past, perhaps because most of the pertinent materials for these conversations reside there. Although the offices are the default origination places by precedent—F feels the need for a back-up if the usual does not hold true, essentially saying, "If you are not in your office when I call, give me your cell phone number to call so we can re-coordinate." Some aspects are communicated simply by their salience in the field of joint activity of the authors. For instance, the "paper" itself is never mentioned. The authors may have communicated by e-mail about all sorts of things: friends and family, an academic meeting they will be attending together in a month, a workshop they are designing on some topic. Yet, somehow, all the prior interaction between the parties has conspired to make the "paper" the salient focus of this exchange in time and space.

The Joint Action Ladder

Clark uses the concept of a Joint Action Ladder (1996, p. 152) to describe the successive inferences that we make during a conversational interaction (Figure 2).

When someone sends a message to a team member, as in the earlier example of the e-mail that F sent to K, four things have to happen for the exchange to be successful in advancing the overarching joint activity—in this case writing the collaborative academic paper. First, when F sends the note to K, K must *attend* to it. This step seems trivial, but it is not. A virus may have temporarily disabled K's e-mail system, leading F to erroneously assume that K has received the message. Second, while K may attend to the note (i.e., he sees that it has arrived), he must also *perceive* its contents. He may fail to do so. For example, he may be in a rush, have hundreds of new email items, open F's message to see if F is writing about the paper or about some other topic, and never read the detail. Third, while the recipient may attend to and perceive the note, he must also *understand* it. Fourth, the recipient must *act* on the message, by doing a number of additional things, among them acknowledging that he has "climbed the action ladder" or indicating how far he has been able to climb it (e.g., "I got your message with the attachment

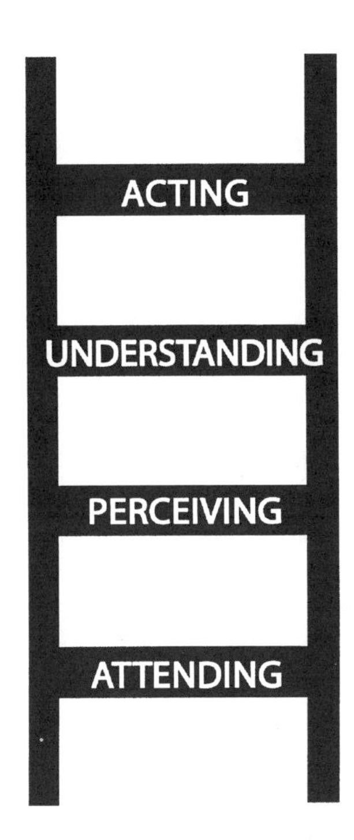

FIGURE 2. Joint Action Ladder.

but for some reason I can't open the attachment"). In the best case, he states what he will do next as a result of receiving the note, which indicates that he has attended to it, perceived it, and understood it.

Certain aspects of the ladder may not need to be addressed directly because of the principle of "downward evidence" (Clark, 1996, p. 148): lower rungs on the ladder are subsumed by higher ones. For example, the initial response by K in our example interchange ("Yes, F, I've been feeling the same way. I'm free all tomorrow afternoon. How's that for you?") lets F know that K has attended to and perceived F's message, and also suggests that K understands the note and intends to act on it to continue the joint work.

Although the interchange in our example of joint activity involved language, this need not always be the case. Anything that carries meaning for the participants—sign, signal, gesture, posture, actions, and so forth—can be used to coordinate joint activity. Indeed, even vertebrate animals participate in simple forms of joint activity, using various forms of signaling and display (Feltovich et al., in press; Smith, 1977, 1995).

The choreography of joint activity does not just cover instances in which the transmission of the joint action signal is more or less direct, face-to-face. For example, the signal can be delayed, mediated by a machine, or mediated by another person. Also, the "cycle time" of any joint action can vary greatly, from the almost instantaneous interactions of two waltz partners to the days or weeks of two authors creating a research paper together. There may be a defined leader or conductor of the joint activity (who decides, for instance, where to move next—commonly the role of the "attending" surgeon during a surgical procedure) or there might not be.

Coordination Costs in Choreography

The effort expended to choreograph joint activity is one type of coordination cost. Fundamental to coordination is the willingness to do additional work and to narrowly downplay one's immediate goals in order to contribute to the joint activity. Consider the example of a relay race. Each handoff takes its toll in terms of time lost. One typical arrangement for a relay race is to use four runners to cover 400 meters. It would be inefficient to have 24 runners stationed around a 400-meter track. The coordination costs would outweigh advantages of the freshness and energy of each new runner.

Schaeffer (1997) and Klinger and Klein (1999) have identified types of coordination costs that are incurred by joint activity: synchronization overhead (time wasted in waiting for one entity to complete its work before the next one can begin); communication overhead (effort to manage the handoff); redirection overhead (wasted time and energy in going in the wrong direction after a new direction is called out but before all entities can be told to change course); and diagnosis overhead (the additional burden of diagnosing a performance problem when multiple moving parts are involved).

Consider also how unwieldy and frustrating a conversation with many participants can be. The coordination costs of choreographing the phases, the time wasted in trying to sort out how to enter a comment into the flow of discussion—each of these increase as the number of participants becomes greater. Sometimes the wait time is so long that when people do get to speak they forget what they wanted to say.

As common ground is lost, coordination costs can rise. Effort spent in improving common ground will ultimately allow more efficient communication and less extensive signaling. On the other hand, failure to invest in adequate signaling and coordination devices, in an attempt to cut coordination costs, increases the likelihood of a breakdown in the joint activity. The next section describes some typical ways that such breakdowns occur.

THE FUNDAMENTAL COMMON GROUND BREAKDOWN

Many different factors can lead to breakdowns in joint activity. In this section, we discuss one type of breakdown that seems to arise repeatedly in the instances we examined. We call this the Fundamental Common Ground Breakdown. We also discuss a form of this breakdown that is particularly troublesome—when one party defects from the joint activity and leaves the other parties believing that the Basic Compact is still in force.

The Logic of the Fundamental Common Ground Breakdown

We assert that, no matter how much care is taken, breakdowns in common ground are inevitable: no amount of procedure or documentation can totally prevent them. During a transaction, while common ground is being established in some ways (e.g., through the record and activities of a conversation or through physical markers), it is simultaneously being lost in others (e.g., as each party has differential access to information and differential interpretation of that information). That is why Weick, Sutcliffe, and Obstfeld (1999) have asserted that high reliability organizations and work processes are marked by a continual mindfulness—a continual searching for early indicators of problems, including indications of a loss of common ground.

Why do teams lose common ground? There are a number of typical reasons for the loss of common ground. These have been identified by Klein, Armstrong, Woods, Gokulachandra, and Klein (2000). In the incidents Klein et al. studied, they found that common ground is continually eroding and requiring repair. As a situation changes, people are likely to make different interpretations about what is happening, and what the others know. So the baseline state is one in which people are detecting and repairing problems with common ground—and not attempting to document all assumptions and duplicate the contents of each person's mind.

Our research has shown that teams tend to lose common ground for the following reasons:

- Team members may lack experience in working together;
- They may have access to different data;
- They may not have a clear rationale for the directives presented by the leader;
- They may be ignorant of differences in stance (e.g., some may have higher workload and competing priorities);
- They may experience an unexpected loss of communications or lack the skill at repairing this disruption;
- They may fail to monitor confirmation of messages and get confused over who knows what.

This last reason, confusion over who knows what, is sufficiently frequent that it deserves more careful examination. We refer to this confusion as the Fundamental Common Ground Breakdown.

The script for the Fundamental Common Ground Breakdown is illustrated by the following example:

Example 2. The Viennese Waltz.

Gary gave a presentation on decision making to a conference in Scotland. One of the attendees requested that Gary give the same presentation in Vienna a few months later. Gary had a schedule conflict but arranged for his close colleague, Dave, to give the presentation in his stead. Dave was willing to make the trip to Vienna, but did not have the time to put together a presentation. Gary assured him that this was no problem—all he had to do was give the same presentation that Gary used in Scotland.

Gary asked his production associate, Veronica, to give Dave the Scotland presentation. She did so.

The day before Dave was going to leave for Vienna, Gary stopped in to talk, and noted that Dave was feeling pressured from the workload because he still had to pull together materials. This surprised Gary. He verified that Dave had gotten the presentation from Veronica. Dave said that there were still some big gaps to fill.

Gary asked to see the presentation, and Dave opened it up on his computer and they went through it, slide by slide. Gary then asked where the exercise materials were, and Dave said he didn't know anything about exercise materials—that was what he had to pull together.

Gary went back to Veronica, and they discovered that Veronica had only sent Dave a file containing the presentation. She didn't know she was also supposed to send a text file containing the handouts with the exercise.

In this incident, Gary assumed that Dave had gotten everything he needed because Gary had requested that from Veronica, Veronica had complied, and Dave had received the file Veronica had sent. The disruption occurred because the word "presentation" meant one thing to Gary (i.e., everything he used during his talk in Scotland) and another to Veronica (i.e., the file containing the presentation—a text file is not a presentation). As a result, Dave almost went off to Vienna without the critical materials needed to run a decision making exercise.

We present this example to illustrate the Fundamental Common Ground Breakdown:

- Party A believes that Party B possesses some knowledge.

- Party B doesn't have this knowledge, and doesn't know he is supposed to have it.

- Therefore, he or she doesn't request it.

- This lack of a request confirms to Party A that Party B has the knowledge.

As a result, they fail to catch the mismatch in their beliefs about what Party B knows and is supposed to know. Further, Party A interprets Party B's subsequent statements and comments with the assumption that Party B possesses the critical knowledge, thereby constructing a cascading set of incorrect inferences about Party B's beliefs and actions.

Too often, people discover a serious loss of common ground when they experience a coordination surprise. When something happens that doesn't make sense in terms of their beliefs, that event may trigger a deeper inquiry. Sometimes, if the parties are sensitive to each other, they will catch the error early enough to repair it, as in the Viennese Waltz example. Too often the discovery is made following an accident or performance failure. (For a discussion of coordination surprises, see Patterson, Woods, Sarter, & Watts-Perotti, 1998; Sarter, Woods, & Billings, 1997.).

The Fundamental Common Ground Breakdown is depicted in Figure 3. Person A assumes that person B knows item "X," whereas person B doesn't have this information. As subsequent events occur, person A interprets B's beliefs and actions in a way that diverges from person B's actual beliefs and rationale for actions. The discrepancy grows greater until they encounter a coordination surprise.

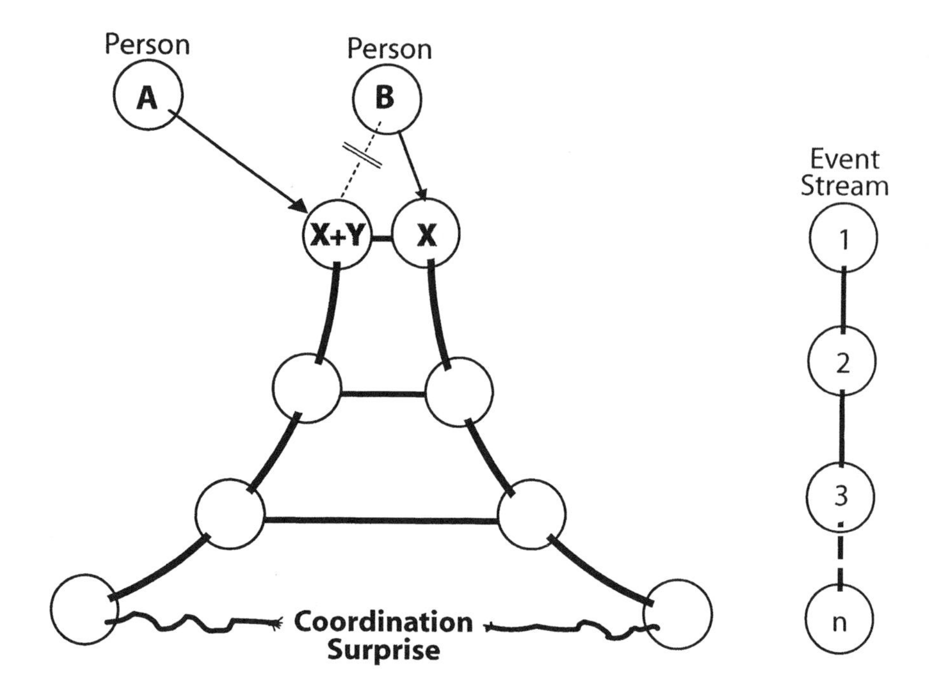

Figure 3. Fundamental Common Ground Breakdown.

This account is different from a case where the parties just don't have a shared understanding of the situation. The breakdown is more than the lack of a shared understanding. The breakdown arises because person A is making incorrect assumptions about what person B knows, and also because of the ripple effects as person A has to explain away the initial anomalies and prediction failures by coming up with more and more elaborate accounts which just end up adding to the distortions.

Careful inspection will show that the Fundamental Common Ground Breakdown is actually a corruption in the execution of the normal Joint Action Ladder sequence. The breakdown occurs when people wrongly assume that the ladder has been scaled. The joint action ladder has four rungs: attending to a signal, perceiving its contents, understanding it, and acting on it. While downward inferences are usually benign (if people act on a message, they have probably read it), upward inferences are very risky (e.g., if a person sends a message then he or she cannot safely assume the recipient has read it).

In the Viennese Waltz, Gary assumed that Dave had gotten and appreciated all the materials because Gary asked Veronica to send them, and Dave said he received them. Unfortunately, Veronica failed to understand what Gary had wanted to request (because of Gary's carelessness in making the request), and therefore failed to act in the way Gary expected. Dave's confirmation that he had received materials from Veronica reinforced Gary's mistaken belief that the Joint Action Ladder had been scaled successfully, and that Veronica had understood his intentions.

In the Fundamental Common Ground Breakdown, activities lower in the ladder are used to hypothesize that the ladder was scaled. We do not, and cannot, take the effort to ensure that every message has been read and properly understood. We often have to make these assumptions, just to get anything done. Then we have to be on the lookout for evidence that the assumptions were unwarranted, although when we hit these coordination surprises we usually think, "Is he stupid, or crazy?"

However there are practices that can be engaged in proactively in order to make it easier to detect a breakdown before a possibly harmful surprise. One of these practices is to elaborate—to the extent possible—all the steps of the Joint Action Ladder during coordination events. The recipient of a message can report back to the sender some indication of the recipient's understanding of the message (in addition to the fact that it has been attended to and perceived). In this case Veronica may have said to Gary's original request: "I got your message about helping Dave, and as a result I have sent him the presentation file named "Scotland," and dated..." This may have prompted Gary to ask about the auxiliary materials for the talk. Even so, we are dubious about the general effectiveness of a reporting back procedure. Veronica would likely only report back in this manner if she was unsure about whether to send the text file—and in that case, she would simply have asked Gary whether she should do so. The breakdown in The Viennese Waltz could best have been avoided if Gary had explained his overarching intent to Veronica—that he wanted Dave to carry out the same program in Vienna that Gary had accomplished in Scotland.

Here are some additional examples of the Fundamental Common Ground Breakdown, taken from Klein et al. (2000):

Example 3. The Fuel Leak.

In a flight simulation to study the decision making of commercial pilots, researchers introduced a malfunction into an otherwise standard scenario: a leak from the fuel tank in one of the wings. This malfunction was patterned after an actual event. Nevertheless, it is an unlikely malfunction, and the aircrew doubted the problem after they spotted it. They thought it was likely that the fuel gauge was malfunctioning, and they spent some time trying to determine if the leak was really happening.

The captain realized that if there were a real leak, the affected wing would be getting lighter, changing the handling characteristics of the craft. He turned to the first officer (who was flying the plane) and said, "Are you having any trouble?"

The first officer answered, "No." That seemed to argue for a gauge problem, and it took another ten minutes to establish that the fuel was really being lost.

Afterwards, during the debriefing, the crew was shown the videotape of their attempts to diagnose the problem. Upon seeing the first officer say "No," the captain and flight engineer seemed to find the culprit. But the first officer asked for the tape to be rewound. He argued that the question he was asked was problematic. Yes, the airplane was handling poorly. He was so busy trying to keep it straight and level that he lost track of their discussion, while assuming that the captain and flight engineer knew about the leak. So, when asked if he was having trouble, he answered honestly that he wasn't. He was able to keep the plane leveled out. He should have been asked if the plane was imbalanced.

The following example of the Fundamental Common Ground Breakdown involves air traffic controllers. In some ways, this example is like a relay race, with the aircraft serving as the baton passed between air traffic controllers. Unhappily, this baton was dropped.

Example 4. Handing off an Airplane.

In aviation, airline dispatchers and flight crews work with air traffic controllers (ATC) to adapt flight plans in the face of constantly changing conditions. Weather can rapidly and unpredictably turn unpleasant, creating obstacles in the planned flight path of an airplane, and air traffic densities can suddenly ebb or flow, changing the assumptions of the plan. Hence, plans must have the ability to be modified to accommodate these changes.

An example of a potentially significant breakdown in communication occurred during a flight from Dallas/Ft. Worth to Miami (Smith et al., 1998). The airline dispatcher in charge of the flight filed an initial flight plan. However, as part of his further responsibility to the flight plan, the dispatcher was also required to warn the pilot of any hazardous weather en route. In this case, after the flight was on its way, the dispatcher noticed a line of thunderstorms that he felt could potentially endanger the safety of the flight and, with the Captain's concurrence, issued a reroute instruction to the aircraft. During this process, the Captain was informed about the situation and the conditions of weather that prompted the reroute. The reroute was coordinated with the ATC and approved.

The dispatcher in charge of the flight, who originally noticed the bad weather, was also responsible for about 30 other planes. Once he had filed the reroute, briefed the Captain, and obtained permission from the first ATC center, he felt that the issue had been resolved and devoted his attention to other airplanes and tasks in his charge.

Once the flight was underway, however, the Receiving Center, (i.e., the ATC that was taking on responsibility for the flight as the aircraft transitioned from one ATC sector to another) rejected the reroute and put the airplane back on its originally filed flight plan. The Captain assented, assuming that the receiving ATC also knew about the weather front.

However, following the original flight plan trapped the aircraft south of the line of thunderstorms. The weather system forced the aircraft to circle, waiting for a break. As the aircraft began to run low on fuel, the crew decided to fly through the weather front to land in Miami (the only usable airfield in the area at the time).

In this case there was a breakdown in common ground between most of the dyads involved. The Captain and the second ATC had trouble modifying the flight plan due to a disrupting event. The Captain knew about the disrupting event, assumed that the ATC did as well, and thought that the ATC had complete, detailed weather information about the relevant portion of the airspace. The Captain assumed that the Joint Action Ladder had been scaled in his interaction with the second ATC, but that ATC did not understand the situation in the way the Captain expected. As a consequence, neither the Captain nor the ATC took the appropriate action.

The second ATC rejected the modified plan and did not take the weather conditions into account. The ATC had a different set of tasks, a different focus of attention (north-south routes on the east coast of Florida), and did not have complete data about the weather situation in the region the pilot was concerned about. Neither party realized that they were on different wavelengths and neither made any inquiries to verify that their assumptions about the other's knowledge, goals, and activities matched their own.

The third party to the plan modification, the dispatcher, had reached closure on this issue and did not realize that it had been reopened. Thus, the dispatcher was unable to participate by elaborating the intent behind the original modification.

Upon modifying the plan following the disruptive event, each of the three parties communicated with others and assumed that all were operating on a common assessment of the situation, that all had examined the issue at hand, and that all were working to the same subset of goals. Yet each of the parties misunderstood the others' perspectives, miscommunicated, and miscoordinated, leaving the flight crew to deal with the obstacle of the weather system—an obstacle that the replanning process had anticipated and tried to avoid.

Defections from the Basic Compact

When the partners in a joint activity agree to work together, they usually are not making a legally binding contract of infinite duration. At any point they may find that they have to abandon the activity, and often this is reasonable. Breakdowns occur when the abandonment is not signaled, with one set of partners continuing to believe that the Basic Compact is still in force while the party who has abandoned the joint activity believes it has become void.

Sometimes the partners will explicitly discuss the fact that they are abandoning the joint activity. But in most cases, the agreement itself was tacit and the partners depend on more subtle types of signaling to convey that they are no longer going to participate in the joint activity. When this signaling fails, the result is a form of Fundamental Common Ground Breakdown.

In the relay race, whether or not participants are remaining in the Basic Compact is visible and fairly unambiguous. The fact that a runner has abandoned the Basic Compact would most often be fairly obvious when, for example, he does not accept the baton at handoff.

Remaining in compact during a conversation is also highly visible in the process of accepting turns, relating understandings, detecting the need for and engaging in repair, displaying a posture of interest, and the like. When these sorts of things are not happening, it may be inferred that one or more of the parties is not wholeheartedly engaged.

On the highway, evidence of breakdown in the compact is often cued by deviations from norms or rules related to the safety of all, such as tailgating, cutting in too fast after passing, or not providing appropriate signals. In this sense, breakdown of the compact on the highway is perhaps easier to verify than its maintenance—the latter being consistent with "nothing unusual."

In football games, there are so many hidden coordination activities (e.g., messages and signals communicated across telephones, or spoken or displayed among players in the throes of the game) embedded in complex activity that knowing the state of the basic compact across the many parties can be quite difficult. The tighter the coupling—the degree to which one's action is dependent on the other—the more likely that lapses in the Basic Compact will show negative results and be noticed. For example, contrast the degree of interdependence between a quarterback and center to that between the two offensive tackles.

We have finished our discussion of the primary factors that affect joint activity. These factors include both support for coordination and the reasons for a loss of common ground and resultant breakdowns in coordination. In the next section we explore the implications of these factors for the design of automation.

MAKING AUTOMATION A TEAM PLAYER

The concept of automation—which began with the straightforward objective of replacing whenever feasible any task currently performed by a human with a machine that could do the same task better, faster, or cheaper—became one of the first issues to attract the notice of early human factors researchers. These researchers attempted to systematically characterize the general strengths and weaknesses of humans and machines (Fitts, 1951). The resulting discipline of *function allocation* aimed to provide a rational means of determining which system-level functions should be carried out by humans and which by machines.

Over time it became plain to researchers that things were not that simple. For example, many functions in complex systems are shared by humans and machines; hence the need to consider synergies and conflicts among the various performers of joint actions (Hoffman, Klein, & Laughery, 2002). Also, the suitability of a particular human or machine to take on a particular task may vary by time and over different situations; hence the need for methods of function allocation that are dynamic and adaptive (Hancock & Scallen, 1998). Moreover, it has become clear that function allocation is not a simple process of transferring responsibilities from one component to another (Boy, 1988). Automated assistance of whatever kind does not simply enhance our ability to perform the task: it changes the nature of the task itself (Christoffersen & Woods, 2002; Feltovich, Hoffman, Woods, & Roesler, 2004; Norman, 1992). Those who have had a five-year-old child help them by doing the dishes know this to be true—from the point of view of an adult, such help does not necessarily diminish the effort involved, it merely effects a transformation of the work from the physical action of washing the dishes to the cognitive task of monitoring the progress (and regress) of the child.

The ultimate desire of researchers and developers is to make automation a team player (Christoffersen & Woods, 2002; Malin et al., 1991). A great deal of the current work to determine how to build automated systems with sophisticated team player qualities is taking place within the software and robotic agent research communities, albeit in many forms and with somewhat divergent perspectives (Allen, 1999; Allen et al., 2000; 2001; 2002; Bradshaw et al., 1997; 2003; 2004a, b, c; in preparation; Christoffersen & Woods, 2002; Clancey, 2004; Cohen & Levesque, 1991; Grosz, 1996; Jennings, 1995; Loftin et al., 2005; Tambe et al., 1999). In contrast to early research that focused almost exclusively on how to make agents more autonomous, much of current agent research seeks to understand and satisfy requirements for the basic aspects of joint activity, either within multi-agent systems or as part of human-agent teamwork.

Clancey (2004) argues that the kind of coordinated joint activity we see in multi-agent systems or joint human-agent activities is of such a shallow nature—being based on artificially-construed goals rather than the rich realm of activities—that it is inappropriate to apply the terms "collaboration" or "teamwork" to them. He prefers to restrict his characterization of agents to the use of the term "assistant" and argues that true collaboration requires the kind of consciousness that allows them to have a personal *project,* not just a job, task, or problem. Few researchers would disagree with his point that human-machine interaction is currently of a very different nature than joint activity among humans. Although we are sympathetic to such arguments, we have decided to use these terms in this chapter because the concepts of teamwork and collaboration may be helpful in guiding automation development so that it becomes less likely to result in coordination surprises.

Researchers in human-agent teamwork have used the term in two broad ways: 1) as a conceptual analogy for heuristically directing research (e.g., to build systems that facilitate fluent, coordinated interaction between the human and agent elements of the system as "team players"), and 2) as the subject matter for research (e.g., to understand the nature of teamwork in people). The first activity focuses on practical engineering of useful systems through application of human-centered design principles, empirical studies of the use of these systems, and often a limited commitment to studying teamwork among people. The second activity is explicitly framed as a scientific study, and may have two angles: 1) providing information relevant to the design of successful human-agent systems, and 2) independent of application, understanding the nature of cognition, communication, and cooperation in people and animals. These researchers see the latter activity as essential for achieving the ultimate goals of artificial intelligence. Because the authors of this chapter are drawn from all of these traditions, our perspective attempts to reflect sensitivity to both: neither undervaluing the independent study of social and cognitive aspects of human teamwork, nor slavishly imitating superfluous aspects of natural systems in the development of artificial ones—like an engineer who insists that successful airplane designs must necessarily feature flapping wings because all birds have them (Ford & Hayes, 1998).

Given the widespread demand for increasing the effectiveness of team play for complex systems that work closely and collaboratively with people, a better understanding of the state of current agent research and observed shortfalls is important. In the remainder of this section we build on our previous examination of the nature of joint activity to describe six common ground requirements that are required to some degree for automation to be a team player. Automation in general, and agent implementations in particular, should be designed and evaluated with a view to their relative sophistication in satisfying each of these requirements.

The Basic Compact

To be a team player, an agent must fulfill the requirements of a Basic Compact to engage in common grounding activities. The Basic Compact is an agreement to work together in a coordinated fashion and to communicate events and changes in status that the other parties need to know in order to coordinate. Not only does the agent need to be able to enter into such a compact, it must also understand and accept the joint goals of the enterprise, understand and accept its roles in the collaboration, be capable of signaling if it is unable or unwilling to fully participate in the activity, and be capable of understanding other team members' signals of their status and changes in participation.

In the limited realm of what software agents can communicate and reason about among themselves, there has been some limited success in the development of theories and implementations of multi-agent cooperation. Teamwork has become the most widely accepted metaphor for describing the nature of such cooperation between software agents. The key concept usually involves some notion of shared knowledge, goals, and intentions that function as the glue that binds the software agents' activities together (Cohen & Levesque, 1991; Jennings, 1995; Tambe et al., 1999). By virtue of a largely reusable explicit formal model of shared "intentions," multiple software agents attempt to manage general responsibilities and commitments to each other in a coherent fashion that facilitates recovery when unanticipated problems arise. For example, a common occurrence in joint action is when a software agent fails and can no longer perform in its role. General-purpose teamwork models typically entail that each team member be notified under appropriate conditions of the failure, thus reducing the requirement for special-purpose exception handling mechanisms for each possible failure mode. In this way researchers have been using computational analogues of human joint activity and the Basic Compact to coordinate multiple software modules acting in parallel.

Addressing human-agent teamwork presents a new set of challenges and opportunities for agent researchers. No form of automation today or on the horizon is capable of entering fully into the rich forms of Basic Compact that are used among people. Thus, agents cannot be full-fledged members of human-agent teams in the same sense that other people are—a basic coordinative asymmetry between people and automata. By stretching, we can imagine in the future that some agents will be able to enter into a Basic Compact with diminished capability (Bradshaw, et al., 2004b, c; in preparation). They may eventually be fellow team members with humans in the way a young child can be—subject to the consequences of brittle and literal-minded interpretation of language and events, inability to appreciate or even attend effectively to key aspects of the interaction, poor anticipation, and insensitivity to nuance.

Consider the activity of a traveler who calls up navigation directions from a web-based service. These services have developed extensive maps to enable travelers to select a route. They can recommend a route and even let a traveler pick the fastest route, the shortest route, or a route that avoids highways. That seems so helpful. But a person needing directions may have a different goal—the least confusing route. The computerized aids don't know what that means. "Least confusing" depends on common ground. To provide a "least confusing" route to someone, it is necessary to appreciate what that person already knows of the area and what landmarks the person is likely to be able to recognize. It is necessary to engage with that person, to appreciate that person's perspective, to draw on common ground. Without this kind of capability software agents will not be able to enter fully into a Basic Compact with humans in these sorts of situations. In such situations the human's ability to appreciate nuance and understand the shortcomings of the machine agent will have to compensate for those aspects of joint activity that are missing in the interaction of machine agents (Clancey, 2004; Woods, Tittle, Feil, & Roesler, 2004).

Interpredictability

To be a team player, an agent has to be reasonably predictable, and has to have a reasonable ability to predict the actions of others. Similarly, one aspect of the

Basic Compact is a commitment to attempt to be predictable to others (i.e., to act neither capriciously nor unobservably), as in Grice's maxims of cooperation (Grice, 1975) and to be sensitive to the signals that others are sending (which often contribute to *their* predictability).

Thus, agents have three challenges for supporting interpredictability: 1) acting predictably and being directable, 2) signaling their status and intentions, and 3) interpreting signals that indicate the status and intentions of other team members. As with any team coordination arrangement, the signaling and interpretation aspects of interpredictability are reciprocal parts of the effort to increase predictability.

Acting Predictably and Being Directable. The intelligence and autonomy of machine agents directly works against the confidence that people have in their predictability. Although people will rapidly confide tasks to simple deterministic mechanisms whose design is artfully made transparent, they naturally are reluctant to trust complex software agents to the same degree (Bradshaw et al., 2004c). On the one hand, their autonomy and intelligence grants agents the flexibility and additional capability needed to handle situations that require more "wiggle room" than traditional software. On the other hand, their blindness to the limits of their competence, their non-transparent complexity, and their inadequate directability can be a formula for disaster (Billings, 1997).

In response to these concerns, agent researchers have increasingly focused on developing means for controlling aspects of agent autonomy in a fashion that can both be dynamically specified and humanly understood—directability. Policies are a means to dynamically regulate the behavior of a system without changing code or requiring the cooperation of the components being governed:

- Through policy, people can precisely express bounds on autonomous behavior in a way that is consistent with their appraisal of an agent's competence in a given context.

- Because policy enforcement is handled externally to the agent, malicious and buggy agents can no more exempt themselves from the constraints of policy than benevolent and well-specified ones can.

- The ability to change policies dynamically means that poorly performing agents can be immediately brought into compliance with corrective measures.

Policy-based approaches to regulation of machine agents have explicitly bound wiggle room, making those aspects of agent behavior about which people have special concern more predictable (Bradshaw et al., 2004a, b, c; Kagal, Finin, & Joshi, 2003; Myers & Morley, 2003).[2]

Signaling Status and Intentions. Agents in general must be able to signal their own status, including, for example, their current goals, stance, state of knowledge and upcoming actions to coordinate with others. Wiener (1989) has commented on the fact that the highest levels of automation on the flight deck of commercial jet transport aircraft (Flight Management Systems or FMS) often leave commercial pilots baffled, wondering what the automation is currently doing, why it is doing that, and what it is going to do next. These confusions demonstrate a failure to establish adequate interpredictability. In essence, the machine agent needs to make its own targets, changes, and upcoming actions externally accessible as people manage and re-direct these systems. This challenge runs counter to the advice that is sometimes given to automation developers to create systems that are barely noticed. We are asserting that people need to have a model of the machine as an agent participating in the joint activity. People can often effectively use their own thought processes as a basis for inferring the way their teammates are thinking. But this self-referential heuristic is not usually available in working with agents. In this regard, there is a growing concern that agents in human-agent teams should do a better job of communicating their current states, capacities, and intentions (e.g., Feltovich et al., in press; Norman, 1992).[3]

Consider this example from a study of pilot interaction with cockpit automation (Sarter & Woods, 2000).

Example 5. Having an Altitude Problem.

A pilot prepares his descent into his destination airport and receives an initial ATC clearance for an instrument landing approach to runway 24 L together with a number of altitude constraints for various waypoints of the arrival. The pilot programs these constraints into the flight management automation.

Shortly after the entire clearance has been programmed, an amended clearance is issued to now make an instrument landing approach to runway 24 R. (The change in runway was made because of an equipment failure). When the pilot changes the runway in the instructions to the automation, the automation signals it understands the runway change and begins to act based on this new target.

---

[2] Feltovich, Bradshaw, Jeffers, Suri., and Uszok (in press) argue that policies have important analogues in animal societies and human cultures that can be exploited in the design of artificial systems.

[3] To be sure, this hoped-for gain in adaptivity would mean some loss in predictability. Moreover, second-order issues of limited competence would no doubt now emerge at the level of the component doing the adjusting. Policies governing social aspects of agent interaction can be used to help assure that humans are kept properly appraised of agent state and intentions (Bradshaw 1997, Bradshaw et al., 2004a, b, c).

Question: Does the automation continue to use the altitude constraints *that still need to be respected*? The pilot assumes that the automation will remember the previously entered altitude constraint, but the system is designed to delete the previous entry when a new entry is made.

If the pilot were working with a human team member, it would be reasonable to expect that person to continue to apply the constraints—to "remember" the previous intent and know that it is still relevant to the situation. Both team members know that the altitude constraints apply to the current situation; and that the change in runways is not relevant to the altitude constraints. However, in this instance the automation doesn't behave this way—the software drops all of the altitude constraints entered for the originally planned approach. In addition, the only signal that these constraints have been dropped is through the disappearance of two indications that would be present if the constraints were still in the program.

The pilots may not understand or anticipate that the automation does not remember the constraints following a change but rather reverts to a baseline condition. This lack of interpredictability (and the lack of observability in the displays of automation activity) creates the conditions for an automation surprise—the automation will violate the altitude constraints as it flies the descent unless the human pilot notices and intervenes (in the simulation study 4 of 18 experienced pilots never understood or noticed the automation's behavior and 12 of 14 who noticed at some point were unable to recover before the constraints were violated). This illustrates how Fundamental Common Ground Breakdowns occur in human-automation interaction unless special measures are taken in design.

Interpreting Status and Intention Signals. Agents must be able to appreciate the signals given by human teammates. The ideal agent would grasp the significance of such things as pauses, rapid pacing, and public representations that help to mark the coordination activity. Olson and Sarter (2001) have described the way Flight Management Systems may obscure the downstream implications of current decisions. Pilots may believe they are making a choice at point A and not realize that their choice will have unexpected consequences later on, at point B (see also Feltovich, Hoffman et al., 2004). These kinds of automation surprises are the antithesis of team coordination. An ideal team member would instead identify the kinds of outcomes that a pilot might not anticipate, and alert the pilot to these implications.

Few agents are intended to read the signals of their operator teammates with any degree of substantial understanding, let alone nuance. As a result, the devices are unable to recognize the stance of the operator, much less appreciate the operator's knowledge, mental models, or goals.

With respect to the design of automation that monitors human state and changes based on the machine's assessment of that state, Billings (1997) and Woods (2002) have elucidated many fundamental concerns and have voiced their general skepticism about the value of this line of research. More generally, they are concerned about the basic asymmetry in coordinative competencies between people and machines. Given that asymmetry, the design of human-agent teams will always be difficult.

A few researchers are exploring ways to stretch the performance of agents in order to overcome this asymmetry, such as exploiting and integrating available channels of communication from the agent to the human, and conversely sensing and inferring cognitive state through a range of physiological measures of the human in real time so they can be used to tune agent behavior and thus enhance joint-human machine performance (Bradshaw et al., 2004a; Forsythe & Xavier, in press; Kass, Doyle, Raj, Andrasik, & Higgins, 2003; Raj, Bradshaw, Carff, Johnson, & Kulkarni, 2004). Similarly, a few research efforts are taking seriously the agent's need to interpret the physical environment (e.g., Rybski & Veloso, in press). Efforts such as these can help us appreciate the difficulty of this problem. For example, by making the machine agents more adaptable, we also make them less predictable.

We see this tradeoff with approaches to "adjustable autonomy" that enable policies to be adjusted as circumstances change, without requiring a human in the loop, essentially amounting to an automated way to "wiggle the bounds of the wiggle room" (e.g., Bradshaw, et al., 2004b, c; Falcone & Castelfranchi, in press; Maheswaran, Tambe, Varakantham, & Myers, 2003; Scerri, Pynadath, & Tambe, 2002). As Klein (2004) has pointed out, the more a system takes the initiative in adapting to the existing working style of its operator, the more reluctant operators may be to adapt their own behavior because of the confusions these adaptations might create. This is another example of how the asymmetry between people and agents creates a barrier to forming a full-fledged Basic Compact.

Goal Negotiation

To be a team player, an entity has to be able to enter into goal negotiation, particularly when the situation changes and the team has to adapt. Agents need to convey their current and potential goals so that operators can participate in the negotiations. Agents need to be readily re-programmable, to allow themselves to be re-directed, and to improve the resilience of the people who work with them to adapt to unexpected events.

Unlike the aviation automation described by Sarter and Woods (2000) and Olson and Sarter (2001), agents have to clearly announce their current intent and permit the operator to easily anticipate the consequences of making a change, the way any human team member would convey the implications of departing from a game plan. If agents are unable to readily represent, reason about, or modify their goals, they will interfere with common ground and coordination.

Traditional planning technologies for software and robotic agents typically take an *autonomy-centered* approach, with representations, mechanisms, and algorithms that have been designed to ingest a set of goals and output as if they can provide a complete plan that handles all situations. This approach is not compatible with what we know about optimal coordination in human-agent interaction. A *collaborative autonomy* approach, on the other hand, takes as a premise that people are working in conjunction with autonomous systems, and hence adopts the stance that the processes of understanding, problem solving, and task execution are necessarily incremental, subject to negotiation, and forever tentative (Bradshaw, et al., 2003; 2004d). Thus, a successful approach to collaborative autonomy will require that every element of the autonomous system be designed to facilitate the kind of give-and-take that quintessentially characterizes natural and effective teamwork among groups of people.

Allen's research on a Collaboration Management Agent (CMA) is a good initial example. It is designed to support human-agent, human-human, and agent-agent interaction and collaboration within mixed human-robotic teams (Allen et al., 2000; Allen et al., 2001; Allen & Ferguson, 2002). The CMA interacts with individual agents in order to: 1) maintain an overall picture of the current situation and status of the overall plan, as completely as possible based on available reports; 2) detect possible failures that become more likely as the plan execution evolves and to invoke replanning; 3) evaluate the viability of proposed changes to plans by agents; 4) manage replanning when situations exceed the capabilities of individual agents, including recruiting more capable agents to perform the replanning; 5) manage the re-tasking of agents when changes are made; and 6) adjust its communications to the capabilities of the agents (e.g., graphical interfaces work well for a human but wouldn't help most other agents). Because the agents will be in different states based on how much of their original plan they have executed, the CMA must support further negotiation and re-planning among team members while the plan is being executed. These sorts of capabilities should provide a foundation for more ambitious forms of goal negotiation in future agent research.

Coordination Phases

To be a team player, agents have to partner with humans in carrying out all of the coordination phases. In many cases, the human-agent coordination involves the handoff of information, from the human to the system or from the system to the human.

Clark described coordination as embodying phases that can be thought of as event patterns. This pattern consists of an entry, a body of action, and an exit. The exit portion of the phase is particularly interesting in considering automation issues; each member of the team has to provide evidence to the other members that a particular phase has been completed. Agent-based systems are sometimes lax when it comes to providing feedback to humans, especially about completing or abandoning tasks.

Common ground is created or lost during handoffs between team members (Patterson, Roth, Woods, Chow, & Gomes, 2004). Schrekenghost and her colleagues (Schreckenghost, Martin, Bonasso et al., 2003; Schreckenghost, Martin, & Thronesbery, 2003) have attempted to address some of these problems in their work on agent support for teams of operators. Their vision of future human-agent interaction is that of loosely coordinated groups of humans and agents. As capabilities and opportunities for autonomous operation grow in the future, agents will perform their tasks for increasingly long periods of time with only intermittent supervision. Most of the time routine operation is managed by the agents while the human crews perform other tasks. Occasionally, however, when unexpected problems or novel opportunities arise, people must assist the agents. Because of the loose nature of these groups, such communication and collaboration must proceed asynchronously and in a mixed-initiative manner. Humans must quickly come up to speed on situations with which they may have had little involvement for hours or days (Patterson & Woods, 2001). Then they must cooperate effectively and naturally with the agents. Schrekenghost's group has developed interesting capabilities for managing notification and situation awareness for the crewmembers in these situations. Also relevant is work on generic teamwork phases as it is being developed in an ongoing manner in Brahms and KAoS (Sierhuis et al., 2003; Bradshaw et al., 2004d).

Attention Management

In coordinated activity, team members help each other direct their attention to signals, activities and changes that are important (Moore & Dunham, 1995). Similarly, machine agents should refrain from working autonomously and silently, because this places the burden on the operators to discover changes. The other extreme doesn't help either—the case where a partner generates a large volume of low-level messages with little signal but a great deal of distracting noise.

The Basic Compact in part shows how responsible team members expend effort to appreciate what the other needs to notice, within the context of the task and the current situation. Similarly, open shared workspaces provide the means for one agent to notice where another's attention is directed, the activity ongoing, and their stance toward that activity (Carroll, Neale, Isenhour, Rosson, & McCrickard, 2003; Patterson & Woods, 2001; Woods, 1995).

To see the issues, take one example of a coordination breakdown between crews and flight deck automation—bumpy transfer of control (Sarter & Woods, 2000). Trouble begins and slowly builds, for example asymmetric lift worsens as wing icing develops or trouble in an engine slowly reduces performance. Automation can compensate for the trouble, but silently (Norman, 1990). Whether the automation is acting at all, a little, or working more and more to compensate is not visible (an example of a private workspace). Crews can remain unaware of the developing trouble until the automation nears the limits of its authority or capability to compensate. The crew may take over too late or be unprepared to handle the disturbance once they take over, resulting in a bumpy transfer of control and significant control excursions. This general problem has been a part of several incident and accident scenarios.

In contrast, in a well-coordinated human team, the active partner would comment on the unusual difficulty or increasing effort needed to keep the relevant parameters on target. Or, in an open environment, supervisors could notice the extra work or effort exerted by their partners and ask about the difficulty, investigate the problem, or intervene to achieve overall safety goals.

Notice how difficult it is to get the machine to communicate as fluently as a well-coordinated human team working in an open visible environment. The automation should signal (or its activities should be open and visible) to other agents so they can see:

- When it is having trouble handling the situation (e.g., turbulence);
- When it is taking extreme action or moving towards the extreme end of its range of authority.

These are quite interesting relational judgments about another agent's activities including: How do we tell when an agent is having trouble in performing a function, but not yet failing to perform? How and when does one effectively reveal or communicate that they are moving towards a limit of capability?

Adding threshold-crossing alarms is the usual answer to these questions in the design of machine agents. But this is a design dilemma as the thresholds are inevitably set too early (resulting in an agent that speaks up too often, too soon) or too late (resulting in an agent that is too silent, speaking up too little). For example, designers have added auditory warning that sounds whenever the automation is active but practitioners usually remove them because the signal is a nuisance, distraction, or false alarm. Sarter and her colleagues (Ho, Waters, & Sarter, in press; Sklar & Sarter, 1999) have developed a successful way to overcome these difficulties by designing tactile feedback to signal automation activities. The tactile cues are non-disruptive to other ongoing activities, yet allow the human partner to stay peripherally aware of automation changes and to focus only on those that are unusual given the context or the practitioner's assessment of the situation, thus reducing coordination surprises.

An important aspect of cooperative communications is gauging the interruptability of other practitioners. In a study of an emergency call-in center, Dugdale, Pavard, and Soubie (2000) found that directing another's attention depended on being able to see what the other agent is doing in order for one agent to be able to judge when another was interruptible. In other words, interruptability is a joint function of the new message and the ongoing activity. This requires one agent being able to see the activity of the other in enough detail to characterize the state of the other's activities—what line of reasoning are they on? Are they having trouble? Does their activity conform to your expectations about what the other should be doing at this stage of the task? Are they interruptible?

For example, Patterson, Watts-Perotti, and Woods (1999) observed the role of voice loops in mission control at Johnson Space Center. They noticed that controllers gauge the interruptability of practitioners outside their immediate team before communicating with them. When a controller needed to communicate with another controller working on a different subsystem, the controller would first listen in on that controller's voice loop. By listening to that loop, she could estimate the controller's current workload to judge how interruptible the controller would be in terms of the criticality of the issues that she or he is addressing. Using this strategy reduced the number of unnecessary interruptions and allowed controllers to judge the priority of their item against the ongoing work context. This reduced the occurrence of situations where a controller was forced to direct her attention away from current tasks in order to receive information about a lower priority item.

The Department of Defense has tried to develop a means of attention management in command posts described as a "common operating picture." The concept was to provide a single map display that showed everything on the battlefield location of friendly and enemy forces, aviation assets, topography, and so forth. The intent to provide common ground by presenting a single shared picture misses the active nature of building common ground.

To illustrate how a single picture is not the same as common ground, consider an exercise in which a Brigade Commander acted like an aide, in order to ensure that a staff member had seen a key piece of information on the display. During the exercise, a critical event occurred and was entered into the large screen display. The commander heard about it on his radio, and noted the change in the display. But he was not sure that one of his staff members had seen the change to the large screen display. Therefore, the commander made a radio call to the staff member to point out the event, as marked on the display. The commander made the call because he felt it was so important to manage the attention of his subordinate, and because the technology did not let him see if the staff member had noticed the event.

The common operating picture eliminated the feedback portion of the information handoff. As a result, the Brigade Commander could not infer whether the staff member had scaled the Joint Action Ladder, and he had to intervene personally to get confirmation. The Brigade Commander appreciated how the technology created a potential for a Fundamental Common Ground Breakdown and sought to prevent a coordination surprise.

A common picture is not necessarily a barrier to common ground. It can serve as a useful platform that enables distributed team members to calibrate the way they understand events. However, it does not automatically produce common ground. Even worse, the use of a common picture can create the conditions for a Fundamental Common Ground Breakdown, as illustrated by the example involving the Brigade Commander, because of the way it can interfere with attention management.

Controlling the Costs of Coordinated Activity

The Basic Compact commits people to coordinating with each other, and to incurring the costs of providing signals, improving predictability, monitoring the other's status, and so forth. All of these take time and energy. These coordination costs can easily get out of hand, and therefore the partners in a coordination transaction have to do what they reasonably can to keep coordination costs down. This is a tacit expectation—to try to achieve economy of effort.

Achieving coordination requires continuing investment and hence the power of the Basic Compact—a willingness to invest energy and accommodate to others, rather than just performing alone in one's narrow scope and sub-goals. The six challenges are ongoing investments. Coordination doesn't come for free; and coordination, once achieved, does not allow one to stop investing. Otherwise the coordination breaks down.

Agents can pose difficulties in serving as team members, and they can also impose penalties when team members are trying to coordinate with each other. Here, interaction with agents can increase the coordination costs of an information handoff. To find out if someone received an e-mail we might request a receipt, but in so doing we incur and inflict coordination costs of five or ten seconds per transaction from each person. To gauge the effect of an anesthetic on a patient, a physician may have to watch a computer monitor for thirty seconds after the initial physiological reaction becomes measurable. To trick the system into showing critical measures, the physician has to endure coordination costs for what used to be a simple information seeking action.

Keeping coordination costs down is partly a matter of good human-computer interface design. But it takes more than skillful design: the agents must be able to actively seek to conform to the needs of the operators, rather than requiring operators to adapt to them. Information handoff, which is a basic exchange in coordination phases involving humans and agents, depends on common ground and interpredictability. As we have seen, agents have to become more understandable and predictable, and more sensitive to the needs and knowledge of people.

The work of Horvitz (1999; Horvitz, Jacobs, & Hovel, 1999) provides one example (among the many that could be provided) of explicitly taking the cost of coordination into account in planning interaction with the user. For example, he has used measures of expected utility to evaluate the tradeoffs involved in potentially interrupting the ongoing activities of humans.

Concluding Remarks About the Challenges

The six challenges we have presented can be viewed in different lights. They can be seen as a blueprint for inspectors who want to evaluate automation. These challenges can be viewed as requirements for successful operation, so that the design of software and robotic agents and other kinds of automation are less likely to result in coordination breakdowns. The six challenges can also be viewed as a cautionary tale, ways that the technology can disrupt rather than support coordination. They can be taken as reminders of the importance of shared experience and expertise with regard to common ground. Simply relying on explicit procedures, such as common operating pictures, is not likely to be sufficient. The challenges can also be used to design team and organizational simulations that capture coordination breakdowns and other features of joint activity. Finally, the six challenges can be viewed as the underlying basis for human-agent systems.

SUMMARY

Based on our review of alternative analogies and previous results, we identified a set of central concepts for coordination in joint activity. The foundation for coordinated activity is the Basic Compact or intent to work together to align goals and to invest effort to sustain common interests. The Basic Compact reflects reciprocal commitment that is renewed and sustained. From this foundation, the criteria for engaging in a joint activity include the interdependence of the work performed by the participants, and the expectation that the joint activity will be resilient and adaptive to unexpected events. If these criteria are to be met, then the participants must meet requirements for making their actions predictable to each other, for sustaining common ground, and for being open to direction and redirection from each other as the activity unfolds. The choreography for carrying out these requirements involves coordinating a series of phases, and it is accomplished through employing various forms of signaling and the use of coordination devices, all of which incur coordination costs. Participants have to signal each other, and they also have to direct each other's attention to ensure that the signals are received, and to ensure that public events are noticed.

One of the key aspects of joint action is the process of sustaining common ground to enable coordination. Common ground is not a binary or constant feature—it is both continuous in its degree and constantly changing over time (cf. Feltovich et al., 1989; 2004, regarding the "reductive bias"). This includes the role of shared experience and expertise, and poses a limitation to those who think coordination can be manufactured through procedures and explicit guidelines.

Key aspects of common ground include:

- The types of knowledge, beliefs, and assumptions that are important for joint activity, including knowledge of roles and functions, standard routines, and so forth;
- Mechanisms for carrying out the grounding process: to prepare, monitor and sustain, catch and repair breakdowns; and
- The Basic Compact committing the parties in a joint activity to continually inspect and adjust common ground.

Common ground is likely to become degraded during team interactions, unless effort is put into calibrating the perspectives of the team members. We described the Fundamental Common Ground Breakdown as a paradigm for how team members can lose their calibration without knowing it, and continue to interact until they run into a coordination surprise.

Common ground is reflected in the amount of work needed in order to manage the communications for a joint activity. As common ground builds mutual knowledge, beliefs, and assumptions, participant's communications become coded and abbreviated, and economical. As common ground improves, the effort needed to clarify and explain should decrease. That is why effective teams are so careful to establish as much common ground as they can in advance of critical periods of activity. Furthermore, effective teams have also learned where to expect coordination breakdowns, and in preparation they elaborate common ground in these areas: e.g., clarifying the roles and functions of the participants, the goals and goal tradeoffs, the skills and competencies of the participants, and the preexisting differences in mental models.

Our examination of joint activity and coordination did not particularly focus on the use of technology to facilitate team coordination—the field of Computer Supported Collaborative Work (CSCW). Others are pursuing the application of Clark's work on joint activity and common ground to CSCW (e.g., Carroll et al., 2003). Since the costs of coordinative communication can be considerable, particularly for a large organization, another topic for additional investigation is how these workload costs are managed relative to the benefits of establishing and sustaining common ground.

This chapter should provide the reader with a deeper appreciation of the nuances of coordination, and some ideas about how to evaluate coordination mechanisms in an organization and in technologies. The concepts that we have discussed in this chapter should be useful for researchers who need to observe teams in action. The chapter may also be helpful for modelers trying to describe and simulate teamwork - perhaps as elements of organizational simulations - by identifying important variables, especially those pertinent to coordination. Finally, our account of joint activity and coordination should be informative to developers of automation, by serving as guidelines they can use for improving coordination in mixed human-agent work configurations.

ACKNOWLEDGEMENTS

We would like to thank Joan Feltovich for reading drafts and making suggestions that improved the manuscript. John Flach generously provided us with the results of his observational study of high school football coaches. Herb Bell, Donald Cox, and Dave Klinger provided us with very valuable and constructive critiques of earlier drafts. This research was supported by contracts to Klein Associates, Ohio State, and to IHMC through participation in the Advanced Decision Architectures Collaborative Technology Alliance, sponsored by the U.S. Army Research Laboratory under cooperative agreement DAAD19-01-2-0009.

REFERENCES

Allen, J., Byron, D. K., Dzikovska, M., Ferguson, G., Galescu, L., & Stent, A. (2000). An architecture for a generic dialogue shell. *Journal of Natural Language Engineering, 6*(3), 1-16.

Allen, J. F. (1999). Mixed-initiative interaction. *IEEE Intelligent Systems, 14*(5), 14-16.

Allen, J. F., Byron, D. K., Dzikovska, M., Ferguson, G., Galescu, L., & Stent, A. (2001). Towards conversational human-computer interaction. *AI Magazine, 22*(4), 27-35.

Allen, J. F., & Ferguson, G. (2002). Human-machine collaborative planning,. In *Proceedings of the NASA Planning and Scheduling Workshop*. Houston, TX.

Billings, C. E. (1997). *Aviation automation: The search for a human-centered approach*. Mahwah, NJ: Lawrence Erlbaum Associates.

Boy, G. A. (1988). *Cognitive function analysis*. Norwood, NJ: Ablex.

Bradshaw, J. M. (Ed.). (1997). *Software Agents*. Cambridge, MA: The AAAI Press/The MIT Press.

Bradshaw, J. M., Acquisti, A., Allen, J., Breedy, M., Bunch, L., Chambers, N., Galescu, L., Goodrich, M., Jeffers, R., Johnson, M., Jung, H., Lott, J., et al. (2004d). Teamwork-centered autonomy for extended human-agent interaction in space applications. In *Proceedings of the AAAI Spring Symposium* (pp. 136-140). Stanford, CA: The AAAI Press.

Bradshaw, J. M., Beautement, P., Breedy, M., Bunch, L., Drakunov, S. V., Feltovich, P. J., Hoffman, R. R., Jeffers, R., Johnson, M., Kulkarni, S., Lott, J., Raj, A., Suri, N., & Uszok, A. (2004a). Making agents acceptable to people. In N. Zhong & J. Liu (Eds.), *Intelligent Technologies for Information Analysis: Advances in Agents, Data Mining, and Statistical Learning*. Berlin: Springer Verlag.

Bradshaw, J. M., Boy, G., Durfee, E., Gruninger, M., Hexmoor, H., Suri, N., Tambe, M., Uschold, M., & Vitek, J. (Ed.). (in preparation). *Software Agents for the Warfighter. ITAC Consortium Report*.

Bradshaw, J. M., Feltovich, P. J., Jung, H., Kulkarni, S., Taysom, W., & Uszok, A. (2004b). Dimensions of adjustable autonomy and mixed-initiative interaction. In M. Klusch, G. Weiss & M. Rovatsos (Eds.), *Computational Autonomy*. Berlin, Germany: Springer-Verlag.

Bradshaw, J. M., Jung, H., Kulkarni, S., Allen, J. Bunch, L., Chambers, N., Feltovich, P., Galescu, L., Jeffers, R., Johnson, M., Taysom, W. & Uszok, A. (2004c). Toward trustworthy adjustable autonomy and mixed-initiative interaction in KAoS. Proceedings of the AAMAS 2004 Trust workshop, New York City, NY, July.

Bradshaw, J. M., Sierhaus, M., Acquisti, A., Feltovich, P., Hoffman, R., Jeffers, R., Prescott, D., Suri, N., Uszok, A., & Van Hoof, R. (2003). Adjustable autonomy and human-agent teamwork in practice: An interim report from space applications. In H. Hexmoor, C. Castelfranchi & R. Falcone (Eds.), *Agent autonomy* (pp. 243-280). Boston, MA: Kluwer Academic Press.

Brennan, S. E. (1998). The grounding problem in conversations with and through computers. In S. R. Fussel & R. J. Kreuz (Eds.), *Social and cognitive psychological approaches to interpersonal communication* (pp. 210-225). Mahwah, NJ: Lawrence Erlbaum Associates.

Cahn, J. E., & Brennan, S. E. (1999). A psychological model of grounding and repair in dialog. In *Proceedings of AAAI Fall Symposium on Psychological Models in Collaborative Systems* (pp. 25-33). North Falmouth, MA: American Assoc. for Artificial Intelligence.

Carroll, J. M., Neale, D. C., Isenhour, P. L., Rosson, M. B., & McCrickard, D. S. (2003). Notification and awareness: synchronizing task-oriented collaborative activity. *International Journal of Human-Computer Studies, 58*, 605-632.

Christoffersen, K., & Woods, D. D. (2002). How to make automated systems team players. *Advances in Human Performance and Cognitive Engineering Research, 2*, 1-12.

Clancey, W. B. (2004). Roles for agent assistants in field science: Understanding personal projects and collaboration. *IEEE Transactions on Systems, Man, and Cybernetics--Part C: Applications and Reviews, 32*(2), 125-137.

Clancey, W. J. (1997). The conceptual nature of knowledge, situations, and activities. In P. J. Feltovich, K. M. Ford & R. R. Hoffman (Eds.), *Expertise in context: Human and machine* (pp. 248-291). Menlo Park, CA: AAAI/MIT Press.

Clark, H. (1996). *Using language.* Cambridge: Cambridge University Press.

Clark, H. H., & Brennan, S. E. (1991). Grounding in communication. In L. B. Resnick, J. M. Levine & S. D. Teasley (Eds.), *Perspectives on socially shared cognition.* Washington: D.C.: American Psychological Association.

Cohen, P. R., & Levesque, H. J. (1991). Teamwork. *Nous, 25*, 487-512.

Cook, R. I., & Woods, D. D. (1994). Operating at the Sharp End. In M. S. Bogner (Ed.), *Human Error in Medicine.* Mahwah, NJ: Lawrence Erlbaum Associates.

Dismukes, K., Young, G., & Sumwalt, R. (1998). Cockpit Interruptions and Distractions: Effective Management Requires a Careful Balancing Act. *ASRS Directline.*

Dugdale, J., Pavard, B., & Soubie, J. L. (2000). A Pragmatic Development of a Computer Simulation of an Emergency Call Centre. In *Proceedings of COOP 2000, Fourth International Conference on the Design of Cooperative Systems.* Cannes, France.

Falcone, R., & Castelfranchi, C. (in press). Adjustable Social Autonomy. In J. Pitt (Ed.), *The Open Agent Society.* New York: John Wiley & Sons.

Feltovich, P. J., Bradshaw, J. M., Jeffers, R., Suri, N., & Uszok, A. (in press). Social order and adaptability in animal and human cultures as analogues for agent communities: Toward a policy-based approach. In A. Omacini, P. Petta & J. Pitt (Eds.), *Engineering societies in the agents world IV (Lecture Notes in Computer Science Series).* Heidelberg, Germany: Springer-Verlag.

Feltovich, P. J., Hoffman, R. R., Woods, D., & Roesler, A. (2004). Keeping it too simple: How the reductive tendency affects cognitive engineering. *IEEE Intelligent Systems, 19*(3), 90-94.

Feltovich, P. J., Spiro, R. J., & Coulson, R. L. (1989). The nature of conceptual understanding in biomedicine: The deep structure of complex ideas and the development of misconceptions. In D. Evans & V. Patel (Eds.), *Cognitive science in medicine: Biomedical modeling.* Cambridge, MA: MIT Press.

Feltovich, P. J., Spiro, R. J., Coulson, R. L., & Feltovich, J. (1996). Collaboration within and among minds: Mastering complexity, individually and in groups. In T. Koschmann (Ed.), *CSCL: Theory and practice of an emerging paradigm* (pp. 25-44). Mahwah, NJ: Lawrence Erlbaum Associates.

Fitts, P. M. (Ed.). (1951). *Human engineering for an effective air navigation and traffic control system.* Washington, D.C.: National Research Council.

Ford, K. M., & Hayes, P. J. (1998). On computational wings: Rethinking the goals of Artificial Intelligence. Scientific American. *Scientific American. Special issue on "Exploring Intelligence", 9*(4), 78-83.

Forsythe, C., & Xavier, P. (in press). Cognitive models to cognitive systems. In C. Forsythe, M. L. Bernold, & T. E. Goldsmith (Eds.), *Cognitive Systems: Cognitive Models in System Design*. Hillsdale, N.J.: Lawrence Erlbaum.

Grice, H.P. (1975). Logic and conversation. In P. Cole & J. Morgan (Eds), *Syntax and Semantics* (Vol. 3: Speech Acts). NY: Academic Press.

Grosz, B. (1996). Collaborative Systems. *AI Magazine, 2*(17), 67-85.

Hancock, P. A., & Scallen, S. F. (1998). Allocating functions in human-machine systems. In R. Hoffman, M. F. Sherrick & J. S. Warm (Eds.), *Viewing Psychology as a Whole* (pp. 509-540). Washington, D.C.: American Psychological Association.

Ho, C.-Y., M., N., Waters, M., & Sarter, N. B. (in press). Not now: supporting attention management by indicating the modality and urgency of pending task. *Human Factors*.

Hoffman, R. R., Klein, G., & Laughery, K. R. (2002). The state of cognitive systems engineering. *IEEE Intelligent Systems, 17*(1), 73-75.

Horvitz, E. (1999). Principles of mixed-initiative user interfaces. In *Proceedings of the ACM SIGCHI Conference on Human Factors in Computing Systems (CHI '99) held in Pittsburgh, PA*. Pittsburgh, PA: ACM Press.

Horvitz, E., Jacobs, A., & Hovel, D. (1999). Attention-sensitive alerting. In *Proceedings of the Conference on Uncertainty and Artificial Intelligence (UAI '99)* (pp. 305-313). Stockholm, Sweden.

Jennings, N. R. (1995). Controlling cooperative problem-solving in industrial multiagent systems using joint intentions. *Artificial Intelligence*, 75, 195-240.

Kagal, L., Finin, T., & Joshi, A. (2003). A policy language for pervasive systems. In *Proceedings of the Fourth IEEE International Workshop on Policies for Distributed Systems and Networks, June*. Lake Como, Italy.

Kass, S. J., Doyle, M., Raj, A. K., Andrasik, F., & Higgins, J. (2003). Intelligent adaptive automation for safer work environments. In J. C. Wallace & G. Chen (Co-Chairs) (Eds.), *Occupational health and safety: Encompassing personality, emotion, teams, and automation. Symposium conducted at the Society for Industrial and Organizational Psychology 18th Annual Conference*. Orlando, FL, April.

Klein, G. (2001). Features of team coordination. In M. McNeese, M. R. Endsley & E. Salas (Eds.), *New trends in cooperative activities* (pp. 68-95). Santa Monica, CA: HFES.

Klein, G. (2004). *The power of intuition*. New York: A Currency Book/Doubleday.

Klein, G., Armstrong, A., Woods, D. D., Gokulachandra, M., & Klein, H. A. (2000). *Cognitive wavelength: The role of common ground in distributed replanning* (Final Technical No. Report No. AFRL-HE-WP-TR-2001-0029). Wright-Patterson AFB, OH: United States Air Force Research Laboratory.

Klein, H. & McHugh, A.P. (2005). National Differences in Teamwork. In W.B. Rouse & K.R. Boff, Eds., *Organizational Simulation: From Modeling and Simulation to Games and Entertainment*. New York: Wiley.

Klinger, D. W., & Klein, G. (1999). Emergency response organizations: An accident waiting to happen. *Ergonomics In Design, 7*(3), 20-25.

Koschmann, T. D., LeBaron, C., Goodwin, C., & Feltovich, P. J. (2001). Dissecting common ground: Examining an instance of reference repair. In *Proceedings of the 23rd Conference of the Cognitive Science Society*. Mahwah, NJ: Lawrence Erlbaum Associates.

Loftin, R. B., Petty, M.D., Gaskins, R.C., & McKenzie, F.D. (2005). Modeling Crowd Behavior For Military Simulation Applications. In W.B. Rouse & K.R. Boff, Eds., *Organizational Simulation: From Modeling and Simulation to Games and Entertainment.* New York: Wiley.

Maheswaran, R. T., Tambe, M., Varakantham, P., & Myers, K. (2003). Adjustable Autonomy challenges in Personal Assistant Agents: A Position Paper. In *Proceedings of Computational Autonomy--Potential, Risks, Solutions (Autonomy 2003).* Melbourne, Australia.

Malin, J. T., Schreckenghost, D. L., Woods, D. D., Potter, S. S., Johannesen, L., Holloway, M., & Forbus, K. D. (1991). *Making intelligent systems team players: Case studies and design issues* (NASA Technical Memorandum 104738). Houston, TX: NASA Johnson Space Center.

Malone, T. W., & Crowston, K. (1994). The interdisciplinary study of coordination. *ACM Computing Surveys, 26*, 87-119.

Moore, C., & Dunham, P. (Eds.). (1995). *Joint attention: Its origins and role in development.* Hillsdale, NJ: Lawrence Erlbaum Associates.

Myers, K., & Morley, D. (2003). Directing agents. In H. Hexmoor, C. Castelfranchi & R. Falcone (Eds.), *Agent Autonomy* (pp. 143-162). Dordrecht, The Netherlands: Kluwer.

Norman, D. A. (1990). The "problem" with automation: Inappropriate feedback and interaction, not "over-automation." *Philosophical transactions of the Royal Society of London, 327*, 585-593.

Norman, D. A. (1992). Turn signals are the facial expressions of automobiles. In *Turn Signals Are the Facial Expressions of Automobiles.* (pp. 117-134). Reading, MA: Addison-Wesley.

Olson, G. M., Malone, T. W., & Smith, J. B. (2001). *Coordination theory and collaboration technology*. Mahwah, NJ: Lawrence Erlbaum Associates.

Olson, W. A., & Sarter, N. B. (2001). Management by consent in human-machine systems: When and why it breaks down. *Human Factors, 43*(2), 255-266.

Patterson, E. S., Roth, E. M., Woods, D. D., Chow, R., & Gomes, J. O. (2004). Handoff strategies in settings with high consequences for failure: Lessons for health care operations. *International Journal for Quality in Health Care, 16*(2), 125-132.

Patterson, E. S., Watts-Perotti, J. C., & Woods, D. D. (1999). Voice loops as coordination aids in Space Shuttle Mission Control. *Computer Supported Cooperative Work, 8*, 353-371.

Patterson, E. S., & Woods, D. D. (2001). Shift changes, updates, and the on-call model in space shuttle mission control. *Computer Supported Cooperative Work: The Journal of Collaborative Computing, 10*(3), 317-346.

Patterson, E. S., Woods, D. D., Sarter, N. B., & Watts-Perotti, J. C. (1998). Patterns in cooperative cognition. In *COOP '98, Third International Conference on the Design of Cooperative Systems*. Cannes, France.

Raj, A. K., Bradshaw, J. M., Carff, R. W., Johnson, M., & Kulkarni, S. (2004). An agent based approach for Aug Cog integration and interaction. In *Proceedings of Augmented Cognition-Improving Warfighter Information Intake Under Stress, Scientific Investigators Meeting*. Orlando, FL 6-8 Jan 04.

Rybski, P. E., & Veloso, M. M. (in press). Inferring human interactions from sparse visual data. In *Proceedings of the Autonomous Agents and Multi-Agent Systems Conference (AAMAS 2004)*. New York: ACM Press.

Salas, E., & Cannon-Bowers, J. A. (2001). The science of training: A decade of progress. *Annual Review of Psychology, 52*, 471-499.

Sarter, N., & Woods, D. D. (2000). Team Play with a Powerful and Independent Agent: A Full Mission Simulation. *Human Factors., 42*, 390-402.

Sarter, N. B., Woods, D. D., & Billings, C. (1997). Automation surprises. In G. Salvendy (Ed.), *Handbook of human factors/ergonomics* (Second ed.). New York: John Wiley & Sons, Inc.

Scerri, P., Pynadath, D., & Tambe, M. (2002). Towards adjustable autonomy for the real world. *Journal of AI Research (JAIR), 17*, 171-228.

Schreckenghost, D., Martin, C., Bonasso, P., Kortenkamp, D., Milam, T., & Thronesbery, C. (2003). Supporting group interaction among humans and autonomous agents. Submitted for publication.

Schreckenghost, D., Martin, C., & Thronesbery, C. (2003). Specifying organizational policies and individual preferences for human-software interaction. Submitted for publication.

Schaeffer, J. (1997). *One jump ahead: Challenging human supremacy in checkers*. New York: Springer-Verlag.

Shalin, V. L., Geddes, N. D., Bertram, D., Szczepkowski, M. A., & DuBois, D. (1997). Expertise in dynamic, physical task domains. In P. J. Feltovich, K. M. Ford & R. R. Hoffman (Eds.), *Expertise in context: Human and machine* (pp. 195-217). Menlo park, CA: AAAI/MIT Press.

Shattuck, L. G., & Woods, D. D. (2000). Communication of intent in military command and control systems. In C. McCann & R. Pigeau (Eds.), *The human in command: Exploring the modern military experience* (pp. 279-291). New York: Plenum Publishers.

Sierhuis, M., Bradshaw, J. M., Acquisti, A., Van Hoof, R., Jeffers, R., & Uszok, A. (2003). Human-agent teamwork and adjustable autonomy in practice. In *Proceedings of the Seventh International Symposium on Artificial Intelligence, Robotics and Automation in Space (i-SAIRAS), 19-23 May*. Nara, Japan.

Sklar, A. E., & Sarter, N. B. (1999). "Good Vibrations": The Use of Tactile Feedback In Support of Mode Awareness on Advanced Technology Aircraft. *Human Factors, 41*(4), 543-552.

Smith, P., Woods, D., McCoy, E., Billings, C., Sarter, N., R., D., & Dekker, S. (1998). Using forecasts of future incidents to evaluate future ATM system designs. *Air Traffic Control Quarterly, 6*(1), 71-85.

Smith, W. J. (1977). *The Behavior of Communicating*. Cambridge, MA: Harvard University Press.

Smith, W. J. (1995). The biological bases of social attunement. *Journal of Contemporary Legal Issues, 6.*

Spiro, R. J., Feltovich, P. J., Coulson, R. L., & Anderson, D. K. (1989). Multiple analogies for complex concepts: Antidotes for analogy-induced misconception in advanced knowledge acquisition. In S. Vosniadou & A. Ortony (Eds.), *Similarity and analogical reasoning* (pp. 498-531). Cambridge, England: Cambridge University Press.

Tambe, M., Shen, W., Mataric, M., Pynadath, D. V., Goldberg, D., Modi, P. J., Qiu, Z., & Salemi, B. (1999). Teamwork in cyberspace: Using TEAMCORE to make agents team-ready. In *Proceedings of the AAAI Spring Symposium on Agents in Cyberspace.* Menlo Park, CA: The AAAI Press.

Weick, K. E., Sutcliffe, K. M., & Obstfeld, D. (1999). Organizing for high reliability: Processes of collective mindfulness. *Research in Organizational Behavior, 21*, 13-81.

Wiener, E. L. (1989). *Human factors of advanced technology ("glass cockpit") transport aircraft* (No. NASA Report 177528). Moffett Field, CA: Ames Research Center.

Woods, D. D. (1995). The alarm problem and directed attention in dynamic fault management. *Ergonomics, 38*(11), 2371-2393.

Woods, D. D. (2002). *Steering the reverberations of technology change on fields of practice: Laws that govern cognitive work*, Proceedings of the 24th Annual Meeting of the Cognitive Science Society [Plenary Address] url: http://csel.eng.ohio-state.edu/laws

Woods, D. D., Tittle, J., Feil, M., & Roesler, A. (2004). Envisioning human-robot coordination in future operations. *IEEE SMC Part C, 34*(2), 210-218.

Zalesny, M. D., Salas, E., & Prince, C. (1995). Conceptual and measurement issues in coordination: Implications for team behavior and performance. In M. D. Zalesny, E. Salas & C. Prince (Eds.), *Research in personnel and human resources management* (Vol. 13, pp. 81-115). Greenwich, CT: JAI Press Inc.

CHAPTER 7

MODELING TEAM PERFORMANCE: THE BASIC INGREDIENTS AND RESEARCH NEEDS

EDUARDO SALAS, JOSEPH W. GUTHRIE, JR., KATHERINE A. WILSON-DONNELLY, HEATHER A. PRIEST AND C. SHAWN BURKE

ABSTRACT

Organizational simulations have been used and applied for some time; however, the modeling of team performance is still in its infancy. Although increasingly important, modeling team performance is more difficult than modeling individual performance because of the complexities and dynamics inherent in team performance. As such, this chapter provides a heuristic to aide practitioners and researchers in determining the 'must have ingredients' as well as those components which may be 'nice to have', but are not necessarily essential in modeling of team performance. We provide an extensive literature review of many of the factors that affect team performance, including individual characteristics, team characteristics, task characteristics, work structure, team competencies and environmental influences. In addition, we provide a brief overview of some of the mathematical and statistical methods that can be used to model team performance. Next, we discuss the impact that team performance modeling has on three areas, human system integration, scenario-based training and decision support systems. We conclude the chapter by identifying research areas that are critical to the progress of team performance modeling.

INTRODUCTION

Organizational simulations have been used for some time as a way to gain insights into the processes that occur within organizations, to identify any bottlenecks, and to improve the quality and efficiency of the organization (Fridsma & Thomsen, 1998). Organizational simulations have been useful in business, manufacturing, and engineering design tasks. For example, in the medical community organizational simulations have been used to determine performance effects of variables such as organizational structure, work processes and people (Fridsma & Thomsen, 1998).

While the earliest simulators, dating as far back as 1910, were basic and addressed more skill-based issues (http://www.bleep.demon.co.uk/SimHist1.html), simulation has evolved into a multi-dimensional training and assessment tool. From flight trainers first primarily used during World War II to present day virtual environments used to train US military ground troops, simulations are a critical aspect of organizations. Through the use of simulations and team training strategies, organizational teams (e.g., aviation cockpit crews, production teams), have become more efficient, productive and safe. For example, simulations used in the military today provide a wide range of mission based scenarios that help to equip soldiers with the proper competencies (i.e., knowledge, skills, and attitudes -- KSAs) to make difficult decisions in stressful situations.

Organizational simulations were originally used to model individual components within the organization. However, as the use of teams to achieve organizational goals has grown dramatically in the past decade, there has been a movement to use simulations to model team processes within organizations. Furthermore, the globalization of organizations has increased the use of both distributed and cross-cultural teams, changing the dynamics of team performance. Although much has been said about team performance, modeling and simulation in the literature, there is still much to be learned about the implementation of team performance models in organizations.

While using organizational simulations to model team performance is becoming increasingly important, to our knowledge, there is no single source that provides the key constructs, elements or ingredients to consider when modeling team performance. As such, the purpose of this chapter is four-fold. First, we will present a brief review of teams and team performance in organizations. Next, using a heuristics framework, we will discuss the "ingredients", -- those aspects that are needed or must be considered --, if one wants to model team performance. Third, we will discuss the impact and importance of modeling team performance on three areas: human system integration, scenario-based training, and decision support systems. Finally, we will conclude with some critical research needs.

WHAT DO WE KNOW ABOUT TEAMS AND TEAM PERFORMANCE?

Over the past two decades, a plethora of literature has been published regarding teams and team performance in organizations (Guzzo & Dickinson, 1996). Teams are being used to improve the quality and quantity of products, safety, national defense and a host of other organizational issues. But since all teams are not created equal, we start with a definition.

What Is a Team?

After decades of research, a widely accepted definition of a team has emerged. A team can be defined as consisting of two or more individuals working interdependently to achieve a shared goal (Salas, Dickinson, Converse, & Tannenbaum, 1992). Team members must interact dynamically, interdependently, and adaptively in order to perform effectively. The key characteristic that separates teams from groups (we argue) is the interdependence of its members. In order for a team to complete a task, the team must: (a) exchange information and resources dynamically among one another, (b) coordinate their task activities (e.g., active communication), (c) make adjustments as task demands require, and (d) organizationally structure its members. Finally, the element of time must be considered when looking at teams. While most teams have a limited life span of several hours (e.g., aircrews, surgical teams), others can last for days or months (e.g., military ground troops, advertising teams). Teams are fluid entities that engage in episodic performance requests (Marks, Mathieu & Zaccaro, 2001). But, what is teamwork? We discuss that next.

What Is Teamwork?

Teamwork is a complex, yet elegant phenomenon. Teamwork can be defined as a set of interrelated behaviors, actions, cognitions and attitudes that facilitate the required taskwork that must be completed. In order to coordinate and cooperate, team members must take their individual proficiencies and meld together to develop team-based KSAs. These team-based KSAs make up the team competencies that are needed for effective teamwork.

Teamwork and team performance is a direct result of team members holding required competencies. All team members and teams have competencies that enable them to perform effectively. These competencies are a set of interrelated, multi-level KSAs necessary for team function (Cannon-Bowers, Tannenbaum, Salas & Volpe, 1995). Simply put, these competencies are what team members "think, feel and do" to function in a team setting (Salas & Cannon-Bowers, 2000). Team competencies are the key to modeling team performance because they provide a glimpse into the inner workings of teams-- meaning that they are the engines that drive team performance. We will discuss in greater detail the specific team competencies needed for team performance modeling later in the chapter.

What Is Team Performance?

If team competencies are the engines that drive team performance, then team performance must be considered to be the outcomes and outputs of those processes. While there is not a clear definition of team performance, Hackman (1990) discussed three primary elements of team performance: (1) whether or not the team accomplishes its goals, (2) the satisfaction of members with the team and commitment to the team's goals and (3) the ability of the team to improve different facets of team effectiveness over time. These elements of team performance should be considered when modeling team performance.

A team's ability to accomplish its goals is not only dependent on whether or not they produce an output, product or a change at the prescribed time, but also whether or not that output/product meets the goals and expectations of the organization. Examples of other performance outputs include less team error, more team products, better decision-making and enhanced quality. These are easy outcomes of team performance to model. And these are the ones that can be quantified and measured.

Another outcome of team performance that is less tangible but just as important is team member attitudes. Team members can have a positive attitude towards the team, its members, its goals and the organization. This positive attitude or satisfaction will help to enhance each member's commitment to the team and its goals. As team members become more satisfied, for example, new levels of motivation are achieved and new efficient norms are established.

Finally, team performance is dependent upon the team's ability to improve over time (Marks et al., 2001). A high performing team will continue to develop better chemistry between its members, which will allow the team to improve in specific areas of performance that may have been lacking in the beginning, such as back-up behavior or closed-loop communication (McIntyre & Salas, 1995). Team performance is dynamic and constantly changing. That is why team processes are key ingredients in modeling team performance. We discuss these in more detail later. Beyond team processes, the moment-to-moment behaviors cannot be ignored in modeling.

What Is Team Performance Modeling?

Modeling human performance is a relatively new phenomenon that has resulted from the technological advances in computers and computer simulation. Up until the 1950's when computers became popular for commercial use, work on human performance modeling was restricted to quantitative models developed by researchers at universities (Zachary, Campbell, Laughery, Glenn & Cannon-Bowers, 2001). Since that time, the majority of research has focused on modeling individual human performance. These individual models of human performance have primarily focused on the underlying mechanisms of performance (Zachary et al., 2001), such as knowledge, problem solving and decision-making. Unlike the work done with individual models of human performance, the small amount of research that has focused on modeling performance at the group and team level has focused on the "constraints and structure on processes" (p. 204).

We believe that the research conducted on team performance modeling should follow a similar path to that of modeling individual performance in that we should focus on the underlying mechanisms that impact team performance (e.g. individual characteristics, task characteristics, work structure, team characteristics and team competencies). When determining the best methods to model team performance, it is important to consider both the individual and team components related to teamwork. The determination of which individual and team components are most important to a team's success is based on several factors. For example, the type of team (e.g. advisory or production, co-located or distributed) helps determine which characteristics of the individual and which components of the team will impact team performance. The type of task (e.g. additive or disjunctive) will also help to select which individual and team components will impact team performance. Also, work structure (e.g. work assignment, team norms and communication structure) helps to establish the individual and team components necessary for exceptional team performance. So, given the myriad of factors that affect team performance, what are the key ingredients for modeling team performance? We describe some of the key ingredients necessary for modeling team performance next. We offer these ingredients as essential elements. We are well aware that the state of the art in modeling team performance is still in its infancy. These elements are then a guide, a reference, a point of departure where the modeler can select what ingredients are to be included based on need.

A FRAMEWORK FOR MODELING TEAM PERFORMANCE

Thus, team performance is complex, dynamic, and elusive. As is evident by the number of variables that impact team performance, the modeling of such aspects is challenging. This chapter is organized around key ingredients needed to effectively model team performance in organizations. The organizing framework — a heuristic-- provided in Figure 1 serves to guide the remainder of this chapter. First we will discuss the factors that serve as inputs to a team performance model. Specifically, we will discuss factors that relate to the individual (i.e., what each member brings to the team), the team (i.e., competencies and processes), the type of task being performed, the work structure, and the modeling techniques available. Finally, we offer three examples of how team performance modeling can help organizational simulations—human system integration, scenario-based training, and decision support systems (see Figure 1). Using the framework developed to guide this chapter, we will next discuss the key ingredients for modeling team performance.

Individual Characteristics

We will first explore some of the characteristics that individuals may hold that will impact the way they interact and perform within teams; culture, attitudes, personality, cognitive ability, motivation, mental models and expertise. It should be noted that many of these factors are related and therefore multiple factors may need to be considered during the modeling process. Individual characteristics are a critical component of team performance modeling because they impact team performance as well as interactions between team members. As such, each of these factors should be considered when modeling team performance.

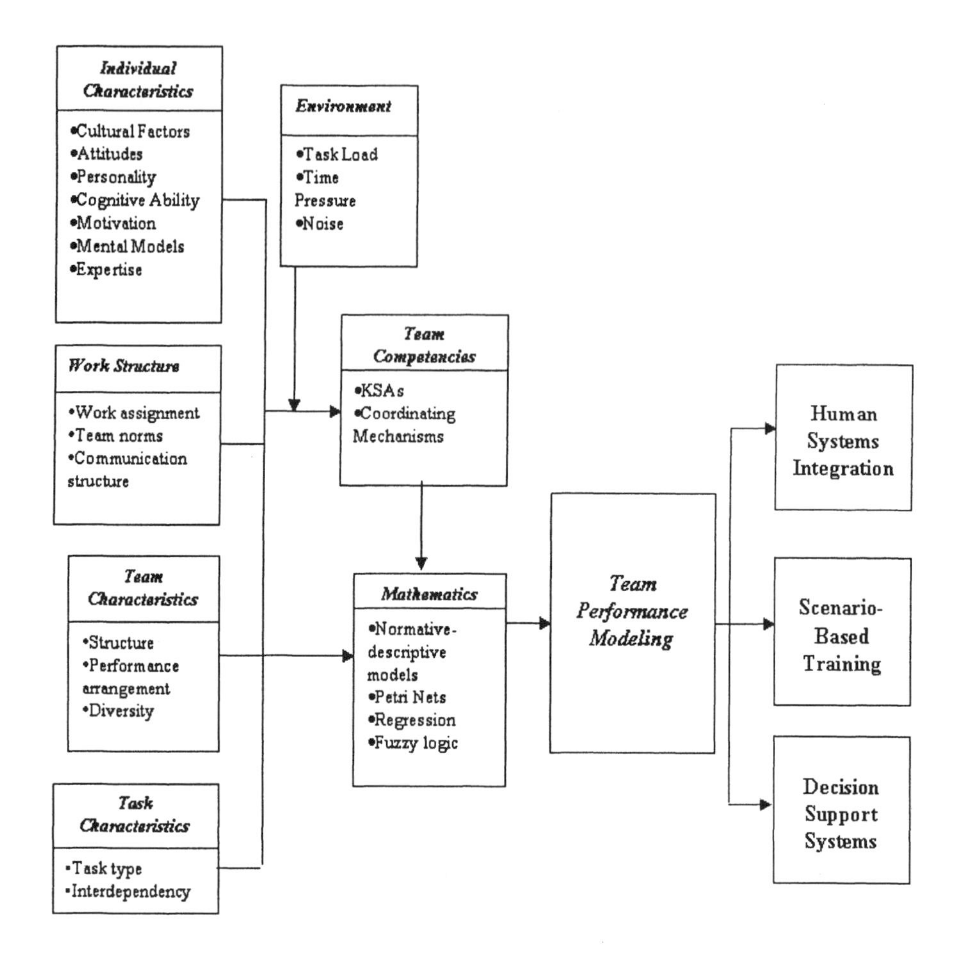

FIGURE 1. Elements for Team Performance Modeling

Cultural factors

It has been suggested that national culture impacts social behaviors such as aggression, conflict resolution, social distance, helping, dominance, conformity, obedience, decision-making, and leadership behaviors (Hambrick, Davison, Snell, & Snow, 1998; Hofstede, 1980; Schneider & De Meyer, 1991; Shane, 1994; Triandis, 1994), ultimately impacting team performance. Furthermore, culture may guide cognitive approaches and influence individual choices, commitments and standards of behavior (Erez & Earley, 1993). A number of cultural factors have been discussed in the literature. For example, Geert Hofstede's (1980) cultural dimensions (i.e., individualism-collectivism, power distance, uncertainty avoidance, masculinity-femininity) have received the greatest attention. However, there are additional cultural factors found in the literature that may impact team performance as well (i.e., field dependence-independence, high-low context, analytic-holistic reasoning). It has been suggested in the literature that the two most influential cultural dimensions on team performance are collectivism and field independence (Earley & Gibson, 2002; Erez & Earley, 1993; Triandis, 1989). However, the lack of research looking at the impact of culture on team performance suggests that others may have an impact as well. While extensive research has examined a number of these cultural dimensions, more research is needed on how variations within cultures influence the expression of certain group norms (Cox, Lobel, & McCloud, 1991). Nevertheless, research investigating the impact of culture on team performance suggests culture does in fact have an impact and therefore is a factor to consider when modeling team performance (Earley & Gibson, 2002). For example, while individuals may take different approaches to complete a task due to their ethnic differences, national cultural differences may not be observed as frequently in teams that are highly populated by one particular nationality (Cox et al., 1991).

Attitudes

Individuals coming into a team will each have differing attitudes, some of which will impact team performance. For example, if an individual had a poor team experience in the past, his/her attitude towards working in a team may not be positive. On the other hand, if his/her previous team experiences have always been positive, the attitudes will likely be more positive as well. These attitudes in conjunction with our beliefs about a situation will influence our expectations for the team (Pew & Mavor, 1998). Furthermore, these attitudes in combination with job characteristics (e.g., rewards, structure) will likely impact the individual's motivation to perform on the team. For example, an individual who has a negative attitude towards teams yet is required to work in a team may show resentment and lack of motivation to perform.

Personality

It has been reported that team members' personalities can have both a positive and a negative effect on team performance (Hackman & Morris, 1975). While some researchers have said that there is not sufficient conclusive evidence regarding the impact of personality on team performance to use it as a factor when developing a team (Kahan, Webb, Shavelson, & Stolzenberg, 1985), it has been shown to impact team performance in a variety of ways. For example, personality variables have been said to influence decision-making (Janis & Mann, 1977) and other behaviors in an organizational context (e.g., Heise, 1992). Five personality traits are commonly used to characterize individual personalities: openness (e.g., creative, curious, broad interests), conscientiousness (e.g., organized, reliable, honest), extroversion (e.g., sociable, optimistic, talkative), agreeableness (e.g., trusting, helpful, forgiving), and neuroticism (e.g., worrying, nervous, insecure). Janis (1989) suggests that three of the five traits (openness, conscientiousness, and neuroticism) are essential for individuals in command-like positions because an individual's scores on these traits can predict whether the individual will fail or not when making executive decisions. This being said, depending on the size of the group (e.g., large or very large) or the nature of the task (e.g., command and control) it has also been suggested that the impact of personalities may be counteracted by other personalities or the constraints of the task, respectively (Pew & Mavor, 1998).

Cognitive ability

Cognitive ability can be referred to as an individual's capacity to process and learn information (Kanfer & Ackerman, 1989). Individuals have differing cognitive abilities and research suggests that cognitive ability has an impact on job performance (e.g., Hunter, 1986) and likely team performance. In addition, individuals with higher cognitive abilities can more readily adapt to novel situations (e.g., Hartigan & Wigdor, 1989), an important factor for teams operating in complex environments. Lepine, Hanson, Borman, & Motowidlo (2000) and others (e.g., Hackman, 1987) suggest that high cognitive ability of team members leads to greater role effectiveness and ultimately higher team performance if the team can integrate their roles in a competent manner. A recent study suggests that teams whose members have higher cognitive ability, (among a few other things) can more easily adapt to changes in their task context (Lepine, 2003).

Motivation

The notion of motivation can be thought of in general terms as "Why does a person do X?" For example, "why does a person participate in the team?" Motivation influences the amount of effort that a team member is willing to exert to accomplish a team task (Ilgen, Shapiro, Salas, & Weiss, 1987). Individual motivation to perform in the team is said to influence team performance (Swezey & Llaneras, 1997).

An individual's motivation to perform is also influenced by the difficulty of the task and team size. Specifically, more challenging task goals leading to higher levels of motivation (e.g., Gladstein, 1984). The size of the team is a complicating factor in that a large team increases the team's resources and thus improves the team's motivation to accomplish its goals (Shaw, 1976). On the other hand, a large team may decrease motivation due to effects such as social loafing (e.g., Karau & Williams, 1993).

Mental models

Also influencing team performance are the mental models that individuals bring to the team. A mental model can be defined as a "mechanism whereby humans generate descriptions of system purpose and form, explanations of system functioning and observed system states, and predictions of future system states" (Rouse & Morris, 1986, p. 360). This system can be either a technological or human (e.g., team) system. Individual mental models are developed through past experiences and influence what we perceive in future situations (Wickens & Carswell, 1997) and how we will act (Senge, 1990). For example, if a team member has participated in a team activity in which one team member dominated the group and others were not encouraged to participate, this will influence how that team member views the new team and his/her participation. As additional experiences are gained, these mental models may be altered or new mental models generated.

Expertise

Expertise has been defined in short as knowing how to do something well (see Smith, Ford, & Kozlowski, 1997). This can be further expanded to suggest that expertise is dependent on detailed knowledge about a specific domain and the ability of the expert to represent and understand problems that occur in that domain. This allows experts to quickly apply solutions or strategies to well-learned, familiar, and routine tasks. Teams in organizations are often comprised of members with varying expertise, requiring input from each of the team members to accomplish the team's goals. As such, team members must rely on one another to supply the requisite knowledge or skills so that performance will not decline. Experts have an increased advantage over novices because they are better able to update their knowledge base, which allows for their mental models to be revised more accurately (Burke, 1999).

Team Characteristics

Beyond individual characteristics, modeling team performance must also include team characteristics. Team characteristics to consider include, but are not limited to, team structure, cohesiveness, team size, team climate and team diversity. These factors should be considered in modeling team performance because they impact the importance of the different team competencies (discussed later) and aspects of work structure and task characteristics. First, we describe different types of teams according to a typology developed by Sundstrom, DeMeuse, and Futrell (1990). The description of types of teams is followed by a discussion of the characteristics of teams mentioned above in order to provide a better understanding of the way teams function.

Types of teams

There are a number of different teams used by organizations to help achieve its goals. While there are a number of team typologies, one of the most widely accepted will be discussed here. Sundstrom and colleagues (1990; 2000) argue that there are several types of teams commonly used in organizations—production and service teams, management teams, project teams, action and performing teams, and advisory teams. As all teams are not created equal, it is important to examine each of these teams individually to better understand their characteristics to more accurately model their performance. We will next briefly discuss each of the teams identified by Sundstrom and colleagues.

Production and Service Teams. Production and service teams are those in which front-line employees produce outputs on a continual basis. While production teams (e.g., assembly line workers) interact with technology to produce tangible outputs (e.g., automobiles, computer microchips), service teams (e.g., flight attendants) interact with people (i.e., customers) and produce both tangible and intangible outputs or services. These teams range in the amount of autonomy they are given. Autonomous or self-managing work teams are defined as teams that give members more responsibility and independence from management in that they "manage themselves, assign jobs, plan and schedule work, make production- or service-related decisions, and take action on problems" (Kirkman & Shapiro, 2001, p. 557). Semi-autonomous teams are those that are led by a supervisor who is higher in rank than other team members (Sundstrom et al., 2000). While some research suggests that higher levels of autonomy may result in greater levels of job satisfaction and organizational commitment (e.g., Cordery, Mueller, & Smith, 1991), the level of autonomy varies by the work required of the team. For example, tasks conducted by coal mining crews are generally led by a supervisor who makes decisions for the team (Sundstrom et al., 2000). On the other hand, some automobile manufacturers allow team members to self-direct themselves with limited supervision from management outside of the team as long as they reach a predetermined quota (Katz & Kahn, 1978).

Management Teams. Management teams consist of members who are at higher levels of the organization including executive or senior-level managers and those managers who report to them (Sundstrom et al., 2000). The expertise of the team members allows management teams to perform tasks critical to the success of the organization (e.g., policy making, staffing). These top management teams are becoming increasingly more common as organizations expand to global markets (Cohen & Bailey, 1997). Based on the decisions made by management teams, direction is provided to those units that are under their jurisdiction.

Project Teams. Project teams generally consist of white-collar professionals (e.g., researchers, programmers, engineers) that have been organized to collaborate on a particular project (e.g., new product development; Sundstrom et al., 1990; 2000). These teams are cross-functional and team members are assembled from various units within the organization. Their work cycles are generally longer than that of service and production teams and their outputs are more complex and "one-of-a-kind" (e.g., research studies; Sundstrom et al., 1990). Project teams are time-limited and, hence, once the project is completed the team will split up.

Action and Performing Teams. Members of action and performing teams are high in domain expertise (e.g., cockpit crews, military teams, surgical teams, negotiation teams) whose roles are specialized yet interdependent (Sundstrom et al., 1990; 2000). These teams receive specific training to improve teamwork performance in the complex environments in which they operate. Additionally, action and performing teams must remain flexible and adaptive to the dynamic and often ambiguous environment surrounding them. Their success is critical as failure can result in catastrophic consequences (e.g., loss of life, monetary losses).

Advisory Teams. Finally, advisory teams are temporary teams assigned by managers to solve a particular problem and offer corrective solutions (Sundstrom et al., 1990; 2000). These teams include quality control circles, selection committees and employee participation groups. The team's lifespan is typically short and/or the amount of time that the team works together is limited (e.g., several hours per month). Advisory teams have become especially popular in organizations since the 1980s (Lawler, Mohrman, & Ledford, 1998).

Team Structure

Each of the types of teams discussed above can be classified as co-located or distributed. With co-located teams, all team members are in the same physical space (e.g. cockpit crews, surgery teams, assembly teams, etc.). Distributed teams, sometimes called virtual teams, are defined as teams whose members are mediated by time, space, or technology (Driskell, Radtke & Salas, 2003). A single team member or the entire team may be physically located in different offices, buildings, time zones or even continents. There a number of benefits of using distributed teams in organizations. Distributed teams are desirable to organizations because they save time and money on transportation to meetings and workplaces, resulting in increased employee satisfaction because they spend less time traveling. In addition, the distribution of employees in different time zones and locations means that someone is always working for the organization. Organizations are also able to get the best people for the job without having to uproot employees so they can be co-located.

Distributive performance arrangements can have a significant effect on performance because of the 'space between' team members (Fiore, Cuevas, Scielzo & Salas, 2002). The distance and lack of familiarity between team members changes how they think, feel, and act. For example, distribution can affect the shared mental models of team members, creating less consensus than between co-located team members. Research has also shown more pronounced problems with conflict, shared identity, leadership, and decision-making between distributed team members (Bell & Kozlowski, 2002; Mortensen & Hinds, 2001). Fiore and colleagues even coined the term 'team opacity' to explain the additional workload distributed teams experience because of a lack of cues, present when teams have additional behavioral and contextual cues (i.e., when the team is co-located or face-to-face). Therefore, team performance models should consider these issues when providing training to distributed team members with additional KSAs that will prepare them to compensate for deficiencies based on distribution and emphasize strengths inherent to these teams.

Cohesiveness

Cohesion was initially defined by Festinger (1950) as "the resultant forces that are acting on the members to stay in the group" (p. 274), but has come to be accepted as "a dynamic process that is reflected in the tendency of a group to stick together and remain united in its pursuit of its instrumental objectives and/or for the satisfaction of member affective needs" (Carron, Brawley, & Widmeyer, 1988, p. 213). Dating back several decades, team cohesion has been found to influence team performance (Seashore, 1955; Stogdill, 1972; Swezey, Meltzer, & Salas, 1994; Wech, Mossholder, Steel, & Bennett, 1998). For example, Barrick, Stewart, Neubert and Mount (1998) positively linked cohesion and team performance, calling cohesion a "general indicator of synergistic group interaction—or processes" (p. 382).

Factors that influence the extent or degree of team cohesiveness include member similarity, external challenges, shared goals, member interaction, team size and team success (McShane & Von Glinow, 2000). If teams are cohesive, members tend to want to remain a part of the team, have strong interpersonal bonds, are more willing to share information, tend to have more enjoyment and less stress, and resolve conflict effectively.

Team Size

The size of the team can also affect how team members perform and interact (Curral, Forrester, Dawson & West, 2001). There is some variability in the research as to whether teams should be small or large. While it was once argued that a team size of 8 to 10 members was ideal, some suggest that a team size of 5 to 7 members is best for complex tasks (e.g., Shaw, 1981). Much of the research suggests that teams that are too large can be detrimental to performance. For example, teams of more than 10 members often result in the team dividing into sub-teams to interact and accomplish the task (Likert, 1977). As a result, larger teams may exhibit process loss, have less clear objectives, lower levels of participation, and less emphasis on quality as team size increases (e.g., Poulton & West, 1999; Steiner, 1972). Furthermore, others suggest that as team size increases, there is a proportional increase in the number of problems that arise (Seaman, 1981). Finally, larger teams, especially when performing simple tasks, may result in social loafing by team members. It is suggested, however, that this effect may be nullified as task complexity increases (Jackson & Williams, 1985) and/or when group size decreases (Harkins & Szymanski, 1988). Conversely, research into small teams does not suggest optimal results either. For example, Jackson (1996) has found that very small teams (i.e., 2 to 3 team members) lack diversity and do not have the different viewpoints that are often seen as benefits of teams. As can be seen, the size of the team will impact the team's effectiveness and thus how performance should be modeled. It is therefore important that modelers consider not only the size of the team but also the task that these teams must perform when developing the model.

Team Climate

Team climate is referred to as an 'ambient' stimulus, meaning that it reflects the nature of a whole group rather than a particular member of the group or team (Hackman, 1992). Simply put, team climate is not based in, or a symptom of, one team member, but is based on overall interaction patterns of the entire team. Therefore, the nature of the climate of any team can have multiple effects at a number of levels. Research has found that team climate has a significant effect on team behaviors because climate is basically a cognitive representation of the team based on members' own individual interactions within the team (e.g., Anderson & West, 1998; Schneider & Reichers, 1983). In addition, Edmondson (1999) found perceptions of team climate to be positively related to changes in self-efficacy. This can lead team members to be more open to new ways of doing their tasks, more open to feedback exchange and interaction without fear of appraisal.

Team Diversity

A team's composition will also influence how effectively they perform. According to the literature there is a potential for both positive and negative impact as a result of team diversity. Team composition is often a confusing and conflictive area of study, which has been referred to as both "conceptually scattered" (McGrath,

1998, p. 594) and "atheoretical" (Levine & Moreland, 1990, p. 594). One reason for this is the varying types of heterogeneity that may exist in a team (e.g., personality, cultural, ability, expertise). Another factor that influences the effects of team heterogeneity is the nature of the tasks. Mohammed and Angell (2003) found that, within heterogeneous teams, performance varied based on whether they were doing a written or oral task. Furthermore, Sessa and Jackson (1995) found the impact of diversity varies with the complexity of the task. Other research found positive influences of diversity including the availability of more diverse information, different viewpoints, richer discussions, and more complete analysis of information (Chatman, Polzer, Barsade, & Neale, 1998; Nemeth, 1986; Simons, Pelled, & Smith, 1999).

Team Competencies

Up to this point we have covered individual characteristics and a number of team characteristics that are important when modeling because they impact the way that teams work together. Next, we consider the competencies that these inputs are acting upon (i.e. teamwork). Team competencies are the knowledge, skills, and attitudes (KSAs) needed to be an effective team member (Cannon-Bowers & Salas, 1997). Thus, teamwork can be defined as the ability of the team to coordinate and cooperatively interact. It has been argued that there are five core components of teamwork—team leadership, mutual performance monitoring, back up behavior, adaptability/flexibility, and team orientation (Salas, Sims, & Burke, 2004). Additionally, there are three coordinating mechanisms—shared mental models, closed-loop communication, and mutual trust. Because these KSAs are a critical component to all teams, they are therefore, critical to successful team performance modeling.

Team Leadership. As the name suggests, part of the team leader's role is to lead the team, however in reality, the team leader does so much more. The team leader also serves as a model of teamwork. While a number of leadership theories exist (Bass, 1990) and research regarding teams is extensive, lacking is a synthesis between the two so as to form a theory of team leadership. Team leaders are particularly important for promoting the dynamic throughput processes, which comprise teamwork and facilitate adaptive performance (Cannon-Bowers et al., 1995; Kozlowski, Gully, Nason, & Smith, 1999; McIntyre & Salas, 1995). For example, team leaders drive team performance and effectiveness by communicating a clear direction, creating enabling performance environments and by providing process coaching (Hackman & Walton, 1986). Furthermore, team leaders serve to train and develop team member's individual and team level competencies (Kozlowski, Gully, Salas & Cannon-Bowers, 1996). In addition, research suggests that the importance of leadership increases as the problems faced increase in complexity (Jacobs & Jaques, 1987; Zaccaro, Rittman, Orvis, Marks, & Mathieu, 2002).

Mutual Performance Monitoring. In order to perform effectively, teams must be aware of how the team is functioning. Team members must monitor the work of their teammates. This is called mutual performance monitoring (MPM). The goal of MPM is to catch any mistakes, slips or lapses before, or at least shortly after, they occur. Specifically, MPM is the ability to, "keep track of fellow team member's work while carrying out their own...to ensure that everything is running as expected and ... to ensure that they are following procedures correctly..." (McIntyre & Salas, 1995, p. 23). In addition, for MPM to be effective, team members must agree that their fellow members can provide effective assistance and feedback. MPM is influenced by other team competencies. For example, high team orientation (discussed in a later section) contributes to team members' abilities and willingness to give and accept feedback and help. Successful MPM leads to team backup behavior, which is discussed next.

Back-up Behavior. Another component of teamwork that must be considered when modeling is backup behavior, defined as "the discretionary provision of resources and task-related effort to another...that is intended to help that team member obtain the goals as defined by his or her role when...there is recognition by potential back-up providers that there is a workload distribution problem in their team" (Porter, Hollenbeck, Ilgen, Ellis, West & Moon 2003, pp. 391-392). Furthermore, this competency has been empirically linked with effective team performance (Porter et al., 2003). The goal of backup behavior is effective performance (1) by providing feedback and coaching to improve performance, (2) assisting the teammate in performing a task, and (3) completing a task for the team member when an overload is detected. However, backup behavior will not be present unless it is preceded by other competencies like shared mental models and mutual performance monitoring. For more information on backup behavior see McIntyre and Salas (1995) and Marks, Mathieu, and Zaccaro (2000).

Adaptability. Another important aspect of team performance is adaptability. This may be the most difficult competency to model because it requires the constant reassessment and evaluation of the environment, team tasks and team processes. However, MPM and backup behavior go a long way in making a team adaptable. Adaptability can be defined as the ability to recognize deviations from expected action and readjust actions accordingly (Priest, Burke, Munim, Salas, 2002) or to "adapt their strategies according to the particular task demands at hand..." (Cannon-Bowers et al., 1995, p. 360). It is important to note, however, that detecting changes and adjusting behavior is not enough to ensure successful performance. Teams must select the appropriate strategies that will help them adjust to the dynamic situation. While adaptability is a goal of teams, and therefore may be considered an output, it is also considered a process in that it enables a team to perform successfully (Kendall, Salas, Burke & Stagl, 2003).

Team Orientation. While the previous competencies have all been behavioral, team orientation is also important to performance modeling. Team orientation is defined as a preference for working with others and the tendency to enhance individual performance through the coordination, evaluation, and utilization of task inputs from other group members while performing group tasks (Driskell & Salas, 1992). Team members who have a team orientation perform more effectively in a team environment than individuals who have an egocentric or individualistic orientation (i.e., they prefer to work alone; e.g., Driskell & Salas, 1992). Research found that team orientation is influenced by several factors, including past experiences in teams, the perceived ability to complete the task, and expected positive outcomes (e.g., Bandura, 1991; Eby & Dobbins, 1997; Loher, Vancouver, & Czajka, 1994; Vancouver & Ilgen, 1989).

Shared Mental Models (SMM). As discussed earlier, behavior and orientation, while important, are not the only necessary competencies of teamwork. Knowledge or cognition at the team level influences how teams perform. A cognitive component of teamwork is shared mental models. Interdependency demands that individual team members coordinate, which includes anticipating and predicting the actions of other team members. In order to coordinate, team members must have a common or shared knowledge structure or understanding of their environment and roles. Shared mental models enable team members to coordinate. Specifically, shared mental models are defined as the shared understanding or representation of team goals, individual team member tasks, and how the team will coordinate to achieve their common goals (Cannon-Bowers, et al., 1995). In addition, team members use shared mental models to encode information (e.g., the dynamics of the environment they are embedded in and the response patterns needed to manage these dynamics, the purpose of the team; Zaccaro, Rittman & Marks, 2001). Shared mental models are essential for effective communication, exhibiting teamwork behaviors, and a willingness to work together in the future (e.g., Cannon-Bowers, Salas, & Converse, 1993; Griepentrog & Fleming, 2003; Marks et al., 2000; Mohammed, Klimoski, & Rentsch, 2000; Rentsch & Klimoski, 2001; Stout, Cannon-Bowers, Salas, & Milanovich, 1999).

Closed Loop Communication. Communication in general is defined as the "exchange of information between a sender and a receiver" irrespective of the medium (McIntyre & Salas, 1995, p. 25). Teams must communicate, either implicitly or explicitly, in order to coordinate. While communication is necessary, it is sometimes not enough to ensure effective performance. Sometimes messages are confused or missed, especially in dynamic, complex environments. Therefore, teams should use a more effective and precise form of communication: closed loop communication. Closed loop communication requires that the sender initiate a message, the receiver receive the message, interpret it, and acknowledge its receipt, and that the sender follow up to insure the intended message was received (McIntyre & Salas, 1995). In addition to the act of relaying messages, communication also promotes the updating of shared mental models and key teamwork behaviors (e.g., backup behavior).

Task Characteristics

In addition to individual and team characteristics and competencies, characteristics of the task will also influence the performance of a team. Here we will briefly describe those task characteristics, task type and task interdependency. The characteristics of the task, in part, determine the type of team. Therefore, the factors discussed below should be considered when modeling team performance. In addition, the type of task and the interdependency involved in the task should influence the individual characteristics of the persons chosen to make up the team.

Task Type

Within organizations, just as there are different types of teams (project teams, production and service teams, etc.) there are also types of tasks that teams are required to perform. Based on these tasks, the needs and goals of the team are determined. There are a number of task typologies that describe the types of tasks teams may have to perform. For example, Shaw (1964) described tasks in terms of their level of difficulty. Tasks could then be categorized into simple (e.g., tasks with limited information processing) or complex (e.g., tasks requiring large amounts of information processing) types. Steiner (1972) suggested that tasks can be categorized based on how team members contribute to the task (i.e., additive, disjunctive, conjunctive). In additive tasks (e.g., number of items inspected at a plant), tasks can be divided into minimal units, tasks are easily accomplished by any team member, and performance is dependent on the summated effort of the whole team. On the other hand, disjunctive tasks cannot be divided into smaller units and can either be described as a judgment task (i.e., choose correct answer of all possible alternatives) or decision-making task (i.e., choose correct answer from a subset of alternatives). The outcome of disjunctive tasks is dependent on the best member of team. Third, conjunctive tasks are those in which the team's performance is dependent on the weakest member of a group. For example, in a production assembly line, a product can only be made as quickly as the slowest member on the line.

A fourth task typology suggests that tasks can be categorized based on the performance processes involved in completing the task (Hackman, 1968). Hackman argued that tasks can be categorized as a production, discussion or problem-solving task. Finally, McGrath (1984) suggests that tasks are divisible into four categories—generating ideas and plans, choosing between alternatives, negotiating conflicts of interest, and executing work. McGrath further argues that these task categories can be separated into behavioral (e.g., executing work) or conceptual (e.g., generating ideas) tasks. A challenge to those modeling team performance is the reality that teams rarely perform just one task (Argote & McGrath, 1993). As such, at any given time the team may switch between tasks. Adding to this challenge is the fact that teams are not only switching between behavioral tasks, but between behavioral and cognitive tasks. As such, it is important that this be taken into consideration when modeling team performance.

Task Interdependence

Task interdependence refers to how much team members must rely on each other to perform their tasks effectively based on the design of their jobs (Saavedra, Early, & Van Dyne, 1993) and is determined in large part from the requirements and constraints inherent in the task (Wageman, 1995). There are several task typologies discussed in the literature (e.g., Saavedra et al., 1993; Devine, 2002; Steiner, 1972; Thompson, 1967; Van de Ven, Delbecq, & Koenig, 1976), indicating different manifestations and causes of interdependencies in teams. These typologies are important for modeling efforts as the nature of the interdependence needs to be taken into account, as this will determine the instrumentality of specific teamwork competencies (discussed later). Saavedra and colleagues (1993) discuss four levels of task interdependence-pooled (i.e., members make separate, interdependent contributions; performance is sum of all contributions), sequential (i.e., members perform tasks sequentially and one's performance is dependent on other's contributions), reciprocal (i.e., members work closely together and share information; performance is dependent on work of others; task order is flexible), and team (i.e., members jointly diagnose, problem solve, and collaborate to complete task; performance is dependent on mutual interactions). For example, if the team task that is being modeled is one in which reciprocal interdependence is present as opposed to team interdependence, then back-up behavior may need to be modeled in terms of verbal as opposed to physical back-up.

Work Structure

Characteristics of the work structure also have an impact on team performance. Work structure can be described as the way a team approaches a task in terms of work assignment, team norms and communication structure (Tannenbaum et al., 1992). These factors are important to team performance modeling because they help to determine the level of interaction between team members.

Work Assignment

Work assignment is the "manner in which the task components are distributed among team members" (Naylor & Dickinson, 1969, p. 167). Work assignment is closely related to team interdependency because it determines which team member does what and when and how each individual component of the task is integrated so that teams work effectively. Team members can be assigned the same task components or different task components depending upon the nature of the task.

Team Norms

Norms are defined as "social standards that describe what behaviors should and should not be performed in any social setting" (Forsythe, 1983, p. 160). Teams have norms that are typically adopted implicitly and are usually accepted gradually regarding their work, with team members aligning their behaviors to match certain "acceptable" standards. Norms within teams can shape team behavior in indirect and direct ways regarding the structure of their work (Caldwell & O'Reilly, 2003). Research has shown that norms affect both specific behaviors like listening (Cialdini, Kallgren, & Reno, 1991; Kallgren, Reno, & Cialdini, 2000; Reno, Cialdini, & Kallgren, 1993) or more general factors like team climate (West and Farr, 1990).

Communication Structure

Communication structure is defined as "the communication interrelationships which exist between team member" (Naylor & Dickinson, 1969, p. 167). Generally, teams develop their own communication structure based on the type of task and work structure of the team. For example, a highly complex task would require that team members communicate more than for a less complex task. In addition, teams in which members are performing the same tasks might require more communication to determine if they are each taking the same steps in completing the task.

Environment

There are a number of environmental stressors that will influence team performance and need to be considered when modeling team performance. Typical performance decrements that occur as a result of stress include accuracy, speed, and variability in accuracy or speed (Salas, Driskell, & Hughes, 1996). The severity of each stressor is dependent on the level in which the operator is exposed to it (Salas et al., 1996). In general, the more extreme the level is, the greater the performance decrement will be. While there are many environmental factors that can affect performance, we will only discuss three: task load, time pressure and noise. We encourage the reader to look at Driskell & Salas (1996) for other potential stressors (e.g., threat, fatigue, group pressure).

Task Load

The first stressor that will be discussed is task load (Salas et al., 1996). It is often necessary for teams to perform more than one task at a time. Using the aviation domain as an example, when landing the airplane, the flight crew is responsible for maintaining a specified airspeed, a constant rate of descent and heading, monitoring all instruments in the cockpit, listening for and making radio calls, completing all checklists, and scanning for other aircraft in the vicinity. Although these tasks can be divided between the crewmembers, each crewmember is still required to conduct several tasks at once. If the crew is also handling an emergency situation, such as an engine failure, additional tasks (e.g., shutting off the fuel supply to the failed engine and making necessary flight corrections to continue the flight safely) will be added.

Research suggests that when a second simultaneous task is added (e.g., talking to the control tower), accuracy on the primary task (e.g., manipulating the flight controls) may be impaired (Bowers & LaBarba, 1991). It has also been shown that when similar tasks are performed concurrently, task load has a more negative effect on performance (Atwood, 1971; Parkinson, Parks, & Kroll, 1971). More specifically, it is more difficult for an individual to perform two verbal tasks (e.g., listening to someone speak and trying to speak simultaneously) than it is for him/her to perform a verbal and a spatial task (e.g., listening to someone speak and watching playing a video game). In some organizations, automated technology (e.g., flight automation in aviation) has been introduced to minimize the effects of task load on teams. However, we caution organizations implementing technology to reduce task load because in some cases it can lead to lowered alertness and increased workload.

Time Pressure

In addition to task load, time pressure, or the time restriction required for task performance, has also been shown to impact team performance. Often the result of organization (e.g., project deadlines) and public demands (e.g., on-time flight departures in aviation) (Prince & Salas, 2000), time pressure has been cited as the most detrimental stressor (Salas et al., 1996). This is especially true when teams are required to perform under extreme time pressure. Furthermore, time pressure has been cited as a contributing factor in several well publicized accidents including the USS Vincennes incident (Fogarty, 1988, as cited in Collyer & Malecki, 1998) and the Space Shuttle Challenger launch (Vaughan, 1996).

Noise

A third stressor influencing team performance in organizations is noise (i.e., any unwanted sound by a listener; Cohen & Weinstein, 1981). While noise can have long-term effects such as permanent hearing loss, noise can also have short-term effects on performance such as narrowing of attention (Endsley, 1995). Narrowing of attention can thus lead to task cues being ignored and performance decrements occurring, especially for complex tasks (Salas et al., 1996). Additional effects of noise are loss of information relevant cues (e.g., inability to hear alarms or auditory cues), increased difficulty in hearing (e.g., resulting in missed clearances from ATC), and increased difficulty in ability to speak (i.e., difficulty in communicating with crew members).

However, some exposure to noise may serve as an arousal and may actually increase performance on vigilance or simple tasks. While this may be the case sometimes, the effects of noise are usually negative. For example, noise often causes a person to focus on only the primary task and cause the secondary task to be ignored (i.e., attention narrowing or decreased situation awareness). This can have serious consequences if a pilot focuses only on flying the aircraft while ignoring other tasks like fuel management (Endsley, 1995). The effects of noise are more severe as the intensity or loudness of the noise increases and as the noise becomes increasingly intermittent and random (e.g., warning alarms). Research suggests that as experience with the task increases, the severity of noise effects decreases. While most say that exposure to noise is debilitating, some also point out that people can habituate to the noise after long periods of exposure (e.g., pilots may habituate to the roar of the engines), which minimizes performance decrements.

Modeling Tools

Given the individual characteristics, team characteristics, task characteristics, team competencies and work structure critical to good team performance, one can then choose the best method to model team performance given the needs motivating the modeling effort. Here we describe several options that have been used previously in organizational simulations and team performance modeling, normative-descriptive models, Petri nets, regression and Fuzzy Systems Modeling.

Normative-Descriptive Models

Over the last 25 years researchers have used normative-descriptive models in order to effectively and realistically model team performance. The normative modeling approach allows researchers to use mathematical formulas to predict the best possible behavior or decisions of individuals and teams under a specified set of conditions and situations. Typically, mathematical theories such as estimation theory, decision theory or dynamic programming (Domenech & Messons, 1975; Dreyfus, 1957) have been used to make predictions of optimal human performance and decision making. After data is collected from individuals or teams, the actual performance data is compared to the theoretical prediction of behavior to gauge overall performance. Differences between actual performance and predicted performance are quantified and incorporated into the normative model to create the normative-descriptive model that more realistically predicts actual human performance.

Petri Nets

The use of Petri nets to model decision-making is another alternative. Proponents of using Petri nets look to the "capability for modeling and understanding team decision making from various perspectives and at multiple levels of abstraction, thus enhancing our understanding of the variables and processes important under various conditions of team decision making" (Coovert & McNelis, 1992, p. 248).

At its most basic level, the Petri net is nothing more than an abstract model of relationships among processes and events (Coovert & McNelis, 1992). Therefore, meaning must be attached to its parts in order for it to be useful. This allows organizations to simulate many different scenarios simply by changing the meaning of the Petri net. Petri nets have three primary advantages that make them usable for organizations. First, Petri nets can model both conflict and concurrency. Second, Petri nets can model at different levels of abstraction. Third, Petri nets can be analyzed in several ways to validate the model (for more thorough explanation see Coovert & McNelis, 1992) The results of the implementation of a decision support system such as developed for the TADMUS (Tactical Decision Making Under Stress) program, show increased situation awareness, lower workload, more confidence in the decisions made and more effective performance.

Regression

A more traditional method to mathematically or statistically model human performance and decision-making is to use regression (linear and nonlinear). Although using regression to predict human performance is simpler than using more complex mathematical models, there are several drawbacks that must be noted. The majority of research utilizes linear regression although many of the presumed linear relationships related to human performance have not been tested using nonlinear models (Dorsey & Coovert, 2003). However, using nonlinear regression models has its own set of troubles. Generally when modeling using nonlinear regression, researchers must state a priori the nature of the nonlinear relationship and the interactions between the variables. The nature of nonlinear relationships in the areas of human performance and decision-making are mostly unknown making it difficult to accurately identify a realistic model a priori.

Fuzzy Systems Modeling

In contrast, using fuzzy logic when modeling human performance and decision-making does not require making a priori hypotheses about the relationships among performance variables that may have a nonlinear relationship (Dorsey & Coovert, 2003). Fuzzy logic uses the ideas first proposed by Zadeh (1964) that allows researchers to create categories that are not clearly defined. This is important because many real-world situations are not clearly defined but instead have a great deal of uncertainty and ambiguity. Another strength of fuzzy systems modeling is the ease of building models, which is very time and cost effective (Bolton, Holness, Buff & Campbell, 2001). In addition, Dorsey and Coovert (2003), reported results from a group of studies that found fuzzy systems models to produce better model fits than those based on linear and nonlinear regression.

Although fuzzy systems modeling has several positive attributes over both normative-descriptive models and using regression as a modeling procedure, it has several weaknesses that are inherent in all mathematical models. First, it tends to require a large amount of data to be collected which requires researchers to choose an area that allows participants to make many decisions in a short amount of time (Bolton et al., 2001). In addition, mathematical models will produce the same predictions of performance when cues are correlated, making it difficult to choose which model to use. Finally, mathematical models are difficult to use in cooperation with scenario or event-based training (SBT; EBAT) because they offer only a general strategy of decision-making and do not specify particular decisions given a specific scenario or event.

The previous examples of modeling methods are just a few of many tools available to model team performance (see Zachary et al., 2001 for a more complete list of modeling methods). As we are not as knowledgeable about many of the methods used for modeling, we call on other researchers who have expertise in these areas to provide a more thorough knowledge base of when to use a particular modeling method over another and a better understanding of the pros and cons associated with different modeling methods.

CHOOSING COMPONENTS TO MODEL

The method used to determine what components of team performance to model in an organizational simulation can be difficult because of the many factors affecting team performance as well as organizational constraints such as time and money.

Therefore, the framework we have provided and the components described in this chapter are meant to be used as a guideline for determining how to best model team performance in organizations under ideal conditions. While all of the components we have provided certainly impact team performance, they do so to varying degrees, and in most, if not all, organizations, modeling is rarely done under ideal conditions, making it impossible to model all of the components of team performance.

Table 1 provides a breakdown of components into categories of "must be modeled", "should be modeled", and "would like to model". This differentiation between components of team performance can provide a step-by-step method for the modeling process. The ingredients that we argue must be modeled have been categorized as such because we believe that these are necessary for any basic model of team performance. The "should be modeled" ingredients are those where adding them to the model would improve the realism and complexity of the model. Finally, there are some ingredients where data are lacking as to how to model them and therefore they have been categorized as "would like to model". Many times it will be difficult to identify those components that "should be modeled," and those that one "would like to model." Therefore, we have provided a grouping of questions to guide the thinking process through the steps to model team performance (see Table 2). This table provides general questions to ask about all of the factors that affect team performance in order to aid the modeler in determining the specific components and subcomponents to model.

The components that "must be modeled" are the characteristics of team performance that we feel must be present in all models of team performance. For example, the type of team (e.g. advisory, action or project) is a critical component to begin modeling team performance. In addition to the type of team, the type of task and the level of interdependency as well as the personality and cognitive ability of individuals provide the basis for modeling team performance. These factors were chosen as the most critical because the other components of team performance vary in importance based on the differences in team, task type, interdependency and individual personality and cognitive ability.

The next step in the modeling process is to determine which of the components from the "should be modeled" section to include. The components that should be modeled will vary based on the type of team, the type of task, the level of interdependency and individual personality and cognitive ability. After determining the components of team performance that "should be modeled," the next step involves including other components that one "would like to model" if resources and time provide that opportunity. These are the factors of team performance that either there is not enough data to determine the impact on team performance or are unlikely to be modeled because of technology and resources constraints.

Components that "Must be Modeled"	
• Individual Characteristics o Cognitive Ability o Personality • Team Characteristics o Team Type o Team Structure o Team Size • Task Characteristics o Task Type o Interdependency	• Work Structure o Work Assignment o Communication Structure • Environment o Task Load o Time Pressure • Noise
Components that "Should be Modeled"	
• Individual Characteristics o Expertise o Mental Models o Cultural Factors o Motivation o Attitudes	• Team Characteristics o Team Diversity
Components one "Would Like to Model"	
• Team Competencies o Team Leadership o Mutual Performance Monitoring o Back-up Behavior o Shared Mental Models o Team Orientation o Adaptability o Closed-loop Communication	• Team Characteristics o Cohesiveness o Team Climate • Work Structure o Team Norms

TABLE 1: Important Components of Team Performance Modeling

Input Factors	Questions to Ask When Modeling	

Individual Characteristics	• What is the cultural background of individuals on the team?
	• How do individuals feel about working in a team?
	• How do individuals feel about the tasks to be performed?
	• What are the personalities of individual on the team?
	  o How will these impact team interactions?
	  o How will they impact the task?
	• Do individuals have high or low cognitive ability?
	  o How will these impact team interactions?
	  o How will this impact the task?
		• Are team members motivated?
		  o To work in a team?
		  o To perform the task?
		• What is the expertise of members?
		  o Is expertise shared?
		• What previous training have members received?

Task Characteristics	• Is type(s) of task(s) are teams performing?
	  o Is the task simple or complex?
	  o Is the task additive, disjunctive, conjunctive?
	  o Does the task involve problem solving, discussion, or production?
	  o Does the task involve generating ideas and plans, choosing between alternatives, negotiating conflicts of interest, or executing work?
		• At what level is task interdependence?
		  o Does the task require pooled, sequential, reciprocal, or team task interdependence?

Team Characteristics	• How many individuals make up the team?
	  o Is the team large or small?
	  o How is this expected to impact performance?
	• Have the members worked together before?
	• Are teams co-located or distributed?
		• Does the team have a diverse makeup?
		  o How do team members differ (e.g., expertise, cultural background, sex)?
		  o Is there one token member or are a number of different cultures, sexes, etc. integrated?

Work Structure	• How are the tasks distributed to the team members?
	  o Do the team members perform the same subtasks or are each assigned separate subtasks?
	• What organizational norms are present that should be developed into the team?
		• How do team members communicate?
		  o Do team members communicate face-to-face?
		  o How much communication is required by the task?

Team Competencies	• Which competencies are required by the task? • Are teams co-located or distributed? • Which competencies are most important for the task?	

TABLE 2: Guiding Questions Through the Modeling Process

The final step in modeling team performance is to choose the method with which to model performance. The examples of methods provided in this paper include both rule-based and linear-based models of team performance. Each method has positives and negatives associated with its use in modeling team performance. Therefore, it is important to consult with other experts and researchers when choosing a specific modeling tool.

IMPACTS OF MODELING TEAM PERFORMANCE

Up until this point we have discussed the key ingredients needed to model team performance. We will now discuss the impact of modeling team performance on three areas important to organizational simulation—human-systems integration, scenario-based training, and decision support systems. We are sure that modeling team performance has a number of implications for other areas of organizations; however, we have chosen to limit our discussion to those we have deemed most important.

Human Systems Integration

Human Systems Integration is a philosophy that espouses focusing on the human elements in complex systems will dramatically increase both performance and productivity (Booher, 2003). The process of HSI requires the integration of human factors engineering (HFE); manpower, personnel, training (MPT); health hazards; safety factors; medical factors; personnel (or human) survivability factors; and habitability (http://fac.dtic.mil/hsiac/HIS.htm). Implementing an HSI approach for an operational system requires that organizations prescribe training requirements early on to ensure that all necessary training capabilities are built into the system (Buff, Bolton, & Campbell, 2003). The implementation of HSI programs such as the Army's MANPRINT and cognitive engineering of the digital battlefield science and technology objective (CE STO) and the Navy's Tactical Decision Making Under Stress (TADMUS) have been highly successful in reducing the expanding gap between technological advances in military operations and humans' limited ability to integrate large amounts of information into their decision-making process (Booher, 2003; Pierce & Salas, 2003).

Despite the success of HSI programs, Pierce and Salas (2003) describe eight problems facing HSI integration into information systems operations (Table 3). Through team performance modeling many of the issues can be addressed and allow researchers to more accurately design proper decision making methods for teams to use in similar situations in the future. For example environmental factors that influence team performance (e.g. time pressure, noise, uncertainty) can be

Issues in Information Systems
• Learning *how* to think, not *what* to think
• Leader mindsets constraining flexibility
• Difficulty managing uncertainty
• Degraded situation awareness
• Problems with team coordination
• Inadequate information filters
• Abuse, misuse, and disuse of automation
• Inadequate human performance assessments

TABLE 3. Human Performance (Pierce & Salas, 2003)

included in a team performance model so that teams can receive feedback as to better courses of action in ambiguous situations. As previously stated, team leadership, team coordination and technological considerations are vital factors that must be modeled to provide a realistic and effective model of team performance.

Scenario-Based Training

Practice is essential to the translation of knowledge into actual skill, but it should not be a random event, nor should it be unguided (i.e., practice, practice, practice does not make perfect). The most effective method is the event-based approach to training where scenarios are defined a priori based on the training objectives identified in the training needs analysis. This instructional strategy is typically known as simulation/scenario-based training (SBT) or event-based training (EBAT) (Prince, Oser, Salas, & Woodruff, 1993; Oser, Cannon-Bowers, Salas, & Dwyer, 1999). As shown in Figure 2, use of this strategy begins with the development of a skill inventory abstracted through the results of a job and/or task

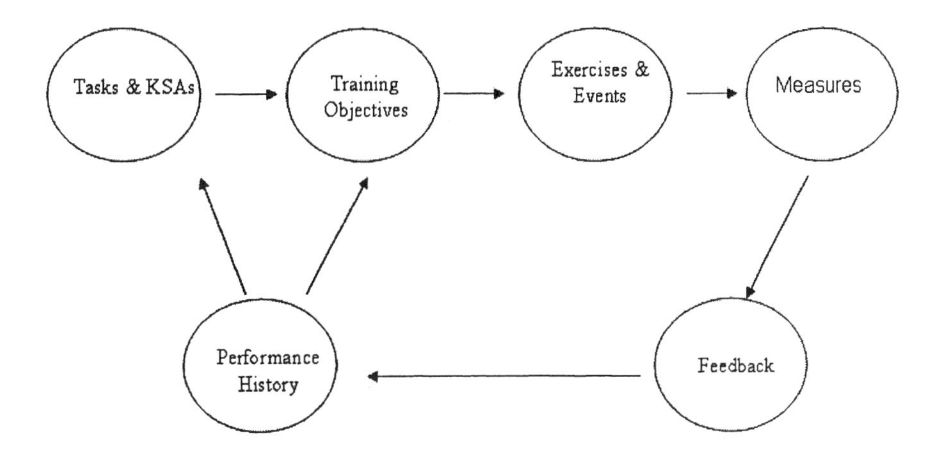

FIGURE 2. Steps for Implementing SBT
(Adapted from Cannon-Bowers, Burns, Salas, & Pruitt, 1998)

analysis. From this skill inventory training objectives and competencies are identified. These training objectives are then turned into learning objectives, which serve to drive the development of the actual training scenarios and associated scripts. These training scenarios contain embedded a priori "trigger" events at several points. Ideally multiple events should vary in difficulty, should be specified for each objective and should be introduced at several points during the exercise. These events serve as known opportunities for trainees to exhibit those competencies targeted in training. After events and the associated scenarios are created, measurement instruments are developed that will be used to assess task performance during each a priori defined event. Measurement tools allow the assessment of both process and outcome level feedback, both of which are important for teams within organizations. Specifically, measurement should allow the assessment of whether targeted competencies are learned (i.e., outcomes) along with why performance occurred as it did (i.e., process) by being tied to learning objectives and scenario events. This instructional strategy explicitly links learning objectives, exercise events, performance measures, and the associated feedback, but also allows for standardized measurement and training and reduces the workload for those in charge of observation and collection of performance data. As performance measures are tied to previously defined events, the observer does not have to observe every instance of behavior (reducing workload) and the form of the measurement instrument has also been argued to allow near real time feedback to trainees.

The development of human behavior representations (HBRs) that allow full interaction between the human user(s) and the computer-simulated image is one example of how advances in technology have increased the usability of team performance models implemented through SBT. Zachary, Santarelli, Lyons, Bergondy and Johnston (2001) discuss the Synthetic Cognition for Operational Team Training (SCOTT) that trains team skills by substituting team members with HBRs. The HBRs in the SCOTT program verbally interact with the trainee in real time scenario-based environments. The HBR is also able to give the trainee feedback to increase training efficiency. In addition to verbal interactions, the HBR is designed so that errors in decision-making occur, thereby better simulating performance under different levels of stress.

In addition to the SCOTT program, several groups are now coordinating to develop Synthetic Teammates for Realtime Anywhere Training and Assessment (STRATA; Bell, 2003). STRATA is specifically developed for pilot training. The purpose is to create a simulation that uses both human team members and HBRs that will interact verbally on scenario-based flight missions. By using HBRs, the STRATA program will be better equipped to train pilots in a variety of real time scenarios without the need for a full complement of team members.

Decision Support Systems

Decision support systems (DSS) are designed to improve and support human decision-making (Brody, Kowalczyk & Coulter, 2003). The design and implementation of the Navy's DSS associated with TADMUS focused on the usability of the DSS as well as its structure (Zachary, Ryder & Hicinbothom, 1998). Organizations can take the general standards and principles outlined by Zaklad and Zachary (1992), and model more specific principles based on the needs of the organization or the team. For example, a general standard for DSS is that they should be designed to support coordination among team members. The Navy took that general guideline and made specific principles related to TADMUS, such as, "the sharing of mental models of team activities...the transmission and acknowledgment of intentions across the AAW team because of the prominent role that such communications play in the timely triggering of cognitive tasks, and helping each operator in the team know when it would be appropriate to notify or ask for permission from higher authorities (Zachary et al., 1998, p. 340)." The results of the implementation of DSS such as developed for TADMUS, show increased situation awareness, lower workload, more confidence in the decisions made and more effective performance

The example above provides strong support for the use of team performance modeling in organizational simulations. As discussed previously, the modeling of teamwork components (e.g. shared mental models, team coordination) is critical to team performance modeling and can be structured and implemented into DSS with specific principles related to the needs of the team.

Intelligent tutoring systems (ITS) are another example of a type of decision support systems (Ong & Ramachandran, 2003). Woolf, Beck, Eliot and Stern

(2001) discuss some of the abilities of various ITS. In modeling team performance, ITS can teach a variety of knowledge domains; however, it requires an extensive knowledge of the subject, as well as, strategies for error diagnosis and decision-making and examples and analogies of relevant topics. This requires a great deal of time and work to design and program.

For organizational simulations used by employees who are already trained, ITS do not have to be equipped with the same set of information or provide as much guidance, because the persons already have a knowledge of the domain. For example, the Tactical Action Officer (TAO) developed for the Navy allows the ITS to analyze the actions of the users to help teach tactics during scenario-based training exercises.

HSI, SBT and DSS are three examples of how team performance modeling has been utilized in organizational simulations. Although we have discussed each separately here, many times the three are used in combination when applying team performance modeling to organizational training. As stated above with the TAO program, ITS can be used with specific planned scenarios to better teach team coordination skills during particular events.

PROSPECTS

As noted earlier, this field is in its infancy. Much more work needs to be done before modeling team performance can be credible, reliable and where the results are of use to organizations. We know a lot about teamwork and team performance. However, the body of knowledge cannot (yet) suggest robust models of team performance. So, what do we need? First, we need more multi-disciplinary team performance research. Modeling team performance does not only belong to team researchers, but to computer scientists, mathematicians, modelers and cognitive scientists. It is multi-disciplinary. To make great strides, we need more and better collaboration between several disciplines. Only then can all benefit from it. Second, while collaboration is needed among the different interested disciplines, we all need to understand each other's requests for modeling team performance. That is, we all need to articulate what we need in terms of data, performance, findings and knowledge to each other. Without this communication, the work will be piece-meal and integration for modeling purposes will be cumbersome and ineffective. Third, we need better simulation tools that allow us to model the complexities of team performance. Our tools are too static and linear. We need more fluid, flexible and adaptable tools that help capture realistic team performance embedded in organizations. Finally, we need more precision in our findings. All those involved in conducting team performance research must be more precise in reporting findings. The more precise we are, the better our modeling; the better our modeling the better the benefit to the organizations

CONCLUSION

As teams are increasingly being used in organizations, we must find ways to better understand how they work and their impact on organizational goals and outcomes (e.g., human system integration, scenario-based training, and decision support systems). Organization simulations in conjunction with team performance models are one means to do this. As such, the purpose of this chapter was to examine the team performance literature to determine the "key ingredients" needed when modeling teams and their performance in organizational settings. We have presented a framework by which those interested in modeling team performance can follow to improve this process. We highlight the importance of factors outside of the team itself that exert influence -- the individual, task, work, and environment. We realize that it may not (currently) be feasible to model all of the ingredients presented for financial, technological or time constraints. And even then maybe we could only model all variables in one simulation in a perfect world. Furthermore, as the community's knowledge regarding some areas may be weak (i.e., some areas of mathematics), we encourage experts from this area to educate the modeling community so that we can better understand the area and more effectively model team performance. It is our hope that the information presented here will inspire more research in this area.

REFERENCES

Anderson, N.R., & West, M.A. (1998). Measuring climate for work group innovation: Development and validation of the team climate inventory. Journal of Organizational Behavior. 19, 235-258.

Argote, L., & McGrath, J.D. (1993). Group processes in organizations: Continuity and change. In C.L. Cooper & I.T. Robertson (Eds.), International review of industrial and organizational psychology, (Vol. 8), New York: John Wiley & Sons.

Atwood, G.A. (1971). An experimental study of visual imagination and memory. Cognitive Psychology. 2, 290-299.

Bandura, A. (1991). The changing icons in personality psychology. In J. H. Cantor (Ed), Psychology at Iowa: Centennial essays (pp. 117-139). Hillsdale, NJ, England: Lawrence Erlbaum Associates, Inc.

Bass, B. M. (1990). From transactional to transformational leadership: Learning to share the vision. Organizational Dynamics, 18(3), 19-31.

Barrick, M. R., Stewart, G. L., Neubert, M. J. & Mount, M. K. (1998). Relating member ability and personality to work-team processes and team effectiveness. Journal of Applied Psychology, 83(3), 377-391.

Bell, B. (2003). On-demand team training with simulated teammates: Some preliminary thoughts on cognitive model reuse. TTG News, 6-8.

Bell, B. S. & Kozlowski, S. J. (2002). A typology of virtual teams: Implications for effective leadership. Group & Organization Management, 27(1), 14-49.

Bolton, A. E., Holness, D. O., Buff, W. L., and Campbell, G. E. (2001). An application of mathematical modeling in training systems: Is it a viable alternative to cognitive modeling? Proceedings of the 10th Annual Conference on Computer Generated Forces and Behavioral Representation. 497 -505.

Booher, H.R. (2003). Introduction: Human systems integration. In H.R. Booher (Ed.), Handbook of human system integration (pp.1-30). Hoboken, NJ: John Wiley & Sons.

Bowers, C.A., & LaBarba, R.C. (1991). Secondary task assessment of workload: Neuropsychological considerations for applied psychology. Perceptual and Motor Skills. 73, 487-496.

Brannick, M.T. & Prince, C. (1997). Overview of team performance measurement. In M.T. Brannick, E. Salas, and C. Prince (Eds.), Team Performance Assessment and Measurement: Theory, Methods, and Applications (pp 3-16). Mahwah, NJ: Lawrence Erlbaum Associates.

Brody, R.G., Kowalczyk, T.K., & Coulter, J.M. (2003). The effect of a computerized decision aid on the development of knowledge. Journal of Business & Psychology. 18(2), 157-174.

Buff, W.L., Bolton, A.E., Campbell, G.E. (2003). Providing an integrated team training capability using synthetic teammates. Paper presented at the ASNE Human Systems Integration Symposium, Tyson's Corner, VA.

Burke, C. S. (1999). Examination of the cognitive mechanisms through which team leaders promote effective team processes and adaptive team performance. Unpublished doctoral dissertation, George Mason University, Virginia.

Caldwell, D.F., & O'Reilly, C.A. (2003). The determinants of team-based innovations: The role of social influence. Small Group Research. 34(4), 497-517.

Cannon-Bowers, J.A., & Salas, E. (1997). Teamwork competencies: The interaction of team member knowledge, skills, and attitudes. In H.F. O'Niel (Ed.), Workforce readiness: Competencies and assessment (pp. 151-174). Mahwah, NJ: Lawrence Erlbaum Associates.

Cannon-Bowers, J.A., Salas, E., & Converse, S.A. (1993). Shared mental models in expert team decision-making. In Individual and group decision making (pp. 221-246). Hillsdale, NJ: Lawrence Erlbaum Associates.

Cannon-Bowers, J.A., Tannenbaum, S.I., Salas, E., & Volpe, C.E. (1995). Defining team competencies and establishing team training requirements. In R. Guzzo, E. Salas, & Associates (Eds.), Team effectiveness and decision making in organizations (pp. 333-380). San Francisco: Jossey-Bass.

Carron, A.V., Brawley, L.R., & Widmeyer, W.N. (1998). The measurement of cohesiveness in sport groups. In J.L. Duda (Ed.), Advances in sports and exercise psychology measurement (pp. 213-226). Morgantown, WV: Fitness Information Technology.

Chatman, J.A., Polzer, J.T., Barsade, S.G., & Neale, M.A. (1998). Being different yet feeling similar: The influence of demographic composition and organizational culture on work processes and outcomes. Administrative Science Quarterly. 43, 749-780.

Cialdini, R., Kallgren, C., & Reno, R. (1991). A focus theory of normative conduct: A theoretical refinement and reevaluation of the role of norms in human behavior. Advances in Experimental Social Psychology. 24, 201-234.

Cohen, S. G. and Bailey, D.E. (1997). What makes teams work: Group effectiveness research from the shop floor to the executive suite. Journal of Management, 23(3): 239-290.

Cohen, S. & Weinstein, N. (1981). Nonauditory effects of noise on behavior and health. Journal of Social Issues, 37(1), 36-70.

Collyer, S.C., Malecki, G.S. (1998). Tactical decision making under Stress: history and overview. In: Cannon-Bowers, J.A., Salas, E. (Eds.),. Making Decisions Under Stress: Implications for Individual and Team Training. American Psychological Association, Washington, DC.

Coovert, M.D., & McNelis, K. (1992). Team decision making and performance: A review and proposed modeling approach employing Petri nets. In R.W. Swezey & E. Salas (Eds.), Teams: Their training and performance (p. 247-280). Norwood, NJ: Ablex Publishing Corporation.

Cordery, J.L., Mueller, W.S., & Smith, L.M. (1991). Attitudinal and behavioral outcomes of autonomous group working: A longitudinal field study. Academy of Management Journal. 34, 464-476.

Cox, T.H., Lobel, S.A., & McLeod, P.L. (1991). Effects of ethnic group cultural differences on cooperative and competitive behavior on a group task. Academy of Management Journal. 34(4), 827-847.

Curral, L.A., Forrester, R.H., Dawson, J.F., & West, M.A. (2001). It's what you do and the way you do it: Team task, team size, and innovation-related group processes. Europe Journal of Work and Organizational Psychology. 10(2), 187-204.

Devine, D.J. (2002). A review and integration of classification systems relevant to teams in organizations. Group Dynamics. 6(4), 291-310.

Domenech, I., & Messons, J.U. (1975). Introduction to estimation theory and decision theory in statistics. Anuario de Psicologia. 12, 65-95.

Dorsey, D.W., & Coovert, M.D. (2003). Mathematical modeling of decisionmaking: A soft and fuzzy approach to capturing hard decisions. HumanFactors. 45(1), 117-135.

Dreyfus, S.E. (1957). Computational aspects of dynamic programming. Operations Research. 5, 409-415.

Driskell, J.E., & Salas, E. (1992). Collective behavior and team performance. Human Factors. 34, 277-288.

Driskell, J.E., & Salas, E. (Eds.) (1996). Stress and human performance. Hillsdale, NJ: Lawrence Erlbaum Associates.

Driskell, J.E., Radtke, P.H., & Salas, E. (2003). Virtual teams: Effects of technological mediation on team performance. Group Dynamics. 7, 297-323.

Earley, P.C., & Gibson, C.B. (2002). Multinational work teams: A new perspective. Mahwah, NJ: Lawrence Erlbaum Associates.

Eby, L., & Dobbins, G. (1997). Collectivistic orientation in teams: An individual and group-level analysis. Journal of Organizational Behaviour, 18, 275-295.

Edmondson, A. (1999). Psychological safety and learning behavior in work teams. Administrative Science Quarterly. 44, 350-383.

Endsley, M.R. (1995). Toward a theory of situation awareness in dynamic systems. Human Factors. 37(1), 32-64.

Erez, M. and Earley, P. C. (1993). Culture, self-identity, and work. New York: Oxford.

Festinger, L. (1950). Informal social communication. Psychological Review. 57, 271-284.

Fiore, S. M., Cuevas, H. M., Scielzo, S., & Salas, E. (2002). Training individuals for distributed teams: Problem solving assessment for distributed mission research. Computers in Human Behavior, 18, 729-744

Forsythe, D.R. (1983). Group dynamics (2nd Ed.). Pacific Grove, CA: Brooks/Cole Publishing Company.

Fridsma DB, Thomsen J. (1998). Representing medical protocols for organizational simulation: An information processing approach. Computational and Mathematical Organization Theory, 4(1), 71-95.

Gladstein, D.L. (1984). Groups in context: A model of task group effectiveness. Administrative Science Quarterly. 29, 499-517.

Griepentrog, B. K. & Fleming, P. J. (2003). Shared mental models and team performance: Are you thinking what we're thinking? Symposium presented at the 18th Annual Conference for the Society for Industrial Organizational Psychology, Orlando, FL.

Guzzo, R.A. & Dickinson, M.W. (1996). Teams in organizations: Recent research on performance and effectiveness. Annual Review of Psychology, 47, 307-338.

Hackman, J.R. (1968). Effects of task characteristics on group products. Journal of Experimental Social Psychology. 4, 162-187.

Hackman, J.R. (1987). The design of work teams. In J.W. Lorsch (Ed.), Handbook of organizational behavior (pp. 315-342). Englewood Cliffs, NJ: Prentice Hall

Hackman, J.R. (1990). Groups that work (and those that don't). San Francisco: Jossey-Bass.

Hackman, J.R. (1992). Group influences on individuals in organizations. In M.D. Dunnette & L.M. Hough (Eds.), Handbook of industrial and organizational psychology (Vol. 2, pp. 199-267). Palo Alto, CA: Consulting Psychological Press.

Hackman, J.R., & Morris, C.G. (1975). Group tasks, group interaction process, and group performance effectiveness: A review and proposed integration. In L. Berkowitz (Ed.), Advances in experimental social psychology (Vol. 8, pp. 45-99). New York: Academic Press.

Hackman, J. R., & Walton, R. E. (1986). Leading groups in organizations. In P. S. Goodman (Ed.), Designing effective work groups. San Francisco: Jossey-Bass.

Hambrick, D.C., Davison, S.C., Snell, S.A., & Snow, C.C. (1998). When groups consist of multiple nationalities: Towards a new understanding of the implications. Organization Studies. 19, 181-205.

Harkins, S.G., & Szymanski, K. (1988). Social loafing and self-evaluation with an objective standard. Journal of Experimental Social Psychology. 24, 354–365.

Hartigan, J. A. & Wigdor, A. K. (1989). Fairness in employment testing: Validity generalization, minority issues, and the General Aptitude Test Battery. Washington DC: National Academy Press.

Heise, D.R. (1992). Affect control theory and impression formation. In E. Borgotta & M. Borgotta (Eds.), Encyclopedia of sociology (Vol. 1, pp. 12-17). New York: MacMillan.

Hofstede, G. (1980). Culture's consequences: International differences in work related values. Beverly Hills, CA: Sage Publications.

Hunter, J.E. (1986). Cognitive ability, cognitive aptitudes, job knowledge, and job performance. Journal of Vocational Behavior. 29, 340-362.

Ilgen, D.R., Shapiro, J., Salas, E., & Weiss, H. (1987). Functions of group goals: Possible generalizations from individuals to groups (Tech. Rep. No. 87-022). Orlando, FL: US Naval Training Systems Center Technical Reports.

Jackson, S. E. (1996). The consequences of diversity in multidisciplinary work teams. In M. A. West (Ed.), The Psychology of Groups at Work, (pp. 53-76). Chichester, England: John Wiley & Sons.

Jackson, J.M., & Williams, K.D. (1985). Social loafing on difficult tasks: Working collectively can improve performance. Journal of Personality and Social Psychology. 49(4), 937–942.

Jacobs, T.O. & Jaques, E. (1987). Leadership in complex systems. In J.A. Zeidner (Ed), Human productivity enhancement: Organizations, personnel and decision making (p. 7-65). Orange, NJ.

Janis, I.L. (1989). Crucial decisions: Leadership in policymaking and crisis management. New York: The Free Press.

Janis, I.L., & Mann, L. (1977). Decision making: A psychological analysis of conflict, choice and commitment. New York: The Free Press.

Kahan, J.P., Webb, N. Shavelson, R.J., & Stolzenberg, R.M. (1985). Individual characteristics in unit performance. Santa Monica, CA: Rand.

Kallgren, C., Reno, R., & Cialdini, R. (2000). A focus theory of normative conduct: When norms do or do not affect behavior. Personality and Social Psychology Bulletin. 26, 1002-1012.

Kanfer, R., & Ackerman, P.L. (1989). Motivation and cognitive abilities: An integrative/aptitude-treatment interaction approach to skill acquisition. Journal of Applied Psychology. 87, 657-690.

Karau, S.J., & Williams, K.D. (1993), Social loafing: A meta-analytic review and theoretical integration. Journal of Personality and Social Psychology. 65, 681-706.

Katz, D., & Kahn, R.L. (1978). The social psychology of organizations (2nd Ed.). New York: Wiley.

Kendall, D.L., Salas, E., Burke, C.S., Stagl, K. C. (2003). Understanding team adaptability: A conceptual framework. Submitted to Journal of Applied Psychology.

Kirkman, B.L., & Shapiro, D.L. (2001). The impact of cultural values on job satisfaction and organizational commitment in self-managing work teams: The mediating role of employee resistance. Academy of Management Journal. 44(3), 557-569.

Kozlowski, S.W.,J., Gully, S.M., Nason, E.R., & Smith, E. M. (1999). Developing adaptive teams: A theory of compilation and performance across levels and time. In D. E. Ilgen & E. D. Pulakos (Eds.), The Changing Nature of Work Performance: Implications for Staffing, Personnel Actions, and Development, 240-292. San Francisco: Jossey-Bass.

Kozlowski, S.W.J., Gully, S.M., Salas, E., & Cannon-Bowers, J.A. (1996). Team leadership and development: Theory, principles, and guidelines for training leaders and teams. In M. Beyerlein, S. Beyerlein, & D. Johnson (Eds.), Advances in interdisciplinary studies of work teams: Team leadership (Vol. 3, pp. 253-292). Greenwich, CT: JAI Press.

Kozlowski, S.W.J., Toney, R.J., Mullins, M.E., Weissbein, D.A., Brown, K.G. & Bell, B.S. (2001). Developing adaptability: A theory for the design of integrated-embedded training systems. In E. Salas (Ed.), Advances in human performance and cognitive engineering research (Vol. 1, pp. 59-123). Amsterdam: Elsevier Science.

Lawler, E.E., Mohrman, S.A., & Ledford, G.E. (1998). Strategies for high performance organizations: Employee involvement, TOM, and reengineering programs in Fortune 1,000 corporations. San Francisco: Jossey-Bass.

Lepine, J.A. (2003). Team adaptation and postchange performance: Effects of team composition in terms of members' cognitive ability and personality. Journal of Applied Psychology. 88(1), 27-39.

Lepine, J.A., Hanson, M.A., Borman, W.C., & Motowidlo, S.J. (2000). Contextual performance and teamwork: Implications for staffing. In G.R. Ferris & K.M. Rowland (Eds.), Research in personnel and human resources management (Vol. 19, pp. 53-90). Stanford, CT: JAI Press.

Levine, J.M., & Moeland, R.L. (1990). Progress in small-group research. Annual Review of Psychology. 41, 585-634.

Likert, R. (1977). The nature of highly effective groups. In J.L. Gray & F.A. Starke (Eds.), Readings in organizational behavior (pp. 100-129). Columbus, OH: Merrill.

Loher, B. T., Vancouver, J. B., & Czajka, J. (1994). Preferences and reactions to teams. Paper presented at the Ninth Annual Conference of the Society for Industrial and Organizational Psychology, Inc., Nashville, TN.

Marks, M. A., Zaccaro, S. J. & Mathieu, J. E. (2000). Performance implications of leader briefings and team interaction training for team adaptation to novel environments. Journal of Applied Psychology, 85, 971-986.

Marks, M.A., Mathieu, J.E. & Zaccaro, S.J. (2001). A temporally based framework and taxonomy of team process. Academy of Management Review, 26, 356-376.

McGrath, J.E. (1984). Groups: Interaction and performance. Englewood Cliffs, NJ: Prentice Hall.

McGrath, J.E. (1998). A view of group composition through a group-theories lens. In M. A. Neale, E.A. Mannix, & D.H. Gruenfeld (Eds.), Research on managing groups and teams (pp. 255-272). Greenwich, CT: JAI Press.

McIntyre, R.M., & Salas, E. (1995). Measuring and managing for team performance: Emerging principles from complex environments. In R. Guzzo & E. Salas (Eds.), Team effectiveness and decision making in organizations (pp. 149-203). San Francisco: Jossey-Bass.

McShane, S.L., & Von Glinow, M.A. (2000). Organizational behavior. Boston, MA: Irwin/McGraw-Hill.

Mohammed, S., & Angell, L.C. (2003). Personality heterogeneity in teams: Which differences make a difference for team performance? Small Group Research. 34(6), 651-677.

Mohammed, S., Klimoski, R., & Rentsch, J.R (2000). The measurement of team mental models: We have no shared schema. Organizational Research Methods. 3(2), 123-165.

Mortensen, M., & Hinds, P. (2001). Conflict and shared identity in geographically distributed teams. International Journal of Conflict Management. 12(3), 212-238.

Naylor, J.C. and Dickinson, T.L. (1969). Task structure, work structure, and team performance. Journal of Applied Psychology, 53 (3), 167-177.

Nemeth, C.J. (1986). Differential contributions of majority and minority influence. Psychological Review. 93, 23-32.

Ong, J., & Ramachandran, S (2003). Intelligent tutoring systems: Using AI to improve training performance and ROI. Retrieved February 10 from, http://www.shai.com/papers.

Oser, R.L., Cannon-Bowers, J.A., Salas, E., Dwyer, D.J. (1999). Enhancing human performance in technology-rich environments: Guidelines for scenario-based training. In E. Salas (Ed.), Human/technology interaction in complex systems (Vol. 9, 175-202). Greenwich, CT: JAI Press.

Parkinson, S.R., Parks, T.E., & Kroll, N.E. (1971). Visual and auditory short-term memory: Effects of phonemically similar auditory shadow material during the retention interval. Journal of Experimental Psychology. 87, 274-280.

Pew, R.W., & Mavor, A.S. (Eds.). (1998). Modeling human and organizational behavior: Application to military simulations. Washington, DC: National Academy Press.

Pierce, L.G., & Salas, E (2003). Linking human performance principles to design of information systems. In H.R. Booher (Ed.), Handbook of human system integration (pp.1-30). Hoboken, NJ: John Wiley & Sons.

Porter, C. O. L. H., Hollenbeck, J. R., Ilgen, D. R., Ellis, A. P. J., West, B. J. & Moon, H. (2003). Backing up behaviors in teams: The role of personality and legitimacy of need. Journal of Applied Psychology, 88(3), 391-403.

Poulton, B.C., & West, M.A. (1999). The determinants of effectiveness in primary health care. Journal of Interprofessional Care. 13, 7-18.

Priest, H.A., Burke, C.S., Munim, D., & Salas, E. (2002, October). Understanding team adaptability: Initial theoretical and practical considerations. Proceedings of the 46th annual meeting of the Human Factors and Ergonomics Society. 561-565.

Prince, C., & Salas, E. (2000). Team situational awareness, errors, and crew resource management: Research integration for training guidance. In M.R. Endsley & D.J. Garland (Eds.), Situation awareness analysis and measurement (pp. 325-347). Mahwah, NJ: Lawrence Erlbaum Associates

Prince, C., Oser, R, Salas, E., & Woodruff, W. (1993). Increasing hits and reducing misses in CRM/LOS scenarios: Guidelines for simulator scenario development. International Journal of Aviation Psychology. 3(1), 69-82.

Reno, R., Cialdini, R., & Kallgren, C. (1993). The transsituational influence of social norms. Journal of Personality and Social Psychology. 64, 104-112.

Rentsch, J.R., & Klimoski, R.J. (2001). Why do "great minds' think alike?: Antecedents of team member schema agreement. Journal of Organizational Behavior. 22(2), 107-120.

Rouse, W.B., & Morris, N.M. (1986). On looking into the black box: prospects and limits in the search for mental models. Psychological Bulletin. 100, 349-363.

Saavedra, R., Earley, C., & Van Dyne, L. (1993). Complex interdependence in task-performing groups. Journal of Applied Psychology. 78(1), 61-72.

Salas, E., & Cannon-Bowers, J.A. (2001). The science of training: A decade of progress. Annual Review of Psychology. 52, 471-499.

Salas, E. & Cannon-Bowers, J.A. (2000). The anatomy of team training. In S. Tobias & J.D. Fletcher (Eds.), *Training & Retraining: A Handbook for Business, Industry, Government, and the Military* (pp312-335). New York: Macmillan Reference USA.

Salas, E., Dickinson, T.L., Converse, S., & Tannenbaum, S.I. (1992). Toward an understanding of team performance and training. In R.W. Swezey & E. Salas (eds.), Teams: Their training and performance (3-29). Norwood: Ablex.

Salas, E., Driskell, J.E., & Hughes, S. (1996). Introduction: The study of stress and human performance. In J.E. Driskell & E. Salas (Eds.), Stress and human performance (pp. 1-45). Mahwah, NJ: Lawrence Erlbaum Associates.

Salas, Sims, & Burke (2004). Towards a taxonomy of teamwork: The core elements. Manuscript in Preparation.

Schneider, S.C. & De Meyer, A. (1991). Interpreting and responding to strategic issues: The impact of national culture. Strategic Management Journal. 12, 307-320

Schneider, B., & Reichers, A. (1983). On the etiology of climates. Personnel Psychology. 39, 19-39.

Seaman, D.F. (1981). Working effectively with task-oriented groups. New York: McGraw-Hill.

Seashore, S.E. (1955). Group cohesiveness in the industrial work group. Ann Arbor MI: Survey Research Institute for Social Science.

Senge, P. M. (1990). The Fifth Discipline: The Art and Practice of the Learning Organization. New York: Doubleday Currency.

Sessa, V.I., & Jackson, S.E. (1995). Diversity in decision making teams: All differences are not created equal. In M.M. Chemers, S. Oskamp, & M.A. Costanzo (Eds.) Diversity in Organizations, Newbury Park, CA: Sage.

Shane, S. (1994). The effect of national culture on the choice between licensing and direct foreign investment. Strategic Management Journal. 15, 627-642.

Shaw, M.E. (1964). Communication networks. In L. Berkowitz (Ed.), Advances in experimental social psychology (Vol. 1), pp. 111-147). New York: Academic Press.

Shaw, M.E. (1981). Group dynamics: The psychology of small group behavior (3rd Ed.). New York: McGraw-Hill.

Shaw, M.E. (1976). Group dynamics: The psychology of small group behavior. New York: McGraw-Hill.

Simons, T. Pelled, L.H., & Smith, K.A. (1999). Making use of difference: Diversity, debate, and decision comprehensiveness in top management teams. Academy of Management Journal. 42, 662-673.

Smith, E.M., Ford, J.K., & Kozlowski, S.W.J. (1997). Building adaptive expertise: Implications for training design strategies. In M.A. Quinones & A. Ehrenstein (Eds.), Training for a rapidly changing workplace: Applications of psychological research (89-118). Washington, DC: American Psychological Association.

Steiner, I.D. (1972). Group processes and productivity. New York: Academic Press.

Stogdill, R.M. (1972). Group productivity, drive, and cohesiveness. Organizational Behavior and Human Performance. 8, 376-396.

Stout, R.J., Cannon-Bowers, J.A., Salas, E., Milanovich, D.M. (1999). Planning, shared mental models, and coordinated performance: An empirical link is established. Human Factors. 41(1), 61-71.

Sundstrom, E.D., De Meuse, K.P., & Futrell, D. (1990). Work teams: Applications and effectiveness. American Psychologist. 45, 120-133.

Sundstrom, E., McIntyre, M., Halfhill, T., & Richards, H. (2000). Work groups: From the Hawthorne studies to work teams of the 1990s and beyond. Group Dynamics. 4, 44-67.

Swezey, R.W., Meltzer, A.L., & Salas, E. (1994). Some issues involved in motivating teams. In H. F. O'Neil Jr. and M. Drillings (Eds.), Motivation: Theory and research (pp. 141-169). Hillsdale, NJ: Lawrence Erlbaum Associates.

Swezey, R.W., & Llaneras, R.E. (1997). Models in training and instruction. In G. Salvendy (Ed.), Handbook of human factors and ergonomics (pp. 514-577). New York: Wiley.

Tannenbaum, S.I., Beard, R.L., & Salas, E. (1992). Team building and its influences on team effectiveness: An examination of conceptual and empirical developments. In K. Kelley (Ed.), Issues, theory and research in industrial/organizational psychology (pp. 117-153). Amsterdam: Elsevier.

Thompson, J.D. (1967). Organizations in Action. New York: McGraw Hill.

Triandis, H. C. (1989). The self and behavior in different cultural contexts. Psychological Review. 96, 506-520.

Triandis, H. C. (1994). Cross cultural industrial and organizational psychology. In H. C. Triandis, M. D. Dunnette, and L Hough (Eds.), Handbook of Industrial and Organizational Psychology, (2nd Ed, Vol. 4; pp. 103-172). Palo Alto, CA: Consulting Psychologists Press.

Vancouver, J. B. & Ilgen, D. R. (1989). The effects of individual difference and the sextype of the task on choosing to work alone or in a group. Journal of Applied Psychology, 74, 927-934.

Van de Ven, A., A. Delbecq, R. Koenig, Jr. (1976) Determinants of coordination modes within organizations. American Sociological Review 41, 322-338.

Vaughan, D. (1996). The Challenger launch decision. Chicago: The University of Chicago Press.

Wageman, R. (1995). Interdependence and group effectiveness. Administrative Science Quarterly. 40(1), 145-180.

Wech, B.A., Mossholder, K.W., Steel, R.P., & Bennett, N. (1998). Does work group cohesiveness affect individual's performance and organizational commitment?: A cross-level examination. Small Group Research. 29(4), 472-494.

West, M.A. and Farr, J.L. (1990). Innovation at work. In J. L. Farr (Eds.) Innovation and creativity at work: Psychological and organizational strategies (pp. 3-13). Oxford, England: John Wiley & Sons.

West, M., & Anderson, N. (1996). Innovation in top management teams. Journal of Applied Psychology, 81, 680-693.

Wickens, C.D. & Carswell, C.M. (1997). Information Processing. In G. Salvendy, (Ed.), The handbook of human factors and ergonomics (2nd ed.; pp. 89-129). New York: John Wiley.

Woolf, B.P., Beck, J., & Eliot, C. & Stern, M. (2001). Growth and maturity of intelligent tutoring systems: A status report. In K.D. Forbus & P.J. Feltovich (Eds.), Smart machines in education: The coming revolution in educational technology (pp. 99-144). Cambridge: The MIT Press.

Zaccaro, S. J., Rittman, A. L., Orvis, K. L., Marks, M. A., & Mathieu, J. E. (2002). Leadership processes in multi-team systems. In J.C. Ziegert & K.J. Klein (Chairs), Team leadership: Current theoretical and research perspectives. Presented at the 19th Annual Conference for the Society of Industrial and Organizational Psychology, Toronto, ON.

Zaccaro, S.J., Rittman, A.L., & Marks, M.A. (2001). Team leadership. Leadership Quarterly. 12(4), 451-483.

Zachary, W.W., Ryder, J.M., & Hicinbothom, J.H. (1998). Cognitive task analysis and modeling of decision making in complex environments. In J.A. Cannon-Bowers & E. Salas (Eds.), Making decisions under stress: Implications for individual and team training (pp. 313-344). Washington D.C.: American Psychological Association.

Zachary, W. Santarelli, T., Lyons, D., Bergondy, M. and Johnston, J. (2001). Using a Community of Intelligent Synthetic Entities to Support Operational Team Training. In Proceedings of the Tenth Conference on Computer Generated Forces and Behavioral Representation. 215-224.

Zachary, W.W., Campbell, G.E., Laughery, R., Glenn, F., and Cannon-Bowers, J.A. (2001). The application of human modeling technology to the design, evaluation, and operation of complex systems. In E. Salas (Eds.), Advances in Human Performance and Cognitive Engineering Research, (pp. 201-250). Amsterdam: Elsevier Science.

Zadeh, L. A. (1964). Concept of State in System Theory. In M. D. Mesarovic (Ed.), Views on General Systems Theory (pp.39-50). New York: Wiley.

Zaklad, A., & Zachary, W.W. (1992). Decision support design principles for tactical decision-making in ship-based anti-air warfare (Tech. Rep. No. 20930.9000). Springhouse, PA: CHI System.

CHAPTER 8

NATIONAL DIFFERENCES IN TEAMWORK

HELEN ALTMAN KLEIN AND ANNA PONGONIS MCHUGH

ABSTRACT

National groups vary in how they engage in organizational collaboration and teamwork. These variations are important to consider when undertaking coalition operations, predicting adversary actions, and facilitating/impeding technology transfer as well as developing organizational simulations. Many descriptions of effective teamwork, such as the "Big Five" model (Sims, Salas, & Burke, 2003), emphasize competencies such as Mutual Performance Monitoring, Back-up Behavior, Adaptability/Flexibility, Team Orientation, and Leadership that promote effective interaction. They look at mechanisms such as Shared Mental Models, Closed-Loop Communication, and Mutual Trust. These team competencies and mechanisms are compatible with Western organizations. They are not consistent with the cognition and interactive patterns that characterize collaboration and organizational functioning in other regions of the world. Therefore, if we want to describe and model multinational organizations and collaboration, we must move beyond our Western research base and incorporate the dynamics of collaboration found in non-Western nations.

INTRODUCTION

The purpose of this chapter is to explain why we cannot generalize from Western models of teamwork and coordination to account for the nature of collaboration in non-Western nations. By Western, we refer mostly to those from English-speaking and Western European nations. Our thesis is simple: the nature of collaboration is strikingly different when we cross national boundaries. Therefore, our preconceptions, our theories, our best practices, all have the potential to mislead us. This is true whether we are designing simulations, establishing doctrine, preparing negotiations, or trying to improve cooperation.

Because we live in a time of globalization, national differences in the nature of collaboration are important in a wide range of organizations and domains. Where businesses serving American interests used to be based primarily in one state or even one city, many, such as DaimlerChrysler, now span continents. Big Macs are flipped and served around the world with the backing of vast multinational management, sales, and distribution organizations. Civil aviation must manage an increasingly multinational work force. An air traffic controller at an international hub is typically supporting pilots from multiple nations. The flight crews may include several nations as expatriates help staff the airlines of many developing nations. Where scientists used to look to colleagues down the hall, they now network with others around the world. The Internet has provided a vital tool for international communication. A variety of services are being developed to provide virtual meetings for organizations working in different countries and different hemispheres.

National differences in collaboration are also important for the military. Military organizations have always been interested in anticipating the actions and reactions of adversaries. In attempting to model the nature of these decisions, analysts have often been concerned with national differences in doctrine, tactics, techniques, and procedures (DTTP), technology, and the command decision processes (CDPs) that might be encountered during conflicts. With the increased visibility of asymmetrical warfare, there is also considerable interest in the organizational structure of smaller groups and cells from relatively unstudied nations and cultures.

Military analysts have needed to expand their concern beyond historical adversaries to include new adversaries. They have to look beyond nation states to non-conventional adversaries including terrorist organizations. Globalization has also expanded the nature and extent of interactions with allies. There has been increased prevalence of coalition operations and humanitarian missions. While the model of the past was one of independent national action, perhaps in alliance with others, national military organizations are now more likely to work in integrated units with counterparts from around the globe. The United Nations as well as regional organizations, such as the North Atlantic Treaty Organization (NATO), work together in many theaters. In Bosnia-Herzegovina, for example, troops from over 30 nations staff NATO's Stabilization Forces Headquarters. Their leaders work together in teams to assess conditions, plan, coordinate actions, and manage emergencies.

The increasingly complex and dynamic tasks facing multinational corporate, scientific, and military organizations often require multinational collaboration. This is important for four reasons:

- Complex tasks demand multiple players in order to complete the required work in a timely fashion.
- Complex tasks can demand expertise in domains beyond the competence of one person or one nation. A corporate decision about marketing, for example, may need regional specialists, product designers, logistics coordinators, and MIS specialists.
- Multinational commerce requires buy-in so that all partners see the big picture and work toward organizational goals.
- Contributions from nationally diverse participants may contribute to adaptability and creativity, as multinational teams draw from a broader base of ideas and constructs.

Multinational collaborations allow the leveraging of expertise, effort, commitment, and adaptability toward a set of common goals. Teams are one important tool for collaboration.

The growing role of multinational collaboration places pressure on practitioners to cultivate effective multinational teams and on researchers to understand and model multinational teams in order to enhance performance. The challenge is to make these collaborations successful despite a range of cultural differences. This chapter describes some of the key cultural differences that can disrupt teamwork. By understanding these potential barriers, practitioners should be better able to overcome the problems created by cultural differences. They will also be able to incorporate different forms of collaboration into organizational simulations to reduce current Western-centered concepts.

Researchers have studied collaboration for many years in their effort to improve the performance of organizations and work groups within organizations. These research efforts have generated descriptive models of organizational performance, and these models generate guidelines and principles of effective team interactions. However, this research and these models are all based on Western populations, primarily United States teams. Will these ideas about effective teamwork generalize to other national groups and cultures? Our claim is that they will not generalize and that attempts to apply Western concepts of effective teamwork to other cultures can result in confusions and misunderstandings. We believe that existing Western-centric models need to be supplemented with alternative concepts of collaboration.

TEAMWORK: CURRENT MODELS AND LIMITATIONS

Current Models

A review of current concepts of teamwork provides a starting point for understanding the challenge of global collaboration. The interest is in the teamwork found in organizations working in natural settings. By this, we mean complex and dynamic settings that may include ill-defined and changing goals, incomplete information, time pressure, and uncertainty. There has been a great deal of research devoted to collaboration, teamwork, and organizational functioning (e.g., Swezey, Llaneras, & Salas, 1992; Zsambok, Klein, Kyne, & Klinger, 1993). Substantial progress has been made in outlining the requirements of teamwork and collaboration. It is beyond the scope of this chapter to review the vast literature on teamwork. Instead, we will rely on a synthesis of this literature that was developed to crystallize all that has been learned into a core set of competencies that are needed for effective teamwork. For further discussion on national differences in crowds see Chapter 17 on crowd modeling.

Sims, Salas, and Burke (2003) provided a "Big Five" model of the competencies required of successful teams. Starting from the accumulated research related to team functioning, they propose five core competencies that together integrate existing research in providing a model of teamwork. They also propose three mechanisms for coordination. The competencies are:

Mutual Performance Monitoring. Natural domains place complex and changing demands on team members. The complex and dynamic nature of the tasks can alter the workloads of members and introduce unexpected demands. This may make it difficult or impossible for an individual team member to complete needed assigned tasks. Mutual Performance Monitoring allows individual team members to identify the mistakes, slips, and lapses of others on the team, based on shared knowledge of the task and of individual resources. Monitoring provides an early warning signal for problems and breakdowns during complex operations.

Back-Up Behavior. Monitoring is necessary but not sufficient for effective teamwork. When mistakes or lapses are identified, successful teams can ameliorate the problems. Back-up Behavior, depending on Monitoring, allows a team member to take on tasks and responsibilities of others as needs and problems develop during work. It serves as a mechanism for balancing workload as demands vary over time. Teams function better when members fill in for each other.

Adaptability/Flexibility. At a strategic level, the team competencies of Monitoring and Back-up require Adaptability/Flexibility. Complex work environments often include elements of uncertainty and surprise. These demand ongoing changes in strategies and plans. An effective team must adapt as a unit to change. When surprises occur, adaptability allows for a smooth transition to a new course of action.

Team Orientation. Effective teams are more than a number of individuals doing an assigned task. They must also have a collective orientation. Team Orientation describes this collective tendency. It describes the extent to which team members identify with the accomplishments of the team and not simply with their own work performance. This Team Orientation results in stronger individual performance through coordination, evaluation, and group communication. Each team member functions as a part of an interdependent group. This conceptualizes an organization as a system functioning in a coordinated way for the common good. Coordination is critical for the accomplishment of the complex tasks demanded in natural domains. Team Orientation can be seen as the attitude that forms an interconnected system of individuals.

Team Leadership. Coordinating the complex tasks of a team requires leadership. Team Leadership creates a Team Orientation, maintains shared knowledge, fosters coordination and skill development, and structures team experience. The leader has the responsibility for defining team goals, organizing team resources, and providing guidance for reaching goals. A leader also sets expectations and fosters a climate that leads to successful interaction patterns. While the leadership role can vary, the competencies are critical. The role of leader is critical no matter how the leader has been selected. Leadership is represented in Figure 1 as an increased flow of information and directives as the leader coordinates actions and intent.

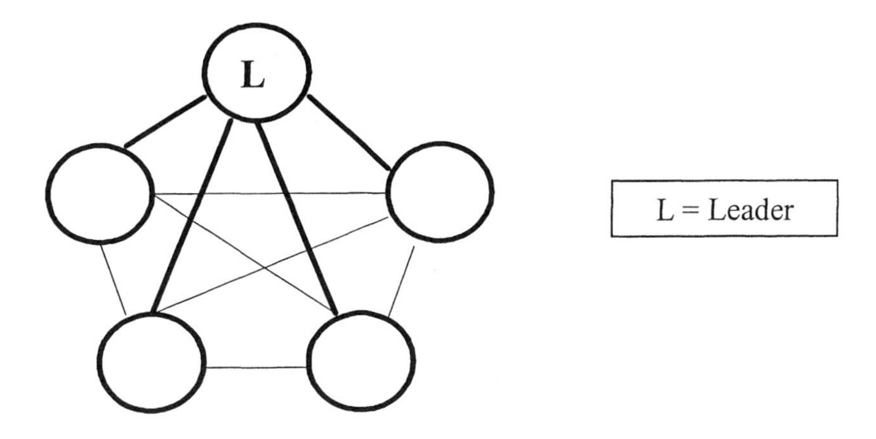

FIGURE 1. Interaction Model.

While task demands can alter the relative importance of these five competencies over the life cycle of a team, Sims et al. (2003) maintain that each competency is important for effective organizations. These five competencies require coordination and three mechanisms for coordination are proposed. Sims and colleagues see these mechanisms as providing a supporting structure for coordination. These mechanisms are:

Shared Mental Model. Shared Mental Models provide each team member with a common understanding of the long-term goals, the nature of the required tasks, the roles, skills, and capacities of team members, and the interrelationship among tasks and people. Shared Mental Models allow the anticipation and prediction of needs and problems. Shared Mental Models support performance monitoring by providing expectations for teams tasks. They allow Adaptation/Flexibility by providing functional understanding of activities and needs.

Closed-Loop Communication. Communication provides ongoing sharing of information about individual functioning and overall status. It gains in importance as complexity increases by allowing the ongoing distribution of new information. Closed-Loop Communication involves the initial sending of a message, the receiving, understanding, and acknowledging of the message, and follow-up to confirm the accurate transmission of information. Closed-Loop Communication insures the exercise of Team Leadership and a Team Orientation.

Mutual Trust. Finally, Mutual Trust allows a sustained commitment to common goals and processes. It provides confidence that others will perform as expected and it conveys that each participant will protect the interests of the others as part of a commitment to their shared goals. Mutual Trust is critical for a Team Orientation. Before team members take on additional duties, they need a sense of common cause and reciprocity.

These and most other models of effective teamwork converge on a set of "best practices," and are overwhelmingly based on research designed by Western scientists with Western samples using Western paradigms.

Describing Applied Domains with Current Models

We will now review several domains that illustrate how national differences call for an expanded description of the ways that organizations work together on complex tasks. These domains are multinational peacekeeping operations, civil aviation, and foreign CDPs. In each domain, we will look at the role of the five competencies in effective organizational functioning. This analysis is based on naturalistic observation, interviews, and research conducted within multinational and non-Western organizations. It suggests that the key five competencies described above as needed for effective team functioning and decision making may be ethnocentric and specific to Western cultures. The "ideal competencies" do not always generalize to non-Western and multinational organizations. Non-Western groups often do not subscribe to these views, and sometimes even find them inefficient. We are not criticizing the work of Sims et al. (2003). We find this work valuable for capturing the essence of Western-based research. Our concern is simply that the factors identified by Sims et al., and by the researchers from whom they draw, cannot be safely generalized to non-Western teams.

Multinational Peacekeeping Operations. Peacekeeping is a growing concern for military training. The Technology Transfer for the Collaborative Technology Alliance program has been focused on this concern. As part of research sponsored by the Army Research Laboratory, we observed top-notch military units from different nations working toward common goals at NATO's Stabilization Force (SFOR) Headquarters in Bosnia-Herzegovina (Hahn, Harris, & Klein, 2003). As personnel from many different nations work together, they sometimes have interesting problems related to the competencies and mechanisms described above.

To learn more about these problems, we interviewed participants to identify and understand the dynamics of their cross-cultural problems. To learn about Back-up Behavior, we asked one non-U.S. soldier what he would do if a coworker was falling behind on his task at a time when he, the interviewee, did not have pressing tasks. We asked if he would fill in for the other, helping with the needed tasks. He shuttered and said, "Of course not. That would be 'eating another man's bread'." He explained that work and performing your assigned work was basic to a man's sense of self and accomplishment. "If you do another guy's job, you would shame him." *Challenge: Does this officer's scorn of Back-up Behavior mean that his nation cannot have effective teamwork? How do they balance workload when demands change?*

We found evidence for this same limitation of Back-Up Behavior in interviews with foreign national graduate students. Teamwork plays a critical role in many Western educational settings. Business and engineering programs often assign group projects to train students for what is viewed as a critical part of professional functioning. While U.S. teams view compensation and Back-up Behaviors as important in organizations, this is not universal.

At SFOR, we also explored the function of Team Orientation (Hahn, et al., 2003). For several of the national groups, optimal organization was seen as having the leader at the center and each member connected to others through the leader. Coordination and communication was accomplished only through the leader. A sense of interdependence was not seen as critical or even part of good group functioning. You work for your commander. There was no role for "Mutual Performance Monitoring." These functions were neither needed nor understood. Instead, the teams appeared to rely on direct dyadic coordination of individual team members with the leader.

Several interviewees reflected on the inefficiency of the U.S., British, and Canadian forces. "They waste a lot of time and effort talking and checking each plan. Our commander knows what is going on and where we are headed. He is in the best position to make the decisions. The Brits are always coordinating because no one is leading." "The more people involved in coordination, the more opportunities for mistakes." The non-Westerners were skeptical of the way that Mutual Performance Monitoring and Back-up Behaviors increase coordination costs.

A different pattern of coordination can be described as a hierarchical pattern or, in the most basic case, a hub-and-spoke pattern (see Figure 2). As shown below, the flow of information, decision making, and coordination is from the team leader to each team member. The hub-and-spoke arrangement is different from the interactive team arrangement assumed by descriptions such as the "Big Five" model. *Challenge: How can communication and coordination be accomplished if the team members are not interacting with each other?*

Finally, our observations revealed limitations with concepts of "adjusting" or Adaptability/Flexibility. "Operation Harvest" in Bosnia has as its goal the collection of guns, explosives, and other weapons. This is to reduce the probability and severity of dangerous clashes between civilians. American and British teams talk about the strength of "staying loose to always be ready for any surprise that comes along." But this is only one approach to a successful operation. Other national groups stress the value of careful, well-defined planning. "The Brits just don't do their work ahead. If you are surprised, you did not do your job. I make sure everything is set. I'm not risking my men's lives." This approach values precision over flexibility. The Westerners say, "They never get out into the field to do the work because they are too busy planning. Then if something really unexpected comes along, they are stumped." Here we see flexibility, a Western preference, over precision. *Challenge: How should we handle differences between national groups in the way precision and flexibility are valued?*

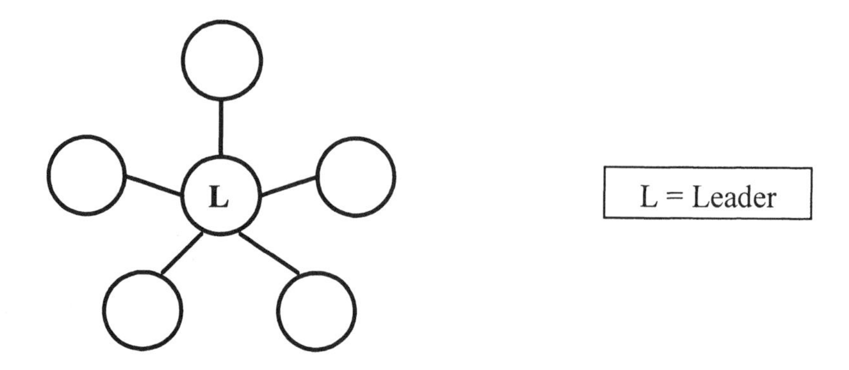

FIGURE 2. Hub-and-spoke model.

Thus, each of these four competencies, as defined in Western terms, runs into trouble. What about Team Leadership? The Western concept of leadership is problematic in two ways. First, several personnel from the former Warsaw Pact nations expressed their dismay that the Western commanders often delegated responsibilities that they felt should not be delegated. "Our commanders do their jobs. They don't ask us to do them." They do not feel it was their role to work directly with others or to make decisions. That was a leader's job. In contrast, personnel from one of the Northern European nations expressed that some leaders did not sufficiently share knowledge or include others in setting goals and organizing actions. "We are highly trained professionals. We work best when we can contribute to decisions." *Challenge: Can people with different leadership models function successfully together on teams?*

Civil Aviation. Next we examined the role of non-Western collaboration in civil aviation. Boeing Aviation sponsored this research. Boeing scientists know that while their equipment, procedures, and training are used worldwide, the safety outcomes vary by region. In an effort to learn more about potential national differences in explaining the differences, Klein, Klein, and Mumaw (2001) interviewed personnel in charge of training and assessing the skills of international pilots. The trainers were all Americans with decades of aviation experience. They used American designed/built planes, flight simulators, and manuals—important tools for training and testing.

Consistent with the model of teamwork presented above, Americans view Adaptability/Flexibility as an important part of training and testing. When American trainers test Americans and other Westerners, they routinely change the orders of simulated incidents and alter their characteristics. They value flexibility in handling routine flight challenges and also non-routine simulator challenges.

Our interviews examined the way Japanese pilots are routinely tested in simulator exercises. When Japanese trainers do this, they use a standard order with no variation. The pilots are required to manage the incidents to achieve rapid, flawless performance. The Japanese, who are exceptionally good on the practiced exercises, are very weak in managing unexpected variations. In contrast to U.S. pilots, those from some other regions were not only less skilled at handling the unexpected, but they also were irritated by the exercises. They made it clear that they trained for precision not flexibility and that they valued precision more highly than flexibility. Their comments reflected the belief that expert pilots flew with precision, eliminating the need for disparate solutions and flexible approaches. *Challenge: How can effective training be provided to crews that strive for precision and those that strive for flexibility?*

Crew Resource Management (CRM), a tool for aviation safety, was developed consistent with an American concept of teams. While CRM has evolved over the years, its underlying intent has always been to use all of the strength available in the cockpit. This means that the pilot does not fly alone but with the aid of the 1st officer's second set of eyes and ears. Crews are trained to attend to this information and to respect it. Considerable effort was required to ensure that this approach was incorporated in Western nations.

It has proven more difficult to incorporate CRM in some non-Western nations. Our belief is that the difficulty stems from national differences in team concepts including Monitoring, Back-Up, Adaptability, Team Orientation, and Leadership. Crew Resource Management has not been easily adopted where it was incompatible with existing team concepts.

Crew Resource Management is based on Western concepts of teamwork, as described by Sims et al.'s (2003) five competencies. For non-Western countries, the CRM guidelines for Mutual Performance Monitoring, Back-up Behaviors, Adaptability/Flexibility, Team Orientation, and the nature of Leadership are simply not relevant. Three examples provide contrasts with typical American performance. As one example Chinese captains receive flight instructions from the control tower. One flight inspector reported how captains would ignore obvious dangers when their judgment conflicted with the instructions they had received. He said he had been on a plane when it flew directly through a severe storm. The Chinese captain had been told to proceed and he did. It had not been his job to assess the weather on route.

The next example occurred during a check ride on a Middle Eastern aircraft. The check pilot told us how he had made several in-flight suggestions to the captain. This is a typical pattern for such flights. After the flight and as soon as the rest of the crew had left the cockpit, the captain told the check pilot that no one could ever criticize him in his aircraft.

A well-documented aviation disaster, VASP, Flight 168, in South America provides a last example of teamwork contrasts. On 8 June 1982, a Boeing 747 was flying near Sierra de Pacatuba, Brazil. The 1st officer asks, "Can you see there are some hills in front?" The captain ignores the 1st officer's comment until it is too late. Then, "Some hills, isn't there?" This was followed by the sound of impact. Here, the 1st officer's attempt to monitor the captain was "inappropriate and so ignored." The aircraft crashed into a 2,500 ft. mountain during an approach in heavy rain and fog. Despite two altitude alert system warnings and the co-pilot's warning of the mountains ahead, the captain continued to descend below the minimum descent altitude. *Challenge: How can a cockpit crew use all of the resources available in a way that is compatible with non-Western as well as Western nations?*

Foreign Command Force Modeling. Military plans try to transform the past actions and decisions of the adversary into predictions and action plans for the future. This foreign CDP comes from a different framework but presents a convergent picture with those of the other domains. During the Cold War, U.S. forces learned a great deal about the Soviet military organization as well as the organizations of U.S. allies. Extended and varied experiences also provided an understanding of the capacities and characteristics of equipment and technology. This knowledge of organizational structure was captured in conceptual models describing the CDP of ground forces. Models described the deliberate planning process—for the offense and defense —for a division commander and staff (Tamucci, Timian, & Burnett, 2000).

During Desert Storm, in 1991, analysts would have benefited from CDP-based models describing Iraqi military actions. Such models would have provided a picture of information flow, command processes, and coordination as well as the nature of plans, risk management, and typical logic. The U.S. and its allies knew that the Iraqis used Soviet equipment and some aspects of Soviet force structure and DTTP, and that they also used aspects of British-based force structure and DTTP. Military analysts attempted to use this information to predict Iraqi patterns. Models of CDP of the deliberate planning process, however, failed to describe the performance of the Iraqi Army. Doctrine-driven models alone could not account for the behavior of a commander and his staff.

Organizational simulations required more than force structure, DTTP, and equipment specifications in order to make good predictions. One missing piece was national differences. Lannon, Klein, and Timian (2001b) described a need to incorporate national differences into any organizational simulation that is intended for use with non-Western cultures. The Iraqi military decision process, for example, does not share British Power Distance or Hypothetical Reasoning. While the British structure was officially hierarchical (see Figure 3), many channels were available for informal communication. The Interaction Model (Figure 1) better helps to understand British leadership and decision making. This interactive element would be lost if the model were to be adopted by nations high in Power Distance or Concrete Reasoning.

Traditionally, military intelligence is focused on doctrine as the primary indicator of how an army will fight. However, doctrine alone cannot account for military behavior. Armies are products of their country's history and culture. The reason that Iraqi forces did not fight exactly like the Soviets is due in part to the fact that the Iraqi and Soviet armies have some differing military structures, but an even more important reason is because the Russian and Iraqi cultures are very different. Even in those cases where Iraq has adopted a Soviet procedure in its entirety, the Iraqi commanders saw that process through an "Iraqi lens" and did not implement that process in a fully "Russian manner" (Lannon, et al., 2001a). Simulation developers cannot simply model formal procedures; they also need to

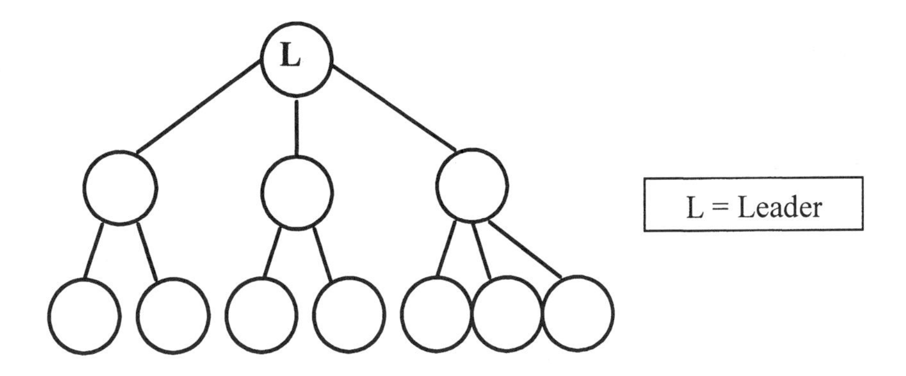

FIGURE 3. Hierarchical Model.

incorporate national differences in the assignment of roles and functions. *Challenge: How can simulations incorporate national differences in teamwork?*

Summary

Researchers have assumed that collaboration is the same across nations because they assume that cognition and the social context of cognition are common across national groups. Yet, differences abound. We see them during multinational peacekeeping missions, and they are a critical contribution to civil aviation crew behaviors. They are a pervasive element in military organizations regardless of formal doctrine. Taken together, evidence from studying the domains of peacekeeping, civil aviation, and command force modeling suggests that current models of organizations and their needed competencies are not universal. Efforts to understand, predict, and simulate organizational action fail when the impact of national differences is not considered. While current models may describe U.S. and probably some Western teams, they do not provide a good picture of collaboration in other regions of the world. This is not surprising because past research is based almost exclusively on research conducted by Western researchers, with Western paradigms, and Western participants. Globalization complicates the task of developing organizational simulations.

In multinational operations, it is necessary to understand and anticipate the teamwork of others. In aviation, it is necessary to design and train for the teamwork of users. In foreign command force modeling, it is necessary to incorporate the teamwork differences of adversaries. We need to anticipate the actions of both our allies and our adversaries from many different nations. Finally, organizational simulations must represent national differences in order to make accurate predictions. To do this we need more than the five competencies proposed as necessary for effective teamwork by Sims et al. (2003).

NATIONAL DIFFERENCES AND THE MECHANISMS OF TEAMS

The five competencies are a necessary consideration in understanding organizations and collaboration. We now focus on how the additional challenge introduced by national differences affects the coordination mechanisms of Shared Mental Models, Closed-Loop Communication, and Mutual Trust. We review national differences in cognition and the social bases of cognition (Klein, H. A. 2004) found in the recent cross-cultural psychology, cultural anthropology, and cognitive psychology literature. We show how these national differences challenge the universality of the coordination mechanisms. If models of Western teams cannot be safely generalized to other national groups, then we need to understand why our Western models are breaking down. By incorporating knowledge of national differences we can provide a more accurate and useful model for understanding the mechanisms for coordination used by multinational teams.

We have already made the case that team competencies vary for different national groups. They also vary in expectations, procedures, reactions to uncertainty, and roles. As a result, members of different groups may be continually surprised and disoriented when faced with the practices of other groups. This reduces predictability, which is a prime requirement for effective coordination. Differing Mental Models make it difficult for members of one national group to anticipate and understand the decisions and actions of other groups. Differences also prevent members of multinational teams from using abbreviated messages and other efficiencies that can cut coordination costs.

Several research traditions describe national differences in cognition and the social context of cognition. We will first describe four of these research traditions and then present specific dimensions that create barriers to effective collaboration in multinational groups. The dimensions influence the creation of Mental Models for cognitive work, judgment, and decision making. They generate different approaches to communication and can interfere with the unimpeded flow of information. To the extent that these dimensions influence people in organizations, they can undermine Mutual Trust within multinational organizations.

- Kluckhohn and Strodtbeck (1961), working in the tradition of cultural anthropology, identified differences among pre-industrial groups. They documented variations in planning, with some groups looking at the weeks ahead while others looked to the long-term needs of their grandchildren. The term "Time Orientation" describes this difference. They also noted that some groups accommodated events in the world while others appeared driven to master them. "Relationship to Nature" describes this difference. Finally, some groups valued work and achievement while others valued people and relationships. They used the terms "Doing" and "Being" to capture this difference.

- Hofstede (1980), a social psychologist, used methods of social psychology and industrial/organizational psychology to identify differences among employees of a large multinational corporation. The research built on earlier research including that of Kluckhohn and

Strodtbeck (1961). Members of different national groups varied in their comfort with uncertainty. Some were comfortable with uncertainty while others worked to reduce it. "Uncertainty Avoidance" describes this difference. Some groups respected and conformed to hierarchical structure while others showed an egalitarian structure. "Power Distance" describes this difference. Recent work confirms the importance of these differences in aviation and medicine (Helmreich & Merritt, 1998). Hofstede (1980) also presented Individual-Collective and Masculinity-Femininity as dimensions of difference.

- Markus and Kitayama (1991) looked at reasoning across national groups from the framework of cognitive psychology. Consistent with earlier notions of individual-collective, they proposed that this social dimension shapes the concept of self. The individual social characteristic is associated with an independent self-concept while the collective person has an interdependent self-concept. These concepts are reflected in cognition. While most people are capable of a range of reasoning, some groups—those that are interdependent—prefer reasoning grounded in concrete reality. Groups that tend to be independent tend to favor more speculative, hypothesis-based reasoning. The distinction is described as "Concrete vs. Hypothetical Reasoning."
- Nisbett and his colleagues (Nisbett, Peng, Choi, & Norenzayan, 2001; Norenzayan, Smith, Kim, & Nisbett, 2002) explored the differences between analytic reasoning—characteristic of people from Western nations—and holistic reasoning—characteristic of east Asian nations. They noted that analytic groups attributed cause to individual dispositional characteristics while holistic groups were more likely to look to situational as well as dispositional contributions (Choi & Nisbett, 1998; Choi, Nisbett, & Norenzayan, 1999; Morris & Peng, 1994). "Attribution" describes this distinction. They also noted that people from analytic nations tend to make decisions by contrasting: seeking distinctions and choosing between options. Those from holistic nations tend to synthesize: seeking commonality. "Differentiation vs. Dialectical Reasoning" describes this difference (Peng & Nisbett, 1999).
- H. A. Klein (2004) formulated a Cultural Lens Model (CLM) to capture the distinctions that appear to be valuable for understanding differences in cognition and the social context of cognition. We will now look more carefully at six dimensions from the CLM that are particularly important for the team coordination mechanisms of Mental Models, Communication, and Mutual Trust. These dimensions include: Relationship-Achievement, Power Distance, Tolerance for Uncertainty, Hypothetical-Concrete Reasoning, Causal Attribution, and Contrasting – Synthesizing. Some of the dimensions are relabeled from the original model for clarity.

Relationship-Achievement

The first of the six dimensions, Relationship vs. Achievement, is parallel to the Being vs. Doing dimension of Kluckhohn and Strodtbeck (1961). This dimension describes the emphasis on social interaction patterns within organizations and has less influence on the underlying logic or reasoning of decisions. It influences the flow of information and the establishment of trust. Those with an Achievement orientation, typical of Westerners, separate their work from their social interactions. They come to work to accomplish goals and view socializing as a waste of time. If it takes being at their desk until late at night, they view it as their job to be at their desk. They define themselves and others by assigned roles. In contrast, those with a Relationship orientation view the social relations formed at work as an integral part of their job. They value the linkages and the social opportunities these create. They view personal relationships and the concurrent Mutual Trust as vital for the long-term success in goals. They know the key people when connections will help. Each of these positions—Relationship and Achievement—carries a different view of teamwork.

At HQ SFOR in Bosnia-Herzegovina, differences on this dimension were reflected in pervasive differences in team functioning. Americans came to work early. They did their job working through lunch if needed. They treated others "professionally" being careful not to waste time in irrelevant chatter. Other groups were more likely to view the work environment as an extension of their social life. They took the time to network and get to know others. They developed working relationships. What may have looked like slacking translated into a strong interpersonal connectedness and a tool for managing task demands. These differences impacted Closed-Loop Communication moderating the flow of information. They influenced the resources each person could tap, the speed with which tasks were accomplished, but also the level of commitment given to tasks. They influenced the time needed to reach productivity.

Both Relationship and Achievement styles can work, but the mechanisms and patterns are different. Each generates a different Mental Model of how team members interact and how tasks should be accomplished. Communication differs because each style suggests different patterns of sharing information. Such differences have a profound impact on Mutual Trust. Nations with similar patterns are mutually respectful and they may look down at others. At SFOR, U.S. staff complained, "A two hour lunch! They waste their time socializing while we work." Italian officers told us, "You can't spend your time at your desk and expect to work on a team. The Americans never figure out how to cooperate with the rest of us." *Challenge*: *How can people work together effectively when they have different expectations for time use and the role of socialization? How can Mutual Trust develop in such an environment?*

Power Distance

Power Distance comes from Hofstede's (1980) work and has been heavily used in business, aviation, and medicine. It has been represented in earlier organizational simulations. This dimension describes the social nature of leadership and the flow of information within an organization. It has less influence on the underlying logic or reasoning of decisions but rather focuses on the social and behavioral constraints that surround organizational processes. It influences the flow of information and the establishment of trust. High Power Distance is consistent with a hierarchical organizational structure. Leaders expect to make decisions and give orders. Other team members want direction from their leader and do not want or have the skills to initiate actions. The assignment of roles and functions is based on place in the power hierarchy. In low Power Distance groups, people at all levels see themselves as sharing in decisions and information.

We hear a great deal about a "military culture," or an "aviation culture," to describe how military officers and pilots and other groups have their own standards and worldviews, regardless of their national origin. Certainly, training in military mission or aviation provides some commonality that transcends national boundaries. However, in observing interactions between military officers and pilots and other groups who claim a distinct professional culture, we are struck by the predictable conflicts and confusions that reflect national origin. Even in military and aviation domains, necessarily high Power Distance organizations, national groups differ greatly in the distribution of power and the flow of information.

The impact of Power Distance differences across national groups can be seen in attempts to export CRM to the cockpits of airlines around the world. Western procedures depend on the free two-way flow of information. Even though the captain has final say, it is expected that the 1st officer tell the captain if there is a problem or if he makes a mistake. Crew Resource Management assumes that information flows freely. The Mental Model demanded for successful CRM is a Western Mental Model of teamwork. The communication patterns demanded are Western communication patterns. When Western procedures are transplanted to high Power Distance nations, they can fail because 1st officers do not provide a useful second set of eyes and ears to detect problems. They do not see it as their role to disagree or contradict the captain. While CRM has proved to be valuable in ritualizing back-up monitoring in Western nations, it has not been as successful in other regions. The next step in exporting CRM to non-Western organizations would be to develop techniques that are compatible with Mental Models and communication patterns stemming from the high Power Distance of non-Western flight crews.

Power Distance is strongly related to leadership. Power can be assigned by competence and training or it can be assigned by family status, kinship affiliation, or group membership. The role may be symbolic with goals, organization, expectations, and interactions carefully defined by a hierarchy or by past practices. Alternately, the role of the leader may be flexible and emerging. When team interactions are minimal, this alters the function of the leader. In interviews, we found that the formality used in selecting a leader varies across Western and Far Eastern cultures. Chinese businessmen described a relatively formal—and less emergent—process for selection. For more important projects, some organizations may use a panel of experts to appoint a leader.

Leaders in some places are clearly identified and "in charge." The leader may sit in a specific place with those of equal or close importance sitting next to him. The farther away you sit, the lesser your importance. One informant reported, "The leader is the only person who speaks in the meeting. The others may converse with him individually. In other groups, the leader may provide input to insiders, but not speak directly to other people at the meeting.

The level of authority that team leaders generally possess may also vary. Chinese MBA students with years of experience working in both private and government-owned companies provided some insight into this. In general, although the leader of a Chinese team is expected to listen to every team member's opinion, it is his responsibility to make the final decision. Once the decision is made in a Chinese business organization, it is highly inappropriate for team members to question it or criticize it. As a team member, you must accept it and support the leader's decision, even if you do not agree with it and even if it leads to a sub-optimal outcome. Adaptation is not an alternative after this point. *Challenge: Can communication be effectively maintained when members have different Mental Models of authority and Power Distance?*

Tolerance for Uncertainty

Tolerance for Uncertainty comes from Hofstede's (1980) Uncertainty Avoidance dimension. The CLM changed the term because the original term was difficult to explain to practitioners. Tolerance for Uncertainty is part of the social context of cognition as well as part of cognition itself. While it moderates the interpersonal interactions within organizations, it is also influential in risk assessment and decision making. This dimension often appears in business and military settings. It was represented in earlier organizational simulations. Tolerance for Uncertainty describes the level of risk, uncertainty, and ambiguity acceptable by members of a group. In low Tolerance for Uncertainty nations, uncertainty is aversive and people work hard to reduce it. In order to accomplish this goal, they emphasize detailed time lines and procedures. Fixed and committed plans are viewed as important while changes are stressful and to be avoided. Care is taken to ensure the accuracy of incoming information. In contrast, high Tolerance for Uncertainty nations prefer flexible planning. They are willing to begin with incomplete information and will change readily when new information becomes available.

Differences in Tolerance for Uncertainty were troublesome during multinational operations at SFOR. Operation Harvest is a disarmament effort in Bosnia-Herzegovina in which weapons are collected in order to lower the risk of aggression. Some national groups would act only when they had complete information, even if it meant fewer collections. They meticulously scripted each home visit and specified many variations of each script. When there was more rain than anticipated, they would cancel a planned collection. They had not worked out the details for the rainier, hence muddier situation. Officers from other nations were appalled by this caution. They wanted a general plan and the flexibility to respond to deviations along the way. Thus, we see two very different Mental Models of appropriate planning.

Differences in Tolerance for Uncertainty decrease Mutual Trust among coalition members. An officer from the low Tolerance for Uncertainty group, uneasy with the "careless" planning of high Tolerance for Uncertainty staff reported, "When we sit down to plan, they drive us nuts! They want to keep everything open. We have to make decisions and we should do it when we have the time to think!" The high Tolerance for Uncertainty officers said, "They are so busy planning, they never get out to collect weapons. Sometimes, you just have to punt! Our men can always figure out what to do."

Business organizations also reflect the Tolerance for Uncertainty characteristic of their nation. Japanese businesses are known for their careful, detailed planning. They devote considerable effort to consider all options and to include all organizational levels. Once the plan has been adopted, it will be carefully followed. They also prefer long-term contracts and commitments to support their planning process. At the same time, Japanese business practice reflects an anxiety with ambiguity. No "shooting from the hip" here! *Challenge: How can team functioning be understood across differences in Tolerance for Uncertainty?*

Hypothetical—Concrete Reasoning

Markus and Kitayama (1991) started with a social concept of self to understand Hypothetical-Concrete Reasoning. This dimension, and the two that follow, describe differences that underlie complex cognitive performance. They are more difficult to represent in simulations, but have the potential for improving predictions of complex operations, including problem identification, planning, and decision making. They are critical for identifying the difficulties that emerge with Mental Models as well as the difficulties that can undermine Mutual Trust. Groups that have an interdependent concept of self ground their reasoning in concrete reality. In contrast, more speculative, hypothesis-based reasoning is associated with an independent self-concept. Parental and educational practices as well as modeling of adult patterns help to develop the reasoning characteristic of a particular national group.

Each of these cognitive patterns leads to a different Mental Model for approaching new situations and challenges. Concrete Reasoning looks to past examples and events as the first guide to understanding and planning. Assessment of situations and planning are both grounded in experience and history. Information needs are high because concrete reasoning strives for precise predictions based on appropriate comparison cases. In contrast, Hypothetical Reasoning requires speculation on the forces at work in a new situation. Engaging in mental simulation of possible actions is part of planning. Those with Hypothetical Reasoning are comfortable thinking about situations they have never experienced. This leads to flexibility in the face of surprises sometimes at the cost of precision.

Differences in reasoning lead to differences in planning and decision making and different information needs and options. A source of mistrust in multinational planning teams at SFOR was the difference in planning generated by differences in Hypothetical-Concrete Reasoning. It is hard for a person who uses Concrete Reasoning to see why others speculate before they have all the needed data. They do not see why plans with potentially serious consequences can be based on "imagination." It is hard for those with Hypothetical Reasoning to see why others refuse to show any creativity in their analysis. They think others should be able to "think outside of the box." Particularly under time pressure, the person with Concrete Reasoning can be seen as an impediment to action. *Challenge: How must the nature of planning change if it is going to be executed by members of cultures who have different Mental Models of reasoning?*

Causal Attribution

Faced with complex information, pressure, or opportunity, people use different Mental Models to attribute causality to their observations. Nisbett and his colleagues (e.g., Ji, Peng, & Nisbett, 2000; Nisbett, 2003; Nisbett, et al., 2001) noted powerful differences among national groups in this attribution (Choi, Dalal, Kim-Prieto, & Park, 2003; Choi, et al., 1999; Morris & Peng, 1994). Attribution focuses attention and narrows the selection criteria for approaches or remedies. Those with dispositional attribution attend to the unique characteristics of the person or object, locating responsibility primarily in the individual (Choi, et al., 1999). In contrast, those with a situational attribution model are more likely to adopt context-dependent and occasion-bound thinking and look to situational and contextual contributions. A dispositional Mental Model of causality is more characteristic of Western nations, and a situational model is more characteristic of East Asian nations.

Attribution provides the initial situational assessment and directs problem identification and problem solving. When organizations or teams encounter anomalies or problems, they must make sense of it before they can make decisions or plan change. In Bosnia, the peacekeepers differed in their attribution. When peacekeepers in Bosnia faced difficulties maintaining services in a refugee camp, they differed in their attribution of cause. Similarly, repetitive aviation equipment failures and unexpected enemy activity brought forth different causal attribution from maintenance personnel and military analysts respectively. Sensemaking among collaborative partners is based on existing Mental Models of causality.

A dispositional attribution demands a plan that addresses the individual characteristic identified as the cause. Training, selection, disciplinary actions, and counseling might be considered appropriate remedy for organizational concerns attributed to individual dispositions. A situational attribution calls forth solutions that may encompass multiple contextual considerations. Those with situational attribution look to the broader context and holistic solutions. They are less uncomfortable with retraining that targets specific individuals. They favor efforts to modify organizations and procedures while placing less weight on selection standards. It is only possible to predict the actions of an adversary or a team member, by using the same Mental Model as the adversary or team member.

There is merit in both approaches to attribution. Multinational organizations, however, have trouble arriving at a solution when they do not share a Mental Model of attribution. As in earlier analysis, a difference in this dimension can reduce Mutual Trust because each party may be critical of the sensemaking and planning of the other. *Challenge: How can people begin to solve problems when they cannot agree on what the problem to be solved is?*

Contrasting – Synthesizing

Contrasting – Synthesizing, taken from Nisbett's concept of Differentiation vs. Dialectical reasoning, describes the difference in how people typically manage inconsistent information and incompatible goals (Peng & Nisbett, 1999). We adopted this new label because it was better understood with users. Contrasters make decisions by seeking distinctions and choosing between options. They understand contradictions by separating and evaluating distinct qualities. This polarization sharpens distinctions by highlighting strengths and weaknesses in order to identify the best option. Synthesizers seek commonality and look for integration rather than sharpening distinctions. They avoid conflict (Chu, Spires, & Sueyoshi, 1999) and believe that all perspectives contain truth. Synthesizers seek harmonious intermediate positions, deny dichotomous descriptions, and retain elements of different perspectives (Peng & Nisbett, 1999; 2000).

Contrasters plan by developing and evaluating the relative merits of two or more plausible alternatives. They may even assign different teams to provide the best case for each alternative. Plans are reviewed with discussion focusing on relative advantages. The best option is selected for implementation although it may be modified to accommodate weaknesses exposed in the decision process. A good leader guides the group to the selection of the best alternative. Among Contrasters, conflict is considered healthy and is even sought out. It is viewed as a way to sharpen ideas and to improve performance.

Synthesizers consider a range of ideas, concerns, and options. The process is directed at integrating as many positive features and contributions as possible. They also try to avoid losing any of the strength. Each person looks for ways to pull ideas and conflicts together and cover up disagreement. A skilled leader would be one who can knit together the seemingly contradictory elements into a functional whole. In synthesizing groups, conflict is viewed as a damaging force and is avoided. People are careful not to offend other team members. If conflict begins to emerge, team members might cover up a disagreement.

This dimension can be seen in many contexts. Aviation maintenance personnel are expected to keep equipment on schedule and, at the same time, ensure the safety of every aircraft. Similarly, a pilot strives to arrive on time, but unexpected weather may make this goal risky. How are the competing goals managed? A manager may want to hire a top-notch scientist and also a dependable, cooperative coworker. How are hiring decisions made when the two characteristics do not appear in the same person?

Contrasters often describe synthesizers as indecisive. "They can never make up their minds! They will do everything in order to avoid making a decision. Their plans are hodgepodges." Synthesizers describe Contrasters as narrow and limited. "They are so eager to find one solution, that they discard a lot of good ideas in the process. It's a weaker plan but they are happy because they value coherence over effectiveness." This lack of a Shared Mental Model reduces Mutual Trust. *Challenge: How can multinational organizations engage in planning when they resolve contradictions and differences in ways that seem incompatible?*

THE PROMISE OF ORGANIZATIONAL SIMULATION

Simulations are important tools for capturing the dynamic functioning of complex systems. Organizational simulations begin with a conceptual model of human organizations and with data on the actual functioning of the types of organizations under study. The simulation may be a complex and large-scale organization or it may abstract a more limited function of an organization. Simulations can use existing data to describe the generation capacity of a power plant or the ground speed of an army in varied terrains. In all cases, a simulation should have the capacity to generate testable predictions with clear analogues in natural organizations. The measurable outcome might be of time to task completion, flow of information, choices, and the like. The closer that prediction is to reality, the greater the confidence in future predictions. Naturally occurring changes or externally induced changes can provide continuing assessment and allow ongoing simulation revision based on observations in context.

As accuracy of predictions increases, we are more comfortable using the simulation outcomes to guide actions. How long will it take for a terrorist cell to recover after the elimination of its leader? What is the optimal staffing size for a surgical team? What external changes might impede a rogue state from rapidly implementing a deadly technology? These questions may carry life-and-death consequences. Comparisons between simulation predictions and actual outcomes expose limitations. Deviations from predictions should be welcomed guides to upgrading simulations. When a simulation makes good predictions about optimal staffing for a small surgical team, but poor ones for a massive procedure such as open-heart surgery, this suggests revisions and extensions that strengthen the simulation.

We make the case that simulations need to incorporate national differences in the nature of collaboration because of their pervasive influence on organizations. They need to incorporate how individuals assume roles and functions in organizations and how they carry out necessary roles and functions. They also must incorporate how information is selected from all valuable information and how it is used in decision making. National differences also influence the product of the collaboration itself—the nature of plans and how they are generated, modified, and executed. Cultural differences influence problem identification, planning, leadership, and coordination. Ignoring national differences leads to errors in understanding allies and adversaries. It can create dissonance and malfunctioning during coalition operations, international business, and other collaborative efforts.

We have well-developed data-based models of teamwork. Why are we proposing something new for multinational teams? While current models may describe American and probably some Western teams, they do not describe collaboration in other regions of the world. We have made the case that non-Western and multinational collaboration can be qualitatively different from the collaboration of Western teams. This is not as surprising as it may at first seem that teamwork models are ethnocentric rather than universal. Current models of teams are based on data almost exclusively from studies in Western nations. They do not incorporate the differences that distinguish national groups. We have detailed the limitations of teamwork models based almost exclusively on Western research. By tapping the research on national differences, we are seeking to provide a more useful model for multinational organizations.

We must not assume that teamwork patterns in other nations mirror teamwork in the U.S. We need to identify how other teams define roles and functions, make decisions, manage conflict, and share information. If organizations do not take national differences into account, they run the risk of inefficiencies, miscommunications, and coordination breakdowns. Given the cognitive differences that have been described, and given the contrast between different team structures (interactive, hub-and-spoke, or hierarchical), we believe that organizational simulations can provide more inclusive and universal models of collaboration. Such models can better inform the design of collaborative information technology to support multinational collaboration. They can also better characterize individual action, craft training interventions, and enhance the productivity and success of multinational teams. We argue that by taking cultural differences in cognition and the social context of cognition seriously, more useful organizational simulations can be designed.

ACKNOWLEDGEMENTS

We would also like to express appreciation to Bianka Hahn, Danyele Harris, Dr. Gary Klein, Dave Klinger, and Mei-Hua Lin for their conceptual contributions to this paper. We also would like to acknowledge the sponsorship of The Boeing Company and the Advanced Decision Architectures Collaborative Technology Alliance sponsored by the U.S. Army Research Laboratory under Cooperative Agreement DAAD-19-01-2-0009 for portions of this work.

REFERENCES

Choi, I., Dalal, R., Kim-Prieto, C., & Park, H. (2003). Culture and judgment of causal relevance. *Journal of Personality and Social Psychology*(84), 46-59.

Choi, I., & Nisbitt, R. E. (1998). Situational salience and cultural differences in the correspondence bias and actor-observer bias. *Personality and Social Psychology Bulletin*(24), 949-960.

Choi, I., Nisbett, R., & Norenzayan, A. (1999). Causal attribution across cultures: Variation and universality. *Psychological Bulletin, 125*(1), 47-63.

Chu, P., Spires, E., & Sueyoshi, T. (1999). Cross-cultural differences in choice behavior and use of decision aids: A comparison of Japan and the United States. *Organizational Behavior and Human Decision, 77*, 174-170.

Hahn, B. B., Harris, D. S., & Klein, H. A. (2003). *Exploring the impact of cultural differences on multinational operations* (Final Report under Prime Contract #DAAD19-01-C-0065, Subcontract No. 8005.004.02 for U. S. Army Research Laboratory). Fairborn, OH: Klein Associates Inc.

Helmreich, R. L., & Merritt, A. C. (1998). *Culture at work in aviation and medicine: National, organizational, and professional influences*. Aldershot, ENG; Brookfield, VT: Ashgate.

Hofstede, G. (1980). *Culture's consequences: International differences in work-related values*. Newbury Park, CA: Sage.

Ji, L. J., Peng, K., & Nisbett, R. E. (2000). Culture, control, and perception of relationships in the environment. *Journal of Personality and Social Psychology, 78*(5), 943-955.

Klein, H. A. (2004). Cognition in natural settings: The cultural lens model. In M. Kaplan (Ed.), *Cultural ergonomics, advances in human performance and cognitive engineering*. Oxford: Elsevier.

Klein, H. A., Klein, G., & Mumaw, R. J. (2001). *A review of cultural dimensions relevant to aviation safety* (Final Report Prepared for Boeing Company under General Consultant Agreement 6-1111-10A-0112). Fairborn: Wright State University.

Kluckhohn, F., & Strodtbeck, F. L. (1961). *Variations in value orientations*. Evanston, IL: Row, Peterson.

Lannon, G., Klein, H. A., & Timian, D. (2001a). *Conceptual modeling of foreign command decision processes.* (Technical Report completed for The Defense Modeling and Simulation Office under Control #61339-01-P-0138 to Sterling Software). Fairborn, OH: Wright State University.

Lannon, G., Klein, H. A., & Timian, D. (2001b). *Integrating cultural factors into threat conceptual models.* Paper presented at the 10th Computer Generated Forces and Behavioral Representation (May 14-17), Norfolk, VA.

Loftin, R. B., M.D. Petty, R.C. Gaskins, & F.D. McKenzie (2005). Modeling Crowd Behavior For Military Simulation Applications. In W.B. Rouse & K.R. Boff, Eds., *Organizational Simulation: From Modeling and Simulation to Games and Entertainment.* New York: Wiley.

Markus, H., & Kitayama, S. (1991). Culture and the self: Implications for cognition, emotion, and motivation. *Psychological Review, 98*(2), 224-253.

Morris, M., & Peng, K. (1994). Culture and cause: American and Chinese attributions for social and physical events. *Journal of Personality & Social Psychology, 67*(6), 949-971.

Nisbett, R. E. (2003). *The geography of thought. How Asians and Westerners think differently... and why.* New York: The Free Press.

Nisbett, R. E., Peng, K., Choi, I., & Norenzayan, A. (2001). Culture and systems of thought: Holistic versus analytic cognition. *Psychological Review, 108*(2), 291-310.

Norenzayan, A., Smith, E. E., Kim, B. J., & Nisbett, R. E. (2002). Cultural preferences for formal versus intuitive reasoning. *Cognitive Science* (26), 653-684.

Peng, K., & Nisbett, R. (1999). Culture, dialectics, and reasoning about contradiction. *American Psychologist, 54*, 741-754.

Peng, K., & Nisbett, R. (2000). Dialectical responses to questions about dialectical thinking. *American Psychologist, 55*(9), 1065-1067.

Sims, D. E., Salas, E., & Burke, C. S. (2003). Is there a "Big Five" in Teamwork?, *Paper to be presented at the 19th Annual meeting of the Society for Industrial and Organizational Psychology*. University of Central Florida.

Swezey, R. W., Llaneras, R. E., & Salas, E. (1992). Ensuring teamwork: A checklist for use in designing team-training programs. *Performance & Instruction.*

Tamucci, M. S., Timian, D., & Burnett, S. K. (2000). *Generic representation of foreign ground forces Paper 9th-CGF-012.* Paper presented at the 9th Conference on Computer Generated Forces and Behavioral Representation.

Zsambok, C. E., Klein, G., Kyne, M., & Klinger, D. W. (1993). *Advanced team decision making: A model for high performance teams* (No. Contract MDA903-90-C-0117 for U.S. Army Research Institute for the Behavioral and Social Sciences). Fairborn, OH: Klein Associates Inc.

CHAPTER 9

HOW WELL DID IT WORK?

Measuring Organizational Performance in Simulation Environments

JEAN MACMILLAN, FREDERICK J. DIEDRICH, ELLIOT E. ENTIN AND DANIEL SERFATY

"We are confronted with insurmountable opportunities..."

—Pogo (Walt Kelly)

ABSTRACT

Immersive virtual simulations offer an opportunity to gain insight and experience in new, innovative organizational structures. Assessing the performance of these new organizations represents a considerable challenge due to the myriad of complex interrelated factors that may contribute to the outcomes observed in the simulation. Theories and models, often in the form of constructive simulations of organizational performance, can guide the development of empirical performance measures by linking detailed behaviors to overall outcomes for organizations. Constructive simulations can be used to create meaningful test conditions for immersive performance measurement, to identify those aspects of performance that are most critical to measure, and to predict the effects of organizational structures on performance. Translating theoretical measures into a form that can guide empirical data collection is a considerable challenge, however. This chapter provides examples of the use of theories and constructive simulations to structure empirical data collection for organizational performance, and discusses the lessons learned from these efforts. The focus is on organizational structures for military command and control, including innovative structures associated with the new and rapidly evolving concept of network-centric warfare.

THE CHALLENGE OF UNDERSTANDING ORGANIZATIONAL PERFORMANCE

The bottom line performance of multi-person teams and organizations, as measured through outcomes such as successful mission completion for military organizations or through profit or productivity measures for business organizations, results from the complex interaction of a myriad of interrelated factors. Organizational performance is a complex, dynamic, stochastic phenomenon. There are a multitude of individual attitudes, behaviors, decisions, and actions—all potentially measurable—that may contribute to successful outcomes for the entire organization (see Klein et al. 2005 and Salas et al. 2005 for a review of these factors). This complexity creates a major challenge for understanding and measuring organizational performance. Without a strong theory to guide the measurement and analysis, a mountain of seemingly unrelated and uninterpretable data can quickly overwhelm the analyst.

Because of the complexity of organizational performance, computational models have come to play a key role in understanding organizational structure and behavior. Hulin and Ilgen (2000) have suggested that computational modeling is a "third scientific discipline" for understanding complex socio-technical systems, supplementing the more traditional approaches of correlational analysis of real-world data and controlled laboratory experimentation. They characterize computer simulations of organizational behavior as "experiments done 'in silica' rather than 'in vitro.'"

Hulin and Ilgen argue that computational organizational models supply an essential capability lacking in other approaches. Controlled experimentation is limited in the number of variables it can consider. Correlational studies are dependent on the existence of a sufficient range and variability of data about the phenomena of interest. Both methods are severely limited in their ability to consider dynamic causal effects over time, depending either on static snapshots of isolated points in time or on arbitrarily selected measurement time periods that are often not dictated by the time cycles of the effects that are to be analyzed.

Models are tools for "illuminating the interface between theory and data" (Hulin and Ilgen, p. 11). Models can act as *dynamic theories* or *dynamic hypotheses* by making testable predictions about how multiple variables will interact to produce measurable outcomes. By making predictions across a theoretical space that includes a wide range of variables and conditions, only a few of which can be tested empirically, models can serve as a bridge between organizational theory and empirical data collection.

Models allow analysts to explore uncharted organizational territory. Computational models can make predictions about the performance of organizations that do not yet exist, under circumstances that have not yet been observed. Further, models can be used to concentrate empirical data collection in the most informative parts of an unexplored organizational space. These capabilities are essential if we wish to create and evaluate new and innovative organizational structures. The need to develop new organizational structures and to understand and predict their effectiveness before they are fully implemented seems especially important at the present time, as information network connectivity changes the face of the organizational landscape.

NETWORK CONNECTIVITY ENABLES NEW ORGANIZATIONAL STRUCTURES

Network connectivity is rapidly changing the information flow and communication constraints that have shaped human organizations for millennia. Physical location is becoming increasingly less critical for information exchange—it has become possible to obtain information from around the world in seconds. New communication technologies make it possible to create "virtual organizations" that can pull together expertise in a variety of locations to address a specific need and then disband as rapidly as they have formed.

Within the military, this change is reflected in a focus on *network-centric warfare* (Cebrowski, 2003; Alberts and Hayes, 2003) that asks how information technology can be used to gain both a strategic and tactical advantage in battle. Network connectivity has removed many of the limitations on information flow that shaped the organizational structure of Napoleon's command center, but a definitive replacement for that structure has yet to emerge. Almost instantaneous information flow seems able to support the formation of flexible, adaptive organizations that can change direction quickly in response to emerging events, but the full potential as well as the limitations of this concept are yet to be explored (see Andrews et al., 2005). At the heart of network centric warfare is the challenge of creating new organizational forms and structures that can use the rapid movement of information to create and maintain a strategic and tactical advantage in military conflicts.

How will we know if these new organizational forms will provide us with the advantages that we seek? How will we know that we have developed organizational structures and procedures that take maximum advantage of new technology for information flow? Organizational models and constructive simulations[1] provide us with a means to test our ideas at a much more rapid pace than the traditional methods of small incremental changes and trial-and-error learning—punctuated by flashes of genius—that have characterized organizational learning in the past. The creation and use of these constructive simulations to test

---

[1] We use the widely accepted term "constructive simulation" in this chapter to describe "third party" models and simulations that do not involve real-time human-in-the-loop participation by live human test subjects. We use the term "virtual simulation" to describe the use of simulations to create an environment in which live humans interact with the simulation and with each other.

new organizational concepts does not free us from the need to collect data through empirical human-in-the-loop testing, however.

HUMAN-IN-THE-LOOP TESTING OF ORGANIZATIONAL STRUCTURES

In the midst of the explosion in information connectivity, human attention spans, perceptual and cognitive capabilities, and processing speeds are unchanged. Although new skills may be developed, fundamental human information processing abilities remain the same. The massive increase in the amount of information available without a fundamental change in the way that human beings attempt to process that information has resulted in the often-cited information overload problem for complex distributed operations such as modern warfare. Information overload has replaced information scarcity as the predominant problem of warfare, and the "fog of war" now results more frequently from too much information than from too little.

Although much progress has been made in understanding and modeling human perceptual and cognitive capabilities, models are still far away from capturing the complex decision-making expertise that underlies a successful military operation and the complex person-to-person interactions that characterize an effective organization. Human-in-the-loop testing with live in-the-flesh decision makers thus remains an essential tool for determining whether humans can use new technology to accomplish their goals and whether they can function effectively in new and innovative organizational structures. Constructive models and simulations can provide insight into how new organizational structures can and should function, and can make predictions about the potential associated with the effective use of new technology, but human-in-the-loop testing is still needed to ensure that abstract ideas about how to use technology and how to alter organizational structures will really play as envisioned when used by real people in the real world.

Virtual simulation environments provide a means to construct artificial worlds in which multiple individuals can experience new technologies, new concepts of operations, and new organizational structures in a realistic interactive environment. These environments can be used for innovation, exploration, and training. The degree of control and observability that is feasible in these environments, while much less than that associated with traditional experimentation, is still much greater than that possible in real-world settings. It is possible, for example, for multiple teams to play the same scenario with different CONOPS (concepts of operation) or technologies, or to test the effectiveness of an organizational structure in circumstances designed specifically to create different types of stress on the organization. Although a complex scenario involving multiple players in a virtual environment does not ever play out in exactly the same way over multiple trials, virtual simulations provide much more testing control and replication than would otherwise be possible.

But how do we know whether a new organizational structure, coupled with new technology, is working as intended when we test it in a virtual simulation environment? What should we measure to gauge success? Overall outcome measures are often not very illuminating. Outcomes may prove to be insensitive to the variations in structure and technology that are being tested within the specific environment and scenario chosen for the test, and it is almost impossible to conduct tests that capture all of the variations in the environment that may affect outcomes. Finer-grained measures are needed that can capture how and why organizational structures and new technologies change the processes and behaviors of the organization in ways that lead to desired outcomes in a specific environment. If detailed process and behavior measures can be linked to desired outcomes and to the specifics of the external events that occur in the scenario, then empirical results can be extrapolated to make testable predictions about other environments and scenarios.

MODELS PROVIDE A FRAMEWORK FOR EMPIRICAL TESTING

It is relatively easy to obtain massive amounts of data from simulation-based exercises, but it can be difficult if not impossible to interpret these data in a meaningful way without a theory, framework, or model that links detailed measurable behaviors to overall results. A framework is needed both to focus the empirical data collection and to aid in finding the patterns of interest in the results.

Executable organizational models serve as dynamic theories to guide the collection and analysis of empirical performance data in virtual human-in-the-loop simulations. Constructive simulations and models can be used to design the scenarios to be played out in virtual simulations; to select the behaviors to be measured, based on predictions about the relationship of those behaviors to outcomes; and to analyze and interpret the empirical data once it is collected. Models make predictions by linking measurable processes to measurable outcomes, and these predictions serve to guide both the collection and the analysis of the empirical data. Further, models "enable surprise" in analyzing complex phenomena.[2] By specifying and instantiating expectations, models allow us to know when we have found the unexpected.

This chapter focuses on analysis of the communication patterns that underlie organizational performance as an example of the use of organizational models to develop process measures that are linked to outcomes, serving to guide the

---

[2] We are indebted to Gary Klein for this apt phrase.

collection and interpretation of empirical data from experiments conducted in virtual simulations. Communication in an organization is easy to observe and measure, but difficult to interpret. For example, communication rates (e.g. message exchange rates between nodes in an organization) are relatively easy to measure quantitatively, but the results can be meaningless without a model or theory that makes predictions about the relationship of these communication rates to the goals of the organization and the outcomes of the mission.

ORGANIZATIONAL MODELING APPROACH

Over the past several years, we have been involved in a series of collaborative studies with the University of Connecticut (MacMillan, Paley, Levchuk, Entin, Serfaty, and Freeman, 1999; Levchuk, Pattipati, and Kleinman, 1998, 1999) to develop models that predict organizational performance and to test and validate those models through empirical experiments. Figure 1 shows the conceptual framework underlying our modeling approach. Organizational models capture the dynamic relationship between the mission to be accomplished (the tasks to be performed), the resources needed to perform those tasks (sensor and weapons systems as well as the information needed for each task), and the human decision makers who use the resources to perform the tasks. The organizational structure is modeled as a three-way matching of humans, tasks, and resources. Based on the interdependency of the tasks in time (some tasks must be completed before others can be undertaken) and the resources required for each task, it is possible to optimally schedule the use of scarce resources to accomplish a mission. Typically there is substitutability among some of the resources (several ways to do the same task), with differences in the projected success of the task depending on which resources are used. As human decision makers are matched to resources and tasks, human capabilities and workload limitations must be taken into account by setting constraints on the number and type of tasks that can be performed simultaneously by each individual. If the workload associated with a task exceeds the capabilities of an individual, then multiple individuals must work together to perform the task. Splitting tasks across individuals, and the interdependencies that exist among different tasks (e.g., one task must be completed before another one can be initiated) create the need for coordination among the individuals in the organization (see Klein et al., 2005 for a more detailed analysis of the nature of coordination and the factors that enable it). The need for coordination spawns the need for communication among the individuals who must coordinate. The organizational structure instantiated in the model thus captures the roles played by each individual in the organization (who does what) and predicts the communication patterns that will be associated with those roles (who talks to whom).

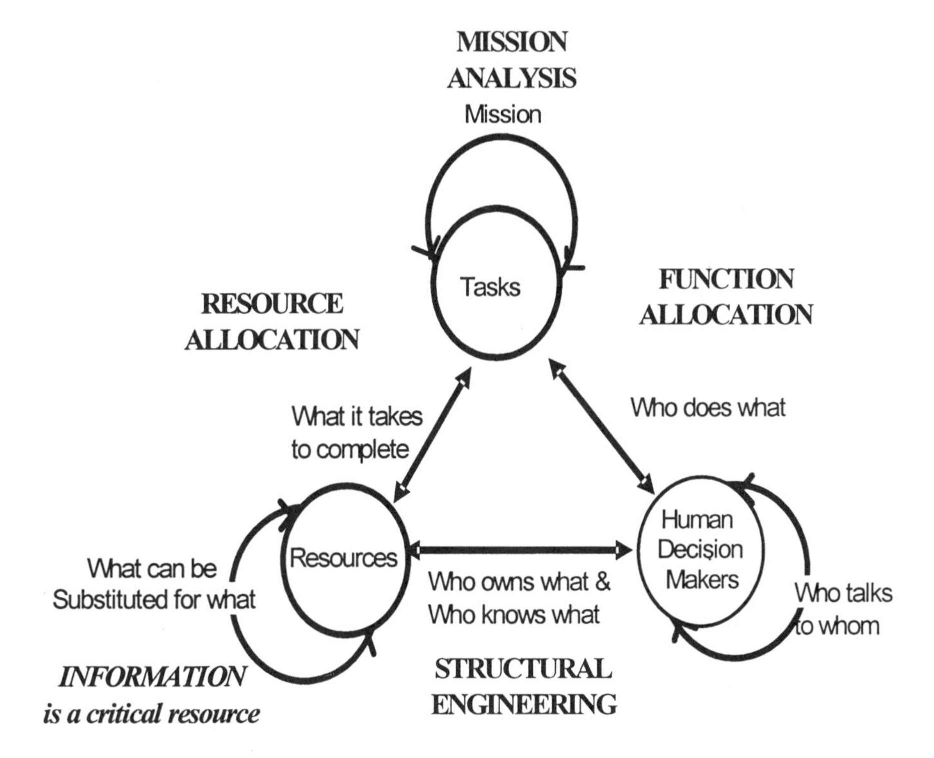

FIGURE 1. Conceptual Framework for Organizational Modeling

COMMUNICATION BEHAVIOR AS A MEASURE OF ORGANIZATIONAL PERFORMANCE

In order for an organization or a team within an organization to act in concert to achieve common goals, the team must have information both about the external situation and about the other team members (Klein et al. 2005 discusses this process in detail). Some information will be relevant to the entire organization, while other information is relevant only to individuals or sub-groups within the organization. The need for shared information depends on the nature of the mission and the specifics of the organizational structure. Although some of the needed information can be shared through common displays or gathered through observation of what other teams members are doing, effective organizational performance almost always requires explicit communication—messages, discussions, directions, and orders—to achieve the necessary sharing of information.

Explicit communication has a cost, however—an overhead associated with the exchange of information among team members. Communication requires both time and cognitive resources, and, to the extent that communication can be made less necessary or more efficient, team performance can benefit as a result. Reductions in the communication overhead associated with coordinated action have been found to result in more efficient and more effective organizational performance (Entin and Serfaty, 1999; MacMillan, Serfaty, and Entin, 2004).

The volume of communication in an organization, taken in a vacuum without considering any other factors, has no natural directionality. Is more communication good? Or is less communication good? The answer depends on the demands of the tasks to be performed, the organizational structure, and the situation in which the organization finds itself. The effective team or organization should *communicate as much as necessary,* but no more. Constructive organizational modeling can provide insights and predictions concerning *how much communication should be necessary* in order for the team to perform effectively for a specific mission in a specific scenario.

The nature of the mission to be performed and the organizational structure of the team interact to determine the communication patterns that will be associated with successful outcomes. For a military mission, the tasks to be performed will have certain requirements for resources, the amount and types of resources needed will depend on the specifics of the scenario, and the structure of the organization will determine how the needed resources are assigned and controlled.

For example, in a simulated amphibious assault scenario recently used to evaluate alternative organizational structures for a Joint Task Force command team (Entin, Diedrich, and Rubineau, 2003; Diedrich, Entin, Hutchins, Hocevar, Rubineau, and MacMillan, 2003), an assault on an airbase might require both strike assets (e.g., cruise missiles) and the support of a Special Forces team. Following this attack, the Special Forces team may then be needed for an assault on a nearby port, and thus the attack on the airbase must be accomplished in time for the Special Forces to deploy to the port. In this case, the amount of coordination required between commanders or command nodes, and therefore the communication required, will depend on who controls which assets. If the strike assets and the Special Forces team are under the control and direction of a single node (commander) in the organization, then little or no inter-person or inter-node communication will be needed to achieve the assault. If, however, the strike assets and the Special Forces team are controlled by different commanders, then extensive inter-person or inter-node communication will be necessary for the assault to be successful. Hence, the fit or match between the organization and the task requirements influences the coordination and communication needed to achieve a successful outcome.

An organizational model that captures the interdependencies among tasks and the way that these interdependencies interact with the control of resources in alternative organizational structures can thus predict how much communication will be needed for successful completion of a mission under each structure. These predictions can then be tested through empirical data collection in a virtual simulation environment.

EMPIRICAL RESULTS: MODEL-BASED PREDICTIONS OF COMMUNICATION BEHAVIOR

We have conducted a number of experimental studies, working with colleagues at the University of Connecticut and the Naval Postgraduate School, that examine how mission structure and organizational structure interact to produce communication behaviors that are associated with effective performance for a command and control organization. In each of these experiments, the interdependence of the tasks being performed and the team structure—the control of resources and the assignment of task responsibilities—generates the need for coordination. This need for coordination then drives the need for communication among team members. Detailed results for these experiments are reported elsewhere (Diedrich et al., 2003; MacMillan, Entin, and Serfaty, 2004; Entin, 1999) but we review them briefly below to show how constructive organizational simulations can serve as a framework to shape empirical data collection in virtual simulations, can guide the analysis and interpretation of organizational performance results, and can enable us to find and interpret the unexpected in our results.

Experiment 1. Can Organizational Structure Be Optimized for a Mission?

This experiment compared both mission outcomes and organizational processes for two different organizational structures (Entin, 1999). The mission involved an air- and sea-based operation to regain control of an allied country that had been taken over in a hostile invasion by a neighboring country. One of the organizational structures tested was a relatively traditional Joint Task Force (JTF) structure, developed by subject matter experts, in which all resources of a similar type (e.g., air strike assets or amphibious landing forces) were controlled by the same node in the organization without explicit consideration of the need to coordinate the use of those resources in the mission. The second structure tested was optimized for the mission using an approach originally developed at the University of Connecticut (Levchuk, Luo, Levchuk, Pattipati, and Kleinman, 1999; Levchuk, Pattipati, and Kleinman, 1998, 1999) in which a model of the activities required to complete the mission was developed and an organizational structure was designed based on a model of how resources could be most efficiently employed during the mission. The optimized organizational structure developed for the experiment was based primarily on two optimization objectives—simultaneously minimizing the coordination required to accomplish the mission and balancing workload across the team. These multiple objectives act to constrain each other, since the workload-balancing objective prevents the assignment of all tasks to only a few team members in order to minimize the coordination requirements.

The model predicted that teams using the optimized structure would perform the mission more successfully. But—most importantly—the model predicted *how and why* teams using the optimized structure would be more successful. Specifically, the model predicted that teams using the optimized structure would need to coordinate less, and therefore would need to communicate less frequently, because each node in the optimized organization controlled all or many of the resources needed to perform the tasks to which they had been assigned. In the optimized structure, it was predicted that the command nodes would be able to achieve the coordinated use of resources to carry out mission tasks without the need for extensive, explicit communication among nodes.

These predictions were tested in an experiment conducted in a virtual simulation with 10 six-person teams of Navy officers (Entin, 1999). The results were as predicted by the model. The teams using the optimized structure accomplished the mission more successfully as judged by outcome measures such as the successful completion of mission tasks, the speed with which mission tasks were completed, and the efficiency of resource usage. More interestingly, the model also successfully predicted differences in the coordination and communication patterns for the two team structures. Teams using the optimized structure coordinated less and communicated less frequently (lower communication rate). Other differences observed for the two structures were not directly predicted by the model, but could be derived from the differences that were designed into the two structures (MacMillan, Entin, and Serfaty, 2004). Because teams using the optimized structure needed to communicate less frequently to accomplish the mission (lower communication overhead), they were predicted to experience a lower subjective workload during the mission -- measured with the NASA TLX questionnaire (Hart & Staveland, 1988) -- and this prediction was borne out by the data.

We also found an effect not predicted by the models—associated with less communication and a lower workload under the optimized structure was a more accurate awareness of what other team members were doing measured via periodic questionnaires about major activities at the other nodes (MacMillan, Entin, and Serfaty, 2004). This finding that less need for coordination and a lower communication rate was associated with a more accurate understanding of others' tasks was both interesting and unexpected. One might have predicted that teams who communicated more frequently would have a more accurate understanding of each other's tasks and situation. Instead, it seems possible that the workload associated with communication may counteract the expected positive effects of that communication on each team member's awareness and understanding of the activities of others. This idea is supported by the finding that teams using the optimized structure were better able to anticipate each other's needs for information, as measured by an *anticipation ratio* (Entin, 1999; MacMillan, Entin, and Serfaty, 2004) that counts the number of transfers of information relative to the number of requests for information across team members.

This experiment illustrates how an organizational model can be used to shape the collection and interpretation of empirical organizational performance data in two ways. First, the model can make explicit predictions about performance that can then be tested experimentally—in this case, a prediction that teams using an optimized organizational structure would be able to accomplish a mission more successfully while coordinating less and communicating less frequently than teams using a traditional JTF structure. Second, the theoretical structure instantiated by the model supports additional theory development and hypothesis formulation that builds on the model to generate new predictions—in this case, the finding that a lower communication rate was apparently associated with a higher level of mutual awareness and understanding among team members. This effect was not predicted by the model, but can be factored into theory formulation and model development in the future. Thus, models not only help us know what to look for in empirical data collection, they can also provide a structure that allows us to find and recognize the unexpected—they "enable surprise" and help us to understand and interpret that surprise.

Experiment 2. Does the Benefit of Collaborative Planning Vary Under Alternative Organizational Structures?

In another experiment (Price, Miller, Entin, and Rubineau, 2001; Miller, Price, Entin; Rubineau, and Elliott, 2001) we examined the effects of using an electronic collaboration tool—a shared whiteboard—during pre-mission planning, and the subsequent effects of that planning on a team's ability to carry out a coordinated mission under two different organizational structures. The mission to be performed was a humanitarian assistance/airlift mission that required the coordinated delivery of food and medical supplies to refugee sites, combined with the need to use defensive weapons to protect the planes making the deliveries. Teams were able to pre-plan this mission based on information about the delivery sites and their respective needs, but other sites and needs that could not be anticipated in advance emerged during the mission.

Three-person teams planned and carried out the mission under two different organizational structures. In a *functional* structure, each team member controlled assets of only one type—medical supplies, food supplies, or defensive weapons. The simultaneous delivery of the needed supplies to a site thus required all three team members to work together to meet the requirements for the mission. In a *divisional* structure, each team member controlled a portion of all three types of resources. In the divisional structure, each member of the team could perform a successful delivery to a site without coordinating with the other team members if he or she had the quantity of resources that were required by the site. In the scenarios that were used in the experiment, some, but not all, of the sites could be serviced independently by individuals under the divisional structure, while all sites required coordination under the functional structure. Thirty-six university students participated as subjects in the experiment, in 12 three-person teams.

The outcome measures used in this experiment were based on the team's ability to deliver exactly the right amount of resources to each site (task accuracy). For example, percent task accuracy was measured by the percentage of times that the team managed to deliver 100 percent of the needed supplies to each of the refugee sites. One might have expected, based on a model that predicts more effective performance if less coordination is required, that the teams using the divisional structure would perform at a higher level of task accuracy than teams using the functional structure. In fact, the opposite was found. Overall performance was significantly better for the teams using the functional structure.

In this experiment, our model "enabled surprise" by making a prediction that was not borne out by the data. It also gave us an indication of where to look in order to understand the results. The performance problems for the divisional teams arose in those tasks for which coordination *was* required. Teams using the divisional structure needed to coordinate much less frequently than the teams using the functional structure, but when they *did* need to coordinate they did so poorly. For example, a coordination success measure (the percentage of required team members who participated in each task) was significantly higher for the functional structure than for the divisional structure.

These results tell us that a model that simply predicts that "teams who need to coordinate less frequently will perform better" is inadequate. If teams vary in their need to coordinate—sometimes they can act independently and sometimes they need to coordinate—then those teams that know how to coordinate will be at an advantage when coordination is needed. So the overall performance levels for the divisional teams who are usually able to act independently will depend on how frequently coordination is needed.

The experiment also examined the effects of using a shared whiteboard for collaborative planning, with teams in one condition using the electronic whiteboard (with a map background) and teams in the other condition conducting their planning using paper maps. There was an overall positive effect of the use of the collaborative whiteboard during planning on mission success (task accuracy), but, interestingly, use of the whiteboard appears to have been more advantageous to the teams using the functional structure than to the teams using the divisional structure (MacMillan, Serfaty, and Entin, 2004). Teams that used the electronic whiteboard collaborated more intensively during the planning process, as measured by the number of collaborative communications that occurred during the planning phase (Miller et al., 2001), and this collaboration during planning seems to have been especially beneficial to the functional teams that needed to coordinate intensively during the mission.

Neither the more intensive collaboration that occurred when the electronic whiteboard was used for planning nor the advantage that this more collaborative planning gave to the high-coordination functional teams during the mission were directly predicted by models or theories prior to the experiment. However, the theoretical framework that distinguishes functional and divisional organizational structures helped us to understand these unexpected results.

The experiment was designed, based on theory, to create organizational structure conditions that differed in the amount of coordination that was required for successful mission execution. At the time the study was designed, we did not predict the difficulty of coordination for teams that needed to coordinate less frequently. However, these results seem consistent with recent studies showing that teams find it more difficult to move from a divisional to a functional structure than vice versa (Hollenbeck et al., 1999; Moon et al., 2000). Teams who are experienced in acting independently seem to find it difficult to learn to coordinate, while teams experienced in coordination find it relatively less difficult to learn to act independently. Given this expectation, it is not surprising that more intensive collaboration during the planning phase was associated with more successful coordination during mission execution. What remains a "surprise" that requires more investigation is why the use of the electronic whiteboard during planning was associated with more intensive collaborative communication. Theory and models that link organizational structure to performance have thus enabled us to identify the "most surprising" results of the study and the most promising areas for future investigation.

Experiment 3. What Happens When the Structure of an Organization Is Incongruent with Its Mission?

A third, recently completed, experiment illustrates how constructive organizational models can be used both to design scenarios that stress specific organizational structures and to collect and interpret empirical performance data for organizations using those structures (Diedrich et al., 2003). The focus of the experiment was to understand and predict the performance of organizations that find themselves out of synch with the mission they are performing. The fundamental concept is that it is possible to mathematically define the degree that an organizational structure is *congruent* with the mission that the organization is performing, and that organizations using structures that are more congruent with a mission will perform that mission more effectively than organizations using structures that are less congruent.

In order to test this concept experimentally, it was necessary to define the organizational structures to be tested in detail and to develop mission scenarios that would provide congruent and incongruent test conditions for those organizations (Kleinman, Levchuk, Hutchins, and Kemple, 2003). The organizational structures for the experiment were developed to correspond roughly to the theoretical divisional and functional structures discussed above, with adjustments made as needed to create an organizational structure that was credible for a Joint Task Force (JTF) in a military context. Two structures were developed for a six-person JTF command team: (1) a functional structure in which all or most of the assets of a specific type (e.g. strike aircraft) were under the control of one commander, and (2) a divisional structure in which all or most of the assets on a specific platform in a geographical area were under the control of one commander.

A model of the divisional (D) and functional (F) structures was developed and used to develop and refine scenarios for the experiment (Kleinman, Levchuk, Hutchins, and Kemple, 2003). Essentially, the test scenarios for the experiment were reverse engineered by using the model to evaluate the amount of incongruence predicted to exist between the two organizational structures and the two different types of mission scenarios. One scenario type (d) was designed to be "tuned" to the divisional structure but mismatched to the functional structure, and the other scenario type (f) was designed for the reverse effect. The major scenario-design factors manipulated to create congruence and incongruence between structures D and F and scenarios d and f were the amount of coordination required to perform the scenario tasks and the spatial-temporal loading of the individual decision makers.

The focus of the experiment was to develop and test model-based predictions about organizational performance in the congruent conditions (Dd and Ff) and the incongruent conditions (Df and Fd). Overall performance as measured by successful mission outcomes was, of course, expected to be higher in the congruent conditions, but it was also of great interest to be able to predict and understand in more detail the organizational behaviors that were affected by the incongruence between the structure and the mission. The experiment was conducted at the Naval Postgraduate School with eight six-person teams composed primarily of Navy officers.

As expected, performance was higher in the congruent conditions as measured by the mean percentage of tasks completed (Diedrich et al., 2003) as well as by an accrued task gain measure that weighted task completion by the value of the task (Levchuk, Kleinman, Ruan, and Pattipati, 2003). Also, as expected, the subjective workload experienced by participants was lower in the congruent than in the incongruent conditions (Diedrich et al., 2003).

The congruent and incongruent conditions were designed so that the requirements for coordination were greater in the incongruent than in the congruent conditions. Communication rates were therefore expected to be higher in the incongruent conditions because the team's primary mechanism of coordination was through verbal communication. As predicted, there was a significant effect of congruence on communication frequency (Diedrich et al., 2003). However, as shown in Figure 2, there was an asymmetry in the effects of

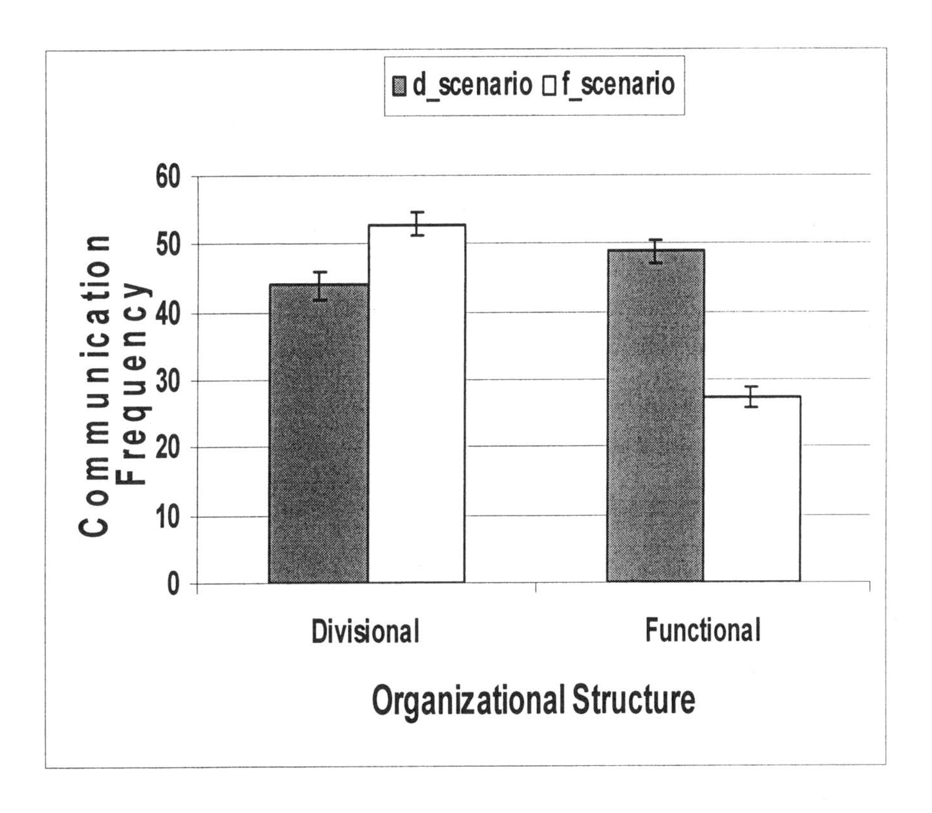

FIGURE 2. Communication Frequency in Congruent and Incongruent Conditions

congruence on communications that had not been predicted by the models. The effect of incongruence on the need to communicate was much greater in the F than in the D organization. Communication rates in the two incongruent conditions (Df and Fd) were roughly the same, while communication rates were much higher in the Dd congruent condition than in the Ff congruent condition.

The unexpected finding that congruence effects on overall communication were quite different for the two organizational structures led us to perform a more detailed analysis of the nature of the communications in the two conditions (Diedrich et al., 2003; Entin, Diedrich, and Rubineau, 2003). Using a categorization of utterances by type, we analyzed changes in the patterns of verbal communication—who talked about what—as the two organizational structures moved between congruent and incongruent scenarios.[3] Teams using the D structure increased their communication in almost all communication categories in the incongruent scenarios as compared to the congruent scenarios. In the D structure, teams talked more overall in the incongruent condition, but there was little change in the *pattern* of those communications. Teams using the F structure, in contrast, showed major changes in their communication patterns between the congruent and incongruent conditions, with some categories of communication increasing greatly while others decreased, and with some positions on the team talking more while others talked less. Team members in the F structure not only talked more in the incongruent condition, they also talked *differently*.

These organizational differences in the amount, type, and nature of communications between congruent and incongruent conditions were not predicted by the model, suggesting that they result from factors not taken into account in the model-based definition of congruence in the experiment. Communications in the functional organizational structure looked quite different in both amount and type in the congruent and incongruent conditions, while communications in the divisional structure increased in volume without changing in pattern. The effect of incongruence on mission outcomes, in contrast, was roughly symmetrical, with both structures suffering about the same decrement under incongruent conditions.

Based on these results, one might argue that the divisional structure was more robust across the congruent and incongruent conditions, if we define robustness as the ability to cope with incongruence without making major changes in organizational communication patterns. The functional structure, in contrast, seems to have been able to take advantage of the congruent conditions to be more efficient (much lower communication rates) and to exhibit more flexibility in changing communication patterns under incongruent conditions. Without more detailed communication-pattern predictions from the model, however, we cannot say to what extent the change in communication patterns shown by teams using the functional structure was necessitated by the coordination requirements that resulted from the combination of the functional structure and the incongruent scenarios. Were the functional teams simply "doing what they had to do" to try to maintain performance under incongruent conditions? Or was there something about the functional structure that resulted in a greater ability to change communication patterns—more adaptability—in response to changing mission requirements? Additional modeling and experimentation is needed to address this issue.

The models in this experiment played a key role in defining the test conditions and in setting the expectations for the experiment, allowing us to be "surprised" by the differences in the effects of incongruence on communication patterns under the two structures. Understanding the implications of this difference in

---

[3] The direction of movement from congruent to incongruent scenarios was counterbalanced for the two organizational structures.

communications under incongruence for the two structures will require further investigation as well as enhancement of the models.

CONCLUSIONS

Our experience in measuring and analyzing organizational performance using a combination of constructive and virtual simulations suggests that simulation technology is entering an innovative period of the "S curve" with regard to understanding organizational behavior (Rouse and Boff, 2005). As the capabilities for virtual simulation of multi-node organizations have advanced, the capabilities for constructive simulation and computational modeling of organizations have advanced in parallel (Rouse, 2005). We argue that these two capabilities are complementary, and, further, that they are both essential for developing new organizational structures that exploit new information technology and connectivity. Organizational performance is complex, and formulating expectations about that performance requires taking into account the interaction of a multitude of variables. Computational models provide a valuable tool for understanding this complexity, allowing us to understand what to measure in virtual simulations, how to construct conditions for data collection, and how to interpret the results of the evaluation.

As technology alters the possibilities for organizational structures, we need new methods for developing effective organizations that move beyond incremental trial and error learning. The ability to combine the capabilities of constructive and virtual simulation puts us at a new threshold for building innovative and effective organizations. Computational models can help us to design new organizational forms that can be experienced in virtual simulations, and can help us evaluate those new organizations by creating model-based expectations that let us know when we have found the unexpected.

The studies reviewed in this chapter suggest that computational organizational models can be extremely useful in creating and using immersive simulation environments to develop and test new organizational forms and structures, as envisioned in the OrgSim environment. Models have three major uses in OrgSim:

- Designing organizational structures to be tested in virtual human-in-the-loop simulations;
- Designing scenarios (external events) for the simulation environment that will stress the organizational structure in predictable ways; and
- Suggesting performance measures for human-in-the-loop testing that will indicate whether an organizational structure performs as predicted.

Constructive models can provide a theoretical framework that ensures that the best possible use is made of experiences in immersive OrgSim testbeds.

ACKNOWLEDGEMENTS

The Office of Naval Research (ONR) and the Air Force Research Laboratory (AFRL) sponsored the experimental efforts reported here. We would like to express our appreciation for the support and review of Dr. Willard Vaughan and Mr. Gerald Malecki at ONR and Dr. Sam Schiflett and Dr. Linda Elliott at AFRL.

REFERENCES

Alberts, D.S. & Hayes, R.E. (2003) Power to the Edge: Command and Control in the Information Age. Washington, DC: Command and Control Research Program Publication Series.

Cebrowski, A.K. (2003) Transforming Defense: The Path Not Taken...Yet. http://www.t2net.org/briefs/dakc/ArthurKCebrowskiv2.ppt

Diedrich, F.J., Entin, E.E., Hutchins, S.G., Hocevar, S.P., Rubineau, B. & MacMillan, J. (2003). When do organizations need to change? (Part I) Coping with incongruence. Proceedings of the 2003 Command and Control Research and Technology Symposium. National Defense University, Washington, DC.

Entin, E. E. (1999). Optimized command and control architectures for improved process and performance. Proceedings of the 1999 Command and Control Research and Technology Symposium, Newport, RI.

Entin, E. E., Diedrich, F. J., and Rubineau, B. (2003). Adaptive communication patterns in different organizational structures. Proceedings of the Human Factors and Ergonomics Society 47th Annual Meeting, Denver, CO.

Entin, E.E. and Serfaty, D. (1999). Adaptive team coordination. Human Factors, 41, 312-325.

Hart, S. G. & Staveland, L. (1988). Development of NASA-TLX (Task Load Index): Results of empirical and theoretical research. In P. A. Hancock & N. Mishkati (Eds.), Human Mental Workload (pp. 139-183). Amsterdam: Elsevier.

Hollenbeck, J.R., Ilgen, D.R., Moon, H., Shepard, L., Ellis, A., West, B., & Porter, C. (1999) Structural contingency theory and individual differences: Examination of external and internal person-team fit. Paper presented at the 31st SIOP Convention, Atlanta, GA.

Hulin, C.L. & Ilgen, D.R. (2000) Introduction to computational modeling in organizations: The good that modeling does. In D.R. Ilgen & C.L. Hulin (Eds.)

Computational Modeling of Behavior in Organizations: The Third Scientific Discipline. Washington, DC. American Psychological Association.

Klein, G., P.J. Feltovich, J.M. Bradshaw, & D.D. Woods (2005). Common Ground and Coordination in Joint Activity. In W.B. Rouse & K.R. Boff, Eds., Organizational Simulation: From Modeling and Simulation to Games and Entertainment. New York: Wiley.

Kleinman, D.L., Levchuk, G.M., Hutchins, S.G. & Kemple, W.G. (2003). Scenario design for the empirical testing of organizational congruence. Proceedings of the 2003 Command and Control Research and Technology Symposium. National Defense University, Washington, DC.

Levchuk, G.M., Kleinman, D.L., Ruan, S., & Pattipati, K.R. (2003). Congruence of human organizations: Theory versus data. In Proceedings of the 2003 Command and Control Research and Technology Symposium. National Defense University, Washington, DC.

Levchuk, Y., Luo, J., Levchuk, G. M., Pattipati, C., & Kleinman, D. (1999). A multi-functional software environment for modeling complex missions and devising adaptive organizations. Proceedings of the 1999 Command and Control Research and Technology Symposium. Naval War College, Newport, RI.

Levchuk, Y., Pattipati, C., & Kleinman, D. (1998). Designing adaptive organizations to process a complex mission: Algorithms and applications. Proceedings of the 1998 Command and Control Research and Technology Symposium. Naval Postgraduate School, Monterey, CA.

Levchuk, Y., Pattipati, C., & Kleinman, D. (1999). Analytic model driven organizational design and experimentation in adaptive command and control. Proceedings of the 1999 Command and Control Research and Technology Symposium. Naval War College, Newport, RI.

MacMillan, J., Paley, M.J., Levchuk, Y.N., Entin, E.E., Serfaty, D., & Freeman, J.T. (2002). Designing the best team for the task: Optimal organizational structures for military missions. In M. McNeese, E. Salas, & M. Endsley (Eds.), New Trends in Cooperative Activities: System Dynamics in Complex Settings. San Diego, CA: Human Factors and Ergonomics Society Press.

MacMillan, J., Entin, E.E., & Serfaty, D. (2004). Communication Overhead: The Hidden Cost of Team Cognition. In E. Salas & S.M. Fiore (Eds.), Team Cognition: Understanding the Factors That Drive Process and Performance. Washington, DC: American Psychological Association.

Miller, D., Price, J.M., Entin, E.E., Rubineau, B., & Elliott, L. (2001). Does planning using groupware foster coordinated team performance? Proceedings of the Human Factors and Ergonomics Society 45th Annual Meeting, Minneapolis, MN.

Moon, H., Hollenbeck, J., Ilgen, D., West, B., Ellis, A., Humphrey, S. & Porter, A (2000) Asymmetry in structure movement: Challenges on the road to adaptive organization structures. Proceedings of the 2000 Command and Control Research and Technology Symposium. Naval Postgraduate School, Monterey, CA.

Price, J. M., Miller, D. L., Entin, E.E., & Rubineau, B. (2001). Collaborative planning and coordinated team performance. Proceedings of the 2001 Command and Control Research and Technology Symposium. Naval Academy, Annapolis, MD.

Rouse, W.B. (2005). Strategic Thinking Via Organizational Simulation. In W.B. Rouse & K.R. Boff, Eds., Organizational Simulation: From Modeling and Simulation to Games and Entertainment. New York: Wiley.

Rouse, W.B. & K.R. Boff (2005). Introduction and Overview. In W.B. Rouse & K.R. Boff, Eds., Organizational Simulation: From Modeling and Simulation to Games and Entertainment. New York: Wiley.

Salas, E., J.W. Guthrie, Jr., K.A. Wilson-Donnelly, H.A. Priest & S. Burke (2005). Modeling Team Performance: The Basic Ingredients and Research Needs. In W.B. Rouse & K.R. Boff, Eds., Organizational Simulation: From Modeling and Simulation to Games and Entertainment. New York: Wiley.

CHAPTER 10

TECHNICAL AND CONCEPTUAL CHALLENGES IN ORGANIZATIONAL SIMULATION

LEON F. MCGINNIS

ABSTRACT

Simulation is a potentially powerful tool for predicting the performance or evolution of complex dynamic systems. In some domains, such as integrated circuits, simulation is a routine, largely automated part of the design process. In organizational settings, such as warehousing, manufacturing, logistics, finance, health care, et cetera, simulation also is widely used, but models are developed in a more *ad hoc* fashion. This chapter explores the ideal use of simulation in analyzing, designing, and understanding organizations, and identifies some fundamental conceptual and technical hurdles in reaching the ideal from today's state of the art.

INTRODUCTION

In organizations, making important decisions is difficult. This is a tautology, because the difficult decisions are those that are important, whether the question is how to respond to a contingency, how to configure the organization, or what should be the organization's mission and goals.

We would like to think better decisions result when decision-makers are better informed—about the problem being addressed, about the options for solving it, and about the consequences of alternative decisions. The emergence of decision-support systems as a focused topic of R&D, and a significant commercial opportunity, is a testament to the perceived value of better informing decision makers.

Over the past forty years, simulation has become a primary tool for informing decision-making in engineering design (e.g., integrated circuits) and in discrete event logistics systems (warehousing, manufacturing, and supply chains). Ostensibly, simulation models accomplish two valuable objectives: (1) they reveal in a controlled way the result of interacting dynamics in complex systems, and (2) they create "synthetic histories" which may reflect the impact of uncertain future events, such as customer demand. These synthetic histories can be analyzed to assess the impact of system design decisions, policies, decision algorithms, or *ad hoc* interventions. Because simulations are computational devices, many different synthetic histories (with different realizations of random processes) can be created, enabling quantitative risk assessments.

A concomitant benefit (some might argue, the larger benefit) from simulation studies is the disciplined process of system description and data gathering. This process alone may reveal information or insight sufficient to guide decision-making.

The idea of using simulation to study organizations is very appealing, and we can easily imagine a range of uses. The VP-Operations, for example, might like to experiment with a range of responses to predictable contingencies, such as a factory destroyed by fire, dramatic changes in fuel prices, or labor disruptions. These are simulation studies not too different from contemporary practice. The CEO might like to experiment with alternative management structures, for example, changing from product oriented divisions to geographic divisions. This kind of simulation study has many elements of contemporary practice, but new issues as well, related to how organizations perform. The VP-Human Resources might like to experiment with alternative labor negotiation strategies, leading to a simulation study quite different from most contemporary practice, because it requires explicitly modeling the behavior of individuals. What we would like to know is how difficult and expensive it will be to construct simulation models for these three distinct kinds of studies.

This chapter explores the concept of organizational simulation from the admittedly narrow perspectives of integrated circuit design and discrete event logistics systems. The goal is to see what conclusions about organizational simulation might be drawn from the history of simulation in these other two domains. I'll begin with a brief discussion of simulation in the other two domains (integrated circuit design and discrete event logistics systems), intending to highlight some relevant points rather than provide a comprehensive overview. Then I'll try to draw relevant parallels between these two domains and organizations, and discuss organizational simulation of a very simple organization—conventional warehouses. Finally, I will discuss the technical and conceptual challenges that emerge, and suggest some conceptual and technical innovations necessary for organizational simulation to become a practical tool.

DISCRETE EVENT SIMULATION

This chapter focuses entirely on the potential for discrete event organizational simulation. Other approaches are possible, and may be more appropriate in particular situations. However, there is a large existing base of expertise and technology for discrete event simulation, which constitutes a significant platform on which to develop organizational simulation.

Two existing application domains for discrete event simulation are discussed below. Integrated circuits are large-scale complex engineered systems. Simulation is an essential supporting technology for integrated circuit design, and is implemented in a way that makes the simulation technology *per se* transparent to the designer. Discrete event logistics systems are large-scale complex socio-technical systems, which operate in an environment of uncertainty, and often embody uncertainty in their own behaviors. Simulation is, today, the only tool that can reveal their complex dynamics, and support "virtual" design and controlled experimentation. However, simulation models of discrete event logistics systems typically are custom-built by simulation experts, who act as interpreters between the model and the decision-maker(s).

Simulating Integrated Circuits

State of the art integrated circuits (ICs) are extremely complex artifacts. For example, the Intel Pentium IV processor released in 2002 contained 55 million transistors, and Intel's stated goal for 2010 is a processor with a billion transistors (Intel, 2002). Note that 55 million transistors translate into many more (perhaps an order of magnitude more) detailed design features, because of the distinct elements of each transistor and the connections between transistors.

In terms of "design features" a contemporary IC is clearly a more complex artifact than could be contemplated, much less created without design automation tools. The fundamental enabling technology for design automation for digital ICs is VHDL (IEEE, 1994), or VHSIC Hardware Description Language [1] and its various implementations and extensions, such as Verilog-HDL (Ormerod and Elliot, 2002). In the following discussion, the term "VHDL" will refer generically to this class of design automation tools.

Fundamentally, ICs with billions of design features can be specified and created because VHDL allows them to be represented at multiple levels of abstraction, and provides tools and methods for elaborating higher level (less detailed) abstractions into corresponding lower level (more detailed) abstractions. As a simple example, consider the function "adder" which takes two inputs, and adds them together. I have just given an *ad hoc* high-level description. VHDL provides a formal way to give a corresponding high-level specification. A graphical way to describe an "adder" would be to use symbols representing Boolean functions, and then connect those symbols. VHDL provides tools that allow a designer to create such a representation directly using a CAD-like interface, or indirectly by "compiling" a functional level specification. Finally, the adder might be described in terms of its silicon instantiation, i.e., the three dimensional configuration of wires and transistors actually created on the silicon wafer. VHDL provides the tools and methods by which a designer can "place and route" these features directly, or allow them to be automatically placed and routed by software which refers to a gate-level specification.

At each level of abstraction supported by VHDL (e.g., functional, gate, and layout), the designer can use a corresponding simulation tool to analyze the behavior of the current specification; all that is required is an appropriate set of simulation inputs. As far as the designer is concerned, the design specification is

---

[1] VHSIC stands for Very High Speed Integrated Circuit, making VHDL a nested acronym.

also, in effect, a specification of the simulation model. What makes this possible is the deterministic nature of integrated circuit behavior. There is a relatively small set of basic functions, the behavior of a particular transistor (in a particular material) is relatively deterministic, a signal is either present or not, and the propagation of a signal on a wire is quite predictable[2]. If you start with a particular input, you get the same output no matter how many times you simulate the behavior. The behavior of a standard "cell"[3] is the same, no matter what is happening elsewhere in the IC.

The output of an IC simulation may take different forms, depending on the level of abstraction. At the lowest level (least abstracted) the outputs represent states of the corresponding physical device that might be observed, e.g., a voltage level over time. At higher levels, the outputs may represent the state of a transistor, or the content of a register.

To summarize, ICs are incredibly complex artifacts, but with a relatively simple architecture that allows a formal descriptive method. In addition, the behavior of an IC is easily determined from the well-understood behaviors of its constituent components and their connectivity. Thus, very precise simulation models can be compiled from functional, logical, or physical descriptions of the IC, and can predict behavior as reflected in well-understood measurements.

Simulating Discrete Event Logistics Systems

The unifying theme in discrete event logistics systems (or DELS) is "flows of materials in discrete quanta." The flow may be single parts in a robotic workcell, patients and staff through a hospital, or cargo containers on a cargo ship crossing the ocean, but the flow is discrete. Every movement involves material (or people, or documents, or digital data, etc), perhaps in a container, with a clearly identifiable start location and time, and end location and time.

Between two movements of a given quantity of material, there may be a conversion or a storage process. Conversion may alter the state of the material (forming, treating, assembling, etc.) or the state of knowledge about the material (its recorded weight, dimensions, characteristics). Storage obviously alters the age of material, but it's also not unusual for the quantity going into storage to be different from the quantity coming out (e.g., put a full box of widgets into a bin, and take widgets out one at a time for some use)[4].

Conversion, movement, and storage are "continuous" processes. For example, a movement from location A to location B will involve a continuous trajectory between the two points. Heat treating a component will take it through a temperature profile. The reason we can speak of "discrete event logistics systems"

---

[2] These statements are roughly true, until transistors and wires become much smaller than current practice.

[3] A "cell" is a standard way of implementing some common function, e.g., the adder discussed previously.

[4] Digital data is slightly different, since it may be stored once and retrieved many times, and it may be erased, so there is not a required "balance of flow".

is that we aren't interested in the intermediate states of the continuous processes, only the starting and ending states. When a continuous process starts, it means the corresponding resource is no longer available, and will not be available until the process ends. Similarly, the material subject to the continuous process is captured by the associated resource, and is not available until the process ends.

Focusing on the discrete events doesn't imply the continuous processes aren't important; clearly they are essential for achieving the desired outcome. Faulty conversion processes, for example, lead to defective product. However, from a system perspective, the conversion processes are *local phenomena* while the discrete events are the *integrating phenomena.*

In DELS, there always is a control system of some type. In a warehouse, for example, the control system will determine where incoming materials should be stored, how customer orders will be grouped for fulfillment operations, the locations from which goods will be picked, how pickers will be routed, etc. In manufacturing, the control system will determine when new work is to be released, and how to sequence work waiting at machines. The control system may be completely computerized, or it may involve people, either in planning or execution decision making.

The material flows within DELS can be conceptualized as a network, where arcs correspond to movement and nodes correspond to conversion or storage. If we assign an identity to each material type, then we will have input arcs corresponding to materials sourced externally and output arcs corresponding to materials delivered to customers. In addition, there will be resources required to support every movement, conversion, or storage.

This conceptual network is useful for understanding in part the behavior of DELS, and also why DELS represent such a difficult design and control challenge. The complexity of DELS increases with the number of distinct materials, the number of storage and conversion operations, the number of arcs in the longest chain from input to output, and the number of resources that must be coordinated. Generally, resources are not dedicated to a specific movement, conversion, or storage, but must be shared among the active material quanta. This usually leads to challenging synchronization and coordination problems.

From a cost perspective, it is important to have high utilization of expensive resources, which may only be achieved by accumulating work for them, leading to storage. From a customer service perspective, it is important to have short material dwell times in the network, which cannot be achieved if material is stored for long periods. Hence there are conflicting objectives, and a fundamental challenge is to provide the right amounts of the right resources along with the right policies for planning and scheduling in order to optimize this trade-off.

Discrete event simulation is the most commonly used analysis tool for evaluating DELS performance. All commercial discrete event simulation languages provide the basic modeling constructs to represent queues, processes, storage, and movement. They all provide modeling mechanisms to represent local material flow control decisions, such as deciding what unit of flow to select next from among those available (called dispatching) and where to send a unit of flow once a process is completed (called routing).

The output of a discrete event simulation model is a set of state variables and the time history of their changes. Often, in addition, a graphic rendering is provided as a visual adjunct, showing in a 2.5 D or 3D rendering, the movement of material and the condition of resources over time. This type of *output animation* may be real-time (displayed as the simulation model is running) or playback (displayed from a log file of the simulation run) (Kelton, *et al*, 2002).

Modeling DELS

In contrast to integrated circuits, simulation modeling of DELS is not a completely formalized process. Conceptually, the modeler or modeling team, through observation, investigation, and experience, accumulates a "catalog of facts" about the target DELS. These facts may be explicit (a listing of resource types, e.g.) or implicit (because of a design flaw, spur #4 receives only 15% of the flow, rather than 25%).

From these facts, a conceptual model is developed. There are no formal tools to support this activity. The resulting conceptual model may correspond to the flow network plus dispatching and routing rules. It may correspond to "event graphs" (Buss, 2002), or to some other conceptualization or framework. The conceptual model is made computational, i.e., implemented, using some simulation language such as AutoMod® (Brooks Automation, 2004) or SIMAN® (Rockwell Software, 2004), or implemented in a general purpose language, such as C++ or Java. Finally, experiments are conducted by providing model input data and analyzing the resulting synthetic histories (Kelton, *et al*, 2002).

This description makes the modeling of large scale DELS seem quite *ad hoc*, and in fact, it is. Two different modelers, even with roughly equivalent experience and skill, may produce quite different models, especially if they work with different simulation languages. Not surprisingly, there is an on-going debate within the discrete event simulation community regarding how to define and measure fidelity and equivalence of models (see, e.g., (Gross, 1999)).

The absence of a formal modeling discipline is one key reason why discrete event simulation modeling is relatively *ad hoc*, but there is another deeper reason. It is relatively easy to observe the material flows in DELS, and, if desired, to construct the corresponding flow network. What is not directly observable, however, is the control system. Even in systems that are completely computer controlled, it simply may not be possible to know precisely how decisions are made. The control systems are themselves large-scale complex systems, difficult to describe, and sometimes not fully documented for users. Thus, simulation modelers often must make assumptions about how decisions are made in order to create fully functional models.

Interestingly, the major research emphasis in discrete event simulation does not address the entire modeling process as described above. Rather, research has focused primarily on the statistical modeling and analysis of input data and time series output data, addressing issues such as significance and technical methods to reduce the variance in estimates of system performance.

Researchers and practitioners have recognized the inefficiency inherent in such an *ad hoc* approach to modeling. There have been some attempts to create "simulators" or "simulation templates" for specific types of DELS problems as a way to streamline the modeling process. Both approaches attempt to create a "standard" conceptual model as the basis for a computational model. In the simulator approach, the resulting computational model is developed once, and for each problem instance, specific data is used to populate the model (see, e.g., (Banks, 2001)). In the simulation template approach, information about the structure of the problem is used to assemble appropriate standardized code modules or templates, which may be populated with problem specific parameters (see, e.g., (Pater & Teunisse, 1997)). There are many examples of each approach, but neither has yet achieved a prominent place in practice.

Simulating Organizations

As of the writing of this chapter, there is not yet a standard definition for organizational simulation, yet we need to distinguish it as a domain of study distinct from DELS. Since I believe most organizations exhibit a flow of some sort—either physical, such as flow of automobiles or breakfast cereals, or information, such as flows of money or music tracks—I will define organizational simulation as subsuming DELS, incorporating explicitly the managerial, financial, and human resource systems that most often are implicit in DELS simulations. This is intended to be an optimistic and expansive view.

The organizational flows that we can contemplate and model are those that are *routine*, i.e., they do not surprise us. The flows need not be repetitive in order to be routine. For example, a military unit in the field will execute maneuvers it has practiced in training, so they are "routine" even though the resulting organizational behavior is not repetitive. If we cannot contemplate an organizational flow (or organizational behavior) then it makes little sense to discuss modeling it or simulating it.

One view of using organizational simulation is the "virtual world" view, i.e., the simulation user has an immersive experience of the organizational simulation using some variation of virtual reality technology. A lower-resolution version of this experience would be similar to that of popular interactive games such as EverQuest (Sony, 2004) or Half-Life (Sierra, 2004). In either case, there is a visual rendering of the "organization" through which the user experiences and interacts with the results of simulation computations.

Arguably, much of the behavior of organizations results from the decision-making and behavior of the people in the organization, either at the executive level (think Bernie Ebbers at WorldCom, Dave Thomas at Wendy's, or Warren Buffet at Berkshire Hathaway), or in operations (e.g., a platoon leader or the waiter at a restaurant). Thus, ultimately, organizational simulation may require the detailed representation of individuals' behaviors. This raises two fundamental issues.

The first of these fundamental issues has to do with the nature of decision-making, and what can be modeled. If we formulate our system boundary to include the organization itself, then we are fundamentally limited in modeling the decision-making of organizational leaders. Leadership (or strategic) decisions are often, if not typically, the result of judgments made about the state of the world exogenous to the organization. If we've drawn our system boundary to exclude explicit representation of those states, then we are, by definition, not prepared to model decisions based on those states, thus we are not prepared to model the strategic decision makers. We can, however, contemplate the modeling of decision makers whose decisions are based on the state of the organization *and on the explicit interactions with the exogenous world.*

The second fundamental issue in representing or modeling individuals is simply the creation of their digital images, including their behavior in the simulation. We are by now somewhat familiar with synthespians™[5] or digitally rendered characters in movies; the infamous Jar Jar Binks from *Star Wars Episode 1: The Phantom Menace*, and Gollum from the Lord of the Rings trilogy are the two best-known examples. What may be instructive in this regard is the near-total contempt for the rendering of Jar Jar Binks (see, e.g., (Barnes, 2000)) and the fact that Gollum is quite literally digital make-up for a human actor (Lamb, 2002). It appears we are still a few technology generations away from having access to high-quality, cost-effective synthespians for organizational simulations.

However, there is room for meaningful organizational simulation with the current state of the art. Clearly, an alternative to synthespians is to have real people engage the simulated organization through an interactive computer interface, making planning and control decisions as if the simulated organization were real. This is the strategy used, for example, in gaming, where there also may be synthespians, but their behavior is largely preprogrammed and predictable. Another alternative is to represent individual behaviors in ways we already know about. For example, the result of individual behavior, such as a decision, a performance time, or a quality of outcome can be randomized in some predetermined fashion. The result alternatively may be modeled as dependent upon some other condition, action, or state, e.g., the productivity of labor may be related to some condition in the workplace that is affected by management decision, such as amount of overtime. This type of modeling is quite common in systems dynamics (see, e.g., (Sterman, 2000)).

The key point is that both these strategies fit within the existing discrete event simulation paradigm. Thus, the technology and expertise from DELS simulation can transfer directly in the creation of large-scale organizational simulations.

Organizational Simulation Issues

For organizational simulation to become a routinely used tool, three issues must be resolved:

---

[5] The term "synthespian" is trademarked (Kleiser, et al, 1996).

- Who is (are) the user(s) and how is the simulation experienced?
- What is the conceptual model, and how is it constructed?
- How is the computational model constructed?

Each of these issues cuts across multiple disciplines, and it is clear that no single discipline's contribution completely addresses any of the issues.

In the remainder of this chapter, I will address only the second issue, which asks a simple but profound question, *viz*, "What is an organization?" If we hope ever to be able to cost-effectively construct computational models of organizations, then we must be able to give a formal answer to this question. Whether the construction of the computational model takes the form of a simulator, or uses a template-based approach to construct the model from modules, it will require a clearly articulated formal model, or meta-model[6].

What I will offer in the remainder of the chapter is not an organization meta-model. Rather, I will discuss the creation of a meta-model for one specific type of organization, from the perspective of material flow, and draw from that some conclusions regarding the conceptual and technical challenges of organizational simulation.

WAREHOUSING AS AN ORGANIZATION TYPE

What we seek is an abstract model of "organization" which may be the basis for creating a conceptual organizational simulation model, and which may be instantiated for a specific organization. This is a challenging task, and I will attempt to illustrate it using the simplest of all industrial organizations, a distribution warehouse or distribution center (DC). My focus will be the representation of a warehouse vis-à-vis material flow.

So, what is a warehouse? How can we begin to construct a formal description of the warehouse as a basis for an organizational simulation? These are fundamentally *modeling* questions, i.e., there is not necessarily a "true" model, only models that are more or less appropriate for particular purposes. In this case, the purpose is to create a "meta-model" of a particular class of organizations, which are referred to as "warehouses". Because specific warehouses would be instances from this class, the corresponding warehouse models would be instances of the class model, or meta-model. The meta-model provides the intellectual foundation for creating instance models, and also the basis for software engineering tools to support the instance-model creation process.

The process of creating such a meta-model is similar to the process of creating a particular simulation model, i.e., we need to assemble a catalog of facts, and then begin to organize those facts in a structured way that supports a formal description

---

[6] In the simulation community, the term "meta-model" is used to refer to a statistical model of an input/output relationship. Here I follow the software engineering community, where meta-model refers to an abstract model of a domain of discourse (see, e.g., (Putman, 2001)).

or model. This may seem a straightforward process, but there are a number of conceptual difficulties.

Warehouse "Facts" Catalog

Distribution centers typically occupy large rectangular building, with dock doors arrayed along two or more sides of the building. Goods come to the DC from factories or other warehouses to fill warehouse replenishment orders, typically in truck load (TL) quantities, although occasionally in less than truckload (LTL) quantities. The goods are received and stored (called the "inbound activities") and then subsequently retrieved, packed and shipped to fulfill customer orders (called the "fulfillment activities"). Typically the quantity of a particular product (called a stock-keeping-unit, or SKU) in a replenishment order is much greater than the quantity in a customer order. Distribution centers employ people and use a variety of equipment to execute the required receiving, put-away, picking, transporting, sorting, accumulating, packing, and shipping functions. Distribution center management must balance concerns for customer service (100% of orders filled correctly on time, for example) with costs (labor, inventory, and capital).

Distribution centers are relatively simple examples of DELS, because the flow network is relatively "shallow"—there are only a few movements and storages (and no real "conversions") between receiving and shipping. Yet there are many decisions to be made, both real-time and off-line. For example, suppose a full pallet of books is received at a publishing warehouse, and must be stored. Where should it be stored? If the entire pallet is shipped as one customer order, then any storage location generally located between receiving and shipping will be adequate, because the shipment will require the equivalent of only one trip from receiving to shipping. However, if the books are shipped in many separate orders, a storage location near shipping is much better, so the total distance traveled to get the books from storage to shipping is minimized. There are many similar examples of decisions to be made, in the design of the warehouse, in the organization of the warehouse, and in its daily operations.

Obviously, there are many more "facts" about warehouses in general or specific warehouses, addressing factors like location, configuration, employees, customers, etc.

Organizing the Facts

A warehouse, like any other organization, consists of <u>*resources*</u> and <u>*processes*</u>. Some resources and processes are common to all instances of warehouses, e.g., every warehouse has dock doors (resource) and receives inbound goods (process). Some resources or processes may be unique, e.g., a warehouse may have a unique, self-developed storage technology, or may use a unique algorithm for assigning orders to order-pickers, but most resources and processes are common within the industry.

If resources and processes significantly influence an organization's behavior, then an organizational meta-model must incorporate them explicitly.

Resources and processes may be distinguished with respect to their *observability*. This is particularly relevant with regard to decision-making; we may observe the results of decision-making, but rarely are able to observe the decision-making process itself. Any organizational meta-model must address this fundamental observability issue, and any technology for organizational simulation must provide a mechanism for dealing with important resources or processes that are not observable.

From the material flow perspective, a warehouse has, at any instant, a set of resources: space (a building), equipment, labor (both permanent and temporary/part time), inventory, and customer orders. Viewing customer orders as a resource may appear counter-intuitive; however a warehouse without customer orders would stand idle. On any given day, some customer orders will be "consumed" and others will be "acquired"; similarly, some inventory will be shipped and additional inventory will be received. The number and hours for temporary and part-time employees may vary on a daily basis as well.

These resources are observable[7], and, in particular, every instance of each resource has a discrete "state". A lift truck, for example, is in a particular place in the warehouse, and a particular condition, e.g., "unloaded and idle at point A," or "unloaded and traveling from point A to point B along path p(A,B)." Each employee, at any instant, is in some location, with some associated "state," such as performing and assigned task, or waiting for task assignment. Every unit of inventory is similarly in a location with a condition. Every storage location is either empty or occupied by one or more units of inventory.

There are two distinct types of processes in a warehouse. The first, and easiest to identify, are *flow activities*; a flow activity is a set of linked state changes in one or more resources that have a meaningful explanation in the context of the warehouse objectives of receiving warehouse replenishment orders or fulfilling customer orders. For example, when a truck parks at a receiving dock, there are flow activities that remove product (e.g., pallets of goods) from the truck and stage them at the receiving dock for subsequent put-away to storage. This kind of flow activity typically will engage one or more pallet jacks[8], one or more employees, and the space dedicated to receiving.

A coherent set of flow activity definitions should encompass every aspect of material flow that could be observed in the warehouse. Every observable physical action in the warehouse has a corresponding flow activity.

The second category of warehouse processes is *control activities*. A control activity is, in essence, a decision-making activity, and initiates one or more flow activities. For example, if there are palletized goods at the receiving dock staging area waiting to be put-away to storage, a control activity may determine which storage location is to be used for each pallet of goods. The resulting decision will

---

[7] Even customer orders in digital form could be observed by appropriate queries on appropriate databases.

[8] A type of equipment used for moving pallets over short distances, typically for floor-to-floor operations.

dictate where each arriving pallet is to be stored. This particular decision might be made by warehouse management system software, or by human lift truck operators.

THE STRUCTURE OF A WAREHOUSE META-MODEL

Given this set of the warehouse facts[9], the next step is to create a formal (conceptual) warehouse meta-model. To do this explicitly and completely would require a much lengthier exposition than is possible in a single chapter. Rather than articulate all the details, I will indicate how such a formal model could be structured and implemented.

Obviously, the meta-model must address both resources and processes. It is not enough, however, simply to catalog all the resource types and process types. For the meta-model to be useful, it also must reflect the explicit organization of resources and processes, and their explicit interactions in operation. Without this essential characterization, it would not be possible to create an organizational simulation, at least not efficiently.

Organizing Concepts

In warehouses, a natural organizing principle for resources and process that can be the basis for a meta-model is the notion of *department*. Some departments are self-evident, such as receiving, storage, and shipping. These departments perform readily identifiable warehousing functions, and have clearly identifiable resources and flow activities. There may be multiple instances of a department type, for example, more than one storage department, corresponding to different storage technologies (e.g., pallets racks versus bin shelving). In contrast to the self-evident departments, there are "transportation departments" which perform warehousing functions associated with moving goods in the warehouse. The resources in a transportation department may include aisles, lift trucks and drivers, conveyors, and other forms of automated material transport.

Note that "department" is a concept distinct from "space". A department will have a space resource assigned to it, as well as equipment and people. Thus, department is an artificial construct, employed to organize our representation of the real, physical resources.

A second organizing concept is *location*, also an artificial construct. As with departments, some locations are self-evident. Every unit of storage capacity (pallet rack opening, shelf, bin, staging location on the floor) has an associated "location ID". Other locations are a representation of the idealized point at which a unit of resource (material, people, and equipment) may be located. In the case of resources in motion (a lift truck traveling with or without a load), we are interested in the locations representing the end points of the trip (origin and destination), or

[9] Recall, this is a limited view, considering only the material flows; a broader organizational view will be discussed briefly later.

perhaps the endpoints of trip segments (two ends of an aisle being traversed, e.g.), and may be willing to ignore the exact (geographic) location in between[10].

All departments contain locations, and all locations are in some department. There must be, in addition, some locations that serve as interfaces between departments. For example, the location where a lift truck stops in order to put a pallet in a rack location is an interface between the transportation department and the storage department.

Meta-Model Components-Resources

The following discussion will describe the essential resource components of a formal warehouse meta-model in general terms. In most cases, the conceptual model is quite straightforward.

Space

The most common representation of space in a warehouse is a plan view, or *layout* drawing, showing the dock doors, the aisles, staging areas, and various functional departments. There are a variety of alternatives for modeling space, and any appropriate mechanism may be selected. The only requirements are that the representation selected enable:

- Space to be defined by its boundary
- A given space to be partitioned into two or more spaces
- Two or more spaces can be combined

Equipment

The major categories of warehouse equipment are vehicles, conveyors, storage systems, automated identification, weighing and measuring, labeling, packaging, and communication and control. The degree of automation can vary significantly within these categories, and between warehouses.

For the purposes of developing a warehouse meta-model, what is required is, in essence, a *dictionary* of warehouse equipment, with appropriate and necessary attributed data. For example, this dictionary might take the form of an object-oriented database. In this case, there are many different ways to define the class hierarchy, and no particular approach is necessarily superior to another. For example, in developing a class hierarchy for lift trucks, one might define a class truck and then subclass first on the type of power used (electric, natural gas, or

---

[10] An approximation often employed is to define shorter path segments, defined by *control points*, in order to refine vehicle location along a path.

diesel, e.g.), and then on type of tires (pneumatic or solid). Alternatively, one might first subclass based on whether the lift truck is counterbalanced or straddle.

Suffice it to say, it is conceptually and technically a straightforward matter to construct an extensible dictionary of warehouse equipment types. The beginnings of such a catalog may be seen in the work of (Kay, 2004) and (Montreuil, et al, 2004).

In creating a specific model of a specific warehouse, the equipment types would be instantiated, and each instance would have an appropriate set of attributes. These attributes might include technical attributes of the equipment, such as load and reach limits, as well as operational attributes, such as operating hours since last maintenance, department assignment, and current task assignment.

People

In a warehouse, people will be assigned to tasks within a department or perhaps shared across multiple departments. For example, a lift truck and driver may do some tasks that are pure "transportation" as well as tasks that are pure "receiving." For the purposes of creating a meta-model, there is a super-class called employee that can be subclassed for particular categories of employee (manager, supervisor, skilled hourly, unskilled hourly, etc). For each person in the warehouse, there is an instance of some subclass of employee. This instance may have instance-specific attributes (SSN, e.g.) as well as operational attributes (current assignment, or overtime hours worked in current pay period, e.g.).

Inventory

A key concept in modeling inventory, or material, in a warehouse is *unit of handling*, which may be pallet, tote, case, or item (sometimes called an "each"). It is not unusual for a particular type of item (a "stock keeping unit" or SKU) to be in the warehouse as several different units of handling. For example, motherboards are received in pallet quantities, ordered by some customers in full cases, but ordered by other customers in single unit quantities. Somewhere in the warehouse, one would see this SKU stored as a pallet, as cases (perhaps on a partial pallet, or perhaps in a case storage area), and as "broken cases" (cases which have been opened to remove some but not all the units).

A warehouse meta-model will include a class sku, used to represent the type of goods in the warehouse. The class sku can be sub-classed for each different SKU, which will have particular attributes such as name, supplier, weight, dimensions, type of storage required (cold, secure), etc. In addition, a particular SKU will have a current quantity in the warehouse, perhaps in different states (received waiting putaway, in storage ready to pick, in fast pick, assigned to customer order, etc). For a particular SKU, there may be goods stored in several or many different locations in the warehouse. So in representing an individual SKU, the model must enable a user to identify all the locations where that SKU may be found and the quantity (and perhaps unit of handling) in each location. Note that, by the same token, the representation of each storage location must enable a user to determine what SKU the location contains and in what quantity.

Representing the types of items in a warehouse and the warehouse inventory are not difficult problems from a conceptual perspective. There are many alternative implementations of the necessary databases, and each implementation will have certain performance characteristics for each type of query that might be needed. The details of implementation are important, therefore, from and end user point of view, but not in the context of developing the meta-model.

Customer orders

A customer order is a relatively simple document. It identifies a customer and terms, and lists the items and quantities the customer requires, and perhaps indicates how they are to be shipped. A class order can be subclassed for categories of customers, if necessary (e.g., if there are retail, wholesale, and mail order customers). An instance of a customer order may have a state, such as received but not released, released but not picked, picked but not shipped, or shipped. The level of granularity achievable in tracking the state of a particular order depends on the level of information technology used in the warehouse.

Departments

A formal meta-model will have a class department and subclasses to represent particular types of departments. Table 1 lists a representative set of department types. It is not necessary that this list be final and complete—if additional department types are identified, they may be added to the list. This is an essential characteristic of an organizational meta-model, i.e., it must admit the possibility of continued elaboration and refinement.

An instance of a department has associated with it instances of the resources that are assigned to that department. Thus, department is a modeling concept that integrates the various resource types in a warehouse model. Figure 1 illustrates the relationships involved. The set of department classes is universal, and a generic department type may have associated with it any of the identified resource types. For a particular warehouse, there is a specific set of department types, and for each department type, one or more department instances. A specific department instance will have associated with it a set of specific resource instances.

Meta-Model Components-Flow Processes

The following discussion will describe the essential flow process components of a formal warehouse meta-model in general terms. In the case of material flow

Dept.	Function
Receiving	Inbound goods are unloaded and verified
QA	Quarantine and inspect goods as necessary
Transportation	Move goods between other departments
Storage	Goods storage; not to include temporary buffer
Forward pick	High demand goods picking area
Sortation	Accumulate picked items and sort for packing
Pack	Pack goods to customer order
Ship	Process customer orders for shipment, load outbound
Value adding	Perform value adding services

TABLE 1. List of Department Types Found in Typical Warehouses

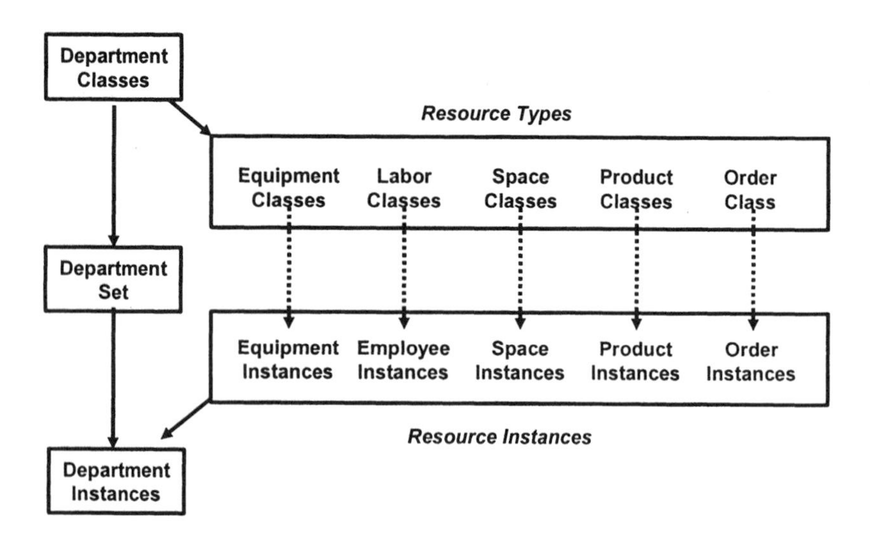

FIGURE 1. Organization of Warehouse Resources

activities, the conceptual model is based on general knowledge of warehouse operations, and is reasonably straightforward.[11]

Material Flow Activities

The material flow activities in a warehouse may be described in terms of department-specific tasks. Table 2 lists a representative, though not necessarily complete list of task types.

Note these are task descriptions that make sense in the context of the warehouse objectives of receiving replenishment orders and fulfilling customer orders. These tasks can be readily observed in an operating warehouse, and could have specific resources assigned for completion. For example, an instance of a transport task will involve a specific handling unit, a specific equipment resource, perhaps a specific labor resource, a specific origin (where the handling unit is initially located), and a specific destination location. It also will have an expected time to complete and, conceptually, could be scheduled. In many respects, these generic warehouse material flow tasks resemble generic process plans in manufacturing.

Flow	Tasks
Receiving	Unload vehicles and verify inbound receipts
Transport	Putaway, retrieval, forward pick replenishment
Pallet pick	Pick pallet, case pick to pallet
Forward pick	Order pick from forward area
Sortation	Sort picked items into orders
Packing	Pack items for a single customer order
Shipping	Load orders to outbound vehicles

TABLE 2. Generic Warehouse Material Flow Task Examples

[11] A reader with knowledge of warehousing will easily recognize the concepts presented; those with no prior knowledge of warehousing may have to take much of this section on faith, and seek additional information in the literature; see, e.g., (Frazelle, 2001) or (Tompkins, et al, 2002).

Material handling tasks correspond either to material flow within a single department (e.g., receiving), or to interactions involving multiple departments (e.g., transport a unit of handling from receiving to storage, which involves three departments).

Warehouse Operations

As defined above, a warehouse task is an aggregate of a number of individual steps, or state changes. For example, unloading a pallet from a truck located at a receiving dock may involve a particular employee walking to a pallet jack, moving the pallet jack into the truck, picking up a pallet, moving the loaded pallet jack to a staging location, and dropping the pallet. These five steps each have a discernable start and end, and each results in a state change for one or more resources. Furthermore, these five steps represent the lowest level at which it makes sense to observe or monitor warehouse tasks. For example, while one might think about each individual stride of the employee walking to retrieve the pallet jack, it hardly makes sense to do so. It is entirely adequate to conceptualize the entire activity in terms of its start location, end location, and duration.

In fact, if one carefully examines all possible warehouse tasks, as defined earlier, one may conclude there is a relatively small set of fundamental warehouse operation types. Table 3 provides an illustrative list.

Operation	Examples
Move	Move from one location to another
Get	Take control of a resource, as "get a pallet jack"
Put	Release control of a resource, as "put a pallet jack"
Store	Place a unit of handling into a location
Retrieve	Remove a unit of handling from a location
Count	Count items in unit of handling
Weigh	Weigh a unit of handling
Measure	Determine one or more dimensions of a unit of handling
Scan	Read a barcode or other label
Communicate	Send or receive information

TABLE 3. Basic Warehouse Operations Examples

A standard form for specifying a basic operation might be:

do *opn-ID* **using** *{res_ID}* **to** *{handling_unit_ID}* **from** *origin_loc_ID* **to** *dest_loc_ID*

where:

opn-ID	indicates what operation type to perform
{res_ID}	indicates which resource(s) to use
{handling_unit_ID}	indicates what is to be operated on, i.e., a container or goods ID
origin_loc_ID	indicates the starting location
dest_loc_ID	indicates the ending location

Any warehouse task instance is a combination of several operations. Moreover, a log of all basic operation instances would allow one to construct a complete state trajectory for all resources in the warehouse.[12]

Figure 2 illustrates the relationships between task classes and instances, and operation classes and instances. There is a generic set of task types, but a given warehouse may not use them all, and will have a specific set of task types. A task instance engages a set of basic operation type instances. Note the symmetry between Figure 1, which addresses resources, and Figure 2, which addresses (material flow) tasks and operations.

There are some important differences between the modeling of resources and the modeling of tasks. In modeling resources, not much judgment is required; a resource is either in a department or it is not. In modeling tasks, however, considerable judgment may be required. For example, removing a pallet from a truck, moving it to a staging area, and scanning it into the WMS could be modeled as one task, or as several tasks. Fewer task types might be desirable from a task identification perspective, but then each task type would be more complex to execute, control, and model in terms of operations.

Another important difference is that resource types will tend to be relatively static in a given warehouse; new technologies, new labor categories, new space, and new product categories are occasionally but rarely added. However, people may change the way they perform functions, perhaps leading to new task types, whenever they see an advantage to doing so.

[12] It is interesting to consider the possibility of defining warehouse tasks by examining a time-stamped log of all basic operations.

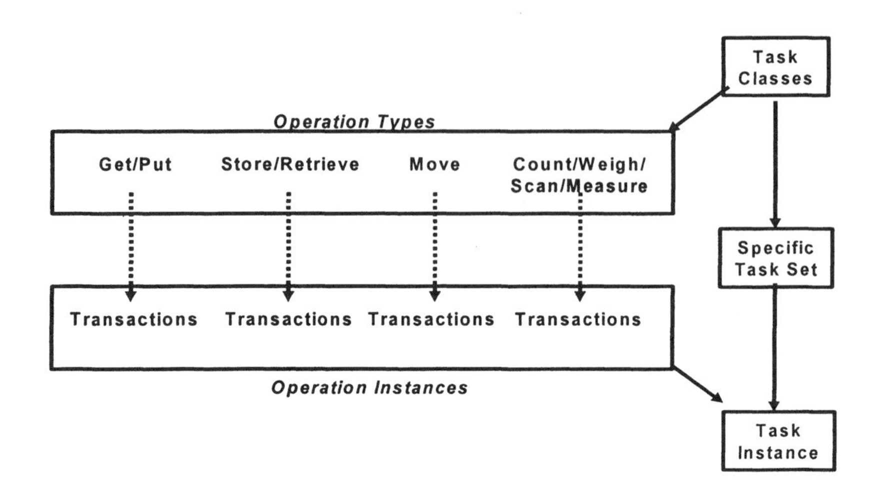

FIGURE 2. Organization of Material Flow Tasks and Operations

Meta-Model Components-Control Processes

Resources and flow processes have a physical manifestation that may be directly translated into modeling components. The major modeling challenge is simply to organize observations into a coherent, manageable modeling structure. Control processes present a very different modeling challenge, because they cannot be directly observed, even when (or especially when) they are automated.[13] For the most part, the modeler must make assumptions about control processes in order to develop a complete model. In other words, the modeler creates a "theory" of control decision making. Two goals should guide the modeler in developing this theory:

- The assumptions should be consistent with what can be observed
- The fewer assumptions, the better

Conceptually at least, the theory is testable, in that results from a simulation can be compared with results from the organization being simulated. Interestingly, there is very little research that explicitly attempts this sort of testing.

In developing a theory of warehouse decision making, we might ask, "What drives a warehouse?" The answer seems straightforward: goods arriving to dock doors drive inbound activities, and arriving customer orders drive fulfillment activities. A little reflection reveals also that the *anticipation* of future inbound or

[13] Just because control processes are implemented in software, one shouldn't assume they are easily described. There are many examples of large scale control software implementations that are not well understood in terms of control decision making.

fulfillment activities also will drive some warehouse behavior, such as scheduling additional part-time labor, or pre-staging particular goods for fast picking.

One approach to modeling warehouse control can be based on the following assumptions:

- Decision making determines task instances to be executed
- Decision making coordinates operations instances required to execute a task
- Decision making is event driven, i.e., decisions are made in response to events

Figure 3 illustrates an inferred structure for control processes in a warehouse. Note that this is a functional representation, not an entity-relationship representation. This is because the implementation of the functions may be distributed across both information technology and human resources. Some event monitoring may be done computationally, by a warehouse management system, and some by the people working in the warehouse. A person assigned to a task will coordinate operations implicitly, for example.

Events

Every resource state change corresponds to an event. We can think of each event generating a message to the warehouse control system, so the information about warehouse state can be updated.

The event messages illustrated in Figure 3 are of three distinct types. External events include the arrival of trucks for unloading, or the receipt of customer orders. Internal events correspond to the initiation or completion of basic warehouse operations (i.e., state changes in resources). Timer events are preplanned events associated with particular times, for example, the execution of a forecasting function at a predetermined time.

Event Monitor

In Figure 3, "event monitor" is a conceptual device created in response to the fact that not all events (state changes) in the warehouse trigger decision making. For example, while the arrival of a truck at a receiving dock triggers decision-making to dispatch resources to unload the truck, the arrival of a pallet jack to remove a particular pallet from the truck does not necessarily trigger decision making. The important events are those that trigger decision making.

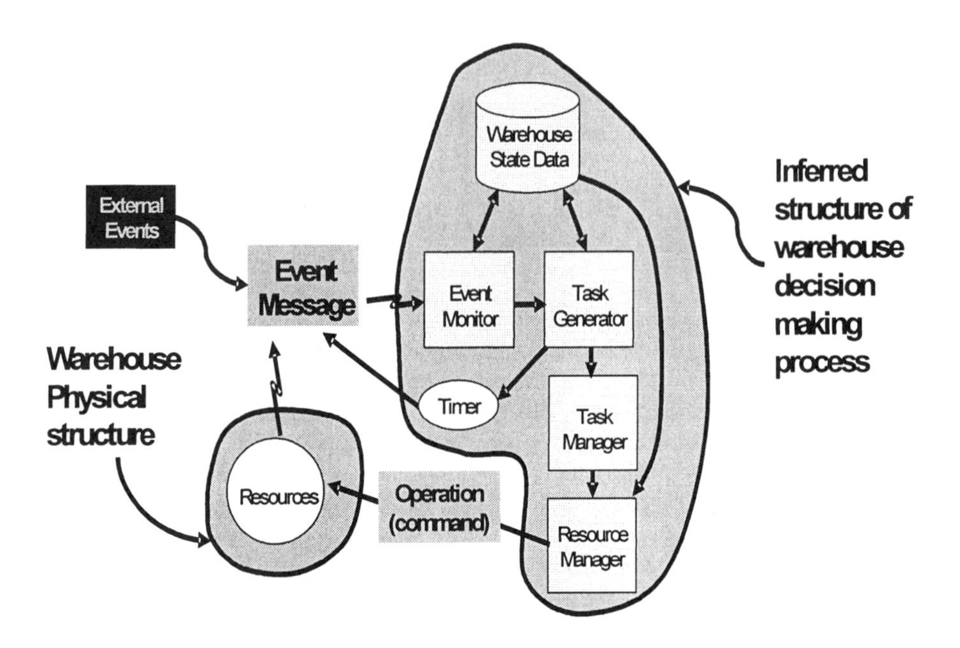

FIGURE 3. Inferred Structure of Decision Making

Task Generator

Implicit in Figure 3 is the assumption that specific event types trigger specific types of decisions. In particular, the generation of a specific warehouse task (as defined earlier) is a response to a specific type of event. So, as an example, the completion of an unloading task (last pallet from the truck is put down in the receiving staging area) may trigger the decision to select the storage location(s) for the inbound goods and generate corresponding putaway tasks.

The task generator is conceptualized as a specific decision making process, i.e., one which creates a specific task type, thus there may be as many different "task generators" as there are task types. Thus, one function of the event monitor is to select the task generator appropriate to the type of trigger event. The task generation process uses warehouse state data, and creates a task and the corresponding set of operations, as defined earlier.

Note that it is not necessary for the task generator to be a computational process. It could as well be a process performed by warehouse employees as part of their routine tasks.

Task Manager

The role of the task manager is simply to insure the operations in a given task are executed in the proper order. One operation in a task sequence cannot be released until its prerequisite operations have been completed. The state of prerequisite operations that have been released for execution can be determined from their corresponding event messages, as reflected in the warehouse state database. In the case where a task generator creates several tasks at once, the task manager also maintains any execution order requirements among them.

Resource Manager

The operations for a particular task may not be completely specified; for example, there may be a number of lift trucks and drivers performing putaway operations, and the particular lift truck chosen to putaway a particular pallet may simply be the "next one available". Since the identity of the resource is not known when the task is specified, it may only be identified by resource type. In this case, a "resource manager" function is required to mediate between active operations competing for the same resource (e.g., all the active move operations that utilize a limited set of lift trucks). When the resource becomes available, the resource manager determines which available operation should claim the resource.

Meta-Modeling Issues

My purpose for discussing a warehouse meta-model has been to establish a framework for discussing organizational simulation meta-models. The most important point to draw from the warehouse example is this: even if we restrict our attention to just the material flow in a warehouse (the simplest of all industrial facilities), modeling the control processes still is quite difficult; we must articulate a theory of control and give it a computational expression.

As the discussion has indicated, it is not inconceivable that a warehousing meta-model can be developed and translated into computational support tools for warehouse modeling and simulation, although there are still significant challenges. Elaborating the models of resources and flow processes seems to be reasonably straightforward, and can, no doubt, be implemented with suitable level of effort in some computational form. The outline of a model for control processes also appears to be amenable to implementation, although it clearly requires more development effort. In particular, those aspects of operational decision making that are managerial, such as how much space to allocate for storing items, or how many part-time employees to bring in next week, may not be easily converted into algorithms. In fact, it may be necessary to incorporate a human interface and have such management decisions made in real-time, thus creating a first generation organizational simulation.

It certainly should be noted that I have not discussed at all the higher-level managerial decision making in warehouses. For example policy issues such as the frequency of replenishment, incentives to modify customer behavior, or the selection of new technologies all are managerial decisions that fundamentally change warehouse operations. Incorporating major resource or policy changes into the execution of a simulation model raises both technical and methodological issues, i.e., how to do it, and how to interpret the results.

CONCLUSION

A very large portion of the organizational simulation problem domain consists of organizations which exhibit discrete flows of materials, people, or information, and whose behavior over time is intimately related to these flows. The modeling of the DELS component of organizations may not yet be at the level of modeling ICs, but the potential is there to achieve similar modeling efficiencies. However, if organizational simulation is to be practical, the other elements of the organization must be modeled with similar efficiency.

We must be able to describe these other systems with at least the same level of precision and rigor as we can describe the discrete event logistics systems. In addition, all these descriptions must be integrated. In other words, if the political systems in an organization impact the flow control processes, then our description of these two systems must allow us to create computational models reflecting these impacts.

The integration of these disparate views of the organization—flows, finances, human resources, authority/responsibility, etc—presents a very significant conceptual challenge. It has proven quite difficult to achieve an integration of the resource, flow process, and control views in the simple case of warehouses, so the problem should not be underestimated.

Some fundamental concepts have proven effective in modeling DELS: resources, flow processes, and event driven control processes. The challenge is to identify the corresponding concepts for the other organizational views, integrate the views, and create the integrated organizational meta-model.

There also are technical challenges in achieving practical organizational simulation. There is the software engineering challenge of constructing large-scale, complex software implementations of organizational simulations. Great strides have been made in recent years in the software engineering world, with software engineering and design tools evolving from UML (Object Management Group, 2004b), to MDA (Model Driven Architecture) (Object Management Group, 2004a), and RM-ODP (Reference Model of Open Distributed Systems) (Putman, 2001). The implementation of organizational simulation may benefit significantly from current developments in the discrete event simulation domain, such as distributed and parallel simulation (Fujimoto, 2004). There is ample reason to conclude that the software engineering challenge is not as difficult as the conceptual challenge of creating the organizational meta-model.

Finally, it is worth repeating that general-purpose simulation languages and tools are often applied to specific problems in an *ad hoc* way that makes them very expensive solutions. Invariably, the applications where simulation is widely adopted are those where an effective meta-model exists and is instantiated in a simulator or in usable templates. The development of organizational simulation is more likely to be successful if it includes a focus on the integrating organizational meta-model.

REFERENCES

Banks, J. (2001). Discrete Event System Simulation. Englewood Cliffs, NJ: Prentice Hall.

Barnes, D. (2000). Attack of the Synthespians. SN&R, October 15, 2000, available at http://www.newsreview.com/issues/sacto/2000-10-05/arts.asp.

Brooks Automation. (2004). Make Better Decisions with Simulation. http://www.automod.com/.

Buss, A. (2002). Component Based Simulation Modeling with SIMKIT. Proceedings of the 2002 Winter Simulation Conference. 243-249.

Frazelle, E. (2001). World-Class Warehousing and Material Handling. New York: McGraw-Hill.

Fujimoto, R. (2005). Distributed Simulation and the High Level Architecture. In W.B. Rouse & K.R.Boff, Eds., Organizational Simulation: From Modeling and Simulation to Games and Entertainment (Chapter 21). New York: John Wiley.

Gerstlauer, A., & Gajski, D. (2002). System-Level Abstraction Semantics. Proceedings ISSS '02. 231-236.

Gross, D. C. (1999). Report from the Fidelity Implementation Study Group, SISO-REF-002-1999. Available at https://www.dmso.mil/public/library/projects /vva/evolvingconcepts/99s-siw-167.pdf.

IEEE. (1994). IEEE Standard VHDL Language Reference Manual, IEEE Std 1076-1993.

Intel. (2002). Expanding Moore's Law: The Exponential Opportunity. Fall 2002 Update, ftp://download.intel.com/labs/eml/download/EML_opportunity.pdf

Kay, M. G. (2004). Material Handling Equipment Taxonomy. Available at http://www.ie.ncsu.edu/kay/mhetax/index.htm.

Kelton, W. D., Sadowski, R.P., & Sadowski, D.A. (2002). Simulation with Arena. New York: McGraw-Hill.

Kleiser, J. (1996). Digital stunt doubles: safety through numbers (panel). Proceedings of the 23rd annual conference on Computer graphics and interactive techniques. 493-494.

Lamb, G. M. (2002). The rise of 'synthespians'. Christian Science Monitor, December 18, 2002. available at http://www.csmonitor.com/2002/1218/p01s01-almo.html.

Montreuil, B., Legare, R., Jr., & Bouchard, J., Jr. (2004). Material Handling Multimedia Bank. available at http://w3.centor.ulaval.ca/MHMultimediaBank/.

Naroska, E. (1998). Parallel VHDL Simulation. Proceedings Design, Automation and Test in Europe 1998. 159-163

Object Management Group. (2004a). The Architecture of Choice for a Changing World. http://www.omg.org/mda/.

Object Management Group. (2004b). UML™ Resource Page. http://www.uml.org

Ormerod, J., & Elliott, I. (2002) Advanced Electronic Design Automation. http://www.ami.ac.uk/courseware/adveda/overview.html.

Pater, A., & Teunisse, M. 1997. The use of a template-based methodology in the simulation of a new cargo track from Rotterdam Harbor to Germany. Proceedings of 1997 Winter Simulation Conference. 1176-1180.

Putman, J., R. (2001). Architecting with RM-ODP, Englewood Cliffs, NJ: Prentice-Hall.

Rockwell Software, Inc. (2004). Arena: Forward Visibility for Your Business. http://www.arenasimulation.com/.

Sierra. Half-Life. http://games.sierra.com/games/half-life/.

Sterman, J. D. (2000). Business Dynamics: Systems Thinking and Modeling for a Complex World. New York: McGraw Hill/Irwin.

Sony Online Entertainment. EverQuest. http://everquest.station.sony.com/.

Tompkins, J. A., White, J.A., Bozer, Y.A., & Tanchoco, J.M.A. (2002). Facilities Planning, 3rd Edition. New York: Wiley.

CHAPTER 11

NARRATIVE ABSTRACTION FOR ORGANIZATIONAL SIMULATIONS

JANET H. MURRAY

ABSTRACT

Story making is a core cognitive and cultural strategy of human beings that contributes to understanding complex phenomena and to social organization. Narratives are abstraction systems, which computers are assimilating through emerging traditions of gaming and interactive story telling. Organizational simulation can be represented as a story-system, allowing us to integrate multiple interpretative and analytical approaches, work at higher levels of abstraction, and make our assumptions more explicit.

NARRATIVE AND SENSEMAKING

As human organizations and our means of analyzing them grow in complexity it is becoming increasingly challenging to describe them in linear formats. Simulations are a more comprehensive strategy, and offer the promise of synthesizing many dimensions of analysis. But the more information a simulation contains, the harder it can be to present the information in a manner that is readable by the observer. A successful interface to a complex simulation would allow the user to vary multiple parameters, to understand the consequences of these variations, and to review and contrast multiple models of past, present, future, and proposed situations. One underexploited resource in designing such interfaces is the fundamental cognitive strategy that allows us to take in information in the form of stories.

Story telling is among the most common of human experiences, one we take for granted most of the time. A story is a causally connected sequence of events with a beginning, middle, and end. Narrative forms a framework in which to assimilate particular information. One of our most basic means of sorting information is through a sequence of events, but a mere sequence, or chronicle is not a narrative. A chronicle takes the form of one thing after another, as in this passage from the first book of the Bible (Genesis 5:5-5:10)

> *5 And all the days that Adam lived were nine hundred and thirty years; and he died. 6 And Seth lived a hundred and five years, and begot Enosh. 7 And Seth lived after he begot Enosh eight hundred and seven years, and*

> *begot sons and daughters. **8** And all the days of Seth were nine hundred and twelve years; and he died. **9** And Enosh lived ninety years, and begot Kenan. **10** And Enosh lived after he begot Kenan eight hundred and fifteen years, and begot sons and daughters.*

To paraphrase E.M. Forster, "The king died and then the queen died" is just a sequence of events. "The Queen died and then the King died of grief" is a plot. It has not just sequence but the subordination of perceived causality.(Forster 1927). Stories select out certain more dramatic events from the stream of human experience, and they arrange the events in memorable patterns that emphasize the most emotionally salient details, such as desires, surprises, triumphs, and defeats. Story patterns often emphasize changes in circumstances, whether that is the emergence of matter from nothingness (as in the Book of Genesis) or the fall of the exceptionally fortunate (as in Aristotle's definition of tragedy). Stories organize and summarize many events into one or more connected sequences. (Martin 1986; Aristotle 1996)

Our ability to produce and process narrative is among our oldest cognitive and cultural strategies for tracing complex causal connections in a situation of data overload. Human culture itself, the shared patterns of meaning and behavior that permeate our existence, rests to a large extent on our ability to capture real and imagined events as sequences of cause and effect (i.e., stories) and to share these sequences – through oral, enacted, and written storytelling technologies. From the ancient proverbs to the latest globally popular films, from the Book of Genesis to the "Big Bang" theory of the cosmos, narrative strategies are essential to the human process of sensemaking. (Bruner 1986; Turner 1996; Bruner 2002)

Storytelling is particularly important to the formation of organizations, starting with the pre-literate clan. Shared stories explain the origin of life, the spiritual truths that guide behavior, the history of the group as distinct from animals and from other human groups. Foundational stories explain to people who they are and how they should behave. History, science, philosophy, religion, and even legal codes can all be seen as specialized forms of storytelling. Stories encode complex relationships, between humans and the cosmos, humans and gods, humans and animals, adults and children, women and men. We may express these relationships in other ways: in legal documents, in mathematical equations, in organizational charts, in symbolic adornments. But stories provide a home truth, a touchstone for what these relationships mean to us experientially as well as cognitively.(Turner 1966; Geertz 1973)

Culture itself is a form of organization, and other organizational changes cannot be viewed apart from culture. For example, the economic changes known as globalization are increasingly recognized as impacting upon and influenced by the individual cultures that bring their own narratives of dominance, resistance, and moral imperatives into the marketplace along with their political and economic circumstances. Events do not "read" the same way to different cultures; they are fit into many master narratives, connected up with the appropriate cause and effect sequences specific to different groups (Bowker and Star 1999). For further discussion on culture's impact on crowd modeling, see Chapter 17.

Storytelling often focuses around cultural behaviors that serve as powerful organizing strategies, such as feuds or journeys. The conflicts that drove the ancient Jews out of Egypt and the subsequent journey through the desert were powerful organizing events for Jewish culture. The telling of the story of the Exodus over and over again as an annual ritual is also an organizational technique, binding the group together from shared remembered experiences, prescribing certain foods and behaviors (e.g., kindness toward strangers).

Storytelling is an organizational tool at the individual and cultural level. The stories that make up the Homeric epics or the Hebrew Bible can be read as a record of organizational strategies for groups of individuals forming themselves into communities with a common culture. The stories of Noah's Flood, Moses' leadership of the Exodus, the destruction of Troy, the voyages of Odysseus preserve considerable information about organizational strategies, both successful and catastrophic. The civic life of Atlanta has moved from one organizational storyline to another. The earlier story line is The Lost Cause, exemplified by the infatuation with Margaret Mitchell's *Gone with the Wind*, and with the film made about the movie, whose premier is still celebrated, for example, in photographs at my local dry cleaners. The newer storyline is The City Too Busy to Hate, exemplified by the life of Martin Luther King, and by the wisdom of the Atlanta white establishment in avoiding the violent confrontation that destroyed the economy of its rival Birmingham (Pomerantz 1996).

History, George Orwell correctly wrote, is written by the winners, who get to impose their narrative on events (Orwell 1944). The front page of today's newspaper is not only the "first draft of history," as it is commonly called, but also the first line of organizational negotiation: of figuring out together, within the constraints of existing power relationships, what to make of the many confusing things happening in the world around us. We tell one another stories in order to understand better where we are and how we should act in the future.

Story telling allows us to compress many kinds of information and many different modes of analysis into an easily recalled and applied master structure. A journey story – from the Odyssey to the Flight of Apollo 13 -- integrates information about places, people, and technologies. A love story – from Romeo and Juliet to Diana and Prince Charles – integrates psychological, sociological, and political information. Some stories, like Aesop's fables, have explicit morals: the grasshopper sings all summer and has nothing to eat in the winter, while the industrious ant has stored up food. Others encode more complex wisdom, about how to face a difficult and dangerous task or the misery that comes from cycles of violence. Stories are memorable patterns of experience that we can use to interpret new events. They form a template and a shorthand for talking about experience. A lengthy and difficult journey is described as an "odyssey," a Palestinian and Israeli couple are called the "Romeo and Juliet" of Jerusalem. Encapsulating experience in stories allows us to apply very complicated patterns to new experiences.

Story-telling also allows us to manage competing views of complex events, to assess them against one another, and to synthesize them, accounting for multiple points of view and multiple strategies of interpretation. Homer gave us an integrated picture of the Trojan War in the *Iliad,* describing life in the besieging army of the Greeks and within the city of Troy. The novelist Tolstoy did something similar in his masterpiece *War and Peace,* which looks at Napoleon's Russian campaign across national and class boundaries. Journalists of the 21st century covering a war, attempt to interview people from both sides of the conflict and from countries neighboring the war zone, and to capture the viewpoints of the generals, the soldiers, and the civilians. We use story-telling technologies of oral recitation, print, moving images and recorded sounds toward the same end: to capture events that are beyond any one person's experience and distant in time or place into a single narrative format. Each of these technologies has developed formats that help us to compare multiple points of view (e.g. successive speeches in Homeric epic; switching narrators in a novel; split screen debates on television) or to move from an overview to a close-up (descriptions of armies followed by descriptions of individual fighters; long-shots to establish a context, followed by closer shots to zoom in on parts of the action).

Narrative making is not the same as fiction making. We do not need to invent incidents in order to make a story. We routinely make our life events into a story through a process of abstraction, taking the relevant details and leaving out the irrelevant ones, emphasizing one part of an event over another. The same is true of public events, and even of scientific observations. We tell one story over another depending on what we decide to observe and to emphasize, and what patterns of causality we understand well enough to apply. The alternation of day and night and summer and winter forms the basis of one story in the Ptolemaic universe, another in the Copernican universe, and a third in the Einsteinian fabric of space/time. Before these were elaborated, human beings had other narratives we now call myths to explain the same phenomena. What we find hardest is to observe without imposing a narrative, to leave phenomena outside the organizing strategy of storytelling.

One of the features of late twentieth century culture is a greater sophistication about narrative, an understanding that the same events can form multiple story patterns depending on the viewpoint or assumptions of the storyteller. In parallel to our greater ability to offer quantitative models of the world, and to describe large phenomena at ever smaller levels of organization, we have developed techniques of storytelling that capture multiple points of view and multiple versions of the same experience. The increased complexity and self-consciousness of storytelling techniques in the 20th century represents a cognitive, cultural, and technical development, independent of the new storytelling technologies (like film, TV, and computers) that have supported it.

In cognitive terms we can now take in more than one moving image of the same event, even simultaneously, as in split-screen coverage of a live news event, or scanning of multiple monitors. We can also assimilate multiple written accounts of the same events, and relish multiple points of view in fictional and non-fictional accounts of dramatic incidents. We watch movies that imagine multiple pasts, presents, and futures (e.g. *Back to the Future, Groundhog Day, Run Lola Run*), and have developed storytelling techniques that help us to keep these multiple versions straight. We have grown suspicious of master narratives and look for the resistant, divergent points of view. When something rivets our attention, we expect to hear it narrated by multiple voices, and we relish the discrepancies and contradictions. In high art and popular culture we are developing forms that juxtapose viewpoints and present variations on the same sequence of events.

Sociologists and anthropologists have recognized the need for more complex forms of narrative to capture the complexities of organizational behavior (Latour 1996; Czarniawska 1998). Descriptions of engineering projects or government structures with changing networks of social actors cannot be captured in a single plotline. Yet narrative remains the most powerful strategy for those within an organization to make sense of their experience. As Karl Weick explains in his study of sensemaking in organizations:

> *If accuracy is nice but not necessary in sensemaking then what is necessary? The answer is, something that preserves plausibility and coherence, something that is reasonable and memorable, something that embodies past experience and expectations, something which resonates with other people, something that can be constructed retrospectively but also can be used prospectively, something that captures both feeling and thought, something that allows for embellishment to fit current oddities, something that is fun to contrast. In short what is necessary in sensemaking is a good story. (Weick 1995).*

The complexities of organizational life are built on interlocking communities of storytellers and story-enactors. But those who study organizations are still actively struggling with a way to represent these multiple shared realities.

Narrative is one of our most ancient and powerful forms of sensemaking, and our narrative sense has been developing with our technologies of representation. Our increasing ability to describe events in words and images, supported by the technologies of drawing, writing, print, theater, painting, sculpture, still photography, moving images, broadcast, etc., has meant an increasing ability to put human experience into narrative form and to share those stories with one another. The invention of the printing press has brought us a richness of detail of inner life of the individual and of the place of the individual life in the social organization because it has made it possible to create and distribute the sustained narratives we know as novels. The invention of the movie camera has continued this trend, deepening our sense of one another's subjective reality and our grasp of the panoramic whole. With the growth of television, we share a repertoire of story patterns, so familiar that they are easily made the object of parody. The computer has assimilated and extended all of these modalities of storytelling. It has also added new capabilities that derive from its own powerful affordances.

COMPUTATIONAL SIMULATION, GAMES, AND NARRATIVE

Simulation is a form of representation that preceded computing but has been brought to new power with the advent of the computer. Simulation is an outgrowth of the study of complex systems, beginning with the cybernetics research of

Norbert Wiener in the mid-twentieth century which pointed out the formal parallels in the complex systems that were instantiated in animals, machines, and human organizations (Wiener 1948; Wiener 1950). The growth of systems theory approaches to the world was made possible by the invention of a machine that could instantiate systems, a machine that was able to behave according to programmed rules that modeled the observed or posited rules of natural or social systems.

At the same time, we have come to understand many real world non-quantifiable behaviors as the result of complex systems of autonomous actors. Psychology has created a systems model of the family, for example, and phenomena from global warming to terrorist attacks have come to be understood as the "unintended consequences" of interventions in systems whose complexity was ill-understood (Tenner 1996; Sandler & Arce 2003). The challenge for the 21st century is to turn our ability to model the world as complex systems with emergent behaviors into an ability to act within that world with greater effectiveness, foresight, and wisdom.

Some, including other authors in this book (Hudlicka & Zacharias, 2005; Klein, H. & McHugh, 2005; Klein, G., Feltovich, Bradshaw and Woods, 2005; McGinnis, 2005), would argue that our ability to model is yet fairly limited. Furthermore, as Sherry Turkle and others have pointed out, we lose track of our assumptions when we move our modeling onto the computer (Starr 1994; Turkle 1995). A fundamental problem is the riskiness of translating conceptual connections of varying degrees of specificity, certainty, and completeness into quantitative and mechanical formats. It is easy to forget that our models are approximations and that the real world has many more variables than we have put into our systems. One way of addressing the issue of the reliability of simulations is to treat them more like story-systems than like calculators.

Simulation, like narrative, is a system for representing cause and effect. The output of simulations can be narrated, and experts can be trained on simulation games that unite computation and live action in a multi-dimensional experience. Simulations can be experienced by these actors as a dramatic story, and often include strong narrative elements such as fictional locales, invented characters, and dramatic incidents. Even when simulations are more quantitative in nature they often include elements of "happy" and "unhappy" endings, and constraints on resources that are similar to those in games. In fact, some of the most successful computer story/games have been based on simulation techniques including *Civilization, SimCity,* and *The Sims.*

Narratives, like simulation, are a means of making sense of a complex world by creating a model with only the most relevant elements of the phenomena being studied. *The Iliad* focuses on the dynamics of the battle for Troy, drawing on other aspects of the two societies only as they throw light on the behavior and fate of the contesting groups. *Hamlet* does not dwell on the inner feelings of Rosencrantz and Guildenstern, but on the wavering consciousness of the prince. Like simulations, narratives can model a huge past event as the sum of many smaller events (e.g. the Napoleonic wars in *War and Peace*) or many versions of an imagined future (e.g. the *Star Trek* saga). Narrative allows us to ask "what if" and to think about the answer in detail, to imagine ourselves experiencing and even acting within alternate realities that bear on our actual world. Narrative also works to construct a causal sequence that explains a current reality, as in a detective story where we work backward from the results of the crime to the facts and causes of the crime.

The computer has enhanced our ability to create simulations because it provides us with a greater power of representation than ever before, including the novel ability to execute models of multi-agent systems. The same unprecedented representational powers have enhanced our ability to create stories, opening up new genres and techniques for practicing the ancient human practice of making meaning by creating significant, discrete sequences of events.

The history of computing in the second half of the twentieth century is studded with examples of pioneer computer users adapting the new representational powers of the computer for narrative purposes (Murray 1997). Joseph Weizenbaum's creation of the first simulated computer character, Eliza, used the procedural and participatory structure of the computer to author a scene that was open to improvisation by the interactor while returning dramatically believable responses. The astonishing achievement of Eliza was that even computationally sophisticated users were fooled into thinking there was a person rather than a program generating the responses. The achievement came not from the complexity of the coding – which was simple pattern matching with ample defaults – but from the cleverness with which the dramatic situation was created, allowing the user to offer free form input that afforded the opportunity for faking understanding, or constructing comically appropriate misunderstandings (Weizenbaum 1966).

Zork (based on the earlier program Adventure) used the spatializing property of the computer to create a landscape that could be navigated, and whose consistency created the illusion of a real, experiential space. It also expanded the expressiveness of the medium by its early use of object-oriented programming techniques, allowing for a flexibility of event structure beyond that of a simple branching narrative (Lebling, Blank et al. 1979).

The development of computer-based narrative overlaps with the development of digital games, which are currently a $25 billion dollar global industry. The first computer games were merely additive creations, taking pre-digital games and porting them to the computer, making use of the ability to automate the game board and even the opponent. Tic Tac Toe (Noughts and Crosses) on the computer could be programmed so that it was impossible to write in a box once it had been written in, and so that the computer took a turn, applying the same strategies as a human player. This was the first computer-based game, and it is a pattern that is still going strong, providing millions of hours of computer solitaire for the ordinary office or home computer users and providing the existential crisis of chess masters defeated by computer programs. The appeal across the spectrum is the same: the game environment focuses our attention on a rule-based, outcome-driven contest in which we can pit ourselves against an external force in an arena limited by space, time, and conventions of engagement.

The combination of narrative and game pleasures has proven to be particularly potent. The first innovative computer game was *SpaceWar!* which was based on

Japanese special effect movies and space explorer science fiction novels. The designers started out with a simple shoot-em-up in space, with two crudely drawn ships that could move and fire at one another. The game was enhanced by the addition of realistic starscapes drawn from an astronomy project, which increased the sense of an immersive adventuresome reality. But the key design feature was the addition of a central star, like the sun within our own star system, with its own gravity. The addition of the star focused all the other aspects of gameplay, and turned the game from one of rapid response and chance into one of strategy, skill and comprehensible physics. The gravity of the star gave it reality outside the actions of the two gamers; the game was more playable because they had something to play against, something that could be used to their own advantage, that obeyed its own rules. (Graetz, 1981).

In a game like baseball, the bat, the ball, the measurements of the playing field, provide this external structure against which the consensual rules – the number of innings and outs, the right way to tag someone, etc. – can be performed. In a card game, this structure is provided by the limits of a single deck of cards. On the computer we have much more flexibility in building the "physics" of the game, in setting up the constraining game resources. We do not have to rely on the real world physics of sports or on the kinds of resources that can be instantiated in atoms. We can bring in elements whose behavior can be scripted based on narrative expectations (like Trolls and enemy aliens), and we can rely on the players' knowledge of narrative elements drawn from fantasy literature to direct their play.

In the decades since the invention of *SpaceWar!*, this love of complex strategy games has continued to be appealing, and the interweaving of strategy and narrative has formed the basis of some of the most successful games in several game genres. For example, the fighting games in the Star Wars series benefit from placing the action in the known storyline of the movies, and from allowing players to position themselves in the army of the Empire or the Rebels. The variety of vehicles within the series is a way of providing variety of strategy and technical challenges. More strategy oriented games like *Civilization* and *SimCity* also exploit story elements to immerse the interactor in the complicated technical environment, adding value to the many parameters within the simulation, and increasing the dramatic investment in the outcome. *SimCity* allows for an open-ended story without win/lose conditions. As its creator, Will Wright, has put it, it is more of a toy than a game. Its latest spin-off, *The Sims*, is in Wright's mind, a "dollhouse," a focus for fantasy play with computer-based characters whose personalities and daily life decisions (to wash, to get a job, to call a friend) determine the state of the simulation, which is an extended interactive representation of bourgeois life and values.

The complexity of digital gaming has increased exponentially since they first came on the scene, 50 years ago. The range of engagement in gaming as well as the volume of games is now so great that there is a separate scholarly organization to study digital gaming. The first studies are now emerging and the general shape of gaming studies is becoming clear. Games are part of our play behavior, which we share with other animals but which is enhanced by the human capacity for symbolic thinking and for the manipulation of symbols. Play has core survival value in helping us to keep a larger repertoire of behaviors and ideas than is needed for immediate survival. Games codify play, and focus it upon particular constrained actions: games are modes of experimentation and specialized training. By demanding the repetitive, coordinated performance of specific skills games allow us the opportunity to model and master situations, and to try the effects of different strategies on the same problem.(Sutton-Smith 1997).

Digital games increase the complexity with which we can model a problem area, because we can bring all the symbolic and computational powers of the computer, all of its speed in tracing dynamic change, to bear in setting up the game arena and in creating opponents to match ourselves against. The plasticity of the computer is also important: it allows us to make things than can be changed without having to engineer them out of atoms. This is an advantage for gaming because it allows for tuning, one of the key ingredients in a successful game. Tuning is the balancing of the challenge, the calibrating of the many individual elements in a game so that they work together for an optimal player experience. Research into games reveals that there is a sweet spot in which the game is not too easy and not hard, but just challenging enough to sustain interest (Loftus & Loftus 1983). Digital games allow for more plasticity, since the parts are all virtual and easily parameterized. Digital games allow us to tune a more complex environment: we can work with an orchestra instead of just a single instrument. The computer is a medium that allows us more representational power than wood, clay, paper, cardboard, pinball springs, etc. in setting up game worlds and in both increasing their complexity and playability.

Narrative is one of the most effective cultural tools available to game designers because narratives script the interactor by setting up expectations and motivating game actions. The elements of narrative exist apart from any particular story. For example, if we were to begin a story with "Once there was a boy..." a listener would immediately anticipate a sequence of good and bad events befalling the boy, and would most likely expect the events to be like those in a fairy tale because of our use of the word "once" as the opening word. Fairy tales are a particular genre of story that are predictable to the extent that they can be mapped into the fairly rigid structures of a simple substitution system (Propp 1928). The same is true of many other kinds of stories, which we call genre fiction, including detective stories, romance stories, and adventure stories. Fairy tales contain narrative elements such as stock characters (a young boy, his brothers, a princess, a helper figure, an evil creature) and stock situations (a capture, a rescue, a flight from evil-doers, a meeting with a helper). Detective stories include familiar characters (a lone detective, a needy client, a skeptical policeman, a criminal) and predictable events (discovery of a body, collecting of evidence, false clues, questioning of suspects, revelation of the culprit), as do other genres.

Games exploit the conventions of genre fiction to signal the player what the game goals and actions should be. When a game is presented as a knight's quest, for example, the interactor will expect swords, dragons, evil knights, treasure, and damsels to be rescued, kings to provide dazzling rewards. When verbal descriptions or images of these narrative elements appear before us we know what to do with them, and we know what kinds of things are likely to happen next. We can experiment with different strategies because we are aware of the kinds of interventions possible, the kinds of obstacles we will face, and the risk/reward structure implicit in the world. In a complex game our original understanding of these basic story elements orients us to the game but they do not exhaust the possibilities of the game. They form the scaffold on which we can hang the particular learning that comes from interacting with the game, from trying out the results of different methods of exploration or social behavior or jousting techniques[1].

The game designer can therefore draw on narrative patterns such as the pursuit of a love-object or a criminal or a reward in constructing the playing field. But so can the player. Simulation games and strategy games often begin with the player setting up the parameters of the world or the heroic attributes of the protagonist. The digital medium gives the players the power to construct the playing field as well as the power to play the games. In *The Sims*, for example, we create the family whose needs and actions we then direct after the game time begins. Some people choose to set up their characters after their own family or friends, or after characters in novels, movies, or TV shows. One of the potential comic plots of a Sims game is the same as the plot of the Neil Simon comedy, *The Odd Couple:* the pairing of a compulsive neatnik with a slob. Other people play out the maturity and marriage plot, educating some single Sims and marrying them off to one another.

A narrative simulation can be played in a focused, goal-driven manner or in an open-ended manner. But the open-ended game is still narrative in that it is full of episodic structure, of small events that have their own causal patterns. My *Sims* character may lose a job after becoming too depressed from not enough fun in her life; or she may enjoy a social evening increasing her friendship score with her neighbors. In order for these events to be memorable they must be readable as narrative. To the extent that they make for little story arcs, I will be likely to repeat the positive patterns and change the negative ones.

Of course, the spectator of a story may just want to "see what happens" rather than "win": the first is a narrative pleasure, the second a game pleasure. In a simulation meant to increase our understanding of complex processes and relationships we would want to encourage people to try out negative behaviors in order to increase their understanding of why and how bad things happen. In playing *The Sims* some people try to kill off the characters or to push them in self-destructive directions. The robustness of the program as a simulation is evidenced by the fact that interesting things happen whether you play positively or negatively.

Computer-based simulation invites us to try out variants, to change parameters in order to see what happens. But when dealing with complex environments it is all too easy to ascribe causality where there is none, or to miss the salient elements that are linked by cause and effect. This is a particular problem in digital

---

[1] The relationship of story forms to game forms is a lively field of critical analysis, see for example (Aarseth 1997; Murray 1997; Aarseth 2001; Murray 2003).

environments where the "Eliza effect" as it is sometimes called causes interactors to assume that the system is more intelligent than it is, to assume that events reflect a greater causality than they actually do. The interactive design principle of visibility – of making clear what the system is doing, and also what it is not doing – is especially important in simulations, if we are to ensure that the interactor connects the right causes with the perceived effects (Turkle 1995). Narrative strategies can be helpful in emphasizing the most salient elements of a complex system, and fostering recall of decision points that have the greatest effects on the outcome. Giving the interactor explicit narrative tasks to perform, allowing them the opportunity to narrate the experience and to produce alternate versions of their narrative can surface causal assumptions that may or may not be accurate.

In creating narratives for organizational simulation the researcher can draw on a rich repertoire of story strategies that have been in development over the past millennia and a growing repertoire of strategies specifically invented for the digital medium. The trend in digital games over the past five decades is to include more and more story patterns and to integrate the story more and more completely into the game. Early games often used the story elements as wallpaper, as the decorative background or costuming for a maze game, for example, or the reward for solving puzzles. "Cut scenes" of animation or live action are often placed in games as rewards for reaching the next level of play. More recent games include rich animation in uninterrupted game flow, and integrate story-appropriate actions throughout the gaming. *The Sims* and *Grand Theft Auto* – two very different story/game environments – both provide richly detailed worlds with a wide range of player actions within an animated world that behaves independently of player action as well as responding to it. The player is more deeply immersed in the game because of this detailing on both the participatory and the procedural levels. These detailed, agent-driven worlds are more realized because they act independent of the player, as well as being more finely responsive to what the player does.

In the research community there have been several efforts to expand the expressiveness of story/games and to increase the range of actions that are available to players. Research into believable characters with deeply modeled inner lives and elaborate social behavior has produced a wide range of technical strategies for simulating interaction, including artificial intelligence techniques that support real time conversations[2].

Computer-based simulated worlds are proving effective for emotional learning, such as overcoming phobias and post traumatic shock (Rothbaum, et al. 1995). The efficacy of these systems is based on the degree of detail in the worlds: they must be close enough to the real experience to invoke the problematic emotional response, but they must be far enough from the actual experience to feel safe. These are "liminal experiences," experiences on the threshold between the

---

[2] See the work of the Oz group at CMU led by Joseph Bates, papers available on line at http://www-2.cs.cmu.edu/afs/cs.cmu.edu/project/oz/web/papers.html, especially Michael Mateas, *Interactive Drama, Art, and Artificial Intelligence*, Ph.D. Thesis, School of Computer Science, Carnegie Mellon University, 2002 and Joseph Bates, "Virtual Reality, Art, and Entertainment." *Presence: The Journal of Teleoperators and Virtual Environments,* 1(1):133-138, Winter 1992.

real and the unreal, the same threshold that is the basis of make-believe, gaming, and the arts (Turner 1966). Sherry Turkle has pointed out that computers themselves occupy this threshold area and are therefore particularly powerful in capturing our attention and holding it (Turkle 1984). Simulated worlds are the most powerful use of the specific affordances of the computer: they are the most powerful uses of the computer's processing power and the most ambitious uses of its participatory property. Narrative itself can be a powerful evoker of the liminal, playful state, the state of openness to new patterns of information. The new kind of learning necessary for coping with multiple possibilities of complex systems would seem to be well suited to an approach that combines computer simulation, virtual environments, and narrative structure.

ABSTRACTION IN NARRATIVE FORMS

A narrative, like a computer program, is a format for the abstract description of objects and behaviors. The oral epic, for example, included specific constraints for describing heroes, including the association of epithets with individuals: Odysseus the wily; Agamemnon, king of men; "swift-footed Achilles," etc. The association of names with epithets is an abstraction system, signaling to us what kind of behavior to expect from the character. The epistolary novel of the 18th English century was told completely in letters, allowing a new exploration of individual consciousness, while the contemporary picaresque novel allowed for an exploration of the variety of the social world, as the hero moved from one place to another, from one adventure to another. Both of these formats allowed an emerging industrial, middle class society to assimilate its experience, to sort out new ways of behaving in changing institutions such as marriage and a changing class structure (Watt 1957). The sitcom TV show assumes unchanging characters and relationships displayed in novel situations in its episode, but the history of the form is evidence of an ongoing effort to abstract and redefine changing racial, sexual, class, and gender relationships in a diverse society (Allen 1992).

Vladimir Propp identified 31 morphemes that make up the Russian fairy tale (Propp 1928). Narrative theorists have identified many other aspects of narrative including genre, point of view, temporal features, and tone (Booth 1961; Chapman 1980; Genette 1983; Bordwell 1985; Bal 1998). Some of these patterns are specific to a particular format (such as jump-cutting in film), while others belong to genres that span multiple media (such as the motif of placing a woman in a sexualized dangerous situation within a gothic story, whether print, film, or theater). The formulaic nature of narrative and the extensive taxonomies of narrative that derive from the critical tradition provide many potential patterns that can be transposed into the universal pattern language of computing (Murray 1997; Murray 1998). In looking at narrative patterns over time and across cultures it is easy to see repeated patterns that seem to reflect some of the core experiences of human life: birth, resilient survival, love, loss, triumph, defeat, sacrifice. Stories are often organized around oppositions and rivalries; they begin with an unstable situation, present a crisis, and reach some resolution.

Stories also reflect ideologies of the cultures in which they flourish. When the European film director George Sluizer remade his original thriller, *Spoorloos* (Netherlands, 1988), about a man obsessed with the kidnapper of his lover, in Hollywood as *The Vanishing* (1993), he found that he could not portray the protagonist as willingly accepting a possibly poisoned cup of coffee from the villain. For a European audience this was understandable as a chosen act of self-destruction, a reflection of the character's devotion to his lost love. For the Hollywood filmmakers the act had to be presented as the result of a "nervous breakdown," a sign of disintegration[3]. The motherless David Copperfield, in the 19th century English ideology of Charles Dickens, is best off when he is sent to school, cleaned up, and cared for by a bossy, caring woman. He is worst off when on his own and unable to go to school. For Mark Twain, reflecting 19th century American ideology, the best place for Huck Finn is on the raft away from the clutches of school and civilizing women, where he is free to be dirty and to care for himself. *The Sims* reflects late 20th century American values, where success is the result of going to school, going to work, giving your spouse backrubs, taking regular showers, being sociable to your neighbors. In all these cases the narrative patterns reflect cultural patterns.

Because narrative patterns are so closely related to cultural patterns, stories change as society changes. In the 19th century a woman in a novel who had sex outside of wedlock wound up either married or dead, often through suicide. In late twentieth century novels, there is more sex and less suicide; marriage is less likely to be a happy ending, and women who are unhappily married at the beginning of a story are often happily divorced by the end of it. As we change who we think we are and how we want to live, we change the kind of stories we tell. Story patterns do not disappear, but they expand to include other ways of looking at the same situation. A woman might still die after having sex out of wedlock in a 21st century American popular media story, but the story would probably include causal sequences such as an abusive boyfriend or a jealous ex-wife, and even her suicide might be an attempt to frame her lover.

Technologies of communication change what can be communicated, and are themselves changed by the demands of new formats of communication. Writing changes the body of oral stories, and literate culture leads to changes in the materials and formats of writing. The printing press changes long repetitive manuscripts into more shapely novels, and the popularity of novels leads to more organized presentation of chapter-based, lengthy stories. Novelists develop techniques for telling picaresque stories, full of adventurous journeys and conflicts, and epistolary stories, made up of letters and recording the interpersonal dramas and subjective emotional turmoil of the literate classes. All of these techniques of storytelling are then available to later novelists who can combine the drama of external events with the emotion of the inner life, the life of the road with the life indoors.

---

[3] His reworking of the scene with the American actor Jeff Bridges was described by the director in a talk at Atlanta's High Museum of Art film series November 14, 2003.

The computer is the inheritor of all the story patterns in novels, theater, film, animation, television, and the other expressive arts. But it adds new possibilities for narrative formats, including:

- Creation of multiple instances of the same object (same character, different choices)
- Creation of multiple variations of the same object (same basic character, different attributes)
- Creation of objects with behaviors (as already discussed)

By the end of the 20th century storytellers in multiple media had developed a number of techniques for representing both interiority and exteriority, for deepening the portrayal of individuals and widening the portrayal of social networks. For example, the novel of the 20th century developed "stream of consciousness" narration and multi-generational epics; films established a visual rhetoric for indicating whose point of view we are seeing things from and whether an event happened earlier, simultaneously, or after another event; television dramas established long-form formats that traced the same characters over multiple episodes, weaving multiple stories into any single episode including one or two that concluded within the episodic boundaries.

At the beginning of the 21st century, interactive storytellers are elaborating a similar range of narrative strategies to take advantage of the representational affordances of the computer. There are at least two main patterns of interactive storytelling: multi-sequential and multi-form stories.

Multi-sequential stories narrate fixed events in different orders. For example, the film *Go* skips around through the time period of the same day, showing events from different characters point of view. A computer-based narrative of this point would give the interactor the power to choose their own path through the story universe, to navigate a fixed story space in multiple ways. This is a form that has been widely explored in hypertext story systems (Landow 1997).

Multi-form stories generate new narratives from the same building blocks. For example, the movie *Groundhog Day* can be thought of as a simulation that runs through all the possible forms that a relationship between the two main characters might take in the course of single day that keeps repeating itself. Storytelling in oral cultures, such as bardic ballads and folktales, is multi-form, because the telling knows many possible variants of the same story patterns, and never generates the same version twice in a row. This is because memorization in an oral culture is used for the building blocks and underlying patterns, the story elements and syntax of the stories, rather than for the exact words of any single telling (Lord 1960). On the computer the interactor develops a similar sense of the underlying building blocks by seeing a story told and retold with significant variants. The key to making the successive tellings memorable is to shape them as a sequence of readable, dramatic, causal events.

What do I mean by "dramatic"? Mostly I mean abstracted and compressed. Events in a simulation are too often over-realistic, which cuts down on their memorable qualities. For example, multi-player online games often require movement through the imaginary space in real time, when nothing much is going to happen between places. In a film, a similar transition would be rendered by a quick cutting between one significant moment and the next. In a well-designed simulation, the system would similarly figure out what events are the dramatic moments, and skip over the more pedestrian action, summarizing it as necessary through appropriate dramatic conventions. In a film we may see someone entering a car and driving off the screen to the right, followed by a cut to another place with different weather or time of day and the same car coming into view again on the left hand side of the screen.

In *The Sims*, we can indicate the passage of time by speeding up the clock, but the characters still have to go to the bathroom, make their meals, and get outside in time for their carpool. There is no way to skip over the mundane events of the day and get to the actions that will truly define their fate, like whether or not they choose to give a backrub to another character after dinner. This has proved to be such an annoyance that future versions of the game will automate the more routine chores. The equivalent of skillful editing for dramatic compression for the computer game may include automating some behaviors so that they take place without explicit user input, or perhaps are scripted by the user in advance. The more complex and detailed the worlds we create the more important it will be to strike the right balance between giving the user control over a richly detailed story world and compressing certain actions so that only the most dramatically important events are enacted.

The creation of multi-form stories has been largely the work of research projects in computer science. The state of the art in such projects is the story system built by Michael Mateas for the story *Façade,* a collaboration with Andrew Stern. *Façade* gives the interactor free play in a dialog-driven graphical world. The interactor is cast in the role of a visitor to a couple who are engaged in a serious marital argument. By typing in conversation that the couple answers, and by moving around their apartment, the visitor influences the unfolding of events. The architecture of the story is based on narrative beats (an abstraction taken from theatrical storytelling), and the input of the user is abstracted into speech acts such as agree, disagree, flirt, and get angry. In this way the system is able to respond and to interpret all input as pushing the story forward. The user has freedom of response, but dramatic compression is maintained (Mateas 2002).

A key design problem in interactive narrative worlds is focusing the user's input on dramatically productive gestures. Games often limit actions too severely for narrative engagement (e.g., kill or be killed) or they provide too much freedom of action with too little dramatic reward (e.g., wander around a space with not much is happening, send a virtual character to the bathroom before they wet the floor). One of the more promising strategies for expanding action without creating overly literal worlds is the use of conventions from the traditions of role-playing games. Massively Multi-player Online Role-Playing Games (MMORGs) are increasingly popular as technology increasingly allows hundreds of thousands of players to share a common graphical reality. MMORGs like *Ultima Online* and *EverQuest* have been successful because they draw on questing patterns and allow users to role-play with one another. The popularity of these environments are making more and more people familiar with the concept of a shared, simulated environment in which players are organized into small teams to accomplish common missions (Turkle 1995). This is a technology and a set of social behaviors that can be readily adapted for organizational simulation (Brown & Duguid 2000).

EXPLOITING NARRATIVE DESIGN FOR ORGANIZATIONAL SIMULATION

As this survey has indicated, narrative is an age-old strategy for information organization, based on one of our core cognitive abilities, the propensity to perceive and describe the world as a sequence of causal actions. Narrative techniques have developed in conjunction with communication technologies, providing us with a rich repertoire of story elements and story formats with which to represent the complex interrelationships of cause and effect that make up our world. Simulation of organizations is a kindred activity, and overlaps with storytelling as our experience with the story/games of the late 20th and early 21st century makes clear. What will it take to maximize the application of these storytelling technologies for the creation of organizational simulation?

We might start by abstracting several knowledge domains in which we understand organizational practices into their core causal and relational patterns. This would be an overwhelmingly large task, of course, since organizations are so varied and they can be understood through so many lenses: management, engineering, economics, social psychology, cultural anthropology, information science, industrial design, military history, etc. We would have to abstract these knowledge bases into the objects and relationships that are most important to their representation of the world. Ideally we should have an ontology for every discipline and a mapping of that ontology onto different varieties of organizations. That would make our overwhelming task exponentially larger. In fact, this approach is so daunting that it would take lifetimes to complete it, and we might never finish describing any single organization within all of these complex data structures, let alone get around to implementing a story structure that would integrate these multiple detailed models.

Suppose we started from the other direction, though? Let us take a specific organizational situation and model it, aiming for a story-system representation rather than a quantifiable model. For example, let us say that we want to study the possible effects of eliminating home kitchens in favor of collective eating halls in a planned development that is being proposed for a future space station, where people will live and work and raise families for periods of up to 20 years. We could create a model of every system and every event that could happen in the space station, or we could abstract away only those elements that would dramatically affect eating arrangements or that eating arrangements would dramatically affect. To find out what these elements would be, we could do an anthropological study of other environments in which unrelated adults people shared eating facilities – colleges, communes, assisted living homes, military bases, monasteries, etc. We could also analyze what characteristics were going to be used to decide who joins the colony, what working life would be like there, what kinds of events will take place on a daily basis or in special circumstances. We would analyze the political and economic structure of the proposed colony.

The results of our analysis would be a set of character traits, locations, and events on which we would build our simulation. As in a novel, we don't have to have a full specification of an individual in order to make a believable character who will serve our plot. We just have to come up with the right abstraction categories to make for action that is significant within our assumptions and the world of the space colony as we currently understand it. For example, if people are going to be living at close quarters then we want to include characters who exhibit claustrophobia, antisocial behaviors, mediation skills, and differing needs for privacy. If we assume with some social psychologists that all human relationships follow familiar patterns of dominance and submission, then we want to create characters whose dominance level can be set, read, and readjusted. If we are including people from different cultures we will want to create characters who exhibit the eating habits of varying cultures (but maybe not bother with their varied musical tastes, religions, or gender roles except as these impact on cooking and eating).

The characters would not be ends in themselves, however, but elements that will help us to generate plot events, because we will want to think at the level of plot. We do not care if people love or hate the collective kitchens except insofar as it impacts behavior and the success of the space colony. The events we care about, then, are events that show a smooth or disrupted community life, a rise or fall in overall morale, a rise or fall in nutrition, a strengthening or weakening of family life. (I'm making the culturally inflected assumption that we want to support family life and community life.) Sexual liaisons, risk-taking behavior by adolescents, competitiveness among high achievers might form the basis of some possible plot events that would involve individuals but impact the larger community. Contingencies of climate, technology, and external conflicts might form larger plot elements that would affect everyone in the colony in some way.

A story-system representation would not have to model the whole physical colony but only those venues that are relevant to the problem at hand. The eating halls would have to be rendered in great detail and in a highly parameterized way: how many families per venue, would children sit with parents, what would the décor be, would there be privileged dining areas, etc. But the workplaces might only exist as relationships among different characters: who was managing whom, who worked together daily, who was passed over for a promotion that benefited someone else, etc.? Venues might also exist as locations for plot events, such as teenage hang-outs or central food supplies that could be contaminated, etc.

The computer-based story-system for the space colony might be used alone or form the basis for live action role-playing scenarios. Its purpose would not be to predict exactly what arrangement would maximize efficiency for the project. Its purpose would be to focus thinking about the many kinds of consequences that might arise from decisions about eating arrangements. Planners would have to specify their assumptions and challenge one another's assumptions in creating the story-world. Challenges to assumptions could take the form of acted out or digitally realized events. Because the simulation exists as a story system that explicitly afforded many outcomes, planners would be discouraged from thinking that they had captured all the variables. But because it could be expanded and altered, the story-system could serve as a way of summing up salient knowledge from many domains. It would provide a shorthand overview of thinking to date on open-ended questions for which there are no single right answers.

I have chosen a fanciful example in order to offer a broad sketch of the benefits of such a system. But story-systems could also work on more immediate planning problems, such as the deployment of rapid-response teams in an emergency. Narrative elements that arise from such situations (e.g. threats to individuals or groups, suffering of victims, loyalty to team-members, rivalry among units, etc.) should be part of the planning process along with resource-allocation issues. Games have been integrating adventure, strategy and resource allocation for quite some time and have developed conventions that could be exploited by organizational planners.

For example, many of the decisions in *Sim City* are budget allocations entered as numbers on a data form, and the city itself is a kind of animated chart, but the game is punctuated by narrative events such as newspaper headlines, cheers of constituents, robberies and fires that reflect and sum up the effects of the quantifiable decisions. Warren Spector, the creator of the 2000 hit shooter-adventure game *Deus Ex,* purposely designed it so that the player can successfully complete the many levels of challenges using stealth and cleverness instead of the more popular method of blasting through it (Spector, 2000). Online multiplayer games, like the table-top *Dungeons and Dragons* games from which they derive, set up problems that require multiple team members with different skill-sets to solve. These strategies have clear applicability to organizational situations.

Live action role-playing games and multi-player online games are divided between pure shooters and story-driven games. In story-driven environments a surprising amount of effort goes into activities that are usually thought of as the least rewarding parts of organizational work, such as team meetings, negotiation meetings, collective ceremonies. These environments have developed conventions for involving role players including elaborate organizational structures, the use of alternate personalities, and the involvement of the player in multiple plots with multiple (and sometimes conflicting) goals and interests. For example, in a live action role playing game, I may belong to the Orculans who are at war with the Tumatans and I may be participating in that war while also trying to marry off my sister to a Tumatan count in order to hide evidence of a swindle from a visiting group of space detectives. Role-playing games are an abstraction methodology for representing complex organizational relationships as a set of interrelated character goals.

Role-playing games offer a methodology for populating a simulated organization with game-master-controlled or computer-controlled characters, called NPCs (non-player characters) in role playing games, and AIs (artificial intelligences, most often characters who shoot at you) in the jargon of videogames. These characters can co-exist with one or more player-controlled characters.

Alternately, an entire world can be made up of computer-controlled characters who can, like the Sims, be individually and sequentially operated by a player; the world can continue without intervention or you can take over one or another of the characters and change the pattern of events. Alternately, the story world can hide the individual characters and only display a map of larger consequences, as in *Sim City* or *Civilization,* where the player acts as non-democratic ruler and the game board displays overviews of the city or empire. Ideally, the interactor should be able to zoom through multiple levels of abstraction, from the overview of the state of the organization to the close-up of the state of subgroups and individual characters.

One of the most salient features of contemporary organizational life is the increasing role of the computer itself. Information technology does not exist in an air-conditioned bubble of electronic circuits. Information has a social life, as John Seely Brown and Paul Duguid (2000) have reminded us. Computer-based story-systems are well suited to modeling the computer itself as a story element. New information systems change the flow of information and it is possible to black box the workings of the system while feeding out the information as it might be fed under actual circumstances, thus capturing the organizational changes and challenges that will come with the change in information structures. It would be useful to bring the potential users of a new system together to test out the information flow in a number of simulated situations (deadline crunches, inventory shortfalls, massive growth, etc.). By combining a simulated information system with scenarios that capture important possible event sequences and actual future users of the system, it might be possible to surface social design issues and secondary effects of the changes before the system is deployed. Story-systems could serve as an extension of the current best practices in task analysis and user testing.

Live action simulation games are currently practiced as part of the routine planning strategies in some organizations including the military. Case-based education is a staple of business schools, and problem-based education is becoming the norm in engineering schools. Multimedia computing has been used as a means of extending this approach, mostly through the use of branching stories that call for multiple choice decisions[4]. The most ambitious application of this approach is underway at USC's Institute for Creative Technologies (http://www.ict.usc.edu) where advanced immersive techniques and Hollywood scripting are employed to approximate a Holodeck experience.

The representational powers of the computer are dazzling in several ways. As the USC project demonstrates, the computer can assimilate the story-telling conventions of all previous forms of narrative. Videogames can grow beyond the

---

[4] A Google search (in 2004) brought over 10,000 hits for "interactive case study" including multimedia, text-based, and live participant cases. Business, law, and medicine seem to be among the most active application areas. Military multimedia simulations have been an active area for two decades. See for example the work of Dartmouth University's Interactive Media Lab (http://iml.dartmouth.edu /index.html). Similar methods have been used for immersive language learning, e.g., (Furstenberg, et al., 1993)

console and PC screen to fill up an entire room. But even larger than the illusion-making machinery is the kaleidoscopic power of the code. Within the computer we can represent worlds with a new plasticity and multiplicity; we can instantiate many versions of the same event, the same situation, the same organization. The computer can deceive us by hiding the basis of our assumptions, by creating a falsely authoritative interpretation of the world. It also can undeceive us, allowing us to make our visions of the future clearer to one another, make our assumptions more explicit, and open our minds to multiple representations of the world. Human beings have used stories in both ways throughout our cultural history: to assert a single reality and to call those assertions into question. By combining this core human activity with our most powerful technology of representation we are creating a medium with the potential to examine our most complex organizational structures and to re-imagine them as more responsive to human needs.

REFERENCES

Aarseth, E. (2001). "Computer Game Studies, Year One." Game Studies **I**(1).

Aarseth, E. J. (1997). Cybertext: Perspectives on Ergodic Literature. Baltimore Maryland, John Hopkins University Press.

Allen, R. (1992). Channels of Discourse, Reassembled: Television and Contemporary Criticism. Chapel Hill, University of North Carolina Press.

Aristotle (1996). Poetics. London,New York, Penguin Books.

Bal, M. (1998). Narratology: Introduction to the Theory of Narrative. Toronto, University of Toronto Press.

Booth, W. C. (1961). The Rhetoric of Fiction. Chicago, University of Chicago Press.

Bordwell, D. (1985). Narration in the Fiction Film. Madison, University of Wisconsin Press.

Bowker, G., & Star, S.L. (1999). Sorting Things Out: Classification and Its Consequences. Cambridge MA, MIT Press.

Brown, J. S., & Duguid, P. (2000). The Social Life of Information. Boston, Harvard Business School Press.

Bruner, J. (1986). Actual Minds, Possible Worlds. Cambridge MA, Harvard UP.

Bruner, J. (2002). Making Stories: Law,Literature,Life. New York, Farrar, Strauss, and Giroux.

Chapman, S. (1980). Story and Discourse: Narrative Structure in Fiction and Film. Ithaca, Cornell University Press.

Czarniawska, B. (1998). A Narrative Approach to Organizational Stuies. Thousand Oaks, London, New Delhi, Sage Publications.

Forster, E. M. (1927). Aspects of the Novel. London, Arnold.

Furstenberg, G., A. Farman-Farmaian, et al. (1993). A la rencontre de Philippe.

Geertz, C. (1973). The Interpretation of Cultures. New York, Basic Books.

Genette, G. (1983). Narrative Discourse: An Essay in Method. Ithaca, Cornell University Press.

Graetz, J.M. (1981). The Origin of Spacewar! Creative Computing Video and Arcade Games. 1(1).

Hudlicka, E. and G. Zacharias (2005). Requirements and Approaches for Modeling Individuals Within Organizational Simulations. In W.B. Rouse & K.R. Boff, Eds., Organizational Simulation: From Modeling and Simulation to Games and Entertainment. New York: Wiley.

Klein, G., P.J. Feltovich, J.M. Bradshaw, & D.D. Woods (2005). Common Ground and Coordination in Joint Activity. In W.B. Rouse & K.R. Boff, Eds., Organizational Simulation: From Modeling and Simulation to Games and Entertainment. New York: Wiley.

Klein, H. & A.P. McHugh (2005). National Differences in Teamwork. In W.B. Rouse & K.R. Boff, Eds., Organizational Simulation: From Modeling and Simulation to Games and Entertainment. New York: Wiley.

Landow, G. (1997). Hypertext 2.0. Baltimore MD, Johns Hopkins University Press.

Latour, B. (1996). Aramis, or the Love of Technology. Cambridge, Harvard University Press.

Lebling, P. D., Blank, M.S., et al. (1979). "Zork:A Computerized Fantasy Game." IEEE Computer 12(4): 51-59.

Loftin, R. B., M.D. Petty, R.C. Gaskins, & F.D. McKenzie (2005). Modeling Crowd Behavior For Military Simulation Applications. In W.B. Rouse & K.R. Boff, Eds., Organizational Simulation: From Modeling and Simulation to Games and Entertainment. New York: Wiley.

Loftus, G. R., &Loftus, E.F. (1983). The Mind at Play: The Psychology of Video Games, Basic Books.

Lord, A. B. (1960). The Singer of Tales. Cambridge, Harvard University Press.

Martin, W. (1986). Recent Theories of Narrative. Ithaca, Cornell University Press.

Mateas, M. (2002). Interactive Drama, Art, and Artificial Intelligence. Computer Science. Pittsburgh PA, Carnegie Mellon University.

McGinnis, L. (2005). Technical and Conceptual Challenges in Organizational Simulation. In W.B. Rouse & K.R. Boff, Eds., Organizational Simulation: From Modeling and Simulation to Games and Entertainment. New York: Wiley.

Murray, J. H. (1997). Hamlet on the Holodeck: The Future of Narrative in Cyberspace. New York, Simon & Schuster/Free Press.

Murray, J. H. (1998). "Building Coherent Plots in Interactive Fiction." IEEE Intelligent Systems.

Murray, J. H. (2003). "Is there a story-game?" First Person: New Media as Story, Performance, and Game. P. Harrington and N. Wardrip-Fruin. Cambridge MA, MIT Press.

Orwell, G. (1944). As I Please. Tribune.

Pomerantz, G. M. (1996). Where Peachtree Meets Sweet Auburn: A Saga of Race and Family. New York, Scribner.

Propp, V. (1928). Morphology of the Folktale. Austin, University of Texas Press.

Rothbaum, B., & Hodges, L.F., et al. (1995). "Effectiveness of Computer-Generated (Virtual Reality) Graded Exposure in the Treatment of Acrophobia." American Journal of Psychiatry 52(4): 626-40.

Sandler, T, & Arce, D.G., (2003). "Terrorism and Game Theory." Simulation and Gaming 34(3).

Spector, W. (2000). Deus Ex, Ion Storm. single player; role playing game.

Starr, P. (1994). "Seductions of Sim: Policy as a Simulation Game." American Prospect 17(Spring 1994): 19-29.

Sutton-Smith, B. (1997). The Ambiguity of Play. Cambridge, Harvard University Press.

Tenner, E. (1996). Why Things Bite Back: Technology and Revenge of Unintended Consequences. New York, Knopf.

Turkle, S. (1984). The Second Self: Computers and the Human Spirit. New York, Simon and Schuster.

Turkle, S. (1995). Life on the Screen: Identity in the Age of the Internet, xxx.

Turner, M. (1996). The Literary Mind: The Origins of Thought and Language. New York; Oxford, Oxford UP.

Turner, V. (1966). The Ritual Process: Structure and Anti-Structure. Chicago, Aldrine.

Watt, I. (1957). The Rise of the Novel. Berkeley, University of California Press.

Weick, K. E. (1995). Sensemaking in Organizations. Thousand Oaks, CA, Sage Publications.

Weizenbaum, J. (1966). "Eliza-- A Computer Program for the Study of Natural Language Communication between Man and Machine." Communications of the Association for Computing Machinery 9(1): 36-45.

Wiener, N. (1948). Cybernetics or the Control and Communication in the Animal and the Machine. Cambridge MA, MIT Press.

Wiener, N. (1950). The Human Use of Human Beings : Cybernetics and Society. Cambridge MA, Houghton-Mifflin.

CHAPTER 12

AGENT-BASED MODELING AND SIMULATION OF SOCIO-TECHNICAL SYSTEMS

ANUJ P. SHAW AND AMY R. PRITCHETT

ABSTRACT

This chapter reviews the current state of the art in agent-based modeling and simulation of socio-technical systems. A socio-technical system is modeled here as being composed of multiple agents, human and technological, each in pursuit of their goals. These agents operate in a rich environment created by their interactions with other agents as well as both physical constraints on behavior, and the more organizational, social and regulatory structures that guide and constrain actions. Implementing these models into a computer simulation also requires consideration of the timing and data-passing considerations that capture the agents' interactions, and of software-engineering considerations which enable best use of agent-based simulation in socio-technical system design. The insights provided by this method, and by its corresponding model of socio-technical systems, as phenomena emerge from agent interactions, are discussed.

INTRODUCTION: OUR "WORLD MODEL"

In simulating socio-technical systems by agent-based simulations, we are interested in emergent phenomena. 'Emergent' is formally defined here as a system property in which system behaviors at a higher level of abstraction are caused by behaviors at a lower level of abstraction which could not be predicted, or made sense of, at that lower level. More informally, our agent-based simulations are not based on any high-level models of the socio-technical; instead, we put agent models in a rich environment, simulate them in a realistic scenario, and see what system behavior emerges.

Our levels of abstractions are the agents (typically humans) and the emergent system-wide behavior. Agent-based simulation provides interesting insights at both levels of abstraction. In addition to the system-wide behavior, in agent-based simulations, the agents respond to their environment and each other. While we can model what the agents' responses would be to a variety of conditions, only simulation can predict what specific conditions they will need to respond to . As such, it is often just as interesting to use simulations to see what activities are demanded of an individual agent when out in the socio-technical system, as it is to see what the system-wide behavior will be in response to changes in agent's capabilities or their environment.

This chapter discusses agent-based modeling and simulation methods suitable for simulating socio-technical systems, i.e., simulating complex systems of interacting humans and technologies, usually for the purpose of engineering specific work domains to meet required standards of performance. These simulations can be run offline for analysis purposes, but can also conceivably run in real-time for visualization and for human-in-the-loop interactions. For example, studies of air transportation typically design specific procedures and technologies for their agents – pilots and controllers – with the intention of realizing system-wide levels of safety and throughput, as will be illustrated by examples in this chapter. In addition, such simulations can contribute to – and capitalize upon – theoretical research into human collaboration and coordination at both the 'micro' (individual) and 'macro' (system-wide) levels. These specific contributions and capitalizations are noted throughout.

BACKGROUND

Agent-Based Modeling and Simulation

Recent developments in software engineering, artificial intelligence, complex systems, and simulation science have placed an increasing emphasis on concepts of agents (Bargiela, 2000; Jennings & Wooldridge, 2000; Parunak, 2000; Parunak et al., 1998; Russell & Norvig, 1995; Weiss, 2000; Wooldridge, 2000). The term *agent* has been used to mean anything between a mere subroutine or object and an adaptive, autonomous, intelligent entity (Franklin & Graesser, 1996; Hayes, 1999; Wooldridge, 2000). This paper uses Hayes' definition of an agent as an entity with (1) autonomy, i.e., the capability to carry out some set of local operations and (2) interactivity, i.e., the need and ability to interact with other agents to accomplish its own tasks and goals (Hayes, 1999).

Historically, agent based modeling concentrated on creating intelligent agents towards achievement of autonomy, an artificial intelligence perspective on emulating humans and designing autonomous technologies (Russell & Norvig, 1995; Wooldridge, 2000). More recently, researchers have also applied "multi-agent" simulation of many interacting (but not necessarily fully autonomous) agents. Such multi-agent simulations have two concerns: modeling individual entities as autonomous and interactive agents, and simulating the system behavior that emerges from the agent's collective actions and interactions. These simulations are increasingly being applied in a wide range of areas including social sciences (Gilbert & Troitzsch, 1999; Goldspink, 2000), telecommunications, manufacturing (Barbuceanu et al., 1997; Shen & Norrie, 1999), business processes (Ankenbrand & Tomassini, 1997; Huang, 2001; Mizuta & Yamagata, 2001), transportation, and military simulations (Tambe, 1997; Tambe et al., 1995).

One type of multi-agent simulation has focused on "closed" systems, i.e., systems in which all aspects of the agents can be specified (Castelfranchi, 2000; Kang et al., 1998; Lesser, 1999). These specifications can include: the agents' goals; the internal capabilities and cognitive primitives that determine their individual actions; and the social primitives such as coordination and collaboration that determine their interactions. Closed systems cover a number of applications, such as team activities, distributed sensor networks, and dedicated applications in which the goals of the agents can be aligned with the goals of the system (Castelfranchi, 2000; Filipe, 2002; Howard et al., 2002; Huhns & Stephens, 2000; Kang et al., 1998; Lesser, 1999; Tambe, 1997). This type of simulation is often based on principles from natural and self-organizing systems, and thus often borrows from research in swarm, group and team design (Castelfranchi, 2000; Kennedy, 2000). Agent-based simulation of natural systems has been used in purely social contexts to validate or illustrate social phenomena and theories or to predict the social behavior of interacting individual entities (Conte & Gilbert, 1995; Conte et al., 1997; Davidsson, 2002). Such domains include actors in financial markets, consumer behavior, people in crowds and animals in flocks, and vehicles and pedestrians in traffic situations (Davidsson, 2000).

Agent-based simulation of natural systems often only includes replications of one or a few homogeneous agent models. As such, use of such homogeneous agents does not allow for simulation of large-scale socio-technical systems that may have agents fulfilling many roles. Agent-based simulation is theoretically capable of representing and simulating socio-technical systems in more realistic manners with a variety of interactions between many different types of agents. However, including heterogeneous agent models requires simulation architectures capable of incorporating the timing and data-passing of agent models of varying complexity and form (Lee, 2002).

In socio-technical systems of realistic complexity, specialization and division of labor is necessary. In such "open" systems the specifications of each agent cannot be controlled exactly, and their behavior and objectives are not fully known. Instead, agents are assumed to take actions contingent on their ability to assess their immediate context (Gibson, 1979; Russell & Norvig, 1995; Wooldridge, 2000). Due to differences in their beliefs, capabilities, and desires, coordination and collaboration among agents becomes a necessity. Coordination refers to temporal management of action, events and tasks amongst agents. Coordination is usually achieved through communication of protocol or event information among agents. Collaboration goes beyond coordination to require agents to share their goals and intentions (Singh et al., 2000). Collaboration, therefore, additionally requires communication of cognitive entities such as goals and immediate intentions. For multi-agent systems such communication is achieved through establishment of common semantics for those cognitive entities needing to be shared between agents. Organizational structures, i.e., distribution of roles and tasks, might be imposed on the agents as an inherent mechanism for coordination (Joslyn, 1999).

Social primitives can be additionally employed to achieve collective behaviors from multi-agent systems (Castelfranchi, 2000; Fitoussi & Tennenholtz, 2000; Lesser, 1999; Singh et al., 2000). For example, agents may be designed to have mutual beliefs and joint intentions so that they inherently work towards the same system objectives and do not need to communicate as often (Barett, 1970; Filipe, 2002).

Most recently, agent-based simulation has also examined the importance of explicitly modeling the agents' environment and the agents' ability to sense and interact with it. This view draws on ideas from situated cognition and ecological psychology in which agents' behavior is seen as explicitly using and responding to their environment (Decker & Lesser, 1994; Vicente, 1990, 1999).

Socio-Technical Systems

The term socio-technical system has been used to identify technology and its social setting, and used in studies of patterns of human use of technology (Norman, 1986). Likewise, cognitive engineering examines systems in which humans and technology interact, and is used in studies of human-computer and human-automation interaction (Norman, 1986; Rasmussen, 1988). The cognitive engineering definition of socio-technical systems (used in this chapter) spans all systems that work through interaction of humans and technology, including any industrial, transportation or service system. These systems are composed of people, technologies, physical surroundings, processes and information.

Socio-technical systems have been modeled in many ways for the purposes of improving system performance and the efficiency and productivity of individuals (Ho & Burns, 2003; McNeese & Vidulich, 2002; Vicente, 1999). In cognitive engineering, modeling approaches for socio-technical systems have shifted from a technology-centered perspective (e.g., feedback mechanisms and system dynamics) to a human-centered perspective. To this end, cognitive engineering has introduced methods to describe and analyze the cognitive aspects of human behavior.

Some of these methods focus on normative models, i.e., models that represent how the system (and the people within it) should behave. For example, task analysis methods often focus on (or start with) the procedures and action sequences that humans are expected to follow within the system (Schraagen et al., 2000). With such models, human behavior can be easily represented by procedure following mechanisms, with the procedures represented by expert systems or rule-bases; these mechanisms also provide a ready benchmark for categorizing observed behavior as correct or erroneous.

Other methods have focused on descriptive models, i.e., models that represent how humans actually behave within the system. Historically, these models were used to predict individual's behavior based on elementary perception, attention, working memory, long-term memory and decision-making models of human behaviors. Thus, the output measures of interest have traditionally included task demands, (mental) workload, task load, information load, attention demands, stress, and procedural timing measures (Anderson, 1993; Eberts, 1997; Laughery & Corker, 1997; Pollatsek & Rayner, 1998). These measures have been validated across domains ranging from helicopter operations, nuclear power-plant control, electronic list design for emergency operations, to advanced aviation concepts (Abkin et al., 2001; Corker, 1999; Corker et al., 2000; Corker & Pisanich, 1998; Gore, 2000; Gore & Corker, 2000; Laughery & Corker, 1997).

Criticism of this type of model has been that they predict input-output behavior in mechanistic terms (Craik, 1947). Other descriptive models have instead viewed human cognition as an interactive, "ecological" response to the properties of the immediate environment. For example, Hutchins (1995) described the boundaries of cognition as extending beyond the human alone to the properties of the environment that enable and constrain action; Hollnagel (1993) additionally identified the different patterns of control exhibited by humans over their environment in response to their immediate resources.

Formative models, i.e., models intended to guide design, also use this ecological perspective. For example, Work Domain Analysis assumes that an expert's behavior is adapted to constraints created by the environment, and thus creates a representation of these environmental constraints at multiple levels of abstraction to describe the range of likely human behaviors (Rasmussen et al., 1994; Vicente, 1999). While this "abstraction-decomposition space" is commonly used as a representation of the structural constraints stemming from the physical environment, recent studies have sought to explicitly incorporate process and cognitive constraints (Nickles, 2004), agent intentions (Hajdukiewicz et al., 1999), and team dynamics (Naikar et al., 2003; Vicente, 1999). Likewise, other methods of representing procedures have been proposed that describe expert adaptation and environmental constraints (Ockerman & Pritchett, 2000).

However, these cognitive engineering methods to date have concentrated on the role of an individual or a small team. For example, Vicente (1999) demonstrated how a surgeon and anesthesiologist work within the same environment, and thus their work can be represented on the same abstraction-decomposition space. However, these methods do not explicitly represent the mechanisms by which the agents interact; as significantly, their focus on individuals and small teams obviates their extrapolation to predict the dynamics of large-scale socio-technical systems.

MODELING SOCIO-TECHNICAL SYSTEMS AS AGENT-BASED SYSTEMS

Building on the ecological perspective of socio-technical systems used in cognitive engineering, socio-technical systems exhibit many of the characteristics also used to describe organizations:

- Involving a number of agents in a range of roles with a variety of objectives and capabilities;
- Purpose or goal oriented;
- Having a knowledge, culture, and established processes; and
- Able to affect and be affected by their environment (Carley & Gasser, 2000).

The first characteristic identifies socio-technical systems as open agent-based systems. The agents themselves act to create the system behavior. They form a heterogeneous set whose variety (of roles) and variability (within roles) must be adequately captured. While some of their attributes can be specified (e.g., minimum criteria for training and selection), many attributes cannot: they will also have their idiosyncrasies, including individual goals and intent. (Hudlicka and Zacharias, 2005).

While the agents may only strive for their own goals, the second characteristic of socio-technical systems concerns designers' intentions that agents' "micro-level" behaviors support some system-wide "macro-level" purpose. Agent-based simulation, unlike methods that focus on either micro- or macro-level behavior, simultaneously creates both. This allows not only for examination of one or the other, but can also enable investigation of the interplay between the two.

The third characteristic identifies aspects of socio-technical systems that need to be included in the models underlying the simulation. Agent-based simulation has the benefit of being *structure-preserving*, i.e., its model form and software implementation can (and should) mirror the structures of the "real" system. In the case of modeling socio-technical systems, the agents and their environment can be represented in the simulation using the same semantics as employed to define the agents' roles in the system: identification of the agents and their environment, and description of their roles, duties and expected capabilities. This representation requires a level of effort commensurate with its implementation in the real system. For example, just as established processes can be codified to be distributed to (or learned by) people within an organization, so can it be distributed to (or learned by) agents in a simulation. By using this representation, specific interventions can be examined, including changes to technology, organizational structure, procedures and information distribution.

The fourth characteristic mirrors the recent emphasis in agent-based simulation on including an explicit representation of the agents' environment. This emphasis on an environment model allows for richer models of agents in which situated cognition, distributed cognition and expert adaptation to the environment can be explicitly represented. Likewise, in classic computational organization theory many environmental attributes are modeled as being distributed among (and coded into) the agents; therefore, these attributes do not extend beyond the lifetime of the agents (Carley & Gasser, 2000). In agent-based simulation the organizational, regulatory and physical aspects of the environment instead provide a structure in which agents can be embedded to generate a complete system behavior. This separate representation facilitates modeling more open socio-technical systems in which a variety of agents are enabled or disabled by their environment.

In summary, agent-based simulation captures well important characteristics of socio-technical systems. However, this approach to modeling and simulation can represent a deviation in practice from other methods of examining socio-technical systems that directly incorporate patterns in system-wide behavior into their models. While the structure of the agents' micro-level behavior is preserved in agent-based simulations, the macro-level behavior must be treated as emergent from the simulation, not pre-specified into the simulation. Likewise, it is typically not suited for focus on one facet of macro-level behavior alone, but instead creates the full system behavior with all its interacting and confounding aspects. In doing so, it provides a prediction of system behavior suitable for socio-technical system design with a fidelity limited only by the validity of the agent models and of the structure within which they interact.

CONSTRUCTING AGEN-BASED SIMULATIONS OF SOCIO-TECHNICAL SYSTEMS

To construct an agent-based simulation of a socio-technical system, three components must be developed. First and most obviously, models of the individual agents must be developed that are capable of emulating the relevant behaviors within the system. Second, as just discussed, a model of the environment must be developed which furnishes the agent models with the information they need about the physical, social and process aspects of their situation. Third, mechanisms must be provided for the agents to act and interact, including mechanisms governing the timing of the simulation and data passing within it.

These developments require both conceptual models and their software instantiation. Often projects place more emphasis on one over the other. Some modelers may be more focused on the concepts behind the models, allowing their instantiation in software that does not inherently mimic the model's structure and purpose, and that may not follow software engineering standards for code re-use and adaptation to simulating other systems. In other cases, use of established software structures (or the desire to streamline some aspect of the software's execution) may constrain the conceptual models to particular forms and functions. For example, Carley and Gasser review a variety of simulations used in computational organization theory, and relate their form to their corresponding applications (Carley & Gasser, 2000).

The following sections will detail development of these three components of an agent-based simulation. Throughout, issues with conceptual modeling and software instantiation will be noted, including solutions that benefit both.

Agent Models

As noted earlier, an agent is defined here, for the purpose of agent-based simulation, as an entity within the system that is accurately described as having some autonomy in acting on its own as well as needing to interact with other agents in the system (Hayes, 1999). Implicit in this definition is the pro-activity of an agent – rather than only being a passive element of the system, an agent must act in ways that change the environment or the actions of other agents, and must interact with other agents (Wooldridge, 2000).

These distinctions provide some guidance in identifying the agents in a socio-technical system. Passive elements incapable of actions that change the state of the environment or other agents are typically best described within the environment model, as later discussed in detail. Pro-active elements that act autonomously without interacting with others do not warrant a full agent-based simulation involving other agents. Pro-active elements that do not act autonomously often indicate too fine a resolution of agent models within the simulation in which the true agents of the system are captured by combinations of smaller models rather than distinguishable agent models.

Within this guidance, selecting what entities should be modeled as an agent within a simulation is not always clear-cut. Physical entities may be modeled as agents, or agents may defined around functional attributes and tasks (Decker, 1994; Odell et al., 2002). In organizational simulation, each agent may represent the behavior of one human in the system, or different agents may handle different tasks involving multiple humans. For example, in an air traffic control simulation, each air traffic controller may be represented by an agent, or teams of controllers performing one function may be represented as an agent, or many controllers may be modeled with one agent for their 'monitoring' activities, one agent for their 'conflict resolution' activities, etc. The final selection of agents is an important design decision in developing an agent-based simulation that should be based on the purpose of the simulation and the required fidelity of each of the agent-level behaviors.

Modeling human performance is a common basis for many agent models in agent-based simulation of socio-technical systems. Several different research communities have created a wide variety of such models: the artificial intelligence and intelligent systems community in computer science; the computational organization theory community; and the human performance model community in cognitive science and human factors. The following sections outline two general approaches for developing such models suitable for agent-based simulation.

Required Agent Capabilities

Agent capabilities can be broadly classified as sensing, cognitive and executive. This distinction is commonly used in the computer science communities focusing on intelligent agents, including studies of robots and natural systems as well as models of humans in socio-technical systems (Russell & Norvig, 1995; Wooldridge, 2000).

An agent's *executive* capabilities describe how it can act upon its work environment. Such actions can include changing its own condition relative to the environment, changing the environment directly in a manner that may or may not impact other agents, and coordinating or collaborating with other agents directly (Russell & Norvig, 1995; Wooldridge, 2000). Agents' actions are only possible within their sphere of influence, i.e., the 'space' within the environment on which they can act (Jennings & Wooldridge, 2000). Purely physical actions, for example, may only be possible on elements of the environment within a certain relative position to the agent. Analogous descriptions can be made for the process and social elements of the environment which an agent can act upon at a certain time (Decker, 1994; Odell et al., 2002). For example, air traffic controllers have authority only over aircraft for which they have accepted a 'hand-off' from another controller; while these hand-offs roughly follow aircraft's transitions into geographic sectors covered by each controller, authority over the aircraft is only established when the hand-off process has been executed. Thus, in many systems, modeling the agents' executive capabilities also requires careful modeling of their sphere of influence and their need to purposefully change their situation to position it.

Correspondingly, an agent's *sensing* capabilities describe how it can perceive its environment. In some studies, these sensing capabilities have focused on the physical environment via models of vision, touch, etc., especially in fields such as robotics (Arkin, 1998; Gibson, 1979). However, sensing may also reference more process and social aspects of the environment. For example, an air traffic controller may "sense" which aircraft are under their control due not only to geographic space but also due to process considerations and regulatory constraints. As with executive capabilities, agents may not be able to sense all aspects of the environment. Their relation to the environment may affect what aspects they can sense.

Cognition is broadly used here to describe an agent's capability to take an intentional stance towards its internal state as well as towards externally-visible executive actions on the broader environment (Singh et al., 2000). Agents may be imbued with cognitive abilities from a vast range of forms and models as appropriate to the purpose of the simulation. Some of these cognitive models have been developed to create intelligent agents (Norling et al., 2000; Russell & Norvig, 1995; Wooldridge, 2000). In emulative simulation of socio-technical systems, however, agents are typically intended to model humans and computational human performance models are appropriate, as detailed in below.

The distinctions between these capabilities are not always exact. For example, sensing can have a strong relationship with cognition when modeling *knowledge*, i.e., a representation of the environment maintained within the agent (Markman, 1998; Singh et al., 2000). This knowledge state is built through sensing of the current state of the environment, through memory of past environmental states, and, in some cases, through explicit sharing of knowledge between agents (Nilsson, 1998). These processes are subjective; therefore, an agent's knowledge is likely to be incomplete and often incorrect or uncertain. These processes also can interact with executive capabilities when the agent is explicitly modeled as shaping their environment context as an aid to task demands (Hollnagel, 1993; Kirlik, 1998).

Likewise, determining an agent's sensing and executive capabilities also depends strongly on the environment model available to them. For example, some environment models provide only a representation of the physical space around the agent, disallowing sensing of and action on process or social aspects of their environment, e.g. geographical models in (Sierhuis et al., 2002). The variety of environmental aspects which can be included in an agent-based simulation are later discussed in detail.

Computational Human Performance Models

Cognitive models suitable for agent-based simulation are increasing in detail and ability to capture relevant aspects of human performance, where human characteristics, based on empirical research, are embedded within a computer software structure to represent the human operator (Corker, 1994; Laughery & Corker, 1997). These models are described as modeling *performance* rather the behavior because of their scope – the current state of the art is better at capturing purposeful actions of a human as generated by well-understood psychological phenomenon, than it is at modeling in detail all aspects of human behavior not driven by purpose.

Simple forms of human performance models may use engineering models to replicate identifiable tasks. For example, Pritchett, Lee and Goldsman (Pritchett et al., 2001) modeled air traffic controllers as using simple "dead-reckoning" navigation filters to predict whether aircraft would lose safe separation and to determine speed commands to resolve such conflicts. While such models are often mechanistic and limited to specific tasks, they can capture well-established patterns of performance; for some applications of agent-based simulation this can be sufficient.

Other "selfish" model forms may view agents as pursuing their goals using optimization or decision making mechanisms to select their actions. (For an example in air transportation simulation, see (Niedringhaus, 2004); for examples using Markov Decision Processes, see (Nair et al., 2003; Xuan, 2004)). If the agents are modeled as pursuing "optimal" solutions relative to their goals, this model form can serve in agent-based simulation to examine the system behaviors arising from optimal agents selfishly pursuing their own goals. Such selfish representations may instead assume that the agents' actions are selected based on bounds in their capabilities. For example, the agents may have "bounded rationality" in which their beliefs about their decision options are incomplete, incorrect, or applied in a manner subject to biases and sub-optimal solutions (Simon, 1982; Tversky & Kahneman, 1974). With such bounded models, agent-based simulation can examine the impact on system performance of agents' bounded rationality in their individual decision making.

A specific form of agents capable of pursuing their own goals uses the cognitive primitives of beliefs, desires, goals, intentions and commitments (Bratman, 1987; Bratman et al., 1988; Rao & Georgeff, 1991). Beliefs, i.e., the world model of the agent, form the agent's knowledge base and are usually represented through modal logic. Desires are usually represented as conditions that the agent wants to achieve. Desires can be inconsistent, and the agent may not be able to achieve its desires. Goals are the subset of desires that can be achieved by an agent. Intentions are usually understood as the states that lie on an agent's plan of action toward its desires.

In addition, selfish model forms can be used in simulations with "learning agents" who constantly seek the course of action that will best meet their goals within the emergent dynamics of the system. Such simulations may be re-run many times from the same initial conditions, with each agent remembering statistics from the previous runs that help them converge on an improved course of action to take in the next. For example, Niedringhaus (2004) detailed a simulation in which, given a fixed level of demand for air transportation and capacity at various airports, learning agents of airlines and passengers can be re-run hundreds or thousands of times to find the flight schedule and fare structure maximizing airlines' profits without exceeding airports' capacities for arrivals and departure. In comparison, simulations without learning agents instead focus on predicting system behavior within any single case, such as the simulations of air traffic control detailed later in this chapter.

Another model form uses normative models of performance, i.e., models based upon the prescribed processes that the agents should follow, as judged by the system designer or other external entity. The agent may be modeled as following its processes exactly, following them with some variation (e.g., with stochastic reaction times fitting observed human behavior), or in more complex manners such as selecting processes in forming intentions that will meet their goals. In some domains these processes may be represented by "procedures"; for example, a pilot-agent may be modeled as following checklists and standard operating procedures. These processes may be codified into the agent model by rule-bases and expert systems. For example, in medicine a clinician may be modeled as using a set of if-then rules when using a set of observed symptoms to make a diagnosis (Shortliffe, 1976).

Even when such agent models follow normative standards exactly and deterministically, agent-based simulation provides two interesting insights. First, when the environment is as expected, executing the simulation will reveal whether satisfactory emergent system behavior is created by agents who are all exactly acting according to standards; if this is not the case, then the agents will be forced to deviate from norms for the system to function successfully. Second, the environment in which the agents operate can be perturbed to examine the impact of likely or potential deviations from normative behaviors, whether arising from unintentional variations in human behavior or from purposeful attempts by an agent to violate a standard in order to achieve its goals.

The most advanced models may include elements of each of these model forms. For example, Air MIDAS, developed by NASA Ames Research Center (ARC) and San Jose State University (SJSU) primarily for aviation-related applications, contains several functions within its model of human performance. Mechanistic models of essential psychological and physiological phenomenon such as vision, attention, working memory and motor skills capture well-understood aspects of human behavior. Domain knowledge serves as pre-established knowledge about the task, often represented as procedures and a rule-base of goals and processes for core tasks. An upgradeable world representation also acquires and maintains knowledge about the current state of the environment. Within this framework, a symbolic operator model maintains queues of tasks waiting to occur, and switches tasks between them according to knowledge and goals (Corker, 1994). To date, this model has been used principally for studies of human performance acting individually or in small teams, so an explicit representation of organizational structures and procedures would require further development.

These models, then, can represent human-agent performance in agent-based simulation. Correspondingly, agent-based simulation brings to these models a dynamic representation of their environment, including detailed models of the physical and technical systems. However, given the complexity of current human performance models, many of which have to date been used largely in modeling the performance of a single human, creating the ability to interact with other simulation models (e.g., communicate, coordinate and collaborate with other agents, and interact with the environment model of an agent-based simulation) can require significant adaptations to current software implementations of these models for agent-based simulations involving strong organizational effects.

Environment Models

Including an environment model in an agent-based simulation requires a slightly different conception of "environment" than that commonly used in systems engineering. Rather than viewing the environment as everything outside the system boundary, in agent-based simulation the environment spans all the passive elements of the system that situate the functioning of the pro-active agents. In other words, the environment of the agents is captured within the system boundary. By definition, the agents' environment can have a dramatic impact on their individual behavior and, as a result, on emergent system performance. The environment is also noteworthy in that its elements (including physical space, new technologies for the participants to use, and procedures and regulations) are often the means by which system-wide change is effected.

As such, the importance of properly modeling the environment has been noted in methods for organizational design (Decker & Lesser, 1994; Lawrence & Lorsch, 1967). However, traditional artificial intelligence approaches to agent-based simulation have concentrated on modeling and manipulating the capabilities of single agents (Anderson, 1993; Arkin, 1998; Russell & Norvig, 1995; Wooldridge, 2000) and of multiple agents working collectively (Barbuceanu, 1998; Durfee, 2000; Weiss, 2000). Only more recently have such studies examined environment models (Decker & Lesser, 1994).

Often, fine distinctions may need to be drawn between the agent models and the environmental model. For example, when agents have limited sensing capabilities, the environmental model can provide an objective representation of their envirment while the agents' internal knowledge established a corresponding subjective representation that may be inaccurate, outdated or incomplete. The objective representation is contained within the environmental model during the simulation; the subjective representations are created with the agent models.

Likewise, the form of the environmental model needs to be tailored to the needs of the agents' models. For example, the perspective of topological psychology concentrated on the *invitation character* of the environment, by which environmental elements are seen as proactively inviting the agent to employ them as relevant to their intention and situation (Marrow, 1969). For the environment to create accurate invitations, it needs to know the agents' intentions. A variant of this perspective views the environment as affording or constraining possible actions and functions to agent. These environmental *affordances* or *constraints* are perceived by the agent, allowing the environment to be passive but requiring it to frame its elements in manner by which its affordances are apparent (Gibson, 1979).

This agent-oriented view of the environment categorizes the environment in a number of ways (Wooldridge, 2000). First, whether the environment is 'accessible' or not, i.e., whether the agents can obtain complete and accurate information about the state of the environment. Second, whether the environment is static or contains its own internal dynamics. Third, whether the environment is deterministic or stochastic. Fourth, whether the agent models are driven by discrete episodes (events) triggered by the environment, or whether other mechanisms, including knowledge of past episodes, are available to the agents.

Within these categorizations, a number of model forms and distinctions have been made in agent-based simulation, as detailed in the remainder of this section. Some are largely ad hoc and tailored to specific agent models as suited to the purpose of the simulation (Sierhuis, et al., 2002). Other studies are examining specific structures suitable for environment models generalizable to a number of agents, environments, and applications of agent-based simulation (Decker, 1994). Many of these structures include not only objects in the environment but also tasks and their relationships. Each such structure captures one dimension of the environment; a combination of structures can create a multi-dimensional environment model for agents to reference. The remainder of this sub-section details some of the environmental structures currently being developed in the agent-based simulation community.

The most common structure models the physical surroundings of the agents through various spatial representations (for example topological maps for robot navigation (Borenstein et al., 1996)). For example, in gaming the objects are spatially located in vector spaces managed through reference frames. In air traffic control simulations, the locations of navigation aids, airports etc. are describing using commonly understood coordinate systems (Pritchett et al., 2002). Spatial structures may also utilize topological relations, for example description of road networks in a city by intersections and road names. Topological structures can be better suited for some agent tasks such as navigation through networks; for example, mobile robots may maintain a topological map of interesting locations in the environment (Arkin, 1998; Borenstein et al., 1996).

Work domain analysis introduced *structural means-end relations* within the physical environment (Vicente, 1999). These structural relationships are hierarchical, i.e., elements of the environment are arranged according to levels of abstraction connected by means-end relations: An element in one layer is an *end* that can be achieved by employing elements in the layer below, and a means to achieve the ends of elements in the layer above. This environmental structure is also referred to as the abstraction hierarchy or abstraction-decomposition space.

Task structures are amongst the first environmental structures used to describe non-physical 'process' aspects of the environment. The applicability of these structures has been specifically demonstrated in choices of contingency-based coordination strategies (Decker, 1994). These process elements include the operating concepts, procedures and regulations that exist in the environment to guide or constrain agents. Procedures define how things may be done, i.e., the tired and tested ways of doing things (Ockerman & Pritchett, 2000), while regulations define what should not be done. For example, in ground transportation, procedures are given for drivers for changing lanes, including indicating their intention using their turn signal. Each agent may internalize such procedures as mechanisms for driving, yet this procedure is a feature of the work environment. Likewise, some process elements of the driving environment proscribe actions, for example, it is forbidden to cross an intersection against a red light. Within a socio-technical system, these process elements can serve as sources of expectation between agents, some agents may serve to 'police' other agents in terms of following procedures and meeting regulations, and the desire of agents to violate environmental process constraints is often an interesting metric when analyzing behavior within a socio-technical system.

One representation of process elements, *task dependence structures*, is based on task analysis methods (Johnson & Johnson, 1991; Schraagen et al., 2000; Shepherd, 2000). Tasks are organized in acyclic graphs called task groups. The root node identifies the main task to be accomplished, and the sub-levels identify sub-tasks and methods that may be executed to accomplish the root task. This hierarchy is further enriched by task dependence relations between subtasks and methods such as enables, facilitates, disables, and delays. Additional relationships are defined in (Decker, 1994). For example, the simulation architecture TÆMS concentrates on task-dependence structures and provides a tool for establishing task-resource relations between the tasks and the resources in the environment (Decker, 1994).

Another representation of process elements, *context-process structures*, is based on the premise that agents take all actions based on their context (Hollnagel, 1993; Kirlik, 1998). The context of an agent is defined within these structures as having two parts: world state and the agent's intention. A context-process structure is therefore organized by and accessed via these two attributes. For example, situation-based structuring accesses an environmental process based on world state, either through examining the objective representation in the environment model alone or by considering the subjective representation internal to each agent. Likewise, intention-based structuring is based on hierarchical intention means-end relations. Each intention in this hierarchy may have environmental processes associated with it that enable or constrain actions towards an intention.

Each of the structures discussed above describes and models one objective aspect of the environment. Many other structures are plausible, such as task-resource structures (relating resources to tasks) and task-skill structures (relating tasks to required skills) (Vicente, 1999). The same environmental element may be represented in a number of these structures and the concerns of any agent may cut across several of them. For a complete model that comprehensively describes a socio-technical system for the purpose of the simulation, then, all necessary aspects of the environment must be modeled.

PUTTING IT ALL INTO A COMPUTER SIMULATION

So far in this chapter we have discussed conceptual models of agents and of their environment. Agent-based simulation requires two further developments: (1) formalizing these conceptual models with sufficient exactitude that they can be implemented as computational software objects and (2) placing these software objects within a larger software architecture which creates and maintains their correct interactions, and which provides the other functions (e.g., data recording and analysis) that link the simulation into broader analysis and design processes.

Many multi-agent simulation architectures have been created. Most common are architectures that mostly work with homogeneous agents in closed systems. For example, StarLogo is a simulation tool that models each agent as a homogeneous entity that works in a physical environment, but can only accommodate a limited range of agent capabilities and functions (Starlogo Website). Given these architectural limitations, it has been primarily used for simulating behavior of natural systems with simple relationships and reactive behaviors. Other simulation architectures have been developed primarily to examine teams (Barbuceanu, 1997; Tambe, 1997; TeamBots Website). These have been used to observe coordination mechanisms through codification of behavior and, for example, assignment of roles and responsibilities. However, they often have focused on specific types of agents (e.g., robots) and thus limit the environment model to physical elements and the corresponding sensing, cognitive, and executive capabilities which the agents can exhibit (TeamBots Website).

Simulating socio-technical systems has particular requirements beyond the simulation architectures used for other domains such as robotics and natural systems. At this time, we know of no single simulator architecture that may be considered 'standard' for organizational simulation. For example, Carley and Gasser describe several simulation architectures used in computational organization theory which each capture some aspects of organizations, but also have a structure that limits them to particular model forms and purposes (Carley & Gasser, 2000).

An intellectual duality exists between the conceptual models' forms and the software architecture such that they work best when their structures are consonant. For example, as discussed in detail below, the 'data passing' and 'timing' concerns of the simulation architecture map tightly to the sensing capabilities of the agents and their methods of updating their internal dynamics. To illustrate these concerns, the following section describes our own agent-based simulator architecture that grew out of the Reconfigurable Flight Simulator (RFS) software suite.

Duality Between Model Form and Software Architecture

Ideally, the software architecture for agent-based simulation should be flexible; i.e., the simulation should not be fundamentally constrained by its basic architecture to one mode of operation or to one level of fidelity. A simulation useful for research and design must also accommodate several user types. The first is that of the general user, who wants to use the simulation as part of his or her day-to-day activities without wanting to interact with source code or recompile software. Such a general user benefits from the ability to configure the simulation through graphical user interfaces or simple configuration scripts; such flexibility is expensive (if not impossible) to create for all possible cases. Therefore, simulations for research and design projects must also provide support the developer, i.e., a member of the design team with some programming knowledge whose task is to keep the simulation functioning as a tool. He or she may not be in a situation to understand the complete workings of all the simulation components, and therefore may not understand the impact of their modifications. Additionally, multiple developers may be distributed throughout the design teams using the simulation software.

As such, simulation software for research and design must be designed to be inherently robust and programmable: robust in that modifications to one part of the software should not have widespread, unanticipated effects in other parts of the software; programmable in that the developer should find the overall framework of the simulation easy to understand, and he or she should quickly be able to find where and how modifications should be made to evoke the desired behavior. For lasting use, simulations must also be inherently extendable; i.e. it should be easy to add new functionality to the simulator through the addition of new components, rather than through fundamental changes to the entire architecture.

To meet these software engineering requirements, Object-Oriented Programming (OOP) has been proposed to create software easy to design, modify and re-use (Alagic et al., 1996; Leslie et al., 1998). OOP principles include abstraction, inheritance, layering, encapsulation, and polymorphism. Abstraction refers to the ability of OOP to define computational structures as independent objects, each containing the functions (methods) and data required to perform the functions of that object. Inheritance refers to the ability, using OOP, to define base interface standards for the behavior of a type of object. Layering allows for smaller, more primitive types of objects to be combined to create larger, more sophisticated objects. Encapsulation is a mechanism by which low-level details about an object are 'hidden' within that object type. Finally, polymorphism is an OOP mechanism by which objects inheriting from a parent class can add new functionality while still meeting the base specifications required of the parent class.

OOP was intended to support software re-use and reduce development costs (Leslie et al., 1998). However, it was subsequently found that, without additional guidance, OOP can result in unworkable software. For example, the first truly large-scale project using C++ and OOP was undertaken in 1988 at Mentor Graphics with the decision to completely redesign their CAD application. The project missed its March 1990 deadline by a year. Beta testing sites reported unusually large numbers of errors, and programmers found it difficult to maintain and correct the code (Lakos, 1996).

To address these issues, Object-Oriented Analysis and Design (OOAD) guidelines and principles have subsequently been developed and tested. OOAD does not add additional mechanisms to those provided by OOP; instead, it specifies effective uses of those mechanisms. These effective uses, as described next, require tight synergy with conceptual modeling efforts.

One principle of OOAD is balancing sources of complexity. An OOP application could theoretically have many simple objects each capable of only one function, or have only one object capable of all their functions; either extreme makes the software appear complex to a developer. In the case of agent-based simulation, then, the proper abstraction is to define agents and the environment as the objects to be modeled. Layering can be used within the agent and environment objects to develop them, but should not be used to create a higher-level structure of the socio-technical system 'above' the agents given the purpose of using agent-based simulation to enable emergent system behaviors.

Another OOAD principle is the elimination of cyclic dependencies between objects, i.e., situations in which their data and functions are sufficiently cross-referenced that they cannot be distinguished. In such a situation, other objects in the simulation cannot be certain of whom to contact to exchange data with an agent. In agent-based simulation, this software engineering principle requires implementing conceptual models that clearly delineate the functions and knowledge of the agents and the environment. Cases where cyclic dependencies do arise in agent-based simulation often indicate that the agent models do not represent the correct level of abstraction (e.g., using multiple objects to do the work of one agent), or has not specified their ability to interact and to act autonomously in a manner capable of proper functioning.

Likewise, OOAD principles advise encapsulating objects to keep their internal functioning private, lest abstractions about objects be easily violated in subsequent developments. Such violations make the software hard to understand because the control logic for a single process can jump from object to object. This makes testing very difficult because objects cannot be tested independently. In agent-based simulation, this software engineering principle again requires implementing conceptual models that clearly delineate the functions and knowledge of the agents and the environment. One important aspect of this principle relative to agent-based simulation is the need to clearly distinguish (and keep private) the knowledge state of agents. If agents are generating a subjective representation of the environment or another agent's state, then the objective and subjective representations should be carefully encapsulated in a manner mimicking reality, and the manner in which data is passed between them should mimic the sensing and coordination/collaboration activities of the agents.

In summary, OOP can provide the much-needed benefits of understandable, readily modified, easily re-used software. However, these benefits can only be fully realized if OOAD principles and guidelines are followed. These OOAD stipulations cannot be met through low-level programming standards that specify only such items as conventions for naming variables – instead, they require the software architecture to be laid out well from its inception. This layout must mimic and support the conceptual models' form so that the translation between conceptual model and software implementation is fluid and conformal – in doing so, the conceptual model must be specific and well defined.

Such conformal mapping between the conceptual model and the software implementation commonly has other benefits. First, many established systems have converged on efficient methods of operation; simulation software mimicking these operations has a better chance to be streamlined and computationally efficient, without 'kludges' and extraneous computations. Second, when decomposing the simulation spatially or temporally to focus on more specific aspects of the simulation (or to distribute it across processors, see the chapter in this volume by Richard Fujimoto), a natural decomposition strategy can be found that does not adversely impact the simulation's functioning or require excessive development. Third, the software's behavior can be easily verified relative to that expected conceptually without translating between or adjusting for different model inputs, outputs and behaviors.

Ideally, the software implementation should follow the structure of well-thought-out conceptual models, and these models should capture the fundamental distinctions in the real system being simulated. Setting up such a simulation architecture requires three main considerations: the interface standards for the software objects, the method of advancing simulation time (and having agents interact at the correct time), and methods of interacting with the larger design process.

Software Interface Standards

To create modular, reconfigurable simulation software, software interface standards should specify the functions within software components and their methods for data passing. In object-oriented development, these interface standards thus define several aspects of an agent-based simulation architecture. The first aspect concerns the general functions that should be internal to the agent models. These generally should be only those that define the autonomous actions of the agents, and each agent's internal functioning during interactions with the environment and other agent. Second, the functions internal to the corresponding environmental model need to be defined. Next, standards need to specify which functions are executed by the simulation architecture (and, correspondingly, not executed by the agents and environment model); these generally reflect those functions arising from the simulation and design processes, including visualization, data recording, and time advance (time advance considerations are detailed below). A fourth aspect concerns the data passing requirements of each component in the simulation, including which data each publishes to the rest of the simulation environment, and specifications of whether the components 'push' data to each other or 'pull' data from each other.

In defining these standards, one major consideration is the extent to which the standards exactly specify the capabilities of the agent and environment models. At one extreme, exact standards could strictly define the methods and data allowable within all models. This extreme would create software that is robust to any developer's attempt to incorporate atypical model forms, and would facilitate creation of tools and procedures for a general user to run the simulation; however, this extreme would also tend to limit the simulation to a narrow range of applications and purposes. At the other extreme, loosely specified standards could allow the developer substantial flexibility in the models used within any simulation. This would allow the same architecture to be applied to a broad range of applications and purposes, including multiple levels of fidelity. However, conformance with the software standards alone would not guarantee that any one component would function with other components used in a simulation, requiring the developer to carefully check all components' mutual compatibility with every reconfiguration of the simulation.

Another consideration is how data should be passed between components. In a general sense, this is commensurate with the information processing view of organizations commonly used by related methods such as computational organization theory - see (Carley and Gasser, 2000). When, in the real system, information transfer is clearly initiated by one entity and received by another, then a clear conceptual basis for the standard exists. Unfortunately, not all information transfer mechanisms are that clearly defined in socio-technical systems at the level of detail required for emulative agent models. For example, in reality an air traffic controller does not need to survey all aircraft in the world to identify those within his or her control sector; however, without special 'radar' models, an air traffic controller agent either needs to 'pull' from the simulation all the aircraft positions to identify those under its control, or the aircraft need to 'push' their information to all controllers. Distinguishing these mechanisms is generally as much a conceptual effort in modeling the 'real' information transfer processes as a software development task.

Additionally, in specifying the data passing requirements, the data types to be passed need to be specified exactly down to the level of units and reference points for relative measures. Often different data types are used by different agents for the same fundamental information. Therefore, resolving the question of which data type should be used for each information element either requires favoring the development of specific models (and requiring other models to 'translate' the data into their format internally), or building mechanisms into the simulation architecture to perform the translation of data types for the components as required (e.g., the reference frame manager used to pass spatial properties between simulation components proposed by (Kalaver & Pritchett, 2004)). In establishing the data types used within the simulation, the databases providing input to the simulation and the data recording requirements of the simulation should also be recognized so that the simulation architecture can link easily with design repositories and operational databases.

Time Advance and Timing Agent Interactions

The heart of every simulation is a timing mechanism that advances simulation time and selects the subroutine or object to be executed next (Law & Kelton, 1999). Time constitutes an important component in the behavior of the agents and their interactions; as such, timing mechanisms are one facet of modeling system dynamics (Ghosh & Lee, 2000). For rigorous control over the time advance of all simulation components, the simulation 'clock' and timing mechanisms are typically part of the central simulation architecture.

In agent-based modeling and simulation, timing mechanisms need to address several issues. First, timing mechanisms need to properly handle heterogeneous agent models, which may have considerably different update rates for their internal dynamics. Second, agents must be timely updated for correct interactions between agents.

Timing mechanisms can be typically defined as synchronous or asynchronous. While synchronous timing methods require all agents in the simulation to update at the same time, asynchronous timing methods allow each agent to update independently. For large-scale or repeated runs of the simulation, synchronous timing methods are usually computationally inefficient since the timing method requires all agents to update at every time step, whether each needs to or not. In theory, synchronous timing mechanisms may provide more accurate modeling and simulation results by updating all agents at very small time steps even though they are computationally inefficient (although these frequent updates may be unrealistic when infrequent updates are uncharacteristic of the real entity). On the other hand, asynchronous timing mechanisms can provide correct modeling and simulation results if all agents are updated timely and accurately for their internal dynamics and interactions with other agents, and can be much more computationally efficient than synchronous timing methods (Logan & Theodoropoulos, 2001; Uhrmacher & Gugler, 2000).

The timing method "asynchronous with resynchronization" documented by Pritchett, Lee, and Goldsman (Pritchett et al., 2001) allows agents to update asynchronously following their own update times, but also estimates when interactions may occur in the future, and requires the relevant agents to jointly update at the times of interaction. The asynchronous with resynchronization timing method allows agents in the simulation to update at their own update times asynchronously and autonomously until an agent specifically requires some or all of the other agents to be synchronized together for an interaction, as shown schematically in Figure 1. To operate within a simulation architecture using this timing mechanism, then, each agent must be able to (1) remember the time of its last update, (2) report the next time it needs to update for its autonomous behaviors, and (3) update when commanded by the architecture's central timing mechanism.

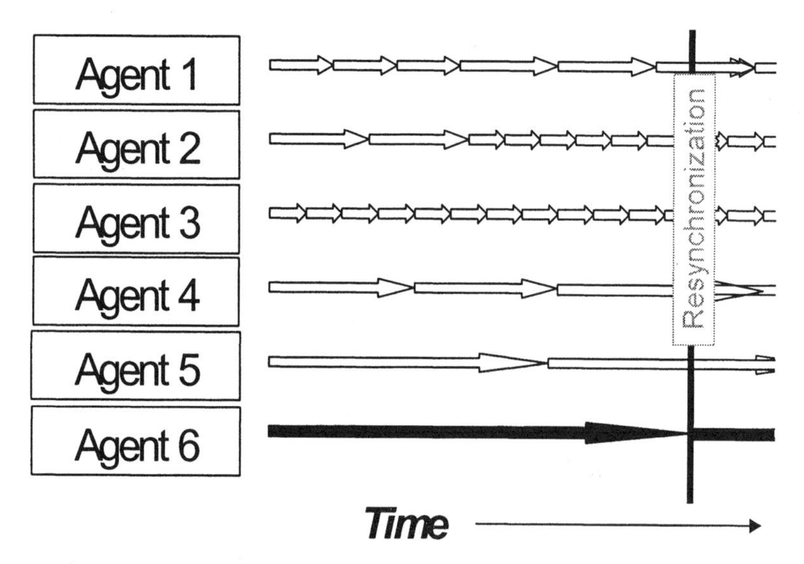

FIGURE 1. Schematic of Timing of Agent Updates Using the 'Asynchronous with Resynchronization' Timing Mechanism

Each agent can be endowed with the ability to monitor and predict its interactions with other agents. However, it can be very difficult and costly to endow agents with more accurate and complex predictive power in a simulation, especially once the simulation contains stochastic elements or when interactions arise due to emergent behaviors and are thus not easily predicted (Lee, 2002). Instead of endowing each agent with the capability of precisely predicting the interactions with other agents, for many applications it is more natural and efficient to develop an object that can monitor and predict the interactions between agents. This object may be conceptualized as providing the monitoring behaviors and physical relationships within the system that trigger interactions between agents.

Accurately capturing these interaction times is important. In socio-technical systems and organizations, interactions with the environment, and collaboration and coordination between agents, are fundamental to system performance. In the real world, many interactions between entities can be easily perceived and detected by themselves or by third entities; for example, in air traffic systems, controllers can easily monitor a spatial radar display to perceive which pairs of aircraft are proximate. In agent-based simulation, on the other hand, this process may require frequent pair-wise comparisons between all pairs of aircraft without the advantage of easy perception of spatial relationships. The example simulation architecture discussed later in this chapter discusses one method for predicting interaction times between agents.

With these reported times – of individual agents' autonomous update times and of predicted interaction times – a central time mechanism can then advance the simulation clock to the next reported time, calling for the corresponding agents to update accordingly. Whether the simulation is running on one or many process. (See the chapter in this volume by Fujimoto for a full description of these timing mechanisms (Fujimoto, 2005)).

One type of timing mechanism is 'optimistic', i.e., simulation processes, including individual agents, are advanced autonomously, resynchronizing across agents (which may span across processors) as required for interactions. Should an interaction be recognized within the simulation only after its occurrence, the simulation clock, environment model and impacted agents then need to be 'rolled-back' to the time of the interaction, either by reverting to recorded descriptions of agent and environment state or by running the models 'backwards'. However, roll-back methods have most commonly been applied to systems with purely discrete dynamics or very simple continuous-time models (Carothers et al., 2000; Jefferson, 1985). Agent-based simulations may be incapable of 'rolling back'. Not only may agent-based simulations use agent models based on legacy code that neither record their state nor can run backwards, but also many aspects of their interactions may also be difficult to record or recreate.

When a simulation cannot be rolled back, 'conservative' timing mechanisms may instead be required. In the context of parallel and distributed simulation, this requires separate processes to advance only forward to a simulation time guaranteed to be before any possible interaction. Additionally, when using asynchronous with resynchronization methods, conservative timing mechanisms require predicting interaction times conservatively, i.e., risking unnecessary updates of agents to see if an interaction will actually occur. While overly conservative predicted interaction times are computationally inefficient due to the unnecessary agent updates they incur, they often do not require computationally complex predictions. On the other hand, more accurate predicted interaction times typically require more computationally extensive calculations; at an extreme, the predictor would need to internally simulate other agents to predict accurately when a problem might occur. As such, the value of better predictions can reach a point of diminishing returns where the additional computations used to predict interaction times more accurately offset any savings in computations by reduced numbers of updates of the agent models.

Many agent-based simulations have continuous dynamics, which thus raises the question of required accuracy in timing. Timing mechanisms have generally been applied to discrete event simulations in which, theoretically, even the smallest error in timing might cause two events to occur in non-causal order resulting in a distortion of the system dynamics not scaled by the magnitude of the error in time. In contrast, in simulations with continuous time dynamics, small errors in update and interaction times might only result in small errors in system dynamics. For example, should an air traffic controller agent interact with two pilot agents slightly later than modeled due to the timing mechanism, the only impact may be that the controller agent commands a slightly stronger collision avoidance maneuver of each of the pilots. Rather than going to significant extremes to make the timing perfect, such variance in the timing mechanism may potentially be capitalized upon to mimic the stochasticity of the timing of behaviors in the real system and to take advantage of any simulated situations in which behavior is not adversely impacted by small errors in timing (Loper & Fujimoto, 2000).

Interacting with the Design Process

Simulation can fit into all stages of research and design. During basic research and conceptual design, low- and medium-fidelity simulations can highlight fundamental issues and constraints on system design. As the design progresses, higher-fidelity models can be added to the simulation so that its output is increasingly detailed and accurate. At the end of design, high-fidelity simulations can serve as a complement to (or replacement for) experimental studies with the actual system, and can serve to train its participants.

However, despite the theoretical utility of simulations throughout design, the time and resources required to develop simulation software for design projects can sometimes limit or prohibit their use. The development of simulation software can take a great deal of time and resources, to the extent that 'rapid' development has been described as weeks to months -- for an aviation example see (Norlin, 1995). Likewise, the development of simulations from scratch requires personnel with substantial skills in many areas, including software engineering and computer programming, computer graphics, and modeling all aspects of the socio-technical system.

Therefore, to serve as an effective design tool, simulation of socio-technical systems must also meet a number of practical software engineering requirements. First, the simulation should be rapidly reconfigurable. As a practical matter, this eases the monetary and time costs of developing a simulation; rapid reconfigurability also enables the simulation to be applied to a range of applications, and to accommodate models of varying form and fidelity as needed for the task at hand. As noted earlier, this concern can be addressed by creation of a simulation architecture conforming to the structures of the conceptual models that will operate within it.

In addition, the simulation should be sufficiently computationally efficient that it can provide a time-effective analysis tool, even when large numbers of runs are required. Computational efficiency can be gained by making the agent-objects computationally efficient and by updating the agent-objects only when needed for correct modeling of system dynamics. As noted earlier, this can be facilitated by using accurate models of the system being simulated such that the relevant dynamic phenomena are propagated as efficiently computationally as they are in reality.

Ideally, the software implementation should follow the structure of well-thought-out conceptual models, and these models should capture the fundamental distinctions in the real system being simulated. If this can be realized, the simulation software can additionally act as a data repository for the design and maintenance of a system. For example, when using simulation to design air vehicles, the vehicle design is often represented by blueprints, circuit diagrams and embedded flight software specifications. Increasingly, vehicle manufacturers are moving to not only allow the simulation to read these specifications directly (rather than laboriously programming them into the simulation), but also making the simulation architecture into the design repository in which designers can store, access and test specifications of their design. An analogous use of simulation in socio-technical systems and organizations can be envisioned in which the change variables (e.g., technology, procedures, regulations and organizational knowledge) are represented in the simulation and presented from there to the system participants.

Finally, simulation can contribute to the analysis and experimentation aspects of a design process. Currently, data analysis is commonly done as a separate post hoc activity acting on data recorded by the simulation. This two-stage process is inefficient in the many cases where statistics could instead be calculated during run-time; in addition, should the data analysis find insufficient or excess observations were taken for accurate statistical comparison, both stages of the process must be repeated.

Incorporating an adaptive control technique, such as a ranking and selection method, offers an additional method of decreasing the effort required to effectively use simulation during design. Ranking and selection methods calculate the number of observations required for statistically sound comparisons between designs. As shown schematically in Figure 2, such adaptive control techniques can rely on statistical estimators embedded in the simulation to calculate the number of observations to make from each simulation (Benson, 2004; Benson et al., 2004). With this and similar techniques, simulation can be made integral to the design process.

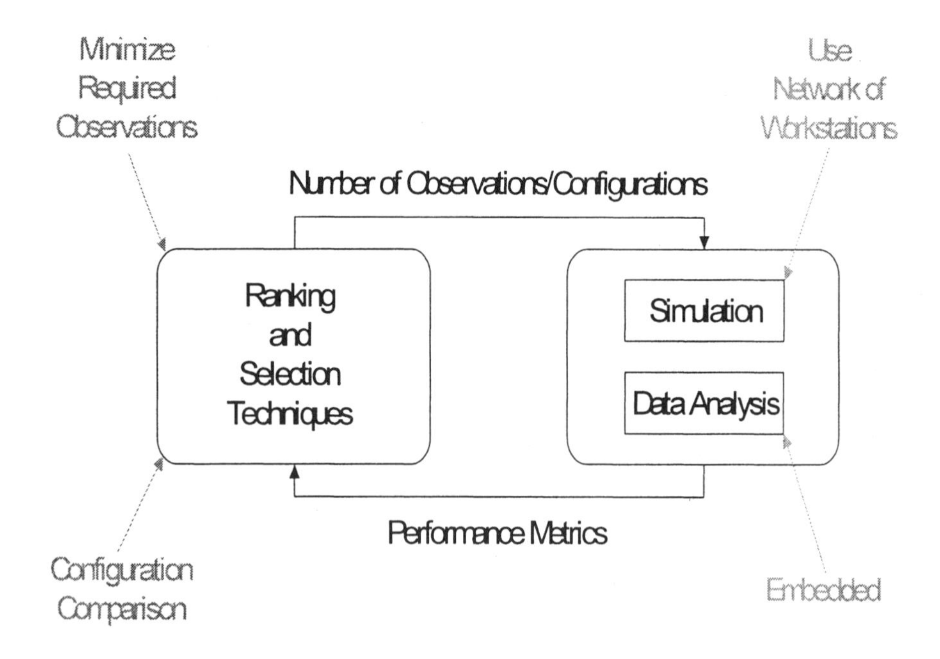

FIGURE 2. Schematic of Automatic Statistical Control of Agent-Based Simulations (Benson, 2004)

EXAMPLE: THE RECONFIGURABLE FLIGHT SIMULATOR AS AN AGENT-BASED SIMULATION ARCHITECTURE

The previous sections discussed conceptual issues with agent-based simulation. This section discusses a particular simulator architecture design. This architecture extends the Reconfigurable Flight Simulator (RFS) software, which was originally designed for real-time human in the loop flight simulation and for fast-time simulation of flight vehicle dynamics (Ippolito & Pritchett, 2000).

RFS Architecture

For modularity, RFS is written in C++ following OOAD principles. The main RFS application does not contain any simulation models but instead provides the run-time support for individual simulation components. This run-time support includes initializing and registering the individual components, and providing communication between them. To provide a dynamic framework, the RFS architecture supports swapping, removing, and loading components during run-time. The architecture also defines the software interface standards for all components. This architecture can be used for real-time or fast-time simulations, human-in-the-loop or autonomous. Many of the components can apply to several types of simulation; for example, aircraft dynamic models have served in flight simulator experiments with pilots as well as representing vehicle dynamics in agent-based simulations of air traffic control systems.

The components are each stored in a precompiled library that can be loaded by the simulator during run-time. Developers extend the capabilities of the simulator by creating new components that the user can select from a library to configure the simulation as desired for any particular run. Because these components are stored in linked lists within the architecture, the number of each which can be included in a simulator configuration at one time is limited only by the computer hardware on which the simulator is being run.

This component-based architecture has several advantages. Developers in different locations and on different projects can create new components and upload them to a central repository; as such, a distributed development environment is possible. Since each module is encapsulated, the developer can work on individual modules without needing knowledge about other components. This facilitates code re-use and reduces the amount of time to tailor the simulation to particular applications. In addition, simulation developers do not require a broad range of expertise. For example, a developer of a human performance model need only modify their component, capitalizing upon established components that provide the visualization, environment model, models of technologies, etc.

The major components of the RFS are shown in Figure 3. Each must fit a base interface for their component type that establishes the data passing requirements judged fundamental to their type of model. The interface for vehicle objects was designed to support continuous-time models of vehicle dynamics. The interface for input-output (IO) objects supports components for visualization, data recording and input functions. Most relevant to agent-based simulation, controller, event and measurement (CEM) objects are less specified, allowing flexibility in the objects' behaviors. As such, these components can range from purely discrete-event models to complex models of human performance; the interface serves to link the model into the data-passing and timing mechanisms of the simulation architecture. In addition, the Environmental Controller and Database (ECAD) object provides a shared simulation environment for all components and provides the foundation for an environment model in an agent-based simulation.

Arrows in Figure 3 represent the access each component has to other objects in the simulation. Cyclic dependencies are avoided by creating a hierarchy within the components; in the dependencies shown, for example, CEM objects can call all the objects in the IO and Vehicles list, and the ECAD.

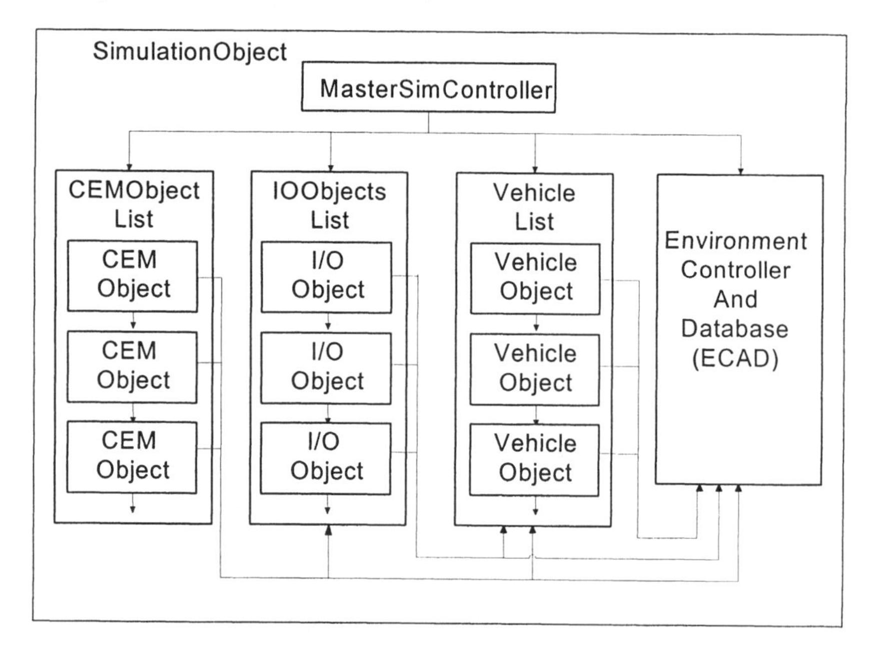

FIGURE 3. Schematic of RFS Simulation Software Architecture

The base objects within the main RFS application define the minimum communication standards estimated to be needed in 90% of the intended applications of each type of component. However, if these standards were the only communication mechanism, then the architecture would not be sufficiently flexible for a wide range of applications. The Object Data/Method Extensions (OD/ME) interface establishes a generic, simulation-wide mechanism for message and data passing between all objects in the RFS. OD/ME adds to the simulation the flexibility to use components with arbitrary data passing requirements. This corresponds to an ability to include components of arbitrarily high fidelity and detail, but also suffers from the dangers noted earlier of developers using OD/ME to circumvent good OOAD practices established by software interface standards. However, because it passes through an interpreter, OD/ME access can be an order of magnitude slower than direct access through the base class interfaces.

To enhance RFS suitability for agent-based simulation, several timing mechanisms have been implemented, including fully synchronous updates of the agents, asynchronous agent updates with all agents resynchronized at the time of any interactions, and asynchronous agent updates with only affected agents resynchronized at the time of their interactions (Pritchett et al., 2001).

Additionally, components have been developed to facilitate predicting the time of agent interactions. Most notably, because predictors based on a model of system dynamics are domain or scenario specific, they incur an obvious development cost for any changes to the simulation configuration. Therefore, an interaction-timing object predicts the time of interactions between agents by learning from the dynamics of the simulation so that scenario-specific predictors of interaction times are not needed. This object uses a back-propagation neural network to learn the system dynamics and predict interaction times between agents. The neural net is trained using a set of training data created through preliminary simulations with a very conservative prediction method. Once the neural network is trained, it can automatically predict interaction times during run-time without expensive computations. When a different interaction between agents needs to be modeled, the neural network can be re-trained.

Application of RFS to an Agent-based Simulation of Air Traffic

The behavior of air traffic systems is created by the interaction of many different elements. Crowded airspace cannot be modeled as a collection of independent aircraft flying simultaneously; instead, controllers (and pilots) are constantly changing their commanded flight paths in response to the actions of others and to changes in the environment. These interactions may meet a number of goals, ranging from time-critical collision avoidance maneuvers to strategic plans for air traffic flow management.

This agent-based simulation was used to investigate the impact of proposed flight deck sensors of clear air turbulence (CAT) (and communication of that information on broadcast air traffic control frequencies) on emergent behavior within the sector and the air traffic controller's task loading -- discussed in detail in Pritchett et al. (2002). This example provided an opportunity to explore a potential safety issue with an expected impact at the system-wide level of the air transportation system. A high-altitude sector in Boston Center (ZBW46) was chosen for its desirable attributes of size, route flexibility, and traffic density. Aircraft were created at random, based on a route-specific exponential distribution of interarrival times with a minimum separation time enforced, derived from the recorded data on a sample day. All flights were given an initial cruise speed of 470 knots, which could change during the simulation in response to controller commands. In this scenario a CAT cell formed in the middle of the sector. Aircraft entered the sector and encountered the turbulence, had the turbulence sensed in advance of their penetration, or overheard pilot reports of turbulence. The experiment was run in three conditions: No sensor, sensor with one-minute look-ahead, and sensor with five-minute look-ahead. Figure 4 illustrates the simulated sector, operational routings, and location of the CAT.

An overview of the simulation is depicted in Figure 5, showing the agents simulated and the information passed among them. The Man-machine Integration Design and Analysis System (MIDAS) human performance model was adapted to model the behavior of pilots and air traffic controllers. MIDAS and RFS operated as separate processes acting as federates joined by an HLA run-time interface.

Baseline behavior for the flight crew agents in the absence of CAT is to follow their commanded route, altitude, and speed. MIDAS was used to model flight crew response to CAT. This agent model uses scripts as its normative standard, generated based upon a focused interview with a captain serving with a major U.S. airline.

It was found that flight crew response to CAT varies with several factors, including: the source of turbulence information; the foreseen spatial extent and altitude of the turbulence; the level (e.g., light, moderate, severe); type (e.g., turbulence versus chop); reported frequency of the turbulence (e.g., occasional, intermittent, or continuous); and actual CAT experience. Pilot responses include: turning the seat-belt sign on and requesting the flight attendants be seated; slowing to turbulence penetration speed; broadcasting pilot reports (PIREPs) on the shared air traffic voice frequencies; and requesting a change in altitude from the controller. MIDAS scripts included these CAT responses as well as standard communication procedures for the various activities required within the sector, including sector entry/exit; ride quality (turbulence) inquiries, requests for altitude change, altitude change clearance acceptance and rejection, etc. (Full details are documented in Pritchett et al., 2002).

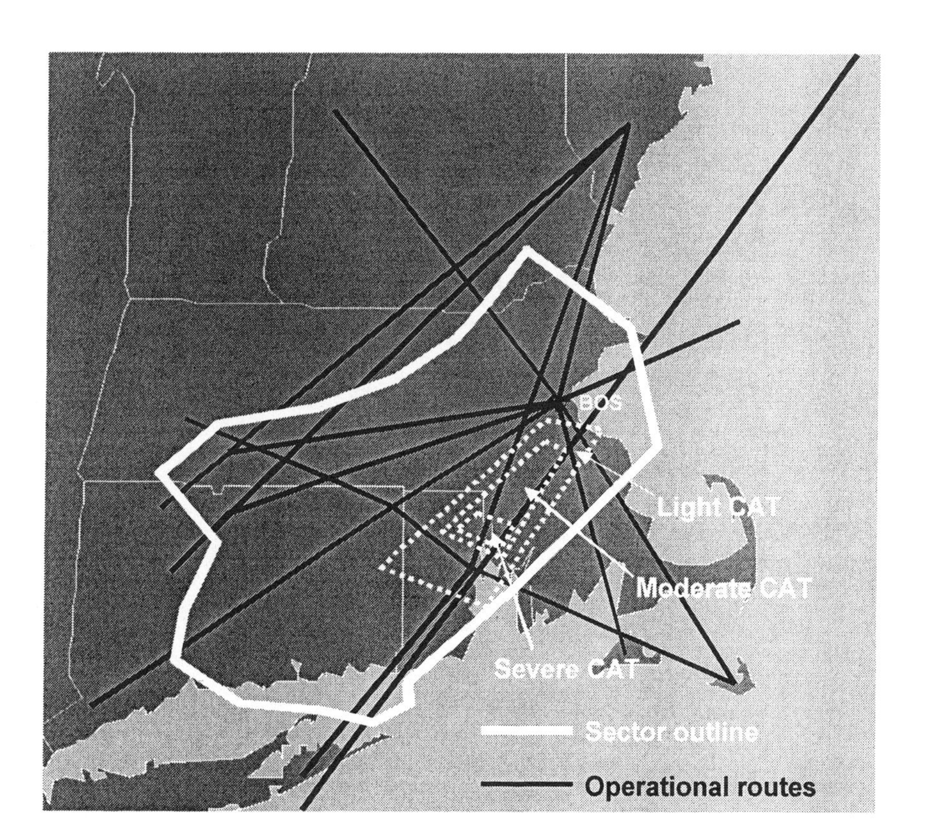

FIGURE 4. Map of Boston High Altitude Sector 46, Showing Airways Simulated and Location of Clear Air Turbulence (CAT)

Similarly, detailed scripts were generated for air traffic controller MIDAS models for sector management (accepting handoffs, initiating handoffs, and monitoring the progress of aircraft) and conflict detection tasks, as well as responses to the CAT-related behaviors scripted for the flight crew. The representation of the cognitive behavior involved in air traffic controller tasks must take account of several aspects, including monitoring and perception, decision making and planning, and associated communications and other actions.

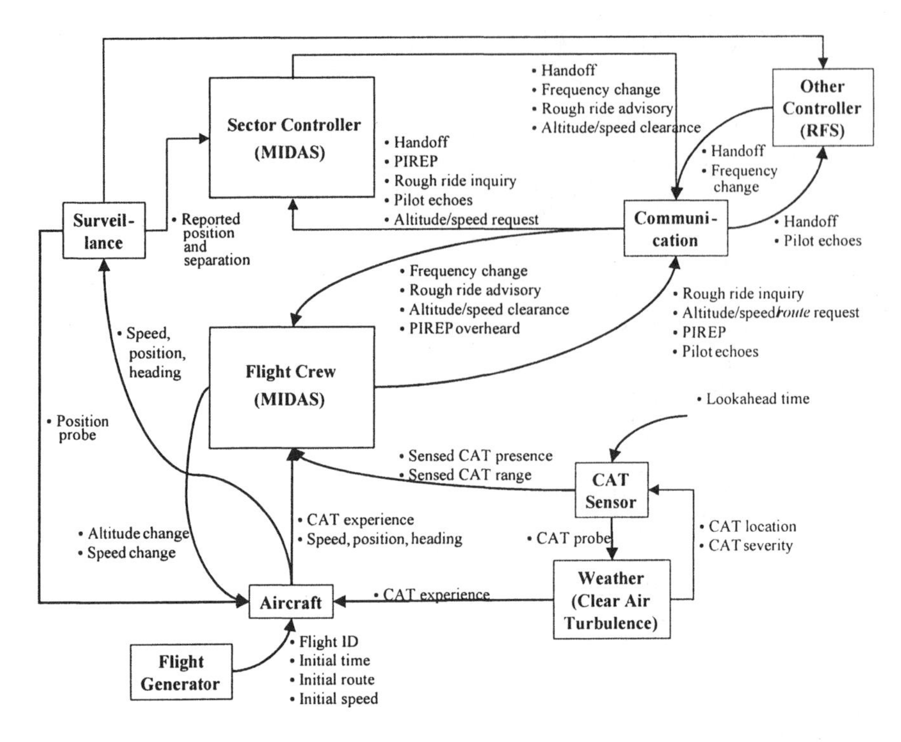

FIGURE 5. Schematic of Agents Included in the Simulation and Their Interactions

Within this framework, controller situation awareness is modeled as a set of data that needs to be constantly updated. As was established by prior efforts to develop a formal representation of controller tasks (Computer Technology Associates Inc., 1987, 1988), controller behavior cannot be represented as a rigid sequence of tasks. Rather, the controller must switch between, suspend and resume, and repeat tasks depending on events. The controller is assumed to maintain a mental representation of the projected flight path of each aircraft, expressed as a sequence of waypoints. The controller needs to maintain a distinction between the commanded and actual flight path, based on the latest radar surveillance information and communications with the flight crew. Controller tasks related to turbulence encounters included: receiving PIREPs about turbulence and maintaining a mental model about it; responding to requests from aircraft about reported ride conditions; evaluating flight crew requests for a speed or altitude change; issuing clearances for conflict-free maneuvers; and responding to requests for an emergency descent to a nearby airport due to turbulence-induced injuries.

The macro-level results are shown in Figure 6. As the number of aircraft in the simulation is increased, the time required for pilot and controller agents to be able to execute necessary responses decreased. This is contrary to the hypotheses that increasing the number of aircraft would yield slower controller response times, not faster response times. Because of some concerns with the timing mechanism used to synchronize MIDAS and RFS, the exact values of these times would require further study with a more conservative timing mechanism maintaining stricter synchronization between the two simulation processes.

Micro-level measures of individual agent behavior and of demands placed on individual agents by the environment included controller task loading measured in four dimensions (visual load, auditory load, cognitive load and psychomotor load) consistent with multiple resource theories of cognitive loading (Corker & Pisanich, 1998; Laughery & Corker, 1997), as shown in Figure 7. Corresponding analysis of demands on the controller due to monitoring the traffic and due to the number of interruptions faced by the controller suggests, first, that putting CAT sensors in the

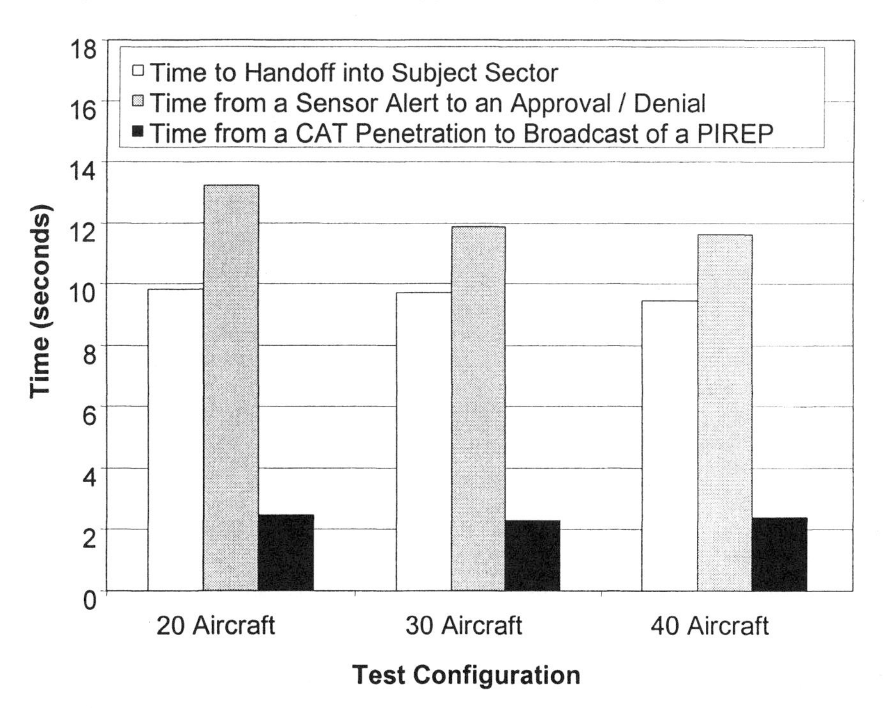

FIGURE 6. Macro-level Measures of System Behavior: Number of Air Traffic Control Communications as a Function of Clear Air Turbulence Sensor

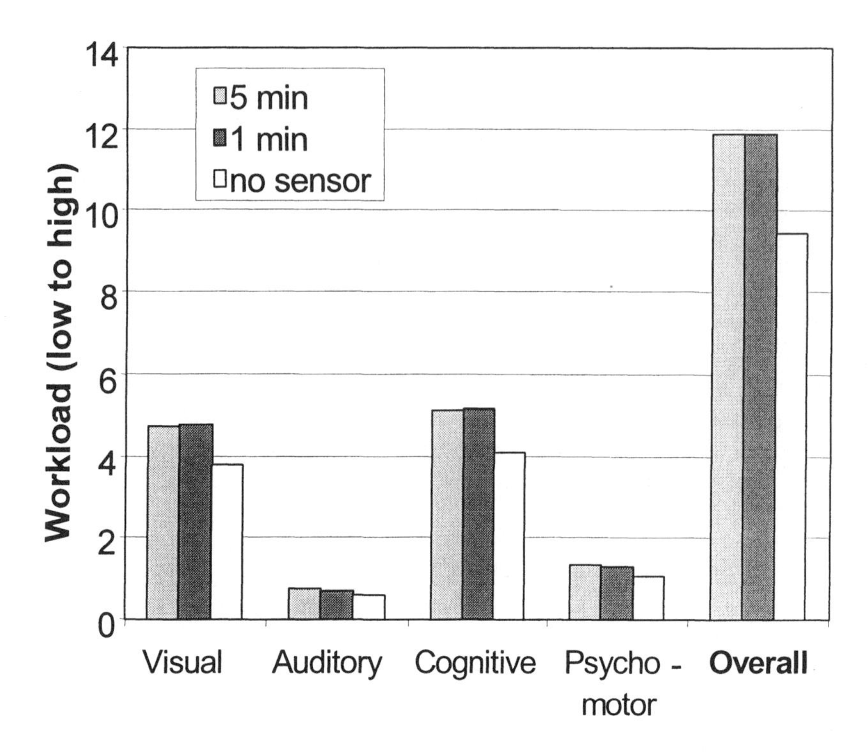

FIGURE 7. Micro-level Measures of Controller Workload as a Function of Clear Air Turbulence Sensor Look-ahead

flight deck impacts the air traffic controller sitting on the ground and, second, that this impact depends on the sensor look-ahead time. Specifically, the short one-minute look-ahead time reduces the time spent on primary controller functions of monitoring and communicating. These results suggest that the one-minute look-ahead time from the sensor and the ensuing calls to the controller focus the controller disproportionally on exploration of alternative routes and other deliberative tasks. The condition of five-minute look ahead has a pattern of response for the controller that is the same as no sensor condition, i.e., the primary controller functions of communicating and monitoring are found to occupy more of the controllers task load.

These results highlight the ability of agent-based simulation to simultaneously provide detailed measures of individual agent performance and of system-level emergent behavior when examining air traffic systems. The individual measures of performance provided insight not only into how the agent will act within (and contribute to) the larger environment, but also highlight the demands of the larger environment on the individual agent.

CONCLUSIONS: QUESTIONS ANSWERED BY (AND RAISED BY) AGENT-BASED MODELING AND SIMULATION OF SOCIO-TECHNICAL SYSTEMS

Agent-based modeling and simulation is a comparatively new method for analyzing socio-technical systems. It builds on those aspects of socio-technical systems that can be directly observed or specified – the work practices and goals of the agents themselves. From these, it predicts the behavior of the socio-technical system as a whole, and the corresponding demands the environment will place on the agents.

In doing so, this method can answer many specific questions about the systems it examines. With simple normative agent models, for example, agent-based simulation can observe whether the system will function as desired when all participants act exactly as procedures, regulations and organizational structures mandate – and highlight areas where individuals' flexibility and creativity are required to advance the system further. Likewise, such normative agents can also record anytime their required behavior conflicts with what they themselves would have done to meet their own goals, highlighting conditions where the real system would be susceptible to non-procedural (or unethical) behaviors. With more descriptive models of human behavior, the simulations could additionally serve to predict the outcome of such unintended behaviors. By simultaneously simulating the 'macro' level system-wide behavior and the 'micro' level agent level behavior, the work environment of agents is also captured, suitable for use in real-time interactive simulations or for computationally estimating the demands imposed on the agents.

This method also illustrates several more general questions whose answers may be subject of debate for some time to come. First, are socio-technical systems emergent? While answering this question comprehensively for all socio-technical systems would be difficult, even the few examples given here highlight insights from agent-based simulations that could be predicted a priori at this level of detail by no other method –for example, the change in air traffic control patterns brought on by the addition of a new sensor to aircraft flight decks. Emergent behaviors appear to be the most relevant when a socio-technical system may be reasonably judged by the product of both individual agents' autonomous actions and their interactions. This is especially relevant when the complexity of and uncertainty within the system obstructs each agent's awareness of the impact of their local behavior on system performance overall.

Second, what are the important aspects of socio-technical systems to include in simulations? Many models focus on specific aspects of socio-technical system behavior. On the other hand, in forming agent-based simulations, accurate prediction of the high-level system behavior requires modeling of many aspects of its low-level structure. The agent models themselves have received the most attention in the research community, and can have a variety of different forms; in this chapter we have argued that simpler agent models can often provide the most definitive results for the purposes of engineering changes to the system. In addition, given our own emphasis on ecological models of cognition, we have argued that a rich environmental model is needed that defines the physical and social constraints on agent behavior, and that establishes the proper mechanisms for agent interactions.

Third, what are the design variables in socio-technical systems – and what are the emergent properties? The historic emphasis of agent-based simulation on analyzing and changing the agents themselves has highlighted the emergence of system behavior from agent behaviors. These insights alone can be useful. In addition it is important to note that, in dealing with the humans in socio-technical systems, we often cannot change the human's cognitive properties directly; instead, we can change the physical and social aspects of their environment, and we can change the mechanisms by which they interact. As such, these design variables can be made manifest in a structure-preserving manner in the environment model. An interesting side-effect of this viewpoint is that changes in the environment model can then create emergent behaviors both in the higher-level system and in the lower-level agent behaviors as they respond and adapt to the demands of the environment.

Finally, given the time and expertise required to establish agent-based models and simulations, is this method practicable for engineering socio-technical systems? This chapter has discussed how a close synergy between the conceptual modeling of agents, the conceptual modeling of socio-technical systems, and the software engineering concerns in establishing computer simulations can provide insight to each other, and thus establish a streamlined analysis and design process. The software engineering concerns center around making simulations that are re-usable and re-configurable. When the agent-based simulation architecture allows for agent and environment models that are structure preserving, the conceptual models should not require translation when codified into a computational representation. Ultimately, by representing the organizational environment in a form that can be specified in vivo to the people in the socio-technical system as well as the simulated agents, the simulation can serve as a design repository in which design variables are not only tested, but also stored and accessed during operations.

ACKNOWLEDGEMENTS

The authors are grateful for the support and direction of Dr. Irv Statler at NASA/Ames Research Center. NASA under Grant NAG2-01291 to Georgia Institute of Technology funded the work reported here. Recognition is also due to S.A. Kalaver and K.M. Corker for their technical expertise and insight.

REFERENCES

Abkin, M. H., Gilgur, A. Z., Bobick, J. C., Hansman Jr, R. J., Reynolds, T. G., Vigeant-Langlois, L., et al. (2001). Development of fast-time simulation techniques to model safety issues in the national airspace system: NASA Contract Report No. NAS2-99072, CY01 Final Report (Phase III) prepared for NASA Ames Research Center.

Alagic, S., Nagati, G., Hutchinson, J., & Ellis, D. (1996). Object-oriented flight simulator technology. Proceedings of the AIAA Modeling and Simulation Technologies Conference and Exhibit, San Diego, CA.

Anderson, J. R. (1993). Rules of the mind: Lawrence Erlbaum Associates.Ankenbrand, T., & Tomassini, M. (1997). Agent-based simulation of multiple financial markets. Neural Networks World, 4, 397-405.

Arkin, R. (1998). Behavior Based Robotics: MIT Press.

Barbuceanu, M. (1997). Coordinating agents by role based social constraints and conversation plans. Proceedings of the Autonomous Agents and Artificial Intelligence.

Barbuceanu, M. (1998). Coordinating with obligations. Proceedings of the Second International Conference on Autonomous Agents.

Barbuceanu, M., Teigen, R., & Fox, M. S. (1997). Agent-based design and simulation of supply chain systems. Proceedings of the 6th Workshop on Enabling Technologies Infrastructure for Collaborative Enterprises (WET-ICE'97).

Barett, J. H. (1970). Individual goals and organizational objectives: a study of integration mechanisms: Braun-Brumfield Inc.

Bargiela, A. (2000). Strategic direction for simulation and modeling. Proceedings of the Plenary Session, Summer Computer Simulation Conference SCSC, Vancouver.

Benson, K. C. (2004). Adaptive control of lasrge scale simulations. Unpublished Doctoral Thesis, Georgia Institute of Technology, Atlanta, GA.

Benson, K. C., Goldsman, D., & Pritchett, A. R. (2004, December 5-8). Applying statistical control techniques to air traffic simulations. Proceedings of the Winter Simulation Conference, Washington, DC.

Borenstein, J., Everett, H. R., & Feng, L. (1996). Navigating mobile robots: Systems and techniques: AK Peters, Ltd.

Bratman, M. E. (1987). Intentions, plans, and practical reason. Cambridge, MA: Harvard University Press.

Bratman, M. E., Israel, D. J., & Pollack, M. E. (1988). Plans and resource-bounded practical reasoning. Computational Intelligence, 4, 349-355.

Carley, K. M., & Gasser, L. (2000). Computational organization theory. In G. Weiss (Ed.), Mutliagent Systems: a modern approach to distributed artificial intelligence: The MIT Press.

Carothers, C. D., Perumall, K. S., & Fujimoto, R. M. (2000). Efficient optimistic parallel simulations using reverse computation. ACM Transactions on Modeling and Computer Simulation, 9(3), 224-253.

Castelfranchi, C. (2000). Engineering Social Order. Proceedings of the ESAW00, Berlin.

Computer Technology Associates Inc. (1987). FAA air traffic control operations concepts (Vol. I). Englewood, Colarado: ATC Background and Analysis Methodology.

Computer Technology Associates Inc. (1988). FAA air traffic control operations concepts (Vol. VI). Englewood, CO: ARTCC/HOST En Route Controllers.

Conte, R., & Gilbert, N. (1995). Introduction: Computer simulation for social theory. In N. Gilbert & R. Conte (Eds.), Artificial Societies: The computer simulation of social life (pp. 1-15). London: UCL Press.

Conte, R., Hegselmann, R., & Terna, P. (1997). Simulating social phenomenon. Berlin: Springer.

Corker, K. M. (1994, October). Man-machine integration design and analysis system (MIDAS) applied to a computer-based procedure-aiding system. Proceedings of the 38th Annual Meeting of the Human Factors and Ergonomics Society, Santa Monica, CA.

Corker, K. M. (1999). Human performance simulation in the analysis of advanced air traffic management. Proceedings of the Winter Simulation Conference, Phoenix AZ.

Corker, K. M., Gore, B. F., Fleming, K., & Lane, J. (2000, June). Free flight and context of control: experiments and modeling to determine the impact of distributed air-ground air-traffic management on safety and procedures. Proceedings of the 3rd Annual Eurocontrol International Symposium on Air Traffic Management, Naples, Italy.

Corker, K. M., & Pisanich, G. (1998, October). Cognitive performance for multiple operators in complex dynamic airspace systems: computational representation and empirical analyses. Proceedings of the Human Factors and Ergonomics Society Meeting.

Craik, K. J. W. (1947). Theory of the human operator in control systems 1: The operator as an engineering system. British Journal of Psychology, 38, 56-61.

Davidsson, P. (2000). Multi agent based simulation: beyond social simulation. In Multi Agent Based Simulation (Vol. 1979): Springer Verlag LNCS series.

Davidsson, P. (2002). Agent-based social simulation: A computer science view. Journal of Artificial Societies and Social Simulation, 5(1).

Decker, K. S. (1994). Environment centered analysis and design of coordination mechanisms. Unpublished PhD, University of Massachusetts, Amherst.

Decker, K. S., & Lesser, V. (1994). The environment centered design of organizations. Proceedings of the Autonomous Agents and Artificial Intelligence.

Durfee, E. H. (2000). Distributed problem solving and planning. In G. Weiss (Ed.), Multiagent Systems: a modern approach to distributed artificial intelligence: The MIT Press.

Eberts, R. E. (1997). Cognitive modeling. In G. Salvendy (Ed.), Handbook of Human Factors and Ergonomics (2nd ed., pp. 1328-1374). New York: Wiley.

Filipe, J. (2002, September). A normative and intentional agent model for organization modeling. Proceedings of the Third International Workshop Engineering Societies in the Agents World.

Fitoussi, D., & Tennenholtz, M. (2000). Choosing social laws for multi-agent systems: minimality and simplicity. Artificial Intelligence, 119, 61-101.

Franklin, S., & Graesser, A. (1996). Is it an agent, or just a program?: A taxonomy of autonomous agents, Third International Workshop on Agent Theories, Architectures and Languages.

Fujimoto, R.M. (2005). Distributed Simulation and the High Level Architecture. In W.B. Rouse & K.R. Boff, Eds., Organizational Simulation: From Modeling and Simulation to Games and Entertainment. New York: Wiley.

Ghosh, S., & Lee, T. S. (2000). Modeling and asynchronous distributed simulation: analyzing complex systems. New York: IEEE Press.

Gibson, J. J. (1979). Ecological approach to visual perception: Houghton Mifflin.

Gilbert, N., & Troitzsch, K. G. (1999). Simulation for the social scientist. Buckingham: Open University Press.

Goldspink, C. (2000). Modeling social systems as complex: Towards a social simulation meta-model. Journal of Artificial Societies and Social Simulation, 3(2).

Gore, B. F. (2000). The study of distributed cognition in Free Flight: a human performance modeling tool structural comparison. Proceedings of the Third Annual SAE International Conference and Exposition - Digital Human Modeling for Design and Engineering, Dearborn, Michigan.

Gore, B. F., & Corker, K. M. (2000). Value of human performance cognitive predictors: a free flight integration application. Proceedings of the 14th Triennial International Ergonomics Association (IEA) and the Human Factors and Ergonomics Society 44th Annual Meeting, Santa Monica, CA.

Hajdukiewicz, J. R., Burns, C. M., Vicente, K. J., & Eggleston, R. G. (1999). Work domain analysis for intentional systems. Proceedings of the Human Factors and Ergonomics Society, 43rd Annual Meeting.

Hayes, C. C. (1999). Agents in a nutshell - a very brief introduction. IEEE Transactions on Knowledge and Data Engineering, 11(1).

Ho, D., & Burns, C. M. (2003). Ecological interface design in aviation domains: work domain analysis of automated collision detection and avoidance. Proceedings of the Human Factors and Ergonomics Society, 47th Annual Meeting.

Hollnagel, E. (1993). Human reliability analysis: context and control. London: Academic Press, Computers and People Series.

Howard, A., Mataric, M. J., & Sukhatme, G. S. (2002). Mobile sensor network deployment using potential fields: a distributed scalable solution to the area coverage problem. Proceedings of the DARS 02, Fukuoka, Japan.

Huang, C. C. (2001). Using intelligent agents to manage fuzzy business process. IEEE Systems, Man and Cybernetics: Part A, Man and Systems, 31(6), 508-523.

Hudlicka, E. & Zacharias, G. (2005). Requirements and Approaches for Modeling Individuals Within Organizational Simulations. In W.B. Rouse & K.R. Boff, Eds., Organizational Simulation: From Modeling and Simulation to Games and Entertainment. New York: Wiley.

Huhns, M. N., & Stephens, L. M. (2000). Multiagent systems and societies of agents. In G. Weiss (Ed.), Multiagent Systems: a modern approach to distributed artificial intelligence: The MIT Press.

Hutchins, E. (1995). Cognition in the wild. Cambridge: The MIT Press.

Ippolito, C. A., & Pritchett, A. R. (2000). SABO: A self-assembling architecture for complex system simulation. Proceedings of the 38th AIAA Aerospace Sciences Meeting and Exhibit, Reno, NV.

Jefferson, D. R. (1985). Virtual time. ACM Transactions on Programming Languages and Systems, 7(3), 404-425.

Jennings, N. R., & Wooldridge, M. (2000). Agent-oriented software engineering. Proceedings of the 9th European Workshop on Modeling Autonomous Agents in a Multi-Agent World: Mutli-Agent System Engineering (MAAMAW-99).

Johnson, H., & Johnson, P. (1991). Task knowledge structures: Psychological basis and integration into system design. Acta Psychologica, 78, 3-24.

Joslyn, C. (1999). Semiotic agent model for simulating socio-technical organizations. Proceedings of the AI, Simulation and Planning in High Autonomy Systems.

Kalaver, S. A., & Pritchett, A. R. (2004). Reference fram management and its application to reducing numerical error in simulation. AIAA Journal of Aerospace Computing, Information and Communication, 1(3), 137-153.

Kang, M., Waisel, L. B., & Wallace, W. A. (1998). Team-Soar: A model of team decision making. In M. J. Prietula, K. M. Carley & L. Gasser (Eds.), Simulating organizations: computational models of institutions and groups: AAAI Press/The MIT Press.

Kennedy, J. (2000). Sociocognitive computation: A new interdisciplinary convergence. Proceedings of the Virtual Worlds and Simulation Conference.

Kirlik, A. (1998). The design of everyday life environments. In W. Bechtel & G. Graham (Eds.), A companion to cognitive science (pp. 702-712): Oxford: Blackwell.

Lakos, J. (1996). Large scale C++ software development. Reading, MA: Addison-Wesley.

Laughery, K. R., & Corker, K. M. (1997). Computer modeling and simulation of human/system performance. In G. Salvendy (Ed.), Handbook of human factors (Second Edition ed.). New York: John Wiley.

Law, A. M., & Kelton, W. D. (1999). Simulation modeling and analysis (3rd ed.). New York, NY: McGraw-Hill.

Lawrence, P. R., & Lorsch, J. W. (1967). Organization and environment: managing differentiation and intergration. Boston: Division of Research, Graduate School of Business Administration, Harvard University.

Lee, S. M. (2002). Agent-based simulation of socio-technical systems: software architecture and timing mehcanisms. Unpublished Doctoral Thesis, Georgia Institute of Technology, Atlanta, GA.

Leslie, R. A., Geyer, D. W., Cunningham, K., Glaab, P. C., Kenney, P. S., & Madden, M. M. (1998). LaSRS++ An object-oriented framework for real-time simulation of aircraft. Proceedings of the AIAA Modeling and Simulation Technologies and Exhibit, Boston, MA.

Lesser, V. R. (1999). Cooperative multiagent systems: a personal view of the state of the art. IEEE Transactions on Knowledge and Data Engineering, 11(1), 133 - 142.

Logan, B., & Theodoropoulos, G. (2001). The distributed simulation of multi-agent systems. Proceedings of the IEEE, 89(2), 174-185.

Loper, M. L., & Fujimoto, R. M. (2000). Pre-sampling as an approach for exploiting temporal uncertainty. Proceedings of the Workshop on Parallel and Distributed Simulation, Bologna, Italy.

Markman, A. B. (1998). Knowledge representation: Lea.

Marrow, F. (1969). The practical theorist: Basic Books Inc.

McNeese, M. D., & Vidulich, M. A. (2002). Cognitive systems engineering in military aviation environments: avoiding cogminutia fragmentosa! : Human Systems Information Analysis Center.

Mizuta, H., & Yamagata, Y. (2001). Agent-based simulation for economic and environmental studies. Proceedings of the JSAI 2001 Workshop.

Naikar, N., Pearce, B., Drumm, D., & Sanderson, P. M. (2003). Designing teams for first-of-a-kind, complex systems using the initial phases of cognitive work analysis: case study. Human Factors, 45(2), 202-217.

Nair, R., Tambe, M., Yokoo, M., Pynadath, D., & Marsella, S. (2003). Taming Decentralized POMDPs: Towards efficient policy computation for multi-agent settings. Proceedings of the International Joint Conference on Artificial Intelligence.

Nickles, G. M. (2004). Work action analysis to structure planning and formative evaluation of an engineering course using a course management system. Unpublished Doctoral Thesis, Georgia Institute of Technology, Atlanta, GA.

Niedringhaus, W. P. (2004). The Jet:Wise model of national air space system evolution. Simulation: Transactions of the Society for Modeling and Simulation, 80(1), 45-58.

Nilsson, N. J. (1998). Artificial intelligence: A new synthesis (1st ed.): Morgan Kaufmann Publishers.

Norlin, K. A. (1995). Fight simulation software at NASA Dryden Flight Reserarch Center. Proceedings of the AIAA Modeling and Simulation Technologies Conference and Exhibit.

Norling, E., Sonnenberg, L., & Ronnquist, R. (2000). Enhancing multi-agent based simulation with human-like decision making strategies. Proceedings of the MABS 2000.

Norman, D. A. (1986). Cognitive Engineering. In D. A. Norman & S. W. Draper (Eds.), User centered system design: New perspectives on human-computer interaction.

Ockerman, J. J., & Pritchett, A. R. (2000). A review and reappraisal of task guidance systems: Aiding workers in procedure following. International Journal of Cognitive Ergonomics, 4(3), 191 - 212.

Odell, J., Parunak, H. V. D., Fleischer, M., & Breuckner, S. (2002). Modeling agents and their environment. In F. Giunchiglia, J. Odell & G. Weiss (Eds.), Agent Oriented Software Engineering (AOSE) III (Vol. 2585, pp. 16 - 31). Berlin: Lecture Notes on Computer Science, Springer.

Parunak, H. V. D. (2000). Industrial and practical applications of DAI. In G. Weiss (Ed.), Multiagent systems: A modern approach to distributed artificial intelligence (pp. 377-421): The MIT Press.

Parunak, H. V. D., Savit, R., & Riolo, R. L. (1998). Agent-based modeling vs. equation-based modeling. Proceedings of the Workshop on Modeling Agent Based Systems (MABS98).

Pollatsek, A., & Rayner, K. (1998). Behavioral experimentation. In W. Bechtel & G. Graham (Eds.), A Companion to Cognitive Science (pp. 352-370): Oxford: Blackwell.

Pritchett, A. R., Lee, S. M., Corker, K. M., Abkin, M. A., Reynolds, T. R., Gosling, G., et al. (2002). Examining air transportation safety issues through agent-based simulation incorporating human performance models. Proceedings of the IEEE/AIAA 21st Digital Avionics Systems Conference, Irvine, CA.

Pritchett, A. R., Lee, S. M., & Goldsman, D. (2001). Hybrid-system simulation for national airspace systems safety analysis. AIAA Journal of Aircraft, 38(5), 835-840.

Rao, A. S., & Georgeff, M. P. (1991). Modeling rational agents within a BDI-architecture. Proceedings of the Knowledge Representation and Reasoning, San Mateo, CA.

Rasmussen, J. (1988). Cognitive engineering, a new profession? In L. P. Goodstein, H. B. Anderson & S. E. Olsen (Eds.), Task, errors and mental models (pp. 325 - 334). London, UK: Taylor and Francis.

Rasmussen, J., Pejtersen, A. M., & Goodstein, L. P. (1994). Cognitive Systems Engineering: Wiley Series in Systems Engineering.

Russell, S., & Norvig, P. (1995). Artificial intelligence: A modern approach: Prentice Hall.

Schraagen, J. M., Chipman, S. F., & Shalin, V. L. (Eds.). (2000). Cognitive task analysis. Mahwah, NJ: Lawrence Erlbaum Associates.

Shen, W., & Norrie, D. H. (1999). Agent-based systems for intelligent manufacturing: A state-of-the-art survey. Knowledge and Information Systems, an International Journal, 1(2), 129-156.

Shepherd, A. (2000). HTA as a framework for task analysis. In J. Annett & N. A. Stanton (Eds.), Task analysis (pp. 9-24). New York: Taylor and Francis.

Shortliffe, E. H. (1976). Computer-based medical consultations: MYCIN. New York, NY: American Elsevier.

Sierhuis, M., Clancey, W., & Sims, M. (2002, January 07-10). Multiagent modeling and simulation in human-robot mission operations work system design. Proceedings of the 35th Annual Hawaii International Conference on System Sciences, Big Island, Hawaii.

Simon, H. A. (1982). Models of bounded rationality (Vol. 2). Cambridge, MA: The MIT Press.

Singh, M. P., Rao, A. S., & Georgeff, M. P. (2000). Formal methods in DAI: logic-based representation and reasoning. In G. Weiss (Ed.), Multiagent systems: A modern approach to distributed artificial Intelligence: The MIT Press.

Starlogo Website. from http://education.mit.edu/starlogo

Tambe, M. (1997). Agent architectures for flexible, practical teamwork. Proceedings of the American Association of Aritificial Intelligence.

Tambe, M., Schwamb, K., & Rosenbloom, P. S. (1995). Building intelligent pilots for simulated rotary wing aircraft. Proceedings of the Conference on computer generated forces and behavioral representation.

TeamBots Website. from www.teambots.org

Tversky, A., & Kahneman, D. (1974). Judgment under uncertainty: Heuristics and biases. Science, 185, 1124-1129.

Uhrmacher, A. M., & Gugler, K. (2000). Distributed, parallel simulation of multiple, deliberative agents. Proceedings of the 14th Workshop on Parallel and Distributed Simulation, Bologna, Italy.

Vicente, K. J. (1990). A few implications of an ecological approach to human factors. Human Factors Society Bulletin, 33(11), 1 - 4.

Vicente, K. J. (1999). Cognitive work analysis: Towards safe, productive and healthy computer based work: Lawrence Erlbaum.

Weiss, G. (Ed.). (2000). Multiagent systems: A modern approach to distributed artificial intelligence: The MIT Press.

Wooldridge, M. (2000). Intelligent Agents. In G. Weiss (Ed.), Multiagent Systems: a modern approach to distributed artificial intelligence: The MIT Press.

Xuan, P. (2004). An MDP approach for agent self monitoring. Proceedings of the The AAAI-04 Workshop on Agent Organizations: Theory and Practice, San Jose, CA.

CHAPTER 13

EXECUTABLE MODELS OF DECISION MAKING ORGANIZATIONS

ALEXANDER H. LEVIS

ABSTRACT

Command centers, whether in a single location or distributed, consist of humans receiving information through display monitors, interacting with other members of the organization through communication systems, making decisions, and sending out information. The overall mission of the command center has been decomposed into tasks that individuals or cells can carry out. This has worked very well in the past. But modern military command centers are faced with a challenge created by the fact that they staffed by a diverse group of professionals provided by the participating coalition partners. This last aspect introduces another set of variables in the design of organizations: cultural differences. The chapter explores a particular computational approach for organizational design especially suited to the class of decision making organizations: coalition command centers. This approach is based on an executable model that can be used both in simulation and in analysis. While some aspects are beginning to be addressed in an exploratory manner, there are many theoretical and technical issues that need to be investigated. Progress in this class of problems will help substantially in designing effective, austere (affordable) command centers that can be deployed easily.

INTRODUCTION

The effort to model organizational behavior with mathematical models has a long history. The groundbreaking work of Marshak & Radner (1972) looked at the communications between organization members; today we would call this connectivity and associated information flows. Drenick (1986) proposed a mathematical organization theory in which a number of fundamental system theoretic ideas were exploited to draw insights for the design of organizations consisting of members who process tasks under time constraints – a form of Simon's (1982) bounded rationality. Levis (1988) and his students developed a discrete event dynamical model and a set of rules that governed the allowed interactions – whether they represented forms of information sharing or of commands. This model, expressed mathematically in the language of Colored Petri

Nets (Jensen, 1990) allowed the design of organizational architectures that could meet accuracy and timeliness constraints while not exceeding the workload limitations of the decision makers. Essentially, the organization members conducted information processing and decision making tasks, often supported by decision support systems in order to reduce workload, while increasing accuracy and timeliness of the organizational response (Levis, 1995).

The basic model (Boettcher and Levis, 1982), that of the single decision maker, evolved over time in order to accommodate more complex interactions and allow for different types of internal processing by the organization members (Levis, 1993). The early focus was on small teams in which several members need to be organized to perform a demanding, time-sensitive task. The objective was to achieve organizational performance without causing excessive workload that would lead to performance degradation.

In the middle 90s, as a consequence of a model-driven experimental program conducted at the Naval Postgraduate School (Levis and Vaughan, 1999; Handley et al., 1999), the focus of the research on organizational architectures changed to the modeling of command centers so that the experiments with human subjects could be simulated. What has been a key consideration from the beginning of this work was the ability to relate structure to behavior. This meant that the structure and attributes of the simulation models must be traceable, in a formal way, to the architecture design. Hence the use of the term "executable" model which denotes that there is a formal mathematical model used for simulation with characteristics that are traceable to the static designs. The mathematical model can also be used for analysis, i.e., properties of the model and performance characteristics can be determined from the mathematical description. A wealth of theoretical results on discrete event dynamical systems, in general, and Colored Petri nets, in particular, can be applied to the executable model.

In the late 90s, a new variant of the problem arose. The armed forces of the US were undertaking operations that ranged from humanitarian assistance and disaster relief all the way to combat operations. And in all cases, this was done as a coalition operation. The coalitions that are being formed are very diverse, going far beyond the set of alliances the US has maintained in the last sixty years. The coalition partners contribute to the goals of the coalition in very different ways – from combat forces to allowing US forces to fly over their territory to conduct operations or re-supply. There are differences in equipment or materiel, differences in command structures, differences in constraints under which they can operate, and, last but not least, differences in cultures. The differences in equipment and in operational constraints can be handled easily in the existing modeling framework. Differences in command structures require some additional work to express these differences in structural and quantitative ways. The real challenge is how to express cultural differences in these, primarily mechanistic, models of organizations.

The command structure of coalitions can take different forms. At one extreme, there can be a functional decomposition of the mission and the assignment of the partitioned tasks to different members of the coalition with each coalition partner having its own command center. A coordinating command center is needed to partition and monitor the tasks at the strategic and high operational level. Each partner's command center operates at the operational and tactical level. This is usually referred to as a functional organization. An alternative partition is a geographic one in which each partner is assigned a geographic sector (e.g., the partition of Germany and Berlin after World War II); this is known as a divisional organization. Again a high-level coordinating command center is needed that determines policy and adjudicates jurisdictional disputes. At the other extreme is a centralized coalition command center staffed by individuals from the coalition partners. Given the diversity of coalition partners, any real coalition combines, centralized aspects with functional and divisional aspects. For example, there can be a centralized coalition command center for conducting air operations with a few of the coalition partners participating; other partners may be assigned a single function, while others may assigned a geographic sector where they conduct multiple functions.

Two other considerations drive the design problem: (a) the tempo of operations has increased substantially, and (b) an approach called "effects based operations" is being used that has the characteristic of requiring the effective combination of multiple resources and the associated functions. The latter consideration requires a different type of modeling to capture the cause-effect relationships. Approaches based on Bayesian networks and their variants have been used to capture such relationships (Wagenhals and Levis, 2002). They will not be considered in this chapter. Instead, this chapter will focus on the ability to introduce attributes that characterize cultural differences into the mechanistic model for organization design and use simulation to see whether these parameters produce significant changes in performance. As stated earlier, the objective is to relate performance to structural features but add attributes that characterize cultural differences. Specifically, the attributes or dimensions defined by Hofstede (2001) are introduced in the model of the interacting decision maker and then different organizations are simulated to observe what kind of performance affecting behaviors these differences induce.

In the second section of this chapter, the modeling approach is described briefly since it has been documented extensively in the literature. In the third section, the Hofstede dimensions are introduced and then applied to the organization model. In the fourth section, a computational experiment is described to show whether this approach is promising. In the final section, advantages and shortcomings of this approach are discussed.

THE DECISIONMAKER MODEL AND ORGANIZATION DESIGN

The five-stage interacting decision maker model (Levis, 1993) had its roots in the investigation of tactical decision making in a distributed environment with efforts to understand cognitive workload, task allocation, and decision making. The five-stage model allows the algorithm in each stage to be defined and makes explicit the input and output interactions of the decision maker with other organization members or the external environment. It also has a well-defined algorithm for characterizing workload. This model has been used for fixed as well as variable structure organizations. Perdu & Levis (1998) described an adaptive decision maker model that used an object class to represent the ability of decision makers to adapt dynamically with local adaptation.

The five-stage decision maker model is shown in Figure 1. The decision maker receives a signal from the external environment or from another decision maker. The Situation Assessment (SA) stage represents the processing of the incoming signal to obtain the assessed situation that may be shared with other decision makers. The decision maker can also receive situation assessment signals from other decision makers within the organization; these signals are then fused together in the Information Fusion (IF) stage to produce the fused situation assessment. The fused information is then processed at the Task Processing (TP) stage to produce a signal that contains the task information necessary to select a response. Command information from superiors is also received. The Command Interpretation (CI) stage then combines internal and external guidance to produce the input to the Response Selection (RS) stage. The RS stage then produces the output to the environment or to other organization members.

The key feature of the model is the explicit depiction of the interactions with other organization members and the environment. These interactions follow a set of rules designed to avoid deadlock in the information flow. A decision maker can receive inputs from the external environment only at the SA stage. However, this input x can also be another decision maker's output. A decision maker can share his assessed input with another organization member. This is depicted as the z" input to the IF stage when the decision maker is receiving a second input. This input must be generated from another decision maker and can be the output of the SA or RS stage. In the CI stage, the decision maker can receive command information as the input v'. This is also internally generated and must originate from another decision maker's RS stage. Thus the interactions between two decision makers are limited by the constraints enumerated above: the output from

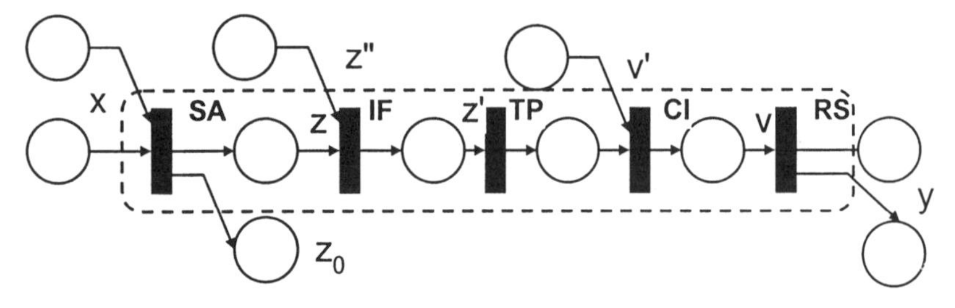

FIGURE 1. The Petri Net Model of the Five-Stage Decision Maker

decision makers are limited by the constraints enumerated above: the output from the SA stage, z, can only be an internal input to another decision maker's IF stage as z_0, and an internal output from the RS stage, y, can only be input to another decision maker's SA stage as x, IF stage as z", or CI stage as v'.

In the Petri net representation of the DM model, the transitions stand for the algorithms, the connectors for the precedence relations between these algorithms, and the tokens for their input and output. The tokens, in the simplest version of the model, are all indistinguishable. A token in a place means simply that an item of information is available to the output transition(s) of that place. A source can be represented by a finite number of tokens x, each one occurring with some probability p(x). However, if the protocols ruling their processing do not vary from one set of attributes to the other, they can be considered as indistinguishable tokens. If the tokens need to be distinct, i.e., carry information, then a Colored Petri Net representation is used. This is described in a later section.

Other organization components can be modeled using the same basic five-stage model, but eliminating one or more of the stages. For example, a processor that receives sensor data and converts it to an estimate of a vector variable can be modeled by a single SA transition, while a data fusion algorithm can be modeled by an IF transition. With this model of the organization member and its variants used to model other components, it is now possible to formulate the problem of designing distributed decision-making organizations.

It was shown in Fig. 1 that a decision maker (DM) can only receive inputs at the SA, IF and CI stages, and produce outputs at the SA and RS stages. These conditions lead to the set of admissible interactions between two DMs shown in Fig. 2. For clarity, only the connectors from DM_i to DM_j are shown; the interactions from DM_j to DM_i are identical.

The mathematical representation of the interactions between DMs is based on the connector labels e_i, s_i, F_{ij}, G_{ij}, H_{ij} and C_{ij} of Fig. 2; they are integer variables taking values in {0, 1} where 1 indicates that the corresponding directed link is actually present in the organization, while 0 reflects the absence of the link. These variables can be aggregated into two vectors **e** and **s**, and four matrices **F**, **G**, **H** and **C**. The interaction structure of an n-decision-maker organization may be represented by the following six arrays: two n x 1 vectors **e** and **s**, representing the interactions between the external environment and the organization:

$$\mathbf{e} = [e_i], \quad \mathbf{s} = [s_i] \quad \text{for i } 1, 2, \ldots, n$$

and four n x n matrices **F**, **G**, **H** and **C** representing the interactions between decision makers inside the organization. Since there are four possible links between any two different DMs, the maximum number of interconnecting links that an n-decision-maker organization can have is

$$k_{max} = 4n^2 - 2n$$

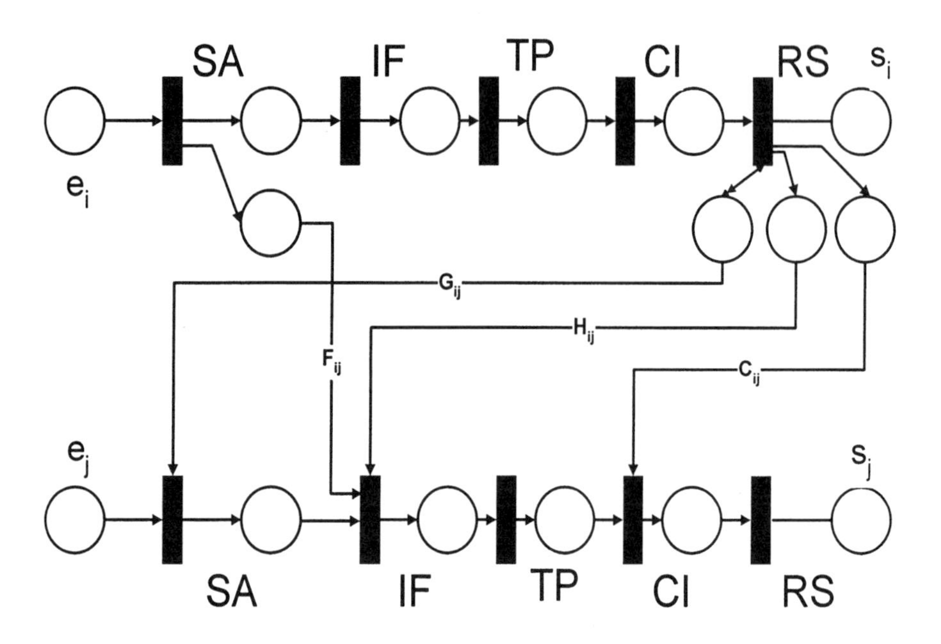

FIGURE 2. Interactions Between Decision Maker i and Decision Maker j

Consequently, if no other considerations were taken into account, there could be $2^{k_{max}}$ alternative organizational forms. This is a very large number: 2^{90} for a five-person organization.

GENERATION OF ORGANIZATIONAL ARCHITECTURES

The analytical description of the possible interactions between organization members forms the basis for an algorithm that generates all the architectures that meet some structural constraints as well as application-specific constraints that may be present. The set of structural constraints that will be introduced rules out a large number of architectures. The most important constraint addresses the connectivity of the organization - it eliminates information structures that do not represent a single integrated organization.

An algorithm has been developed (Remy and Levis, 1988) that determines the maximal and minimal elements of the set of designs that satisfy all the constraints; the entire set can then be generated from its boundaries. The algorithm is based on the notion of a simple path - a directed path without loops from the source to the sink. Feasible architectures are obtained as unions of simple paths. Consequently, they constitute a partially ordered set.

The sextuple {**e**, **s**, **F**, **G**, **H**, **C**} is called a well-defined net (WDN) of dimension n, where n is the number of organization members. A WDN can be represented in two different ways: (a) the matrix representation; i.e., the set of arrays {**e**, **s**, **F**, **G**, **H**, **C**}; and (b) the Petri net representation, given either by the graph or the incidence matrix of the net, with the associated labeling of the transitions. These two representations of a WDN are equivalent; i.e., a one-to-one correspondence exists between them.

Let the organizational structure be modeled as having a single source and a single sink place. Each internal place of a WDN has exactly one input and one output transition. The sink of a WDN has one input but no output transitions, while the opposite stands for the source. If source and sink are merged into one place, Figure. 3, every place in the net will therefore have one input and one output transition.

Considering the source and the sink of a WDN as the same place has no bearing on the internal topology of the organization. The assumption becomes important, however, when the dynamic behavior of a WDN is studied. The merging of source and sink limits the amount of information a given organization can process simultaneously. The initial marking of the place representing the external environment will define this bound. At this stage, a WDN may contain circuits.

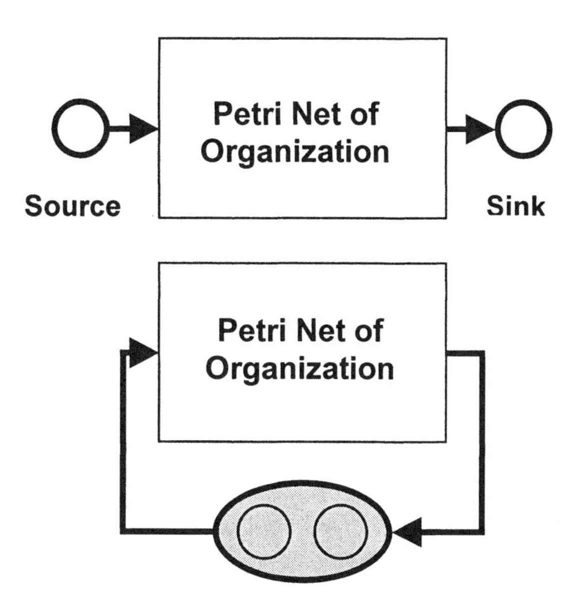

FIGURE 3. Merging the Source and Sink Places

While WDNs constitute the framework within which information structures will be designed, each WDN is not a valid organizational structure. Additional constraints to restrict the set of WDNs to useful information structures are needed. There are some WDNs corresponding to combinations of interactions between components that do not have a physical interpretation; e.g., DMs can exchange information - F_{ij} and F_{ji} can coexist - but commands are unilateral- either C_{ij} or C_{ji} or none, but not both. Those WDNs should be eliminated, if realistic organizational forms are to be generated. The structural constraints define what kinds of combinations of interactions need to be ruled out. A set of four differentstructural constraints is formulated that applies to all organizational structures being considered.

R1 A directed path should exist from the source to every node of the structure and from every node to the sink.

R2 The structure should have no loops; i.e., the organizational structures should be acyclical.

R3 There can be at most one link from the RS stage of a DM to each one of the other DMs; i.e., for each i and j, only one element of the triplet $\{G_{ij}, H_{ij}, C_{ij}\}$ can be nonzero.

R4 Information fusion can take place only at the IF and CI stages. Consequently, the SA and RS stages of each DM can have only one input.

Constraint R1 eliminates structures that do not represent a single integrated organization and ensures that the flow of information is continuous within an organization. Constraint R2, allows acyclical organizations only.[1] Constraint R3 states that the output of the RS stage of one DM or component can be transmitted to another DM or component only once: it does not make much sense to send the same information to the same decision maker at several different stages. Constraint R4 prevents a decision maker from receiving more than one input at the SA stage. The rationale behind this limitation is that information cannot be merged at the SA stage; the IF stage has been specifically introduced to perform such a fusion.

Any realistic design procedure should allow the designer to introduce specific structural characteristics appropriate to the particular design problem. To introduce user-defined constraints that will reflect the specific application the organization designer is considering, appropriate 0s and ls can be placed in the arrays {**e, s, F,**

---

[1] This restriction is made to avoid deadlock and circulation of messages within the organization. It also restricts the marked graphs to occurrence nets, which makes analysis much simpler. Particularly, liveness and safety are easily treated: if the source has initially exactly one token, then each transition fires exactly once and eventually the sink is marked. There is never more than one token in a place.

G, H, C} defining a WDN. The other elements will remain unspecified and will constitute the degrees of freedom of the design. The set of user-defined constraints will be denoted R_x, while the complete set of constraints will be denoted **R**.

A feasible structure is a well-defined net that satisfies both the structural and the user-defined constraints. The design problem is to determine the set of all feasible structures corresponding to a specific set of constraints. Note that this approach is not, by design, concerned with the optimal organizational structure, but with the design of a whole family of feasible structures. By assigning attribute values to the tokens of the Petri net and by specifying the contents of the algorithms embedded in the transitions, we can address the performance problem on the basis of which optimal designs can be obtained. At this stage, we are only concerned with the structure and information flows, i.e., the development of the set of feasible organizational forms. This set will become the admissible set in an optimization problem that will consider performance measures.

The notion of subnet defines an order (denoted $\leq$) on the set of all well defined nets of dimension n. The concepts of maximal and minimal elements can therefore be defined. A maximal element of the set of all feasible structures is called a maximally connected organization (MAXO). Similarly, a minimal element is called a minimally connected organization (MINO). Maximally and minimally connected organizations can be interpreted as follows. A MAXO is a well defined net such that it is not possible to add a single link without violating the set of constraints **R**. Similarly, a MINO is a well defined net such that it is not possible to remove a single link without violating the set of constraints **R**. The following proposition is a direct consequence of the definition of maximal and minimal elements: For any given feasible structure P, there is at least one MINO P_{min} and one MAXO P_{max} such that $P_{min} \leq P \leq P_{max}$. Note that the net P need not be a feasible. There is indeed no guarantee that a well-defined net located between a MAXO and a MINO will fulfill the constraints **R**, since such a net need not be connected. To address this problem, the concept of a simple path is used.

Let P be a WDN that satisfies constraint R1 and whose source and sink have been merged together into a single external place. A simple path P is a directed elementary circuit which includes the (merged) source and sink places. The simple paths of a given WDN are themselves well-defined nets.

The following proposition characterizes the set of all feasible organizational structures: P is a feasible structure if and only if P is a union of simple paths, i.e., P is bounded by at least one MINO and one MAXO. Note that in this approach the incremental unit leading from a WDN to its immediate superordinate is a simple path and not an individual link. In generating organizational structures with simple paths, the connectivity constraint R1 is automatically satisfied.

The algorithm developed by Remy & Levis (1988) generates, once the set of constraints R is specified, the MINOs and the MAXOs that characterize the set of ali organizational structures that satisfy the designer's requirements. The next step of the analysis consists of putting the MINOs and the MAXOs in their actual context to give them a physical instantiation. If the organization designer is interested in a particular (MINO, MAXO) pair because it contains interactions that are deemed desirable for the specific application, he can further investigate the intermediate nets by considering the chain of nets that is obtained by adding simple paths to the MINO until the MAXO is reached.

This methodology provides the designer of organizational structures with a rational way to handle a problem whose combinatorial complexity is very large. Having developed a set of organizational structures that meets the set of logical constraints and is, by construction, free of structural problems, we can now address the performance problem with particular focus on modeling cultural differences. Note that the use of Petri nets has ensured that the models are, by definition, executable. Colored Petri Net software such as Design/CPN™, CPN Tools™ and a plethora of university-developed software can be used to implement the Petri Nets.

MODELING CULTURAL ATTRIBUTES (Handley & Levis, 2001)

Up to this point, the attributes that characterized the decision makers – information structures and the algorithms that processed the data – did not include cultural differences. In order to study interactions that also include heterogeneous decision makers, some type of subjective parameters must be included in the interacting decision maker model. These subjective parameters can represent any factor that constrains the way that decision makers interact or process information. For example, the parameters can represent the differences between services of the same country, for example the differences in procedures between US Army officers and the US Navy ones. The parameters could also represent the differences between members of a governmental organization and a non-governmental organization, such as the US Army and Doctors without Borders. In multi-national coalitions, the parameters can represent national differences that have an effect on the performance of decision makers from different countries working together. Hofstede (1991, 2001) distinguishes dimensions of culture that can be used as an instrument to make comparisons between cultures and to cluster cultures according to behavioral characteristics. Culture is not a characteristic of individuals; it encompasses a number of people who have been conditioned by the same education and life experience. Culture, whether it is based on nationality or group membership such as the military, is what the individual members of a group have in common (Mooij, 1998).

To compare cultures, Hofstede differentiated them according to four dimensions: uncertainty avoidance, power distance, masculinity-femininity and individualism-collectivism. The dimensions were measured on an index scale from 0 to 100, although some countries may have a score below 0 or above 100 because they were measured after the original scale was defined in the 70's. The original data were from an extensive IBM database for which 116,000 questionnaires were used in 72 countries and in 20 languages over a six-year period. Two of these dimensions, power distance and uncertainty avoidance, may affect the interconnections between decision makers working together from different national organizations because different perceptions of hierarchy and ambiguity impact the decision making process.

The power distance dimension can be defined as "the extent to which less powerful members of a society accept and expect that power is distributed unequally" (Hofstede. 1991). An organization with a high power distance value will likely have many levels in its hierarchy and convey decisions from the top of the command structure to personnel lower in the command structure; centralized decision making. Organizations with low power distance values are likely to have decentralized decision making characterized by a flatter organizational structure; personnel at all levels can make decisions when unexpected events occur with no time for additional input from above. Power distance has been used in previous studies to characterize multi-cultural work environments (Helmreich et al., 1996). In studies of aviation and medical teams, when members differed in power distance, problems arose that affected the performance of the team. In aviation, the co-pilot must be willing to speak, interrupt, and correct the pilot and the pilot must be willing to listen, reassess, and change based on the co-pilot's input. Similar results were found in medical settings.

Uncertainty avoidance can be defined as "the extent to which people feel threatened by uncertainty and ambiguity and try to avoid these situations" (Hofstede, 1991). An organization which scores high on un- certainty avoidance will have standardized and formal procedures; clearly defined rules are preferred to unstructured situations. In organizations with low scores on uncertainty avoidance, procedures will be less formal and plans will be continually reassessed for needed modifications. Klein et al. (2000) hypothesized that during complex operations, it may not be possible to specify all possible contingencies in advance and to take into account all complicating factors. "Operators must continually reassess ongoing plans for needed modifications of action. Information may be incomplete and inaccurate but may be the best information available at the time. If the decision is postponed, more information may become avail- able, allowing a better decision, but time and opportunity will be lost" (Klein et al., 2000). The trade off between time and accuracy can be used to study the affect of both power distance and uncertainty avoidance in the model.

Messages exchanged between decision makers can be classified according to three different message types: information, control, and command ones. (Zaidi & Levis, 1995). Information messages include inputs, outputs, and data; control messages are the enabling signals for the initiation of a subtask; and command messages affect the choice of subtask or of response. The messages exchanged between decision makers can be classified according to these different types and each message type can be associated with a subjective parameter. For example, uncertainty avoidance can be associated with control signals that are used to initiate subtasks according to a standard operating procedure. A decision maker with high uncertainty avoidance is likely to follow the procedure regardless of circumstances, while a decision maker with low uncertainty avoidance may be more innovative. Power distance can be associated with command signals. A command center with a high power distance value will respond promptly to a command signal, while in a command center with a low power distance value this signal may not always be acted on or be present.

In the proposed model of a decision making organization in which organization members exhibit cultural differences (heterogeneity), when a message is received by a decision maker, first the message type is checked. Based on the message type, the subjective values of both the sending and the receiving decision maker are compared. Differences in these values across command centers can cause communication and coordination difficulties.

Previously, the algorithms used with the Petri Net representation of an organizational structure were based on operational data and procedures; they did not contain any subjective parameters. Subjective parameters can be included in the model by rule sets associated with each organization member. These rules sets can affect the behavior and performance of the decision making organization in specific situations, such as when messages are being exchanged among heterogeneous decision makers in the same or different command centers. These parameters can influence the organizational response. In order to evaluate the effect of these subjective parameters on the decision process and on the organizational performance, it is necessary to enhance the mathematical model of the five-stage decision maker by recasting it as a Colored Petri net.

THE COLORED PETRI NET MODEL OF THE DECISION MAKER

Colored Petri Nets are an extension of Petri nets (Jensen, 1990). Instead of indistinguishable tokens, tokens now carry attributes or colors. Tokens of a specific color can only reside in places that have the same color set associated with them. The requirements to fire a transition are now specified through arc inscriptions; each input arc inscription specifies the number and type of tokens that need to be in the place for the transition to be enabled. Likewise, output arc inscriptions indicate what tokens will be generated in an output place when the transition fires. A global declaration node of the Colored Petri Net contains definitions of all variables, color sets, and domains for the model. Figure 4 shows the top level of the hierarchical, executable model.

The five transitions that represent the individual stages of the five-stage model are compound transitions; each represents a separate page of the model that contains the functionality of that stage. The places are defined by the different types of messages that are associated with the model; the messages have attributes that are represented by the color sets that annotate the places. The first stage, Situation Assessment, compares the basis for the subjective parameters for the sending and receiving decision makers. The model can be configured to use any basis for comparison and to include the corresponding subjective parameters. In this case, the nationality of the two decision makers is compared, and the power

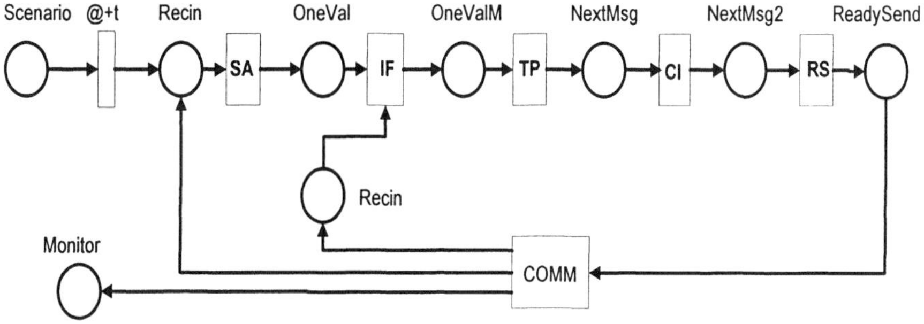

FIGURE 4. Colored Petri Net Representation of Interacting Decision Maker

distance and uncertainty avoidance parameters for each are obtained from a table. To simplify this initial proof-of-concept model, only one set of subjective parameters is retained, based on the incoming message type. If the message type is `command', then the power distance parameters are retained, if the message type is `control', then the uncertainty avoidance parameters are retained.

In order to compare the subjective parameters, a breakpoint must be determined based on the scenario to distinguish low versus high values. In this case a value of 50 was used; 50 is the median value of the Hofstede index. Values below this are considered low (L) and values above this are considered high (H). If both parameters are high, or if both are low, the model's internal parameter is set to high or low, respectively, indicating compatibility between the two decision makers. However, if there is a disagreement in the values, i.e., one value is high and the other is low, the internal parameter is set to undetermined (U), indicating there may be incompatibility on this parameter.

The second stage, the Information Fusion stage, is used when a decision maker carries out information fusion, i.e., he is a fusion node. This occurs when a control signal is required in order to continue processing the task. If the internal parameter for uncertainty avoidance is high, then the decision maker will wait for the control signal before continuing his processing. If however, the internal parameter for uncertainty avoidance is low, the decision maker will proceed without waiting for the control signal in case it has not arrived yet. If this occurs, the accuracy value for this task is decreased by one. If the uncertainty avoidance parameter is undetermined, a toggle is used to decide if the decision maker waits for the control signal before proceeding.

In the Task Processing stage the input message is processed and the response message is generated. If the message requires a degree of redundancy, then more than one response message may be generated. A delay of one time unit is incurred to process this stage of the task and the accuracy value is increased by one to indicate that this step is complete. The following stage, Command Interpretation, is active when the message type is command and the internal parameter, representing power distance, is high. In this case the model incurs a delay of one time unit, representing command confirmation. In the other cases, if the internal parameter is low or the message type is not command, then this stage is a pass through and no delay is incurred. If the internal parameter setting is undetermined, then a toggle is used to determine if a delay is incurred.

The final stage is the Response Selection stage, which prepares the response message to be sent. First the next decision maker is chosen to whom the response message is to be sent. This is done by checking the next billet required in the task processing. Then the nationality of the sending decision maker is appended to the message; this will be used by the receiving decision maker in his Situation Assessment stage. The message is now ready to be sent.

The communications transition (COMM) contains the logic to route messages between decision makers until a task is complete. The input to the communications transition is the output of the decision maker's Response Selection stage. The message is routed to the next decision maker's Situation Assessment stage. When a task is finished, an output is delivered to the message monitor place where the task identification code, the task accuracy, and the task processing time will be used to score the performance of the coalition in completing the task.

SIMULATING ORGANIZATIONAL PERFORMANCE

The consequences of cultural differences in interactions among culturally different decision makers can be illustrated by conducting a virtual experiment. An experimental design was created which simulates the coalition model under different levels of heterogeneity. The organizational design used to populate the coalition model and the task graphs used to create the input scenario were extrapolated from a scenario used for coalition research. This prototype scenario was developed by SPAWAR Systems Center - San Diego to support the Operations Planning Team of the Commander, United States Pacific Command. It is an Indonesian "Rebel Territory" scenario and has been developed to provide a context for development and demonstration of decision support tools and products for coalition operations (Heacox, 1999).

The scenario depicts a situation where growing tensions among multiple ethnic groups has led to armed conflict between a rebel militia group and the host country's military. The rebel group has fled to an enclave of land on the eastern portion of the island nation and has detained a large number of citizens within the rebel-secured territory. Many of these citizens are unsympathetic to the rebels and are considered to be at risk. The host government recognizes that they are unable to maintain peace and that the tide of world opinion has turned against them; the government then asks the US to lead the anticipated coalition operation in an effort ensure aid is delivered to the rebel-secured territory where the food and water supply and sanitation facilities are limited.

When a coalition is composed of one lead country with the majority of the forces, the coalition is not very heterogeneous and its command center is not very integrated. On the other hand, when the coalition is composed of many, equally represented organizations, the coalition is very heterogeneous. In order to represent varying degrees of heterogeneity, three participating coalition partners were chosen based on the coalition scenario: the United States (USA or U), Australia (AUS or A) and the Republic of Korea (ROK or K). The command center billets were assigned as indicated in Table 1, where each column represents a different experimental condition. The scenario is first completed by a homogeneous command centers where there should be no effect of heterogeneous interactions. Then, the coalition command center is equally integrated by assigning billets alternately to officers from two countries. Finally the command center is integrated equally among the three countries present in the scenario. Note that permutations of the order of the mappings did not produce significant differences in performance due to the high interaction of all billets.

The input scenario used in the experiment was modeled after the task blocks sections "Situation Assessment and Preparation" and "Coordinate Operations Across Sectors". Sixteen tasks were identified that require a series of messages among the command center decision makers until an output is generated back to the operation lead. They are composed of command (orders) and control (initiation) messages.

For this experimental design, the independent variables are the organizational configuration, the mapping of command center billets to decision makers of a specific coalition partner, and the scenario used to stimulate the model, the series of tasks to be performed. The dependent variable, the variable to be observed and measured, is the performance of the coalition command center completing the tasks. Monitoring the performance of the command center provides the ability to compare or judge the different levels of heterogeneity. Table 2 indicates the Hofstede values for the scenario countries that were included as subjective parameters. Two measures, timeliness and accuracy, were used to evaluate the coalition output.

Timeliness expresses an organization's ability to respond to an incoming task within an allotted time. The allotted time is the time interval over which the output produced by the organization is effective in its environment. This allotted time can be described as a window of opportunity whose parameters are determined a priori by the requirements of the task. Different task types may have different windows of opportunity. Two quantities are needed to specify the window of opportunity: the lower and the upper bounds of the time interval, t_s and t_f, respectively, or one of the bounds and the length of the interval, e.g. t_s and Δt (Cothier & Levis, 1986). The timeliness of each coalition output was scored based on the task's window of

Country	USA	AUS	ROK	U-A	U-R	A-R	U-A-R
Billet							
CC	USA	AUS	ROK	USA	USA	AUS	USA
CCTF	USA	AUS	ROK	AUS	ROK	ROK	AUS
PSAT	USA	AUS	ROK	USA	USA	AUS	ROK
SJA	USA	AUS	ROK	AUS	ROK	ROK	USA
PAO	USA	AUS	ROK	USA	USA	AUS	AUS
J1	USA	AUS	ROK	AUS	ROK	ROK	ROK
J3	USA	AUS	ROK	USA	USA	AUS	USA
J4	USA	AUS	ROK	AUS	ROK	ROK	AUS
J6	USA	AUS	ROK	USA	USA	AUS	ROK
CMOC	USA	AUS	ROK	AUS	ROK	ROK	USA
NAVFOR	USA	AUS	ROK	USA	USA	AUS	AUS
TRNBN	USA	AUS	ROK	AUS	ROK	ROK	ROK
DJTFAC	USA	AUS	ROK	USA	USA	AUS	USA
AMC	USA	AUS	ROK	AUS	ROK	ROK	AUS
CAO	USA	AUS	ROK	USA	USA	AUS	ROK

TABLE 1. Seven Experimental conditions based on billet assignments

Country	Power Distance	Uncertainty Avoidance
USA	40	46
AUS	36	51
ROK	60	80

TABLE 2. The Hofstede values for the three coalition partners

opportunity; if the response was within the window, it was given a score of two; if it was on the boundary, it was given a score of one; otherwise it received a score of zero.

Similarly, accuracy expresses an organization's ability to make a correct response to an incoming task. The accuracy for each task can be described as an interval that contains the correct response plus or minus a margin of error within which the response is still acceptable. The accuracy value of each coalition output was scored based on the accuracy interval determined a priori for each task. If the value was within the range it was given a score of two, if it was on the limits of the range, it was given a score of one, and otherwise, it received a score of zero.

Timeliness identifies weaknesses caused by delayed information; organizations that have high uncertainty avoidance and high power distance are hypothesized to score lower on timeliness. Accuracy identifies weaknesses caused by incomplete information; organizations that have low uncertainty avoidance and low power distance are hypothesized to score lower on accuracy.

Figure 5 shows the results of the virtual experiment. The results illustrate the trade off between timeliness and accuracy for the different combinations of nationalities. USA and AUS are similar on power distance but differ in uncertainty avoidance. This shows a similar timeliness score, but a difference in the accuracy.

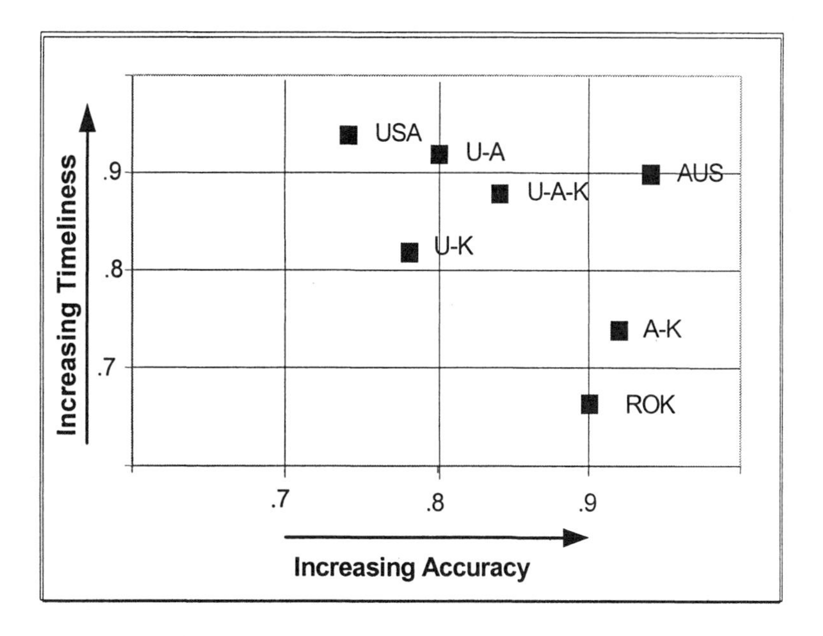

FIGURE 5. Results of Seven Experimental Conditions

AUS and ROK are different on power distance but the same on uncertainty avoidance, thus the accuracy is similar but the timeliness is different. USA and ROK are different on both parameters and this is reflected in their scores. Perhaps the most interesting is the triple combination (UAK), where neither score is maximized or minimized.

This was purely a proof-of-concept experiment. Its purpose was to investigate whether cultural differences could be added to organizations constructed using the five-stage decision maker model. The computational experiment showed that cultural difference parameters that characterize the interactions among decision makers in an organization such as a command center do make a difference in the performance metrics of timeliness and accuracy. By running a virtual experiment using the coalition scenario, the effect of heterogeneity within a coalition command center could be explored.

CONCLUSIONS

The model used in this virtual experiment is a generic model of information processing and decision making organizations which can be populated with data to reflect any type of military or non-military command center, and include any subjective parameters that can be assessed on the different entities in order to explore other effects of heterogeneity. This approach could also be applied to non-military, multinational joint ventures where decision makers from different, distinct organizations are interacting, either directly in a command center type setting such as an executive meeting, or in a more dispersed venue such as through networked exchanges. Other performance measures can also be used that examine other aspects of coalition performance. The effect of interactions among decision makers in the coalition command center is the first step in designing adaptive coalition command centers. One of the characteristics of military coalition command centers is that the staff only serves for finite periods of time and its members are rotated out. Consequently, there is need to develop a design technique, a generalization of the approach introduced in this chapter, that will allow the design of coalition command centers that are adaptive to the changes caused by cultural differences among the decision makers. Since it was shown that cultural difference parameters affect performance, it follows that such parameters should be included in the design algorithms: a generalization of the Lattice algorithm of Remy & Levis (1988) appears to be a promising area of research. Furthermore, computational constraints present at the time the algorithm was developed have disappeared and larger, more complex organizations could be designed.

The problem formulated in this chapter is that of highly structured organizations with well-defined organizational processes. The organization's tasks are essentially planning and the monitoring of the execution of the planned operations: it processes information, makes decisions, and issues commands. The modeling of the organization using Colored Petri Nets allows the application of design algorithms, the use of analysis to identify the behavioral properties of the organization, and simulation to obtain performance characteristics. Consequently, it covers one small but significant part of the OrgSim problem space.

REFERENCES

Boettcher, K. L., & Levis, A. H. (1982). Modeling the interacting decision maker with bounded rationality. IEEE Trans. on Systems, Man, and Cybernetics, SMC-12(3), 334-344.

Cothier, P. H., & Levis, A. H. (1986). Timeliness and measures of effectiveness in command and control. IEEE Trans. on Systems, Man, and Cybernetics, 16(6), 844-853.

Drenick, R. F. (1986). A Mathematical Organization Theory, New York: North-Holland.

Handley, H. A. H. & Levis, A. H. (2001). Incorporating heterogeneity in command center interactions. Information Knowledge Systems Management 2(4).

Handley, H. A. H., Zaidi, Z. R. & Levis, A. H. (1999). The use of simulation models in model driven experimentation. Systems Engineering, 2(2), 108-128.

Heacox, N. J., (1999). Indonesian "rebel territory" scenario for DSSCO (Technical Memorandum). San Diego, CA: Pacific Science and Engineering.

Helmreich, R. L., Merritt, A. C., & Sherman, P. J. (1996). Human Factors and National Culture. ICAO Journal 51(8), 14-16.

Hofstede, G. (2001). Culture's Consequences: Comparing Values, Behaviors, Institutions, and Organizations Across Nations, 2nd Edition, Thousand Oaks, CA: Sage Publications.

Jensen, K. (1990). Coloured Petri Nets, Vols. I, II, and III, Berlin: Springer-Verlag.

Klein, H. A., Pongonis, A., & Klein, G. (2000). Cultural barriers to multinational C2 decision making. Proc. 2000 Command and Control Research Symposium, Monterey, CA.

Levis, A. H. (1988). Quantitative models of organizational information structures. Iin A. P. Sage, (Ed.) Concise Encyclopedia of Information Processing in Systems and Organizations, Oxford: Pergamon Books Ltd.

Levis, A. H. (1993). A Colored Petri Net Model of Command and Control Nodes. In Carl R. Jones (Ed.), Toward a Science of Command Control and Communications, Washington, DC: AIAA Press.

Levis, A. H. (1995). "Human Interaction with Decision Aids: A Mathematical Approach. In W. B. Rouse (Ed.) Human/Technology Interaction in Complex Systems, 7. Greenwich, CT: JAI Press.

Levis, A. H., & Vaughan, W. S. (1999). Model Driven Experimentation. Systems Engineering. 2(2), 62-68.

Marschak, J. & Radner, R. (1972). Economic Theory of Teams, New London, CT: Yale Univ. Press.

Mooij, M. (1998). Global Marketing and Advertising: Understanding Cultural Paradoxes. Thousand Oaks, CA: Sage Publications.

Perdu, D. M., & Levis, A. H. (1998). Adaptation as a morphing process: A methodology for the design and evaluation of adaptive organization structures. Computational and Mathematical Organization Theory 4(1), 5-41.

Remy, P., & Levis, A. H. (1988). On the generation of organizational architectures using Petri nets. In G. Rozenberg (Ed) Advances in Petri Nets 1988, Berlin: Springer-Verlag

Simon, H. A. (1982). Models of Bounded Rationality. Cambridge, MA: MIT Press.

Wagenhals, L. W., & Levis, A. H. (2002). Modeling support of effects-based operations in war games. Proc. 2002 Command and Control Research and Technology Symposium, Monterey, CA.

Zaidi, A. K., & Levis, A. H. (1995). Algorithmic design of distributed intelligence system architectures. In Gupta, M. M., & Sinha, N. K. (Eds.), Intelligent Control Systems: Theory and Practice, New York, NY: IEEE Press, 101-126.

CHAPTER 14

ORGANIZATIONAL DESIGN AND ASSESSMENT IN CYBER-SPACE

KATHLEEN M. CARLEY

ABSTRACT

Organizations are complex dynamic systems composed of an ecology of networks connecting agents (people and intelligent artificial agents), knowledge, resources, and tasks/projects. Organizational management can be characterized as the science of managing the relationships in these networks so as to meet organizational goals such as high performance, minimal costs, and adaptability. Even the most effective managers often have little understanding of the underlying networks and have access only to out of date information on the state of the organization. Consequently, they rely on experience, hearsay, and tradition in designing teams and assessing the vulnerabilities in the organization, often leading to catastrophic errors. Such catastrophic errors can occur as a management change creates either a less than effective team or destroys the competency of an existing team in forming a new team. These problems are exacerbated for organizations in dynamic environments where there is volatility, changing missions, high turnover, and rapidly advancing technology. Advances in network analysis, multi-agent modeling, data-mining and information capture now open the possibility of effective real time monitoring and design of teams, organizations, and games. This enables reasoning about organizations using up-to-date information on the current state of the organization ("actual" or gamed) including who knows who, who knows what, who has done what, what needs to be done and so on. In this chapter, a meta-matrix view of organizations is presented and the process of using measures and tools based on this formulation for team design for "actual" and/or gamed organizations are described and illustrated using data from a dynamic organization. New approaches to assessing organizational vulnerability are described. Finally, the way in which such tools could be linked into existing real time data streams and the values of such linkage are discussed both for actual and gamed organizations.

THE NATURE OF ORGANIZATIONS

The industrial revolution enabled organizations to increase in size, number of divisions (Etzioni, 1964; Fligstein, 1985), level of bureaucracy (Weber, 1947), and level of hierarchy (Blau and Scott, 1962). This increased the demand for information processing. The computer revolution enabled organizations to effectively downsize or outsource (DiMartino and Wirth, 1990), network (Nohira and Eccles, 1992; Miles and Snow, 1995) and form virtual organizations (Lipnack and Stamps, 1997) thus altering where, when and by who work was done. At the same time, changes in the economic climate and longevity increased the likelihood of individuals following several careers, changing jobs, and moving to new locations. Other technological changes enabled the formation of rapidly changing markets, new products, and altering user demand. The upshot is a more complex highly-informated system that needs to adapt rapidly and economically to a volatile and changing environment in situations where there many be insufficient human memory to evaluate possible changes due to personnel transitions.

History has shown that there is no one right organizational design (Mintzberg, 1983, Burton and Obel, 1984). Today, as we move to ubiquitous computing and the world of intelligent agents, previous mechanisms of coordination are increasingly less relevant. Indeed, there can be little doubt that technology and organizations are co-evolving producing new forms of coordination and control. Artificial "smart agents" such as WebBot, robots, e-negotiators, and electronic shoppers are joining humans and groups in the ranks of the smart agents that "work" in and among organizations altering the tempo, quality, and type of work being done (Carley, 2002). Computers are coming to coordinate, control, or participate in the operation of everything from the office and home environment to routine purchases to strategic organizational decisions. As computing becomes ubiquitous with computational capabilities embedded in every device, from pens to microwaves to walls, the spaces around us become intelligent (Nixon, Lacey and Dobson, 1999; Thomas and Gellersen, 2000). Intelligent spaces are characterized by the potential for ubiquitous access to and provision of information and information processing among potentially unbounded networks of intelligent adaptive agents (Kurzweil, 1988) often organized in teams. These teams:

- Might be distributed in space, time and social space, thus increasing reliance on information and telecommunication technology for a) communication, b) reduction in communication barriers, c) coordination, and d) extended team memory.

- Are increasingly distributed across organizational boundaries (due to open source code development, outsourcing, and cross corporate projects) such that many agents working together have a) never previously worked together, b) know their co-workers socially or c) speak the same language.

- Have dynamically changing personnel resulting in possible performance decrements due to the a) exodus of expertise, b) on-the fly training for new members, and c) substitution of artificial agents for humans. Personnel changes, when due to involuntary turnover, can also reduce motivation, inhibit recruitment, exacerbate errors due to the need to use partially trained individuals, increase training costs and increase organizational risk.

- Are strongly instrumented so that humans and technology, including adaptive artificial agents, co-exist and performance is dependent on the integration of the human, intelligent agents and technological systems into an adaptive and evolving system.
- Depend on change and search mechanisms, individual expertise, collective knowledge, and the group's structure to effect performance.
- And can use telecommunication and computational tools to increase the overall effectiveness of the organization by improving the knowledge, communication, "memory", learning, and decision making of people (and their intelligent assistants) in the organization.

The nature of the modern organization poses numerous challenges for organizational effectiveness and risk free operation. We are beginning to have an understanding of how to coordinate, manage, facilitate, or inhibit the performance and adaptability of organizations in which humans and artificial agents work side-by-side. Most of this understanding has been assisted by the use of multi-agent simulation models of organizations (Carley & Gasser, 1999). Nevertheless, this understanding suggests that changes over the past 50 years, and expected changes in the next 10 are producing a management nightmare. There is no one right, or even a small set of right, organizational designs. Even if there were, managers often don't have the experience or up-to-date knowledge of their staff to determine which design is best for the current situation. Finally, even if the manager had such information and experience, the environment is changing so rapidly that the organization needs to be continually redesigned to meet the changing demands. Further, unassisted human minds are just not capable of reasoning about complex non-linear adaptive systems that are fundamental to the nature of the modern organization (Carley, 2001).

Today, collaborative computer games provide both similar challenges and a managerial opportunity. On the one hand, large-scale games in which there are multiple players coordinating as teams to perform various tasks have the same order of complexity as "actual" (that is, non gamed) organizations. The design of such games and the management of the teams in such games are in themselves organizational design challenges similar in kind to those faced by management in "actual" organizations. As such, tools and techniques for designing and managing "actual" organizations are equally valuable in designing and managing gamed organizations. On the other hand, these large-scale collaborative games are a virtual laboratory for reasoning about organizational performance, adaptation, and the impact of new technology. Such games enable the emulation of the social-historical-technological and task environment that has, is, or will be present in an "actual" organization, while still taking advantage of the richness, emotiveness, and innovativeness of the human response to new situations. Such games, as virtual laboratories, become a tool for testing out new coordination schemes and managerial practices. In this context, tools and techniques for designing and managing "actual" and gamed organizations can be evaluated, tested, and improved in this gaming context.

How and when organizations should change, how they will respond to new technology, which organizational personnel should be targeted as early users for new technology, and so on are only some of the questions that managers in actual and "gamed" organizations need to be able to reason about. To answer such questions the manager needs to understand the current organizational design and be able to evaluate possible changes and reduce the negative surprises associated with such changes. In other words, the manager needs a computational toolkit attached to organizational data-streams providing the manager with up-to-date accurate information with associated tools. Such streaming data is even more present and available in gaming environments that in "actual" organizations. Regardless, the organizational design and management toolkit should include tools for monitoring various factors influencing performance such as the social and knowledge networks, project scheduling, database usage and updating, errors and exceptions. In addition there should be situation assessment tools for doing knowledge and risk audits of the organization. And there should be decision aids for doing what if analyses on the potential social and organizational risks and impacts of changes such as new policies, downsizing, hiring, new task assignments, and deployment of new technologies. This chapter lays out a vision for how this can be done and some of the tools and approaches that could form key parts of this toolkit for both "actual" and gamed organizations.

ORGANIZATIONS AS META-MATRICES

The organization, "actual" or gamed, can be usefully characterized as a set of interlocked networks connecting entities such as people, knowledge resources, tasks and groups. These interlocked networks can be represented using the meta-matrix conceptual framework (Table 1) (see also, Krackhardt and Carley, 1998; Carley and Hill, 2001; Carley, 2002). The meta-matrix can be thought of as a conceptual description of the organization and as an ontology for characterizing key organizational entities and the relations among them. There are several key points.

First, in defining a meta-matrix a set of entities are identified. Here, Table 1, these entities are agents, knowledge, resources, tasks and groups. One might imagine more or fewer entities; however, this set has shown to be useful in characterizing teams and organizations in both "actual" and gamed scenarios[1].

[1] In general here, game refers to an on-line game, typically a massively multi-player game or a specifically designed organizational game. For a history of on-line games see King (2002).

	Agents	Groups	Knowledge	Resources	Tasks
Agents	Social Network *Who talks to, works with, and reports to whom*	Group Membership *Who is in which group or groups, or project team or teams*	Knowledge Network *Who knows what, has what expertise or skills*	Resource Network *Who has access to or can use which resource*	Assignment Network *Who is assigned to which task or project, who does what*
Groups		Inter-Group Network *Which groups work with or depend on which*	Knowledge Competency Network *What knowledge is situated in which group*	Resource Competency Network *What resources are associated in which groups*	Divisions *What tasks are done in which groups*
Knowledge			Information Network *Connections among types of knowledge, mental models*	Resource Usage Requirements *What type of knowledge is needed to use that resource*	Knowledge Requirements What type of knowledge is needed for that task or project
Resources				Inter-operability and Co-usage Requirements *Connections among resources, substitutions*	Resource Requirements *What type of resources are needed for that task or project*
Tasks					Precedence and Dependencies - Work Flow *Which tasks are related to which*

TABLE 1. Meta-Matrix Showing Networks of Relations Connecting Key Entities

The term "agent" is used with intention in Table 1. This entity is appropriately thought of as actors with information processing capabilities. As such, it can include corporate personnel, game players, and artificial agents (such as robots or intelligent databases) or actors as in simulated opponents in games. The term "tasks" is a generic placeholder that can be equivalently thought of as events or projects.

Second, in defining a meta-matrix the relations among the entities are identified. Here, Table 1, an illustrative relation is denoted between each pair of entities. One might imagine more or fewer relations depending on the relevant context. For example, in many organizations it is worth considering, as in the agent-x-agent cell, both the authority or reporting structure (formal organizational structure) and the social connections used for advice or friendship (informal organizational structure).

Third, each cell in the meta-matrix described in Table 1 can be instantiated not only in multiple ways, but flexibly. To begin with, the relations can have attributes indicating the strength, frequency or existence of the associated connection among the entities. In addition, we expect the relations embedded in this meta-matrix to be dynamic, changing with changes in personnel, mission (and so tasks) and technology. For example, as new technologies are deployed there will be new areas of knowledge. As another example, personnel turnover impacts the entire first row of the meta-matrix.

Fourth, each individual has his or her own perception of this meta-matrix. This perception, which includes my knowledge of who knows who, who knows what, and so on, can be usefully thought of as the individual's transactive memory. Organizational theorists might usefully think of this as an operationalization of the agent's social capital.

Finally, the social knowledge of the group can be measured as the lossy[2] intersection of the group members' perceptions of the meta-matrices (Carley, 1997). Further, due to the development of social knowledge there is in effect a meta-matrix at each organizational level. That is, there is a view from each individual. In addition, there is an integrated view across all individuals. There is a view "held" by the group, which is itself an intelligent adaptive agent (Carley, 1999).

Thus, the meta-matrix is an extensible flexible structure with multiple instantiations across time and personnel. As such, it is actually a hypercube. As an aside, the reader should note that although it is valuable to characterize the organization as a hypercube, few current visualization tools or organizational metrics can handle hypercube data. This is illustrated graphically in Figure 1.

This meta-matrix serves as an integrating feature of a managerial toolkit. The meta-matrix serves several purposes:

- It provides a way of conceptualizing the set of entities and relations among them that the research and associated tools will focus on

- It brings to the forefront the recognition that the data that is collecting will be not just the attributes of the entities (people, knowledge, resources, tasks and/or projects, and groups or teams) but also the set of relations or ties among them

- It provides an identification of the class of entities and relations that will be used in doing organizational design, analysis and risk evaluation

- It provides a common ontology for talking about and representing organizational information.

---

[2] A lossy intersection is one in which an intersection is formed based on a certain fraction of the items. Imagine that we have three sets (1,2,3) (2,3,4) and (1,4,5) and we use a .67 intersection rule. In this case the TRUE or normal intersection is an empty set; whereas, a lossy intersection at .67 is the set (1,2,3,4).

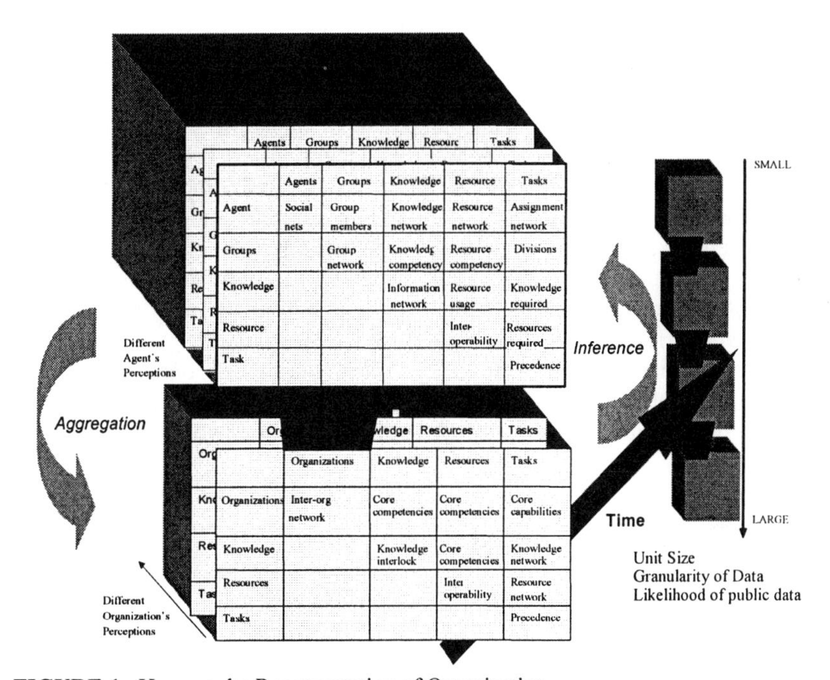

FIGURE 1. Hyper-cube Representation of Organization

The power of this meta-matrix conceptualization comes, in part, from the fact that it integrates organizational design (Table 1, row 1 – agents to “x”), state of art (Table 1, information network, inter-operability), and mission needs (Table 1, resource usage requirements, knowledge requirements, resource requirements, task precedence) into a single conceptual framework for monitoring, assessing and designing teams.

ILLUSTRATIVE ORGANIZATIONS.

Such data characterize the organization’s design and can be used to assess the organization’s human-based vulnerabilities and strengths. One tool for doing this is ORA (Carley & Reminga, 2004)[3]. An organizational can be designed specifying the number of each entity (e.g., number of people and resources) and the relations among the specified entities. Given meta-matrix data, the manager can plan for the future, design new teams, and evaluate the potential impact of new technology using various “*What-if?*” exercises (often referred to as virtual experiments). The

[3] Available at CMU vai http://www.casos.cs.cmu.edu/projects/ORA/index.html

game designer or manager can also design teams or evaluate emergent teams on the fly using this data.

To illustrate these points a data sample consisting of three illustrative and hypothetical organizations will be used. The format for the following data is provided first in Table 2-format. Then, in tables 2-A, 2-B and 2-C the meta-matrices for three different organizational structures are shown. Note, in the three designs (2A, 2B and 2C) the relationships identified in Table 1 have been formalized as a set of binary relations. This is an overall simplification for the sake of illustration. Typically the data might be non-binary and there might even be multiple networks per cell. Since these are extremely small organizations, each with 6 personnel, without loss of generality we define all individuals as being in a single group and so suppress the group column. Further, groups, as emergent phenomena, are derivable from the remainder of the meta-matrix.

Before continuing, let us look at one of these organizations, organization A, in more detail. In Table 2a, 10 cells are shown. In Figure 2, the equivalent graph is shown. Note that the information network is symmetric (see two headed arrows in Figure 2).

	Agents	**Knowledge**	**Resources**	**Tasks**
Agents	Social Network *A 1 indicates that row agent sends messages to column agent*	Knowledge Network *A 1 indicates that row agent has knowledge associated with that column*	Resource Network *A 1 indicates that row agent can use the resource associated with that column*	Assignment Network *A 1 indicates that row agent is assigned to the task associated with that column*
Knowledge		Information Network *A 1 indicates that row knowledge is linked to column knowledge*	Resource Usage Requirements *A 1 indicates that row knowledge is required to use column resource*	Knowledge Requirements *A 1 indicates that row knowledge is required to complete column task*
Resources			Interoperability *A 1 indicates that row resource can be substituted for column resource*	Resource Requirements *A 1 indicates that row resource is required to complete column task*
Tasks				Precedence *A 1 indicates that row task must be finished before column task can begin*

Table 2. Description of Format of Meta-Matrices Shown in Tables 2A, 2B and 2C

	Agents	Knowledge	Resources	Tasks
Agents	0 0 0 0 0 0 1 0 0 0 0 0 0 1 0 0 0 0 0 1 0 0 0 0 0 0 1 0 0 0 0 0 0 1 0 0	1 1 1 1 1 1 1 1 1 1 1 1 0 0 0 0 0 0 0 0 0 1 1 1 0 0 0 0 0 0 0 0 0 1 1 1 0 0 0 0 1 0 0 0 0 0 0 0 0 0 0 0 1 0	1 1 1 1 1 1 0 0 0 0 0 1 0 0 0 0 0 1 0 0 0 0 0 1 0 0 0 0 0 1	0 0 0 0 0 1 1 0 0 0 0 0 0 1 0 0 0 0 0 0 1 0 0 0 0 0 0 1 0 0 0 0 0 0 1 0
Knowledge		0 1 1 0 0 0 0 0 0 1 0 1 0 0 0 0 0 0 1 1 0 0 0 0 0 0 0 0 0 0 0 1 1 0 0 0 0 0 0 1 0 1 0 0 0 0 0 0 1 1 0 0 0 0 0 0 0 0 0 0 0 1 1 0 0 0 0 0 0 1 0 1 0 0 0 0 0 0 1 1 0	1 0 0 0 0 1 0 0 0 0 1 1 0 0 0 0 1 1 0 0 0 0 1 1 0 0 0 0 1 0 0 0 0 0 1 0 0 0 0 1 0 1 0 0 0	0 0 0 0 0 1 0 0 0 0 0 1 1 0 1 0 0 0 0 1 0 1 0 0 1 0 0 0 0 0 0 1 0 0 0 0 0 0 1 0 0 0 0 0 0 1 0 0 0 0 0 0 1 0
Resources			0 1 0 0 0 0 0 1 0 0 0 0 0 0 0 0 0 0 0 0 0 0 0 0 0	1 0 0 0 0 1 1 1 0 0 0 0 0 1 1 0 0 0 0 0 0 1 1 0 0 0 0 0 1 1
Tasks				0 1 0 0 0 0 0 0 1 0 0 0 0 0 0 1 0 0 0 0 0 0 1 0 0 0 0 0 0 1

TABLE 2A. Meta- Matrix for Organization A

	Agents	Knowledge	Resources	Tasks
Agents	0 0 0 0 0 0 1 0 0 0 0 0 1 0 0 0 0 0 1 0 0 0 0 0 1 0 0 0 0 0 1 0 0 0 0 0	1 1 1 1 1 1 1 1 1 1 1 1 0 0 0 0 0 0 0 0 0 1 1 1 0 0 0 0 0 0 0 0 0 1 1 1 0 0 0 0 1 0 0 0 0 0 0 0 0 0 0 0 1 0	1 1 1 1 1 1 0 0 0 0 0 1 0 0 0 0 0 1 0 0 0 0 0 1 0 0 0 0 0 1	0 0 0 0 0 1 1 0 0 0 0 0 0 1 0 0 0 0 0 0 1 0 0 0 0 0 0 1 0 0 0 0 0 0 1 0
Knowledge		0 1 1 0 0 0 0 0 0 1 0 1 0 0 0 0 0 0 1 1 0 0 0 0 0 0 0 0 0 0 0 1 1 0 0 0 0 0 0 1 0 1 0 0 0 0 0 0 1 1 0 0 0 0 0 0 0 0 0 0 0 1 1 0 0 0 0 0 0 1 0 1 0 0 0 0 0 0 1 1 0	1 0 0 0 0 1 0 0 0 0 1 1 0 0 0 0 1 1 0 0 0 0 1 1 0 0 0 0 1 0 0 0 0 0 1 0 0 0 0 1 0 1 0 0 0	0 0 0 0 0 1 0 0 0 0 0 1 1 0 1 0 0 0 0 1 0 1 0 0 1 0 0 0 0 0 0 1 0 0 0 0 0 0 1 0 0 0 0 0 0 1 0 0 0 0 0 0 1 0
Resources			0 1 0 0 0 0 0 1 0 0 0 0 0 0 0 0 0 0 0 0 0 0 0 0 0	1 0 0 0 0 1 1 1 0 0 0 0 0 1 1 0 0 0 0 0 0 1 1 0 0 0 0 0 1 1
Tasks				0 1 0 0 0 0 0 0 1 0 0 0 0 0 0 1 0 0 0 0 0 0 1 0 0 0 0 0 0 1

TABLE 2B. Meta-Matrix for Organization B

	Agents	Knowledge	Resources	Tasks
Agents	000000 100000 010000 010000 001000 000100	111111111 111000000 000111000 000000111 000010000 000000010	11111 10000 01000 00100 00010 00001	111111 111111 111111 111111 111111 111111
Knowledge		011000000 101000000 110000000 000011000 000101000 000110000 000000011 000000101 000000110	10000 10000 11000 01100 00110 00010 00001 00001 01000	000001 000001 101000 010100 100000 010000 001000 000100 000010
Resources			01000 00100 00000 00000 00000	100001 110000 011000 000110 000011
Tasks				010000 001000 000100 000010 000001

TABLE 2C: Meta-Matrix for Organization C

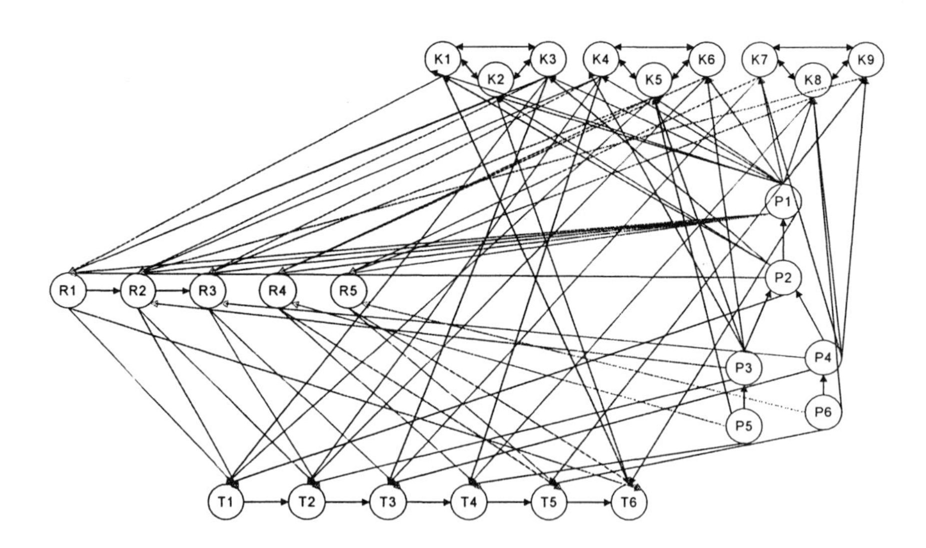

FIGURE 2. Graphical Representation of Organization 2A

It is worth noting that organization B differs from A in having a different social network. A has the classic hierarchical form whereas B has a flatter structure where everyone reports to the same team leader. Similarly, C differs from A only in the task assignment. Where organization A has a unique task assignment for each agent, in organization C all agents work on all tasks. These three designs will be used to illustrate the measures and the proposed approach in the remainder of this chapter.

ASSESSMENT AND DESIGN

Where might a manager or game designer get data like that in Table 2? In point of fact, it can be captured in a number of ways including questionnaires, interviews, observation and so on. However, from a managerial and game design perspective, the optimum approach is to capture the data from live streaming data available in the organization or generated on-line by game players with minimal intervention and minimal effort on the part of the humans in the organizational or the game players. This streaming data goes beyond communication patterns to include various work artifacts such as keystrokes on topics, database entries, billing hours, and so on. This topic of automated data collection will be returned to later. For now, let us assume that such data has been captured. Indeed, we will assume we have captured it as binary data for organizations A, B and C as shown in Tables 2A, 2B, and 2C. This latter assumption is made for the sake of didactic simplicity and we note that in general, the data will not be binary. The data for any single meta-matrix or cube at a specific point in time is simply a snapshot of the organization. In and of itself it is static.

So we have, in Tables 2A, 2B and 2C, snapshots of 3 hypothetical organizations. Analyzing single time period data can provide insight into the causes of current organizational or team performance, vulnerabilities or risks, potential for adaptation, and so on. Such data can be used to assess the current organizational form and to design the future. Such data enables the reconfiguration of teams on the fly as the demand of the environment and changing tasks require. Further, we can compare these snapshots to similarities and differences across organizations.

What Are Key Measures of Vulnerability

Organizational risks and vulnerabilities, from the human or group side, are those structural or psychological attributes that contribute to error, loss of intellectual property, lower performance, reduced adaptability, reduced information gathering and processing capability, and reduced creativity. Even a cursory analysis of the literature reveals a wide variety of measures for assessing organizational risk and vulnerability (e.g., see Wasserman & Fuast, 1994; Ashworth & Carley, 2003;

Thompson, 1967; Borgatti, 2003; Brass, 1984; Carley & Ren, 2001). Such measures vary dramatically in the detail and type of data needed to determine various types of risk and vulnerabilities. These measures span a wide range of vulnerabilities from assessment of who the critical employees are, to the tendency to group think, to the potential for adaptability.

Given the meta-matrix, it is possible to provide a suite of measures and metrics that capture both the organizational design and the possible changes in that design that are likely to result in group think, error cascades, and intellectual property (IP) loss (Carley & Hill, 2001). Given the high potential number of vulnerabilities and risks, what is needed is a framework for evaluating this set of metrics, assessing the value of existing metrics, locating gaps in the existing metrics, developing new metrics as needed, and so providing a more comprehensive guide to which metrics to use when. The ultimate goal would be to develop a tool for organizational risk assessment that is sensitive to variations in the organization design, particularly those centered on the relationships among personnel, knowledge, resources, tasks and groups. As noted earlier, a preliminary version of such a tool exists – ORA (the Organizational Risk Assessment tool).

A large number of metrics for assessing organizational vulnerability and design have been assessed and are being incorporated into ORA. All metrics are based on the meta-matrix and take in to account the relations among agents, knowledge, resources and tasks. These metrics are based on work in social networks, operations research, organization theory, knowledge management, and task management. As metrics are included, care is taken to make them scalable, meaningful, interpretable, and useful for comparing organizational designs. There are a number of techniques that we are using to accomplish these goals. For example, as metrics are incorporated into ORA, if the original version cannot handle binary data, then we are developing a non-binary form of that metric thus extending its capability. As the metrics are added, effort is taken to ensure that the algorithms scale or are at least are using the most optimized algorithm developed to date. As appropriate, both sparse and non-sparse matrix techniques are used thus facilitating data handling. As another example, for most metrics we have created a normalized version that produces values between 0 and 1 to enable comparison across organizational designs. ORA is extensible and there are no limits on the number of nodes of any one type – agents, knowledge, resources, tasks etc – that can be analyzed. However, for some algorithms, the length of processing will vary with the number of nodes. How much so depends on the measure, though most scale as N (the number of nodes) or N^2. Within ORA, help is provided for each metric on how it is measured, what constraints it places on the data, and so on.

In many environments, the manager will want a risk assessment tool that is tailored to the specific corporate concerns. Hence, rather than developing a one-size-fits-all interface we designed ORA in two components the Net-Stat set of metrics and the GUI-Interface. As we go along, we expect that specialized interfaces will be established for the same underlying assessment engine for different applications. For example, one might want a different interface for the intelligence analyst who uses ORA to assess vulnerabilities in covert networks and for the medical CEO who uses ORA to evaluate proposed new organizational designs to reduce medical errors.

ORA can be used to do a risk audit for the organization of its individual and organization risks. Such risks include, but are not limited to, tendency to groupthink, overlooking of information, communication barriers, and critical employees. It evaluates potential organizational risks based upon underlying social, knowledge, resource, and task networks. This tool takes the meta-matrix data at a particular point in time and calculates a series of metrics assessing the team's design, particularly the command and control structure, and the associated organizational risks. ORA has been used to assess risk in, locate more optimized form of, and compare and contrast various organizations in a variety of organizational and government settings including NASA, nursing hospitals, and Joint Task Forces.

The metrics in ORA include social network, task management, and dynamic network metrics. These metrics can be cross-classified by level, type of risk, and data needs. Currently there are two levels of measures: those indicating how a node stands out (node level metrics) and those indicating a characteristic of the entire organization design (graph level metrics). There are seven categories of organizational risk characterized in ORA. These are critical employee, resource allocation, communication, redundancy, task, personnel interaction, and performance. Some metrics are indicative of more than one type of risk. Finally, metrics are cross classified as to whether they need data from one, two or more cells in the meta-matrix. ORA is an extensible system. As the basis for other organizational vulnerabilities are empirically determined, appropriate metrics can be added to ORA. For example, we are currently adding measures to deal with groups or teams within the organization as well as measures appropriate to multiple graphs in the same cell, weighted ties, and so on.

It is beyond the scope of this chapter to present and define all of these measures. Interested readers can turn to Carley and Reminga (2004) for full details. Herein, a small select set of metrics will be described to illustrate the scope and type of vulnerabilities looked at. The metrics described are degree centrality (one of a class of standard social network metrics in ORA) (Wasserman & Faust, 1994), task exclusivity (a multi-cell measure developed to look at organizational risk) (Ashworth & Carley, 2003), cognitive demand[4] (a node level measure derived from the cognitive aspects of workload) (Carley & Reminga, 2004), variance in cognitive demand (a graph level measure), and resource congruence (a graph level measure indicating overall match of the organizational design to its resource needs) (Carley & Reminga, 2004). See Table 3.

---

[4] This measure was previously referred to as cognitive load. The name was changed to more accurately reflect the fact that it was measure what the agent needed to do; i.e., demand. In addition, this avoided confusion with the term workload, which typically reflects physical effort.

Metric	Meaning	Level	Risk	Data Needs
Degree Centrality	In the social network, number of others the person is connected to.	Node	Communication	Single-cell
Task Exclusivity	Detects agents who exclusively perform tasks.	Node	Critical Employee Performance	Single-cell
Cognitive Demand	Measures the total amount of cognitive effort expended by each agent to do its tasks.	Node	Critical Employee	Multi-cell
Variance in Cognitive Demand	Measures the variance in cognitive load.	Graph	Performance	Multi-cell
Resource Congruence	Measures the similarity between what resources are assigned to tasks via agents, and what resources are required to do tasks. Perfect congruence occurs when agents have access to resources when and only when it is needful to complete tasks.	Graph	Resource Allocation, Task	Multi-cell

TABLE 3. Illustrative Metrics

This set of metrics is good for illustration because they cover both standard social network analysis measures (degree centrality) and dynamic network analysis measures (cognitive demand) (Carley, 2003). At least one measure is associated with all entities in the meta-matrix and as such the set shows the power of looking at multi-cell metrics. The measures span a set of different types of human-centered risk that an organization might face. Finally, these metrics have been shown to be related to behavior at the individual, team, and organizational level (Carley & Ren, 2001). In addition to measures of risk or vulnerability are measures of value. Such measures include those for identifying potential change agents (emergent leaders), power brokers, and boundary spanners. Such individuals by virtue of their connection with others are capable of influencing team behavior.

Organizations (or teams) vary on these metrics. In Table 4, we can see how the three hypothetical designs vary. For an organization, a complete analysis of it's meta-matrix using ORA can be thought of as it's design and vulnerability

Metric	Theoretical Min	Theoretical Max	Organization A	Organization B	Organization C
Degree Centrality	0	N	Agent 2 Max 0.30	Agent 1 Max 0.50	Agent 2 Max 0.30
Task Exclusivity	0	1	All equally high Max 1	All equally high Max 1	All equally high Max 0.04
Cognitive Demand	0	1	Agent 1 Max 0.28	Agent 1 Max 0.28	Agent 1 Max 0.50
Variance in Cognitive Demand	0	unbound	0.004	0.004	0.002
Resource Congruence	0	1	0.80	0.80	0.33

TABLE 4. Evaluation of Three Hypothetical Organizations

assessment. Doing such an assessment indicates potential risk factors to which the organizational manager or game designer needs to be alert. Consider risks associated with agents. Imagine the manager finds out, as in organizations A, B and C, that some agent has a much higher degree centrality than others or is higher in task exclusivity. What then? Since individuals who are higher in degree centrality are more likely to get access to and disseminate information, then the high degree centrality person might be someone that the manager wants to secure cooperation from for the dissemination of new technologies. Or if someone is high in task exclusivity, then the organization stands to lose the tacit knowledge held by that person if that person leaves. In this case, the manager, assuming those tasks are likely to remain critical, may wish to find someone to be mentored by the high task exclusivity person. Of course, ORA is only indicating which individuals or groups fall on which dimensions and providing some insight as what this might mean. These are only possibilities; however, not guarantees. Other data should be used to supplement these measures before taking any managerial action.

Now, from a game design perspective, if the designer finds that the proposed organizational design results in an agent having higher than average degree centrality, then the designer can expect that the player in that position may be relied on to send information to the group (or may be instructed to do so). If the design results in a position that has high task exclusivity, the designer may want to add a feature for enabling others to take on that task if that player quits. If one player has excessively high cognitive demand compared to the others, due to the player's structural position in the meta-matrix that might indicate that only experienced players should take on that position, or that the game may be less interesting for other players. In general, pre-assessment of these structural characteristics of individual player positions can provide key guidance as to how the game will play out.

"Actual" and gamed organizations rarely act exactly as expected. Even though one agent should communicate with another due to task dependencies, they may not. Thus, another use for these tools is to contrast the organization as designed and the organization as realized – the game as planned and the game as played. One of the key features of tools like ORA is that they provide a systematic approach for comparison of the designed and the realized.

What Are Key Indicators of Performance and Adaptability

Part of the assessment is to determine how likely it is that a particular organizational design will exhibit the desired level of performance and adaptability. A related issue is estimating how, after a change in the organization's structure, the organization's performance is likely to change. An earlier study by Carley and Ren (2001) found a number of factors that contributed to performance and suggested that there was often a tradeoff between performance and adaptability. For example, organizations in which there was an overall high level of congruence exhibited higher performance; further, a high level of cognitive demand appeared to enable adaptability but at the cost of performance, at least in the short term. Other work has found that high variance in workload and so presumably cognitive demand reduces performance. That is, organizations typically exhibit better performance and have fewer problems with personnel if workload is evenly distributed. Previous work indicated that high performance and adaptive systems tended to exhibit a high level of congruence, or match, between what resources were needed for a task and the availability of those resources (resource congruence) and between who needed to communicate in order to do the task and who actually communicated (Carley, 2002). The factors promoting or inhibiting performance and adaptability are summarized in Table 5. By using ORA to measure the organization's level on these factors, even without data from multiple time periods, general predictions can be made about the ease with which the organization will attain high performance and/or adaptability.

Team Design

A classic question for both managers and collaborative game designers is - are there better designs? That is, is there a better organizational design that, at least in the short run, will lead to better performance or adaptability. There are a number of approaches to answering this question. Two such approaches are optimization and intelligent forecasting. Having meta-matrix type data can support both

Factor	Performance	Adaptability
Need for Negotiation	promotes	promotes
Resource Redundancy	inhibits	promotes
Cognitive Demand	inhibits	promotes
Span of Control	inhibits	inhibits
Size	promotes	promotes
Variance in Cognitive Demand	inhibits	promotes
Resource Congruence	promotes	inhibits

TABLE 5. Factors Enabling Performance and Adaptability

approaches. The first of these, optimization, is supported in ORA, the second is supported in the tool – OrgAhead. This latter approach will be discussed under the team dynamics section.

In the short term, we can think of the mission and technology constrained portions of the organizational design as relatively fixed components of the extant system. In terms of the meta-matrix – this fixed component would be the cells not involving people or groups. In other words, it would include the information network, resource usage, inter-operability and co-usage requirements, knowledge requirements, resource requirements, and precedence and dependencies. These portions of the meta-matrix are fixed in the sense that they represent the current constraints placed by technology, the state of science, fixed costs, or current policy. In contrast, the remainder of the meta-matrix contains relations and entities that are to an extent more malleable. That is – it is "easier" to move people about than create new technology. By viewing the organization's design in terms of this fixed and variable component we open the door for creating and thinking about optimized designs in a systematic fashion.

The organization is optimized if the ties in the variable component of the organization's design are arranged such that they minimize vulnerabilities or maximize congruence. We define a system to have the optimal organizational configuration or design if vulnerabilities due to one or more requisite metrics of concern to the manager are minimized or maximized as needed. There are a number of candidate metrics though which make sense depend on the specific organizational context.

In ORA two heuristic based optimization tools have been implemented – a Monte Carlo procedure and a simulated annealing procedure. There are several advantages to each method. The Monte Carlo method provides a good sampling of the parameter space and theoretically gives the possibility of finding the global optimum. It also allows us to easily keep the logical constraints on the optimization: such as keeping at least one "1" in every row of each of the sub-matrices. Finally, it enables the analyst to simulate the sub-matrices with fixed or randomly distributed densities. Whereas, the disadvantages of the Monte Carlo approach are that it can be slow (if many experiments are selected). Further, because of the random, discrete nature of the search, the global optimum can easily be missed if too few experiments are selected.

In contrast, the simulated annealing method is not "greedy"; i.e., it is not easily fooled by the quick payoff achieved by falling into unfavorable local minima. One of its strengths is that even if it doesn't find the absolutely best solution, it often converges to a solution that is close to the true minimum solution. Finally, it takes less time than Monte Carlo method to get an appropriate solution. Whereas, a disadvantage of the simulated annealing method is that it does not allow one to easily keep the logical constraints on the optimization such as having a 1 in every row of sub-matrices. Further, a generic difficulty in using simulated annealing is that it can be difficult to choose an appropriate rate of cooling and initial temperature for the system that is being optimized. This occurs primarily because of the absence of any rules for selecting them. The selection of these parameters depends on heuristics and varies with the system that is being optimized.

Within ORA, both techniques are available. They can be used to perform single or multi-criteria optimization using either the sum or the product of the measures. The current measures that can be optimized over are:

- Maximizing resource congruence
- Maximizing communication congruence
- Minimizing variation in actual workload
- Minimizing variation in cognitive demand
- Maximizing cognitive demand
- Maximizing resource redundancy

To illustrate the impact of optimization, each of the hypothetical organizations (A, B and C) were optimized for the sum of maximum resource congruence and minimum variation in cognitive demand (Table 6). By maximizing resource congruence the fit of the organization is being improved which should improve performance. Minimizing variation in cognitive demand, allows for individual difference in who is doing what, while setting all effort to be about the same. This should, in an "actual" or gamed organization, improve satisfaction and performance without degrading adaptability or requiring more personnel to get

Metric	Organization A	Organization B	Organization C
Degree Centrality	Agent 6 Max 0.70	Agent 4 Max 0.70	Agent 5 & 6 Max 0.60
Task Exclusivity	Agent 5 Max 0.52	Agent 5 0.52	Agent 5 0.85
Cognitive Load	Agent 3 Max 0.46	Agent 3 Max 0.41	Agent 3 Max 0.57
Variance in Cognitive Load	0.003	0.003	0.008
Resource Congruence	1.0	1.0	1.0
Hamming Distance from Original Design	114	114	119

TABLE 6. Scores for Optimized Organizations

hired or more players needed for the game to achieve the same level of performance.

Examining Table 6, we see that after optimization the three hypothetical organizations are more similar. In part this reflects the fact that they have similar missions in terms of tasks and resources. Further, in all three, resource congruence has increased to its theoretical maximum and in A and B, the variation in cognitive demand has decreased. The internal social structure of these organizations has changed as a result of this optimization. For example, where before agent 1 or 2 was the most connected (highest degree) now the most connected are agents 4, 5, or 6. Further, where before agent 1 had the highest cognitive demand now the highest is agent 3. A limit of this approach is that it does not tell us whether the agents like the change or not. Future work might want to link this to an understanding of personality and differences in the agent's capability to handle work.

It is worth noting that there are potential pathological cases that can arise when doing an optimization, such as eliminating all but one agent. To help eliminate such pathological organizational cases constraints are placed on the optimized form. For example, in this case we have constrained the optimized design such that:

- No agents are added or dropped (thus keeping the number of agents fixed).
- Each agent has at least one area of knowledge (one skill).
- Each agent has at least one resource.
- Each agent is assigned to at least one task.

Once the organization has been optimized one can then ask, how close is the current design to the optimal design? How many changes need to be made? Distance metrics can be used to assess the difference between the original design and the proposed optimized design. Since, in the example, the designs are binary matrices a simple Hamming distance (number of edge differences) is used. More precisely, the distance shown is the Hamming distance divided by the maximum possible distance (number of possible edges, not including diagonal). Note, if there are N personnel, K knowledge areas, R resources and T tasks, then the number of possible edges not including the diagonal is equal to (N+K+R+T)*(N+K+R+T-1) as diagonal linkages are irrelevant. Were the edges weighted, then a Euclidian distance would have been more appropriate. Here we see that organizations A and B are slightly closer to their optimal form than organization C. This should mean that it will require the least effort, though not necessarily the least cost to move from the original design to the optimized one for this organization. As an aside, note, if the object of the game is to find the optimal design, the difficulty of the game can be assessed by the distance from the original configuration to the optimal one.

TEAM DYNAMICS

Just because a particular design is optimal does not mean that the organization will move toward it naturally. In a gaming environment, when players begin in a sub-optimal configuration they may stay in a sub-optimal configuration. Moreover, just because an organization begins with an optimal design does not mean it will stay there. It is worth distinguishing adaptation due to two different causes – factors endemic to the personnel (players) or resources in the organization and managerial controlled (game based intervention) change. The former includes individual learning, individually "chosen" attrition such as dying or moving to a different organization, and resource consumption or growth. In a game setting, this would include players exiting the game permanently or temporarily. In contrast, controlled change includes the hiring, firing and promotion of personnel, re-assignment and re-tasking of personnel, and in the long run changes in mission and so tasks, changes in technology, and so on. In a gaming environment, such controlled change could be designed in exogenous factors (like the chance cards in monopoly), game manager interventions, or predefined game stages.

Were the only change individual learning, then the "actual" or gamed organization would get better and better at the current tasks up to some limit. The limit would not be how good it is possible to do on this task; rather, it would be how good is it possible to do assuming that the entire organizational structure (other than the knowledge network) stayed fixed (Carley & Svoboda, 1996). It is important to note that for different designs, different levels of maximum theoretical performance are possible. That is, there is a structural limit on performance. Up to this limit, agents (that can learn) and the organization will exhibit the typical ojival learning curve. From a gaming perspective, this means that there may be some organizational designs that inhibit the players from achieving a perfect score. In general, for each task and task precedence matrix, the maximum theoretical score for that organizational design needs to be calculated.

Changes in the organizational design, indeed changes in only the variable component of the design can also alter performance. Controlled changes have two effects. First, such changes, since they alter the design of the organization, can alter the structural limit on performance. Second, by altering the design the controlled changes may impact agent strategy by effecting a need to change who reports to who, who does what, who has access to what resources, or are trained to know what. These controlled design changes have the secondary consequence that they impact the effective value of each agent's knowledge enabling the lessons of experience to be used or obviated. This is true at both the task and transactive knowledge level. From a management perspective, obviating the experience of personnel about to retire or who were utilizing obsolete technology may be effective. From a gaming perspective, obviating experience, may re-introduce a level of challenge that some users will find exhilarating and others will find overly complex.

These controlled changes in design can be thought of as structural learning (Carley, 1999). There will be times when structural and individual learning clash leading to major drops in performance (Carley, 1999). From an "actual" organization perspective, such clashes are likely to result in reduction in profits. Further, such clashes may be crippling from a survivability perspective. In a gaming environment, such clashes may alter the game outcome, changing which team is likely to win.

Such clashes are difficult for the manager or game designer to forecast without computational aid. There are numerous reasons, but we will consider only three. First, organizational performance emerges from a complex adaptive system and people are just not good at thinking through the consequence of complex non-linear systems. Second, managers (and people in general) often make their decisions by trying to look ahead, when doing so they rely on information about the current system, such as individual performance. The natural assumption is that tomorrow will be like today, that is, if an individual is a high/low performer they will continue to be so. However, such individual performance is in part a function of the individual's position in the organization. Thus, structural learning will alter that performance despite individual attempts. Moreover, the environment and so tasks to the organization can change thus obviating the value of the look-ahead. Third this look ahead is limited. Consider that in chess even masters rarely look ahead more than a few moves; whereas performance is often based on a longer period of time.

Potential for Dynamic Measures

Multi-time period meta-matrix data is a movie of the observed changes in the organization. Given the frailty of human ability to forecast, computational aids are needed. Data-mining multi-time period data for changes over time can provide guidance for how the organization has and is likely to change. In this approach a sequence of static pictures of the organization – the meta-matrix at multiple periods in time – are collected and analyzed. The advantage of this approach is that it can provide early warning when the organization is beginning to drop in performance or adaptability. The trouble with this approach is that it cannot handle true novelty and it does not lend itself to *what-if* analysis. An alternative approach is to use empirically informed simulation models to conduct a series of *what-if* analyses. Simulation models use as input, not just the current organizational design (the static picture of the organization) but also the set of change-processes that are extant in the organization such as hiring policies. Such models can also take advantage of a sequence of static pictures. The advantage of this approach is that it can suggest potential futures, reduce surprise, and increases the amount of the solution space that is explored. The trouble with this approach is that, given current technology, it is often difficult to provide a detailed explanation of why a particular prediction happens without resorting to the complete model code. It is worth noting that it is even easier to take this approach in a gaming environment where the change rules are more codified. In the gaming context, such pre-simulation can be done either by building synthetic agents to take on the "human" roles, or by building a second simulation of the entire game.

Two such simulation based managerial decision aids are OrgAhead (Carley & Svoboda, 1996) and Construct (Carley, 1990; 1991b)[5]. These are multi-agent network models designed to examine how groups, teams, organizations, networks etc. evolve over time both due to endogenous individual factors and managerial based structural changes. Multi-agent network models are particularly valuable for reasoning about complex adaptive systems such as organizations (Holland 1975; Axelrod and Cohen 1999; Carley and Prietula 1994; Carley 2002). Both OrgAhead and Construct are based on fundamental findings about the nature of individual learning, cognition, and the role of structure in impacting information processing. Both of these tools are interoperable thus enabling the assessment of risk both initially and as the team evolves.

---

[5] Available at CMU via http://www.casos.cs.cmu.edu/projects/OrgAhead/index.html and http://www.casos.cs.cmu.edu/projects/construct/index.htm, respectively.

OrgAhead

OrgAhead (Carley & Svoboda, 1996) is a model of strategic and natural organizational adaptation in which organizations and their members co-evolve. OrgAhead was designed to reflect basic realities of organizational life. That is, in any organization there is a task or set of tasks being done, new tasks arrive over time, a set of personnel, each of whom occupies a particular role in the organization, reports to others, does the tasks and gains experience, and a strategic or management function that tries to anticipate the future, assigns personnel to tasks and determines who reports to whom. In OrgAhead, the organization does a quasi-repetitive classification choice task in a distributed fashion. At the operational level, each individual in the organization works on, at most, part of the task (no individual sees the whole task). Agent learning is thus constrained by the agent's position in the organizational structure.

From a learning perspective, within OrgAhead organizations are characterized at two levels — operational and strategic. At the operational level, the organization is characterized as a collection of intelligent adaptive ACTS agents (Carley & Prietula 1994) engaged in experiential learning; i.e., each agent is cognitively capable, socially situated and engaged in working on a specific task. At the strategic level, the organization is characterized as a purposive actor; i.e., there is a CEO or executive committee which tries to forecast the future and decides how to change the organization to meet that need. At the strategic level, there is structural learning. Each of the three primary components within OrgAhead will be described in turn — the task model, the operational model, and the strategic model.

The basis for OrgAhead is the body of research, both empirical and theoretical, on organizational learning and organizational design. The model has built into it several theories of different aspects of organizational behavior. From the information processing tradition comes a view of organizations as information processors composed of collections of intelligent individuals each of whom is boundedly rational and constrained in actions, access to information by the current organizational design (rules, procedures, authority structure, communication infrastructure, etc.) and by his or her own cognitive capabilities. Organizations are seen as capable of changing their design (Stinchcombe, 1965; DiMaggio & Powell, 1983; Romanelli, 1991), and as needing to change, if they are to adapt to changes in the environment or the available technology (Finne, 1991). Different organizational designs are seen as better suited to some environments or tasks than are others (Lawrence & Lorsch, 1967; Hannan & Freeman, 1977). Aspects of the model have been tuned to reflect the findings of various empirical studies related to these theories. OrgAhead has been validated in a number of settings including hospitals and joint task forces.

Construct

Construct is a multi-agent network model for assessing the evolution, adaptability, and performance of teams where the social and knowledge networks are co-evolving. Construct is designed to capture dynamic behaviors in organizations with different cultural and technological configurations. Construct takes as input the meta-matrix data and then from basic principles of social interaction and task behavior generates expected changes in the associated networks for team design (row 1 in Table 1). In addition, Construct takes as input high-level indicators such as education, tenure and department. Example outputs include changes in the underlying social and knowledge networks, task performance accuracy, task performance efficiency, task performance adaptability, information diffusion, triads, and technology use.

Construct models groups and organizations as complex systems and captures the variability in human and organizational factors through heterogeneity in information processing capabilities, knowledge and resources. The non-linearity of the model generates complex temporal behavior due to dynamic relationships among agents. These dynamic relationships are grounded in symbolic interactionism, information processing, and structuration theory, according to which the social system is continually reconstructed through human interaction based on rules and resources. The changes in the social system are defined and analyzed through the lens of dynamic and social network analysis.

In Construct, the agents communicate, learn, and make decisions in a continuous cycle. Agents select interaction partners based on relative similarity or relative expertise. This selection is dependent upon the perceptions and goals of the individual and the goals and culture of the group. When agents interact they communicate and learn both task knowledge and cognitive knowledge. As they learn their beliefs and perceptions of the world change and they reposition themselves in the network based on these new beliefs and perceptions (reconstruction). Periodically the agents make task-related decisions with the knowledge they have at the time. The knowledge they possess may be sufficient or insufficient for the task at hand as measured by performance.

The Construct model has been scientifically validated several times (Carley, 1990; Carley & Krackhardt, 1996; Carley & Hill, 2001; Schreiber & Carley, 2003). The first validation (Carley, 1990) used Kapferer's Zambia tailor shop data. By combining individual and social considerations, the model was able to predict observed changes in human interactions. The latest validation (Schreiber & Carley, 2003) found significant correlation between communication patterns within real-world organizations and agent interactions within the model. Construct is often used to look at the impact of change in technology.

Construct can be used to estimate projected differences in organizational structures over time under different technological conditions. While there are a number of available measures, two that are particularly valuable are performance and diffusion. One measure of performance is accuracy. In this case, accuracy is measured as the ensemble average of the organization's accuracy on a series of binary choice tasks. In the binary choice task there is a binary string and the organization needs to determine if there are more ones than zeros (Lin & Carley, 2003). What makes this difficult is that not everyone has access to all of the bits in the string. Hence, decisions are limited, as in human organizations, in who knows what. This particular task has been shown to be a reasonable predictor of general performance for organizations in a number of fields as, on average, a large number of the tasks done in organizations are classification choice tasks.

To apply this task in this case, we treat each item of knowledge as a bit in the string. If the agent has that knowledge (a 1 in that location) that means that the agent can "read" that bit in the string. For the bits the agent can see, if the agent sees more 1's than 0's the agent reports to those the agent talks to that the decision should be a 1 else 0. Note, as agents learn they get better at the task. The overall decision is made by combining the individual agent's decisions. Ties are decided at random. For the three hypothetical organizations the results are shown in Figure 3. These results suggest that in fact organization C is designed to have the best performance over time.

Another measure in Construct of particular value is diffusion -- more precisely, the average percentage of people in the group who know a bit of knowledge, averaged across all pieces of knowledge) at a particular time. In Figure 4, the information diffusion results are shown for the three hypothetical organizations. Note that Organization B, which actually exhibited the worst performance, exhibits the fastest rate of diffusion.

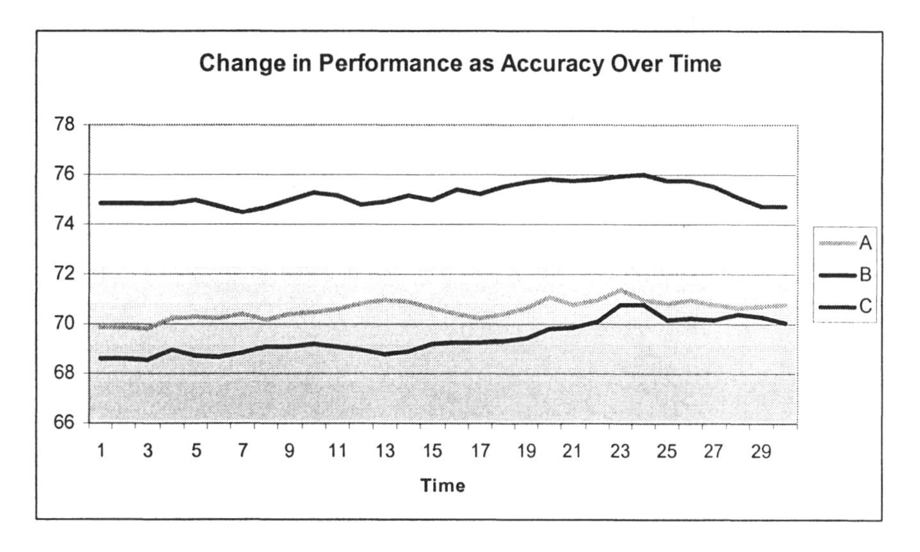

FIGURE 3. Change in Performance Over Time for Three Hypothetical Organizations.

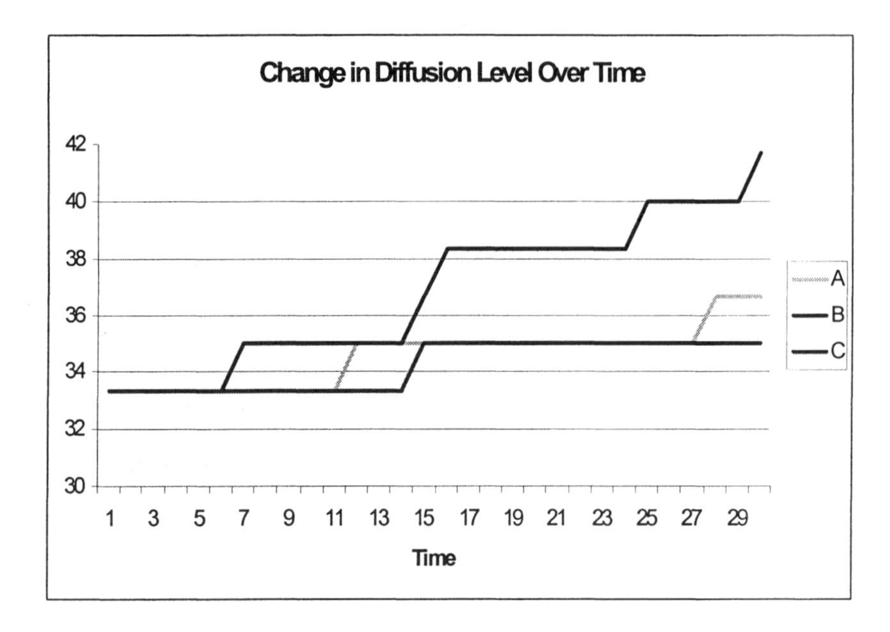

FIGURE 4. Change in the Level of Diffusion Over Time.

Figures 3 and 4 illustrate a key aspect of organizations in particular and complex non-linear systems in general. The point is that complex systems often exhibit unintended consequences due to interactions among the components. In general, in a system where learning leads on average to improved performance on the part of individuals, one would expect it to lead to improved performance on the part of the organization. However, here we see that the team that learns the fastest does not exhibit the best improvement in performance. Why? As individuals learn more, they have, on average a better chance of making the correct decision. However, this is not guaranteed. In particular if the true answer is X and the bit the agent now learns enforces the not-X decision, then the agent's performance is likely to remain the same or degrade. Since, and this is the key point, individuals do not learn randomly, but learn from those with whom they have something in common, incorrect decisions can be enforced, particularly in the short run, as agents continue to send information supporting the wrong decision.

CHALLENGES TO TEAM ASSESSMENT AND DESIGN IN CYBER-SPACE

A key issue is identification of existing data streams for populating the meta-matrix and identifying the "right" level of data compression for risk analysis and team design given usage data (such as blogs, email, web pages, database access hits and so on). Another key issue is determining which of the risks measured by ORA are consistent with or enhance the understanding of managers of the potential vulnerabilities in their organization. It is likely that the relevant set of human-centered risks may be somewhat different depending on the nature of the tasks and resources in the organization. For example, in tasks that require the movement of physical resources, such as building a car or money laundering, an individual who is highly degree central (many people interact with this person on task related issues) is likely to be a bottleneck. Were that person to leave critical skills go with that person and performance should degrade accordingly. In contrast, in a more knowledge intensive task, such as consulting, an individual with high degree centrality is less likely to be a bottleneck and may simply represent someone who likes to talk with others. Were that person to leave, it may be that no critical skills leave and performance may actually improve.

From a gaming perspective, the relevant measures may be still different. There are several factors to consider in this case. First, in massively distributed on-line games it is possible for there to be millions of players. Note, this is 1 to 3 orders of magnitude greater than human organizations. Measures that are interpretable when there are 30, 100, even several thousand agents may have little meaning at this scale, even if they are computable and even if the data can be captured without slowing the game down appreciably or overwhelming available server space. Some studies suggest that standard network measures have little information content and may even be incapable of discriminating between actors for networks of this scale.

These arguments suggest the need for the creation of "measurement sets" – sets of measures that capture the same underlying phenomena but that vary in what type of data is used and the level of fidelity in the data used, but with annotations as to when to use which measure.

What Data Streams are Available

ORA measures could be integrated into control technologies and provided to the team and manager in actual organizations. Tools like Construct and OrgAhead could be linked to these data streams to allow the manager or change management team to examine the potential impact of changes thus enabling the reduction of surprise. This would enable self-monitoring and awareness of potential risks in the organization, resulting in improved performance and adaptability. Such control technologies might include automated keystroke capturers, email, time stamps, project management tools, and so forth.

In a gaming environment, data captured for game management such as communications, player actions, player status, player location, teams the player has worked with, and so on serve as data that can be used by ORA. As previously noted, it is essentially easier to capture such streaming data in a gaming environment. For a game, capturing such data and then analyzing it with ORA provides both a potential source of feedback to the players and feedback to the game designers. Capturing such data, analyzing it, and then providing it in "play-back" mode provides a powerful tool for after-action review and analysis and turns the game in to a training environment.

Overall, this approach to organizational management is potentially quite powerful, but it is data and time intensive. Potentially, gigabytes of data may be collected each day depending on the size of the organization/game and the task/resource complexity. To be effective, we need to be able to collect the data in a non-obtrusive, continual fashion. Moreover, assuming that storage space is not infinite, much of the data will need to be processed on the fly to reduce storage costs. Algorithms for design and analysis of organizations will need to be scalable.

Most organizations collect a number of data streams more or less automatically. To be sure, exactly which streams are collected, how frequently, and in what format varies from organization to organization. Moreover, many organizations implicitly have data on other information that could be collected automatically. By connecting the previously described tools to such data streams we could provide managers with the beginning of a toolkit for real time management not just of tasks, but also of people and organizational risks. By connecting the previously described tools to similar data streams in gaming environments we could provide managers with a virtual test-bed for exploring new organizational designs in a cost-effective and ethical manner.

To illustrate this point a number of potential streams are identified and then listed as to what type of information in the meta-matrix can be derived from which stream or streams in Table 7. This discussion is meant to show the possibilities not to exhaustively enumerate all possible data streams for "actual" organizations. Similar streams for gaming environments are note noted. As a final caveat, new technology will of course make additional data streams available – such as touch sensors on keypads to record typing pattern and body perspiration of the typist as a possible indicator of nerves or health, or mechanism for identifying the user.

What Are The Challenges

There are a number of challenges in moving to this world. These challenges occur at the technical, substantive, human-computer interface and policy level. Only a few of these will be considered. At the technical level, most data streams in organizations contain vast quantities of unstructured data. Collaboration and knowledge management tools would be needed to collect and store such data. The size of the potential data stores, especially when over time data is accumulated, is enormous – requiring larger faster hardware, and faster search mechanisms. However, it may be impractical to store everything thus requiring policies about what to store, for how long, and whom to give access to the data. Additionally, it

Network	Entities	Potential Data Streams
Social Network	People by people	Email, location in floorplan, co-attendance at meetings, chain-of-command, video surveillance of hallways, meeting rooms and lunch rooms.
Knowledge Network	People by knowledge	Web pages, knowledge or skill audit, training records, project role, data base access or contributions, the audio in video surveillance.
Resource Network	People by resource	Resource sign-out or usage records, database access
Assignment Network	People by task/project	Project billing, management plan
Group Membership	People by group	Organizational chart, imputed from other networks
Information Network	Knowledge by knowledge	Percentage of people who have both skills or trained in both areas
Resource Usage Requirements	Knowledge by resource	SME reports, training manual
Knowledge Requirements	Knowledge by task/project	SME reports, % of people with particular skill on that project, project plan, proposals, client requirements
Knowledge Competency	Knowledge by group	Percentage of people in group with that skill, manger characterization
Interoperability	Resource by resource	SME report
Resource Requirements	Resource by task/project	SME reports, percentage of people with that resource, project plan, proposals, client requirements
Resource Competency	Resource by group	Percentage of people in group with that resource, manger characterization
Precedence	Task by task	Gant chart, SME report, project plan, task flow
Governance	Task by group	Organizational chart, charter, SME reports, percentage of people in group assigned to that task
Inter-group Network	Group by group	SME report, organization chart, fraction of social network ties between group members

TABLE 7. Illustrative Data Streams for Capturing Meta-Matrices

may be necessary to develop procedures for automatically determining the "value of old data" in terms of access, technical accuracy, and so on.

Another challenge is automatically creating the list of knowledge areas or skills. Knowledge areas and skills tend to change as organizations evolve, as new technology is created and used, as science progresses, and so on. Automated classification of these areas will require tools for automated or semi-automated ontology creation (Aberer, Cudre-Maurouz & Ouksel, 2004; Agirre, et al. 2004). Knowledge extraction tools, and indeed tools that can extract the entire meta-matrix in an automated or at least semi-automated fashion are needed.

Since data would be coming from live data streams a key issue to address substantively is how to "chunk" the data to contrast previous to current designs. For example, take the issue of when is there a social network tie between two individuals. Does such a tie exist if there is a minimum of some number of communications a day, a week, etc.? Answering this will be a function not only of the substantive issues, but also of managerial needs and privacy considerations. Alternatively, continuous versions of the performance and various individual and organizational risk metrics would need to be created and visualized. Another substantive issue is whether these data streams are accurate enough indicators to get at the data needed for the meta-matrix.

Organizations, both "actual" and gamed, differ, not only in the kinds of tasks they do, but also in culture, the fraction of knowledge work, the tempo of work, and so on. As such, different metrics may be appropriate for assessing vulnerability in different kinds of organizations. For meaningful use, the key metrics central to a particular organization's needs need to be identified and interfaces for these tools tailored to the language of the organization need to be developed. Even as dealing with ledgers led to the development of the field of accounting, dealing with vulnerability audits may lead to a new field of "social accounting." From an individual's point of view this information could be meaningfully employed to help them understand the range of jobs that they are suited for and the path they need to follow to change this. Individuals could engage in *what-if* analyses using the simulations to see if their career goals were in line with the changes being made in the organization. In the gaming environment, the players could use this information to improve game skills.

At the policy level, there are a number of ethical and policy issues. For organizations, understanding the networks in the meta-matrix can facilitate predicting group outcomes such as performance and adaptability. However, when networks within corporations are known it is possible to plan effective take-over strategies, confiscate IP, and plan successful strikes. Individuals can use their networks to locate needed information, pre-determine what products they will like (e.g., when they have similar tastes to their friends), and garner social support which is critical for maintaining health, getting promoted, doing the best possible job, and possible gaining advantage in games. If such data is gathered from email it can be used as part of a system for managing email enable the automatic classification of messages by social group and task; however, such data can be used to explore an individual's personal life in terms of what messages they send to whom on non-work related topics. Managers' knowledge of the social networks can enable them to run more productive organizations with greater satisfaction for the personnel. Game designers' knowledge of the social networks can enable them to develop more effective games. Thus there is a tension between the need to locate networks and the need to selectively reduce the visibility of the networks; between the need to use network information to organization work and the need to retain individual privacy.

Should all of the data collected be shown to all individuals, or should they only see their "ego" network? How can this information be linked to training and personality so as to aid individuals in understanding how to manage change in their lives and attain the job the want? When should knowledge areas, resources and people be de-identified? What does de-identification mean for network data. Which type of data can be used for promotion or demotion? Should a new department be set up in organizations, or should the role of Human Resources be extended, to collect and analyze the data and only report it when there is a problem or answer questions asked by management?

CONCLUSION

In the Gulf War, automated scheduling tools saved almost more money than they cost to develop. Even more money, is spent across the U.S. corporations on re-organizing and recovering from the mismanagement of trying to get new technologies to be adopted than on scheduling. With tools like the ones described in this chapter, managers would be able to pre-assess change, avoid undesirable outcomes, minimize the number of changes, and chart a change path that minimizes cost. Linking these tools to games could reduce costs in training. To do so, a number of challenges in generating real time, dynamic metrics that scale, will need to be met and game designers will need to move beyond key entertainment goals (Mulholland & Hakala, 2002) to include analysis and organizational principles. The boost to the economy and social welfare would be enormous as the billions spent on reorganization and change management are reduced.

ACKNOWLEDGEMENT

This chapter is part of the Organizational Adaptation project in CASOS at CMU. This work was supported in part by the DOD, the Army Research Institute (ARI) DASW01-00-K0018 on Personnel Turnover and the Office of Naval Research (ONR), United States Navy Grant No. N000140210973 on Dynamic Network Analysis under the direction of Rebecca Goolsby and Grant No. N00014970037 on Adaptive Architecture under the direction of Bill Vaughn. Additional support was provided by the NSF IGERT 9972762 for research and training in CASOS and by the center for Computational Analysis of Social and Organizational Systems at Carnegie Mellon University (http://www.casos.cs.cmu.edu). The views and conclusions contained in this document are those of the author and should not be interpreted as representing the official policies, either expressed or implied, of the Department of Defense, the Army Research Institute, the Office of Naval Research, the National Science Foundation or the U.S. government.

REFERENCES

Aberer, K., Cudre-Mauroux, J. P., &. Ouksel, A. M. (2004). Emergent Semantics. Invited paper, The 9th International Conference on Database Systems for Advanced Applications, (DASFAA).

Agirre, E., Ansa, O., Hovy, E., & Martinez, D. (2000). Enriching Very Large Ontologies using the WWW, Workshop on Ontology Construction of the ECAI.

Ashworth, M. & Carley, K. M. (2003). Critical Human Capital. (Working Paper) CASOS, Carnegie Mellon, Pittsburgh PA.

Axelrod, R. & Cohen, M. (1999). Harnessing Complexity. New York: The Free Press.

Blau, P. & Scott, W.R. (1962). Formal Organizations. San Francisco, CA: Chandler.

Borgatti, S. (2003). The Key Player Problem. In R. Breiger, K. Carley, & P. Pattison (Eds.), Dynamic Social Network Modeling and Analysis: Workshop Summary and Papers (pp. 241-252). Committee on Human Factors, National Research Council.

Brass, D. (1984). Being in the right place: A structural analysis of individual influence in an organization. Administrative Science Quarterly. 26, 331-348.

Burton, R.M. and B. Obel, (1984). Designing Efficient Organizations: Modeling and Experimentation. Elsevier Science.

Carley, K.M. & Prietula, M.J. (Eds.). (1994). Computational Organizational Theory. Hillsdale, NJ: Lawrence Erlbaum Associates.

Carley, K.M. (2002). Intra-Organizational Computation and Complexity. In J.A.C. Baum (Eds.) Companion to Organizations. Blackwell Publishers.

Carley, K.M. & Prietula, M. (1994). ACTS Theory: Extending the Model of Bounded Rationality. Iin K.M. Carley & M. Prietula (Eds.) Computational Organization Theory. Hillsdale, NJ: Lawrence Earlbaum Associates.

Carley, K.M. & Reminga, J. (2004). ORA: Organization Risk Analyzer. Carnegie Mellon University, School of Computer Science, Institute for Software Research International, Technical Report CMU-ISRI-04-101.

Carley, K.M. & Gasser, L. (1999). Computational Organization Theory. In G. Weiss (Ed.) Distributed Artificial Intelligence (ch. 7). Cambridge, MA: MIT Press.

Carley, K.M. & Hill, V. (2001). Structural Change and Learning Within Organizations. In A. Lomi & E.R. Larsen Dynamics of Organizations: Computational Modeling and Organizational Theories (ch. 2 pp. 63-92). Boston MA: MIT Press/AAAI Press/Live Oak.

Carley, K.M. & Ren, Y. (2001). Tradeoffs Between Performance and Adaptability for C3I Architectures. In Proceedings of the 2001 Command and Control Research and Technology Symposium, Annapolis, Maryland. Vienna, VA: Evidence Based Research.

Carley, K.M. & Svoboda, D. (1996). Modeling Organizational Adaptation as a Simulated Annealing Process. Sociological Methods and Research. 25(1): 138-168.

Carley, Kathleen, (1990), "Group Stability: A Socio-Cognitive Approach." Pp. 1-44 in Lawler E., Markovsky B., Ridgeway C. & Walker H. (Eds.) Advances in Group Processes: Theory and Research . Vol. VII. Greenwhich, CN: JAI Press.

Carley, Kathleen, (1991b), "A Theory of Group Stability." American Sociological Review ,56(3): 331-354.

Carley, K.M. (1997). Extracting Team Mental Models Through Textual Analysis. Journal of Organizational Behavior. 18: 533-538.

Carley, K.M. (1999). On the Evolution of Social and Organizational Networks. In S.B. Andrews & D. Knoke (Eds.), special issue of Research in the Sociology of Organizations on Networks In and Around Organizations (pp. 3-30). Greenwhich, CN: JAI Press, Inc..

Carley, K.M. (2002). Smart Agents and Organizations of the Future. In L. Lievrouw & S. Livingstone (Eds.), The Handbook of New Media (Ch. 12 pp. 206-220). Thousand Oaks, CA: Sage.

Carley, K.M. (2003). Dynamic Network Analysis. In R. Breiger, K.M. Carley & P. Pattison (Eds.), Dynamic Social Network Modeling and Analysis: Workshop Summary and Papers, Committee on Human Factors (pp. 133-145). National Research Council.

Di Martino, V. & Wirth, L. (1990). Telework: A New Way of Working & Living. International Labour Review, 129(5): 529-554.

DiMaggio, P.J.& Powell, W.W. (1983). The iron cage revisited: institutional isomorphism and collective rationality in organizational fields. American Sociological Review, 48: 147-160.

Etzioni, A. (1964). Modern Organization, Englewood Cliffs, NJ: Prentice-Hall.

Finne, H. (1991). Organizational Adaptation to Changing Contingencies. Futures, 23(10):, 1061-1074.

Fligstein, N. (1985).The spread of the multi-divisional form among large firms, 1919-1979. American Sociological Review, 50: 377-391.

Hannan, M.T. & Freeman, J. (1977). The Population Ecology of Organizations. American Journal of Sociology, 82: 929-64.

Holland, J.H. (1975). Adaptation in Natural and Artificial Systems. Ann Arbor, MI: University of Michigan Press.

King, L. (2002). Game on: The History and Culture of Video Games. New York, NY: St. Martins Press.

Krackhardt, D. & Carley, K.M. (1998). A PCANS Model of Structure in Organization. In Proceedings of the 1998 International Symposium on Command and Control Research and Technology. Monterray, CA. (pp. 113-119). Vienna, VA: Evidence Based Research.

Kurzweil, R. (1988). The Age of Intelligent Machines. Cambridge, MA: MIT Press.

Lawrence, P.R. & Lorsch, J.W. (1967). Organization and Environment: Managing Differentiation and Integration. Boston, MA: Graduate School of Business Administration, Harvard University.

Lin, Zhiang and Kathleen M. Carley, (2003), Designing Stress Resistant Organizations: Computational Theorizing and Crisis Applications, Kluwer Academic Publishers, Boston, MA.

Lipnack, J. & Stamps, J. (1997). Virtual Teams: Reaching Across Space, Time, and Organizations with Technology. New York, NY: John Wiley & Sons.

Miles, R. E. & Snow, C.C. (1995). The new network firm: A spherical structure built on a human investment philosophy. Organizational Dynamics, 23, 5-18.

Mintzberg, H. (1983). Structures in Five: Designing Effective Organization. Prentice Hall Inc.

Mulholland, A. & Hakala, T. (2002). Developer's guide to multiplayer games. http://www.netLibrary.com/urlapi.asp?action=summary&v=1&bookid=70880 An electronic book accessible through the World Wide Web Plano, Tex. : Wordware Pub.

Nixon, P., Lacey, G. & Dobson, S. (Eds.). (1999). Managing Interactions in Smart Environments: 1st International Workshop on Managing Interactions in Smart Environments (MANSE'99). Dublin, Ireland, December .

Nohria, N. & Eccles, R.G. (Eds.). (1992). Networks and Organizations : Structure, Form, and Action. Boston, MA: Harvard Business School Press.

Romanelli, E. (1991). The Evolution of New Organizational Forms. Annual Review of Sociology, 17: 79-103.

Stinchcombe, A. (1965a). Organization-creating Organizations. Trans-actions, 2: 34-35.

Thomas, P. & Gellersen, H.W. (Eds.). (2000). Proceedings of the International Symposium on Handheld and Ubiquitous Computing: Second International Symposium, HUC 2000, Bristol, UK. .

Thompson, J. D. (1967). Organizations in Action. New York, NY: McGraw-Hill.

Wasserman, S. & Faust, K. (1994). Social Network Analysis: Methods and Applications. New York, NY: Cambridge University Press.

Weber, M. (1947). The Theory of Social & Economic Organization, Translated by A.M. Henderson & T. Parsons. New York, NY: The Free Press.

CHAPTER 15

ARTIFICIAL INTELLIGENCE AND ITS APPLICATION TO ORGANIZATIONAL SIMULATION

STEPHEN E. CROSS AND SCOTT FOUSE

ABSTRACT

Artificial Intelligence (AI) has been an exciting and productive field of research for over 50 years. Many useful applications have been realized in a myriad of domains from medical diagnostics, to data analysis in drug research, to detecting and interpreting patterns in marketing data, to automated help in office software applications, to control of household devices. The field has led to many different scientific, philosophical, and engineering views about the possibility and practicality of automating behaviors that seem intelligent when observed in humans. In this chapter, we focus on AI methods and evolving research ideas that can be used to engineer the functionality needed in organizational simulation. We discuss some of the successes of AI, provide an overview of the common methods as well as their limitations and current research to address those limitations, discuss the range of work comprising the field of intelligent agents, and then finally discuss uses of AI in the context of organizational simulation.

INTRODUCTION

In this chapter, we provide a brief overview of the field of artificial intelligence (AI) and discuss several ways that AI methods can contribute to the development of an organizational simulation (OrgSim) capability. We start with definitions of AI and discussion of the goals of AI research, and then briefly review the major methods that have been developed as well as some open research problems, and then discuss the applicability of AI to OrgSim.

Many use the word intelligence when defining AI. This sometimes leads to confusion or ignites intense philosophical debates on whether or not a computer can be as intelligent as a human. For example, Pat Winston (1991) defines AI as "the study of ideas which enable computers to do things which make people seem intelligent." Winston goes on to distinguish between engineering and scientific goals of AI. The engineering goal is to build computers and computer programs that solve hard, real-world problems that are typically solved by humans. The scientific goals are to explain theories of intelligence.

AI is a relatively young field; the term was first coined in 1956, with earlier work evolving from theorem proving research in the 1930s. Turing (1951) proposed a test, the Turing Test, in 1951 as a means to decide if a computer was an intelligent as a human. Basically, the Turing Test involved a human and a computer each trying to convince a judge that they were human. Each communicated with the judge through some medium that disguised the participant's true identify (i.e., teletype). The judge was allowed to ask any question and the participants could respond as deemed appropriate. The judge would conclude that the computer was 'intelligent' if he could not decide which participant was the computer and which was the human.

While Turing's paper did not provide a proof of intelligence, he predicted many of the oppositions to machine-based intelligence and provided much insight into research issues that are still relevant today. A variant of the Turing Test may be applicable to assess the suitability of intelligent software or agents for the task of playing the roles of humans in organization simulation. Such a capability, which will be discussed later in this chapter, would be useful in training humans who have to interact on a team with many other humans. Dedicating the time for humans to help train other humans is often cost prohibitive. In such situations, one would want the automated humans to function in ways that were indistinguishable from the real humans doing similar tasks.

We do not disagree with these definitions or philosophical views of computer versus human intelligence. We acknowledge that they may fuel argument about whether or not a computer can display intelligence that is equivalent with a human. For example, before the middle of the 19th century, humans hand computed navigation tables for the British Navy. This continued until Babbage's Difference Engine showed that such computations could be automated. No one attributed human intelligence to his machine, even though it automated what was previously viewed as a job that required human intelligence. More recently, Deep Blue, the IBM chess program, demonstrated that it could beat grand masters. The program uses methods discovered and refined through years of AI research. Yet it is often claimed that the computer is not intelligent. These arguments will continue and for our purposes are not relevant. We are interested in the use of AI methods that will enable OrgSim functionality, as we will presently describe.

For the purposes of this chapter, we prefer a task-based description of AI that was advocated by James Allen in his American Association of Artificial Intelligence (AAAI) keynote address in 1998 (Allen, 1998). He described AI "as the science of making machines do tasks that humans can do or try to do." He further qualified tasks to be tasks in the real-world that in some way require the machine to sense its environment, select and possibly reason about appropriate actions, and then perform those actions. Allen described this as task-based intelligence. This is similar to arguments presented earlier by Simon (1996). Task-based intelligence can be viewed synonymously as the collection of AI methods that allow one to construct intelligent agents. It is this description and focus of AI that we will use throughout this chapter.

Task-based intelligence requires a machine (e.g., a computer, a computer program, a robot), to sense, reason, and act in a real-world or simulated real-world environment.

Sensing is typically associated with the five human senses – hearing, seeing, feeling, tasting, and smelling. Inherent in all of these 'sensing' capabilities is understanding, that is the ability to derive meaningful descriptions of data in the context of the environment. The most human-like sensing capabilities have been developed in image understanding, speech understanding, and natural language understanding. Significant progress has been made in all three areas. For example, face recognition systems are now employed during high profile security events, such as the Super Bowl, to provide security officials assistance in identifying suspected terrorists. Commercial speech understanding systems are available (Menzies, 2001; Kramer, 2001). For a robot, sensing involves image understanding capabilities as well as the integration and interpretation of data collected from a variety of sensors (e.g., GPS, infrared) tuned to the intended task. For a Softbot[1], sensing might be performed by web services used by a browser or any other services such as *ftp* or *telnet*.

Reasoning has been the 'heart and sole' of AI research for more than 50 years. Initially based on logic theory, early AI researchers investigated automated theorem proving methods to encode deductive and inductive reasoning techniques. Newell's General Problem Solver (GPS) was a key early attempt to automate everyday problem solving capabilities (Newell and Simon, 1963). In the mid-1960's, Buchanan and Feigenbaum at Stanford University began investigations of how human experts solve difficult problems in medical diagnosis. Their underlying philosophy was that it was not the search techniques that made people impressive problem solvers. Rather, it was the domain specific knowledge that they employed to guide and limit search. Today's reasoning systems use a combination of generative and memory-based reasoning capabilities as will be later described. Task-based intelligence sensing methods also extend to data intensive tasks that we do not typically associate with human capabilities. For example, industrial strength data mining capabilities search for interesting patterns in large data stores, a capability routinely used by Wal-Mart to help understand customer buying preferences. Drug Companies, such as Abbott Laboratories, have reduced the number of scientists required on one investigation team from 200 to 6 through the incorporation of AI methods in the search for drug therapies (Port, 2003).

Action is seemingly the most straight forward of the task-based intelligence attributes to automate. Our stereotypical view of action is what we see a robot do to affect change in the real world. For OrgSim, this means the 'agent' exhibiting the intelligent behavior of acting reasonably and intelligibly communicating information to another person or computer program.

---

[1] A softbot is an intelligent agent equivalent of a robot that lives in a software or internet world. The phrase was coined by researchers at the University of Washington in the mid-1990s. For examples, see www.cs.washington.edu/research/projects/WebWare1/www/softbots/softbots.html

We believe that AI, or task-based intelligence has a large role to play in building OrgSim systems. Later in this chapter we will discuss the following six ideas of how AI can support OrgSim functionality.

- Creating novel scenarios
- Role playing parts that otherwise would require humans
- Supporting reuse of simulation artifacts
- Training new team members, including AI-based team members
- Adapting processes and workflows to improve team efficiency and performance
- Discovering and understanding emerging behavior in organizations

In the next section, we present a short overview of basic concepts in AI and then briefly discuss open research issues. Before this review it is interesting to consider the progress that has been made in the past 20 years. Twenty years ago there were intense debates about the best ways to represent knowledge and the best ways to automate problem solving. These debates have transcended into acceptance of multiple methods useful in the development of large software systems. The community no longer talks about pure AI software systems, but the appropriate use of AI methods in operating systems, web search engines, and other forms of commercial software. Today's equivalent of the AI debates of 20 years ago is now centered on architectural approaches to fielding task-based intelligent systems (e.g., agents) and the unsolved problems of AI – learning, common sense reasoning, and computer-based understanding – all capabilities of human intelligence that, if replicable to some degree in computer software, would greatly enhance the role of AI in organizational simulation.

AI METHODS

In this section, we briefly describe AI methods by discussing AI search methods, approaches to knowledge representation, and one example of AI problem solving – planning. AI methods require a problem representation and knowledge of the domain in which the problem is to be solved. Often a problem can be represented as a network (tree structure or state graph). For example, consider a road map. The top node (or root node) is the present location. Each branch represents a highway or road that connects other nodes (locations). The problem solving task known as search requires one to find a path from the root node to some desired location. We will call the desired node the goal. There are many optimal search techniques (i.e., linear programming, dynamic programming) that may provide the best solution given a set of constraints. We can view these techniques as brute force approaches that yield optimal results. The AI approach is to find a solution that is considered 'good enough' without searching the entire network. In this section, we will first review some basic brute force search techniques. Then we will discuss heuristic research. These methods are sometimes called generative methods that use declarative (explicit) representations of knowledge.

Search

Depth-first and breadth-first search are two brute force ways of finding a path between a root node and a goal node. A depth-first search will look down one path until either the goal is found or a dead end is reached (in which case the search backtracks to a fork in the network). A breadth-first search looks at all nodes on one level before going deeper. In breadth-first search, the nodes of the search tree are examined level by level. No node on a given level is examined until every node on the next level is examined.

A breadth-first search can be managed by using a data structure called a queue (as in the waiting line at a bank). Nodes are added to the end of the queue and taken off the front. A breadth-first search starts out with one node in the queue, the root node. The children of the root node are added to the end of the queue. If any of the children are the goal node, the search ends; otherwise, each node's children are added to the end of the queue. The search continues exhaustively until the goal node is found. Breadth-first search will find the shortest path between a root node and goal node, however there may be costs attached to traversing each branch that have not been accounted for.

In depth-first search, the nodes of the search tree are examined in a top down manner (parent to child to child's child, etc.) until either the goal node is found or one is forced to backtrack. In this search technique, the queue becomes a stack and the search strategy is one of 'first on the stack, last off the stack.' A problem with depth-first search is the potential depth of the search tree (depths of 10^{100} are not uncommon). Depth-first and breadth-first searches are the building blocks of other methods. However, in many problems more efficient or elaborate searching is needed.

AI methods may be applicable whenever the tree structure becomes so huge that a computational examination of every branch is not possible. In such cases, it might be possible to apply a heuristic evaluation function that makes an educated guess as to the 'closeness' of a present node to a goal node. Then the nodes that are in the queue or stack can be ordered. A heuristic is a 'rule of thumb' or succinct statement of experience that allows one to make an educated guess. For instance, consider a tree that represents all roads between New York and Los Angeles. If one did not have a pre-planned route (hence the ability to view the tree in a global manner) one would be tempted to drive on roads away from the sun in the afternoon. One would then be using a heuristic of the form 'brave men head west.' Both the breadth-first and depth-first search techniques can be improved markedly by the inclusion of heuristics. There are at least four ways that this can be accomplished.

- The elements in the queue can be ordered based on some guess or knowledge about which nodes are the most likely to be on a solution path.
- One may decide not to add new nodes to the queue because of some *a priori* hunch that it would be a waste of computational resources to search below that node.
- It may be beneficial to prune away search paths that do not appear 'close enough' to a goal after a certain depth is reached.
- The search depth can be varied to allow promising paths to continue a while longer.

There are many variations to the search techniques discussed here. These include minimas, alpha-beta pruning, as well as problems associated with hill climbing. The interested reader is referred to Nilsson (1980) and Winston (1993).

Many problems are too complex for state space (or tree) representations. The state graphs for these types of problems are very large (perhaps infinite) and there are no known heuristics to simplify the search process. More powerful methods are required in order to obtain a solution with a reasonable amount of computational effort.

Humans are very adept at solving difficult problems by decomposing them into sets of smaller problems. Hopefully, each subproblem is easier to solve. Often one will have default methods for solving the subproblems. Consider a simple example. Suppose one wanted to travel from the convention center in Dayton, Ohio to the beach in Honolulu (a very reasonable goal!). The actual path could very possibly be represented in a graph. However, such a graph only gives a path to follow. It tells us nothing about what methods to use to traverse the path. We would naturally divide this problem into a set of more manageable problems: walking to an exit, hailing a cab to the airport, boarding a flight to Honolulu, hailing another cab, walking onto the beach. Each subgoal might have been achieved by many other methods (e.g., hitchhiking to the airport, running to the airport, etc.). Experience plays a large part in problem representation and solving. Powerful heuristics may be applicable to subproblems that were not obvious at a higher level. Much of our current ideas about problem representation can be traced to the 'blocks world' or other micro-problems (Nilsson, 1980). The blocks world consists of a known micro-world; in this case blocks, their locations and attributes, and operators that are useful for changing the world state. The means by which goals and subgoals are achieved is called planning, a topic that is discussed later in this section.

Constraint directed search is an extension of heuristic search where constraints are used to direct the search process. The insight that has led to the development of constraint directed search techniques is that domain knowledge, encapsulated in the form of constraints, can be used to develop a better understanding of the structure of the problem space. This in turn will lead to more efficient problem solvers whose architectures take advantage of known problem space structures and focus of attention methods which reduce the amount of search. Consider the following: If you knew there was a bottleneck resource, you would start scheduling the bottleneck resource first, or if there is an order which had to be delivered by the end of the day, you would schedule that order first, or If you knew too much money had already been spent on producing a particular order, you would choose processes which were least costly. It is clear that constraints provide information that enables the problem solver to focus on the "right" decisions. In other words, constraints lead to a better understanding of the structure of the problem space, which results in the efficient generation of solutions.

Knowledge Representation

AI systems are described as *knowledge-based*, reflecting their ability to apply general knowledge about a domain to solve a wide variety of unique problems, without explicitly pre-programming those problems. Two key features of knowledge-based AI systems give them the ability to compose general knowledge into unique problem-specific solutions and make them fundamentally more powerful than conventional software. First, AI systems separate the representation of programming operators from the data on which they operate. Second, AI systems break the programming model of a rigid sequence of operators, and introduce the ability of the machine to organize its own computations. Exploiting these two features, AI reasoning systems exhibit intelligence by continually evaluating the best operator to apply based, on the current problem solving goal and the current data content. Adaptive ordering of computations lets these systems deal with unique and novel problem situations, composing sequences of rules or operator applications to generate surprisingly complex solutions that often appear extremely innovative.

The features that enable adaptive ordering of computations derive from the distinction between conventional processing at the *data* level vs. knowledge-based processing at the *symbol* level. Knowledge-based systems accept input as collections of asserted facts that map the specific values of a situation onto symbols representing concepts that the machine can understand and manipulate. Using symbol-level manipulation, the AI reasoning mechanism builds complex structures called *semantic networks* that identify the elements that comprise the problem space, and describe the relationships between them. For example, if a new customer entity is added to an enterprise model, numerous new relationships might be established that impact the models, from recognizing co-location relationships impacting workload at enterprise service sites to asserting new types of supplier relationships that impact demands on production and delivery capabilities. The resulting new semantic network structure serves as a refined model of the problem solving situation, and the reasoning system chooses which goals to pursue and which rules or operators to apply by matching the symbols represented in the revised model to the symbols that describe the problem-solving knowledge in the knowledge base. This ability to manipulate and match concepts at the symbol level is fundamental to enabling machine intelligence.

Underpinning the ability of the machine to manipulate symbols is its ability to understand what those symbols mean, and to exploit that understanding in reasoning. Until recently, each AI solution was engineered around a single, unified *domain model* that defined the *semantics* of the domain: the concepts and entity types that needed to be represented, the relationships that connected them to form semantic webs, and the constraints that governed the definition, linkage and application by the system. While the reasoning engine provides a domain-independent capability, it relies on the semantic agreement between the data representation and knowledge representation. The domain model has typically been comprised of the knowledge system engineer's implicit set of semantic definitions which provided that agreement and has been largely hand-encoded in each application design.

More recent AI research has explored the value of making these semantics explicit in a machine-accessible form. By declaring these semantics symbolically, the AI system builder extends the machine's knowledge, and thus its ability to reason, to recognize and build richer semantic network descriptions of a problem space, and to be more adept at recognizing how encoded domain knowledge can be applied to those descriptions. Formalization of machine-understandable semantics has even begun to cross the design barriers between systems, allowing separate knowledge-based systems to share both problem spaces and problem solving knowledge. A key breakthrough in this field has been the emergence of semantic representation languages and tools that let separate applications discover each other, expose their domain models to each other in an understandable form, and apply each other's knowledge on demand. The distributed and constantly evolving nature of the World Wide Web poses a compelling requirement to find and apply the best current source of knowledge. Problem domains, like competitive business-to-business e-commerce, are rapidly driving the reduction of these techniques to common practice.

Ontologies provide the formalization necessary to build machine-usable descriptions of domain models. The simplest ontologies are *taxonomies* that define the concepts that are referenced in describing the domain, and the categories that organize those concepts. Concepts make up the nodes of semantic webs built within the domain, and describe both *concrete* concepts (people, locations, things) and *abstract* concepts (roles, groups, goals, plans, etc.). Categories are typically defined as a hierarchical mapping of types, with more general abstractions at higher levels of the tree and more concrete elements and more detailed concept definitions linked below them in a *specialization hierarchy.* Richer ontologies are defined by extending the concept definitions by defining concept *attributes* (as unary properties on the concept), *roles* (that define legal relationships between concepts) and *constraints* on the values on concept attributes and relationships. The constraints may be as simple as a type dependency that defines the type of concept that can satisfy a role relationship, or as complex as a set of logical assertions that bound the value relationships between related concept values as they form models as graphs of multiple concept nodes.

Various researchers have developed a wider range of languages for representing ontologies in ways that enable knowledge based systems to use the ontology as the basis for inference, classification, and modeling. Nearly all successful work has been based on *description logic*-based languages. Description logic provides a declarative form of representation for the subset of first-order predicate logic most useful for representing and manipulating semantic representations. It provides an underpinning formalism for many common representational frameworks, including object-oriented frame systems, type systems, semantic data modeling and semantic network modeling languages. Reasoning on description logics can be summarized as *subsumption and instance checking* (reasoning about specializations, class membership and inheritance across specializations); *classification* (interpreting a concept from its structure and the structure of its filled roles); and *satisfiability* (determining whether a construct of concepts is consistent with a set of constraints).

Recently, researchers from across the field have contributed to the emergence of a standard description logic-based language for semantic representation. Research through the 1990s library science and philosophy combined with knowledge representation research to provide early ontology languages and tools for ontological interchange. These tools began to address capabilities for semantically grounded interoperability and composition. In the late 90's, DARPA, the National Science Foundation, and the European Commission Information Society Technologies (ITS) and the U.S. World Wide Web Consortium (W3C) all began investment toward a common language. Unifying these efforts was Tim Berners-Lee, the inventor of the WWW, and his vision of a "Semantic Web" where a future World Wide Web made content, knowledge, and services available to any machine intelligence on a global scale.

Intermediate products such as DAML (the DARPA Agent Markup Language) and the W3C Metadata Activity RDF (Resource Description Framework) standard, and the EU OIL (Ontology Interchange Level) have now coalesced into OWL (the Web Ontology Language). The OWL standard is lead by the W3C's Web Ontology Working Group (WOWG). (It would be most accurate to describe OWL as the latest set of contributions to a community that has developed a family of languages, a layer cake of representational capabilities that include XML and XML schema underpinnings; RDF metadata and RDF Schema resource layers; OWL-based ontology languages, and OWL extensions for rules (OWL-R), and Web Services (OWL-S), with surely more to come.). Berners-Lee, Hendler, and Lassila (2001) provide an excellent overview and current related work can be accessed online.[2]

Planning

Search and knowledge can be combined to solve many task-based intelligence problems. One is planning. In planning, the planner seeks to find a sequence of actions it can perform in a real or simulated world to move from a given current state to a goal state. Three distinct styles of planning have emerged: generative planning, constraint satisfaction, and case-based planning (Fowler, 1995).

Generative planning refers to a body of research and techniques which generate sequences of actions that will achieve given goals in a domain complex enough that the appropriateness and consequences of the actions depend upon the

---

[2] See www.w3.org

world states in which they are to be executed. This is a creative, synthetic process, not merely a matter of filling in certain slots with correct values. In particular, the planning system must keep track of and reason about differing world states at different points in time. This feature distinguishes the planning problem from similar problems such as scheduling, and makes planning inherently difficult as it involves the solution of several subproblems. Faced with this overwhelming complexity, practical planning systems must balance epistemological and heuristic adequacy, retaining as much expressive power as is practical, yet making enough restricting assumptions so that a viable, efficient implementation can still be realized.

Because of the complexity of the problem, existing generative planning systems are run off-line and do not react to the world. This is a major drawback in a dynamic world, since the planner's assumptions about the world are becoming invalid as planning proceeds. However, much military planning, for example, is done off- line, months or even years before the plans are executed in a dynamic environment. Clearly, a system embedded in and reacting to a dynamic world must have an architecture that is reactive and responsive. But even in this later case, the system will still need a generative planning component if it is to take reasonable actions from a broad task-level perspective in response to unforeseen situations. Thus generative planning techniques are of crucial importance in military planning and logistics.

Generative planning has been an active area of research within the AI community since the early 1960s. A brief summary of this history will show that significant progress has been made both in understanding the problem, and in developing practical techniques for generating plans. Early attempts at AI planning, for example the robotic systems developed at SRI International during the late sixties and early seventies, made certain assumptions about the planning problem that dominated AI planning research until the mid eighties (Nilsson, 1980). These assumptions include perfect information about the world and about the results of actions, a static world, and a state-based representation of the world.[3]

Another key problem solving technique is constraint satisfaction. One of the key techniques of problem solving is to exploit all the natural constraints possible. It turns out that the world is very well formed and constrained. This helps to simplify many problems. For example, we know that if the edge of a block is attached to the floor, then both ends of that edge must be attached. This may sound simple, but is one of the keys of AI problem solving. Many simple search procedures consider many possibilities that could not exist. For example, if we had six labels for each vertex then we could have six times six possible combinations for a line. But when we consider how the world is formed, only a small subset of these is actually realizable. Constraint exploitation adds knowledge to a problem. By knowing how things can and cannot be, we add constraints to the problem domain. These constraints can reduce a very hard problem to a very easy one. Genetic algorithms and neural networks are variants of constraint satisfaction where constraints are represented numerically.

---

[3] For more recent work, see www.aiai.ed.ac.uk/~oplan/oplan/oplan-doc.html, www.cs.umd.edu/~nau/planning/, or www.ai.sri.com/~sipe

One technology that seems well suited to the task of generating schedules for large-scale transportation problems comes out of the emerging work in case-based planning. Case-based planning is founded on the idea that a machine planner should make use of its own past experience in developing new plans, relying on its memories instead of a base of rules. Complete plans for conjunctive goals are stored in memory for later use. Memories of plan failures are stored so that they can be recalled and thus avoided in later planning. Specific repairs are stored and reapplied when the problems that they solve are anticipated and specific techniques for conjoining plans are saved. In general, planning is a matter of recalling what has worked in the past rather than projecting what could work in the future. Case-based planning suggests that the way to deal with the combinations of planning and projection is to let experience tell the planner when and where things work and don't work. Rather than replanning, reuse plans. Rather than projecting the effects of actions into the future, recall what they were in the past. Rather than simulating a plan to tease out problematic interactions, recall and avoid those that have cropped up before.

In summary, one of the key ideas of AI is representation of knowledge and use of that knowledge in problem solving. Heuristically guided search is the 'work horse' of AI. Variations on search, such as constraint satisfaction systems, genetic algorithms, and neural networks provide additional methods useful for many problems.

Intelligent Agents

Autonomous machines that display task-based intelligence are called agents. Such machines may perform autonomous action in the physical world (robots), in the world of the internet (softbots), or in simulation worlds -- for our purposes, we will call them agents or intelligent agents. Besides having some degree of autonomy, agents also interact with other agents in intelligent ways. Hayes (1999) provides a good overview of agents and we only focus on a few key points in this section.

Figure 1 provides a simple depiction of an agent as a robot. This does not imply that robots are simply machines, only that the agent architecture illustrated here is simple in terms of its cognitive capabilities.

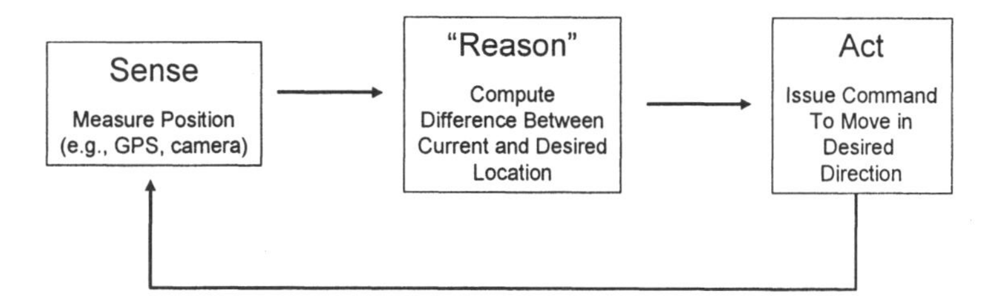

FIGURE 1. Notional (Simple) Robot

In this system, a robot is able to move freely, that is autonomously, in the world. It senses its location in the world through one or more sensors (e.g., an infrared sensor). It is programmed to compute its position based on the sensor outputs in some predefined coordinate system and to then compute the difference between its current and desired location. The desired location is either programmed in or otherwise provided as input from an outside source. The robot then issues movement commands to change its location in the world.

One would have a hard time arguing that the robot shown in Figure 1 is intelligent. But suppose there was a higher form of intelligence that compiled the behavior described above into control laws or other suitable real-time executable software? Rasmussen (1983) describes three levels of human problem solving behavior: skill-based, rule-based, and knowledge-based. The skill level represents "sensory motor performance during acts or activities which, following a statement of intent, take place without conscious control as smooth, automated, highly integrated patterns of behavior." At the rule-based level, a human performs a function when a triggering pattern matches incoming data. The function and situation may be learned from experience and is goal-directed. The performance is within the context of a domain model that may or may not accurately describe the present domain stage. At the knowledge-level, goals are explicitly formulated, based on an analysis of the environment and the overall aims of the person. An important aspect is the maintenance of the domain model.

The robot could exhibit higher forms of intelligent behavior than the skill-level behavior illustrated above. For example, while executing tasks at the skill-based level, it could also be executing rule-based tasks such as navigating around chairs and adapting a plan to guide itself around a room. It could also be executing tasks at the knowledge-level, for example updating its map of its world, monitoring resources such as battery power, and planning routes to 'sweep over' as much of the room as possible before it runs out of power. In fact, vacuum sweepers enabled by such task-based intelligence now exist (Manley, 2003).

In 1988, John Sculley, then CEO of Apple Computer, proposed an example of a robot in the world of software. The *Knowledge Navigator*[4] illustrated a vision of a future computer in which a personal intelligent agent worked collaboratively with a person to solve a variety of problems. The agent was able to communicate in natural language, find and summarize documents, create setup files for simulations, fuse information from various models and simulations, construct visualizations of those simulations, and perform a variety of intelligent assistant tasks such as meeting scheduling. While the cognitive capabilities of such an intelligent agent are still well beyond the capabilities of AI, many of the visionary aspects of the intelligent agent are achievable today (for example, intelligent web search).

The above discussion illustrates two characteristics of intelligent agents. First, they are autonomous. And second, they have sensing, reasoning, and action capabilities based on an architecture that encodes a theory of cognition. By architecture, we mean the ways AI methods and other computing methods are

[4] See weblog.infoworld.com/udell/2003/10/23.html or www.billzarchy.com/clips/clips_apple_nav.htm

organized to achieve these capabilities. There are two fundamental different architectural styles as illustrated in Figure 2.

The first concept is similar to the block diagram in Figure 1 and maps easily into the 'sense-reason-act' model. Each block seeks to replicate a task-based intelligent behavior. Each may involve the use of AI methods to replicate behaviors at the skill, rule, and knowledge levels. For example, much research has been conducted on robotic perception, trying to build systems that replicate human visual understanding capabilities. This is illustrated in the upper half of Figure 2.

A good example of an intelligent agent architecture modeled on this approach is SOAR (Newell, 1990). SOAR is a robust architecture that is both a model of human cognition and a useful framework for constructing intelligent systems that use large amounts of knowledge. It is based on the production system paradigm with engineering improvements that enhance its efficiency. For example, production rules are compiled into a network that accelerates search time by propagating data through the rule network (instead of matching rules against a database) as well as a highly tuned maintenance capability to help maintain the consistency of rules (e.g., if a new rule is entered that counteracts a previous stored rule, a 'truth maintenance system' helps the human engineer understand the inconsistency.

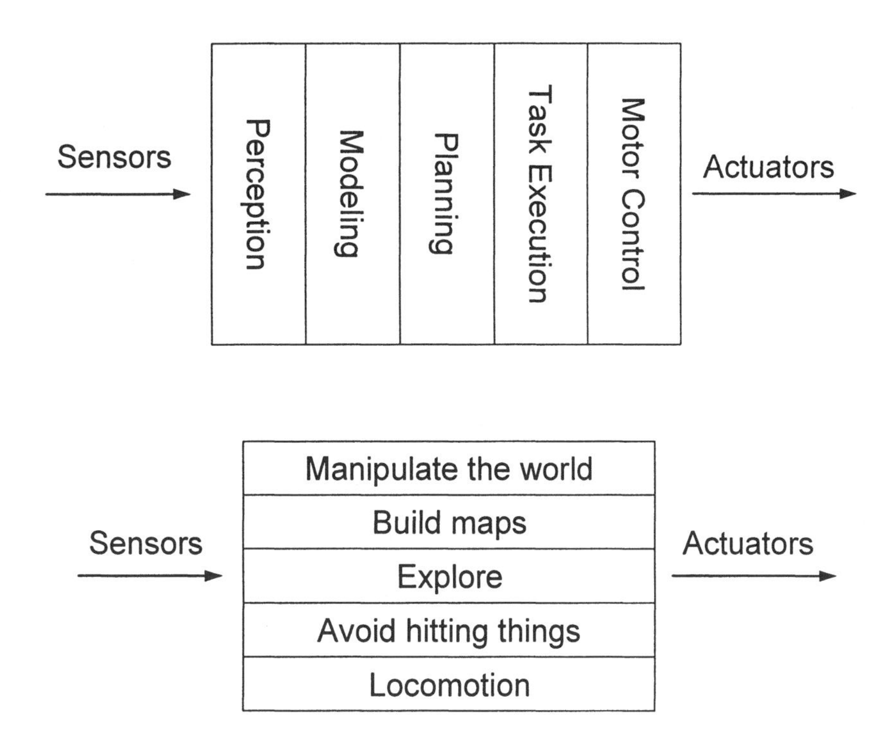

FIGURE 2. Two Architectural Concepts for Intelligent Agents. (Brooks, 1991)

The subsumption architecture (suggested in the lower half of Figure 2) as advocated by Rodney Brooks (1991)[5] is fundamentally different. Task-based intelligence is embedded in behavior-generating modules. For example, the ability to move in the world may involve sense-reason-act capabilities that are encoded at the skill-based level, while exploration or map building behavior models may involve sense-reason-act capabilities at the rule-based level. (See the chapter in this volume on Crowd Modeling).

In this section, we have focused on intelligent architectures for robots. But the same concepts apply for other worlds such as the internet (softbots for intelligent search of web pages) or simulated worlds. There are now many public domain agent systems and agent programming languages[6] as well as standard setting bodies for agents and their use in various applications.[7]

As Hayes (1999) points out in her overview, control issues in large software systems comprised of multiple agents is a difficult problem. Veloso and her research group (Vu, Go, Kaminka, Veloso, & Browning, 2003) at Carnegie Mellon University have researched this issue in a variety of robotic and simulated domains. They have proposed an architecture for multi-agent control "which integrates script- and code-based off-line team programming and team design, with a run-time coordination engine that can execute different team-control designs automatically within a behavior-based framework."

Modern applications need to be highly tailored to the task at hand and yet flexible enough to support rapid, (ideally) automatic reconfiguration to adapt to changes in the situation, currently accessible IT systems and data sources, and available computing power and communications bandwidth.. Software agent technologies facilitate building "smarter" components which can participate in these kinds of applications. Using object-oriented software engineering, knowledge-based reasoning and other techniques, agents can (1) describe their needs, capabilities and interfaces to other agents, (2) find and work with other (lightweight) agent components to accomplish complex tasks in flexible teams, versus a single monolithic application, (3) interact with humans and other agents to accept tasking and present results, and (4) adapt to changes in tasking, the battlespace, the computing environment, other agents, or new opportunities.

A mechanism is needed to enable the dynamic, runtime integration of agent, object, and legacy software components into applications. A key enabler of this vision is the CoABS Grid[8,9], a major research and development effort of the DARPA Control of Agent Based Systems (CoABS) Program. The Grid supports the development of applications for dynamic domains which require the composability, adaptability, and autonomy provided by software agents interoperating in dynamic, mixed-initiative teams with human users. The Grid does this by providing access to shared protocols and ontologies, mechanisms for describing agents' capabilities and needs, and services that support interoperability

[5] The vacuum sweeper described by Manley (Manley, 2003) is based on this intelligent agent architecture.
[6] See www.agentlink.org/resources/agent-software.php or www.fipa.org/resources/livesystems.html
[7] See www.fipa.org or www.omg.org
[8] See www.coabs.globalinfoteck.com
[9] See www.darpa.mil/ipto/programs/coabs

among agents at flexible levels of semantics – all distributed across a network infrastructure.

There are three views of an agent grid (Thompson, 1998). All are, or can simultaneously be, valid views:

- a collection of agent-related mechanisms (brokers, agent communication languages, rules, planners, etc.) that augment object and other technology to increase the ability of developers to model and dynamically compose systems. Many of these mechanisms can be independently studied and could be standardized so that communities that used these standard mechanisms could more easily develop interoperable "agent-based" capabilities.
- a framework for connecting agents and agent systems. The framework could be built using the standard mechanisms as well as existing non-agent technology (e.g., middleware). The benefit of this view is to show how the agent mechanisms fit together to make it easier to construct agent systems of systems, using framework-aware (grid-aware) protocols, resources, and services as well as foreign wrapped agent systems and also legacy non-agent systems. The grid as framework is responsible for providing services and allocating resources among its members. One of the issues with this notion of grid is whether there is just one or several and if the latter how they interoperate.
- model entities that might be used to model an org chart, team or other ensemble. This view assumes there are many grids but they interoperate according to the standard protocols. In this view, a grid might represent a battalion ensemble which has assigned tasks and resources. Lower level units (agents/grids) use the resources to work on the tasks assigned to the battalion.

Although many multi-agent system (MAS) architectures provide some of the interoperability and other services that the Grid provides, each architecture typically supports specialized agent types, communication, and control mechanisms. This specialization is desirable because a particular MAS can use mechanisms appropriate to the problem domain/task to be solved. The Grid is *not* intended to replace current agent architectures but rather to augment their capabilities with services supporting trans-architecture teams of agents. Agent technologies will support semantically rich conversations among these agents to allow them to interoperate outside their local agent "community". An analogue is the Internet's bridging of heterogeneous networks by gateways and protocols. Programmers will make their components "Grid-aware", much as many network applications are now made "Internet ready" or "Web ready" by supporting protocols and languages such as TCP/IP, HTTP, HTML, and XML. Furthermore, programmers will *want* to make their components "Grid-aware" to enable them to participate in dynamic teams that leverage other components discovered at runtime.

How might agents be used in simulation? What engineering issues need to be addressed to incorporate agents into simulation? We propose a three-dimensional space as depicted in Figure 3. Tasks that an agent might support in simulation range from data intensive computation tasks (such as helping to create source files or scenarios to run the simulation) to performing tasks usually accomplished by humans. Such tasks might involve monitoring, planning, and analysis. It is suggested for the purposes of this chapter that the complexity of these tasks is ordered from easiest to hardest. Each task might be executed at different levels of cognitive capability as suggested by Rasmussen's taxonomy. The third dimension relates to how the cognitive capability is acquired by the agent, from being programmed, to being shown (perhaps via a graphical programming environment), to being told, to being discovered by the agent itself.

Based on this simple taxonomy, one could think of progressively more complex agents for use in organization simulation. Three examples are provided.

- Monitoring agent. An agent that is coded by a software engineering or AI expert to track data or outputs from another human or agent with programmed instructions to take specific actions when some threshold value is reached. Such agent behavior is characteristic in smart forms used by logistics planners in the military to track explicit, but cumbersome constraints. Another example that has been proposed for the medical professions is smart form based agents that will check a person's medical history to ensure they are not allergic to a medication prescribed by a physician.

- Planning assistant agent. An agent that acquires its knowledge by being told by an expert planner what it should do in certain situations. "Being told" is a form of machine learning that can be implemented in domain-constrained ways through a visual programming environment or through use of natural language that exploits the ontology of a given domain. An example would be an intelligent scheduling program that generates acceptable transportation plans as a planning service in a large warehouse, supply chain, or military command and control function. Such a capability is described by Mulvehill (2005).

- Analysis agent. Such an agent would remember all past exercises in memory as cases and be able to generate appropriate analogies as proposed solutions to a new given instance. In such cases, the agent discovers novel uses of past knowledge and stores this knowledge for later use. An example would be the agent that can review case studies in a given domain and adapts a course of action that runs counter to what a trainee has previously experienced.

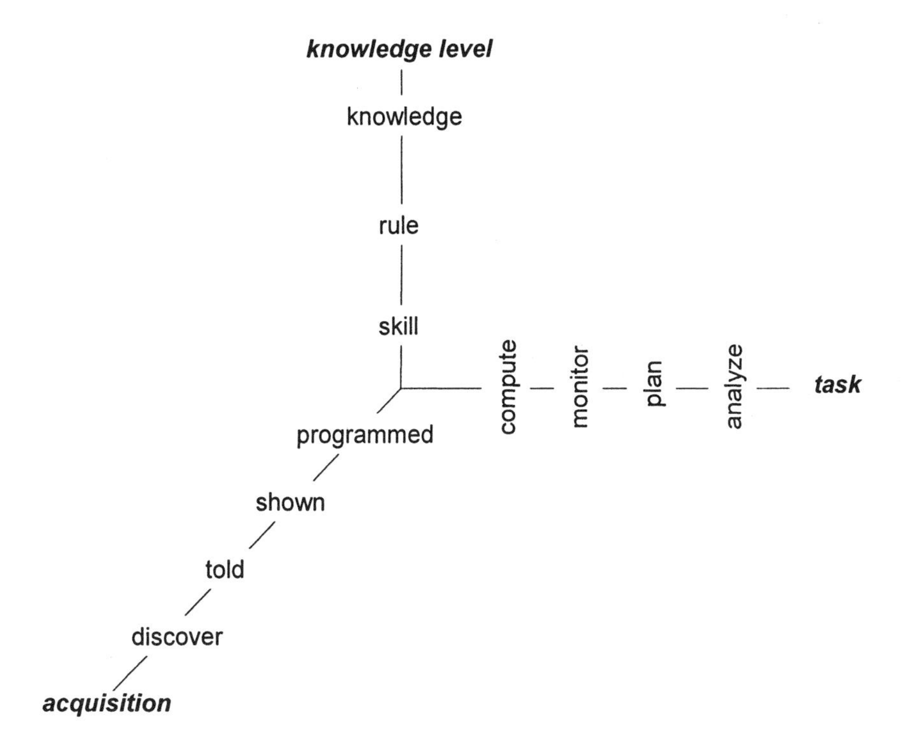

FIGURE 3. Dimensions of intelligent agent capabilities for use in simulation

APPLICATIONS OF AI TO ORGSIM

What if a team could co-evolve better ways of doing work while experiencing what it would be like and feel like to work this way?

What if their legacy IT and communication systems could be integrated at the same time?

What if this could all be done faster than the unfolding events of an unexpected event or crisis?

These questions frame our view of OrgSim. We mean something quite different from the common approach to organization simulation as pursued in the social and organizational behavior community, for example (Carley, 2004). Rather we are interested in immersive environments that can be used to discover new ways of work, to evolve IT systems to support those ways of work, and in the process to train intelligent agents to support that work. In this section we discuss how AI methods might be used to achieve such a capability.

To illustrate the concepts, consider the current popular video game, *Black Hawk Down*. In this game, the human player assumes the role of a U.S. Army platoon leader. He/she leads troops into an urban area and must make real-time decisions about the intent of adversarial forces (e.g., to shoot them because their intent is hostile). What if one could do much more in such an immersive environment? What if one could interview U.S. soldiers or captured enemy soldiers? What if one could change the scenario to reflect the cultural or emotional states of the occupied town? What if any of the 'humans' in the game could be played by real humans or artificial beings whose behavior is learned automatically from observing a human player? What if the maturity or effectiveness of the U.S. team (that is, what is learned) could be measured as a result of playing the game? Such capabilities are goals of researchers in a diverse set of fields including organizational modeling, human-computer interaction, performance measurement, and virtual reality. An OrgSim test bed would enable such diverse researchers to integrate their research software into the test bed and share their results (and software) with a larger research community. Such a test bed would, thereby, support experimental science in the new interdisciplinary field of organization simulation.

We describe six particularly salient ways AI methods can be used to support the development of organization simulation.

Creating Novel Scenarios: Wildberger (2001) observes that simulation and AI are frequently used to assist contingency planning for military operations and their logistical support. In these situations, simulation is most often used to assess the feasibility of human generated plans. AI is most often relegated to the assembling of plans from components or aiding data interpretation tasks. These wargames and military exercises are often based on *a priori* scenarios. These scenarios are developed by humans and as such often reflect the 'last war.' In addition they are data intensive and very time consuming to construct. Wildberger suggests that libraries of past scenarios could be constructed and indexed based on the component size fragments from each scenario and then a computer could construct novel combinations of components from the library in an effort to develop novel scenarios. Just as AI-based data processing is being used so effectively in the drug industry, data intensive tasks for scenario development could also be offloaded to computers utilizing appropriate AI methods.

Role Playing: There has been ongoing effort on using more cognitively capable agents in human intensive exercises. Often these agents are called Semi-Autonomous Forces (or SAFORs). Perhaps the most notable SAFOR is the use of SOAR-based agents in virtual reality flight training (Wray, 2002). The goal of this work has been to create intelligent agents to assume the role of other pilots – both potential enemies and members of one's own unit (or flight) – to provide more realistic training in air-to-air operations. This work highlights requirements for agents in order to play the roles of opponents in exercises:

- Observational fidelity – to be as realistic as possible
- Faster than real-time – to enable exploration of more options

- Behavior variability – to prevent becoming predictable
- Transparency – to be able to explain its actions
- Rapid and low-cost development

To date, the SOAR-based SAFORs have been highly effective in terms of their cognitive capabilities and faster than real-time performance. But the cost of creating such SAFORs has limited their wide spread introduction into more complex simulations (e.g., for air-to-ground training). Cost could be reduced through three means. The first is to limit the cognitive capabilities of the agent to simple, but useful monitoring, planning, or analysis tasks. The second is to focus on the creation and introduction of user-centered tools that allow human users to program agents by showing them or telling them what to do in certain situations. Memory-based AI techniques like case-based reasoning should prove effective in enabling this capability. The third area is to promote greater reuse and sharing. There is related work in reuse of AI methods that may help in the future to lower the cost of introducing AI technology into simulation.

Supporting Reuse: ROSS – the Rand Object Oriented System - (McArthur, 1982) was an early attempt to provide a public domain version of an object-oriented environment tailored for simulation of command and control systems. There have been few examples since. Fu and Houlette (2002), for example, observe the lack of reusable AI software in the game industry. Despite the increased desire for agents and other forms of AI to make games more intelligent and more challenging, most game developers create their own AI software. Griss (2003) describes the use of softbots as the next advance in component-based software engineering as an approach to accelerate software system development with agent-based components. In related work, Riley and Riley (2003) explore the integration of conventional simulation methods with multiple agents to leverage the existing body of simulation methods to produce distributed, repeatable, and efficient simulations.

More Robust Simulation: Another aspect of military exercises is the human dimension. It is not uncommon for hundreds of humans to be involved in training exercises. Such exercises often follow a predefined script. The exercise itself resembles a depth first search through some predefined scenario, allowing no opportunity to explore multiple options or to back up and repeat a course of action with a more robust but high-risk option. An interesting area of research would be the development of a game or simulation controller, an agent that could stop play, backtrack through previous team decisions, and pose additional challenging tasks for a team to explore. This is related to a final possible functionality for OrgSim, adaptive processes and workflows.

Adaptive Processes and Workflows: A process or workflow is a sequence of tasks that are interrelated to accomplish some specific purpose. Each task is performed by one or more intelligent entities – they receive information and add or manipulate that information and produce information or knowledge that is useful to the performer of the next task. Specific tasks could be made the responsibilities of an intelligent agent that is programmed to be competent on that task, for example the automated filling out of a form used to convey an intelligence report from the intelligence staff to the commander in a a command and control scenario. The team could then experiment with a given process and propose changes, discovering new and better ways to work. While agents themselves are a relatively new concept in the study of workflow techniques, there has been some recent research along these lines (Zeng, 2002). For example, Shah and Pritchett (2004) explore the role of work environment in the design of multi-agent simulations.

Emergent Behaviors: As was noted in early knowledge-based simulations (McArthur & Khlar, 1982), interesting behaviors that are not easily anticipated arise from the interaction of intelligent agents during the simulation. Understanding emerging behavior in collaborative networks has been an active area of research at the intersection of many fields including AI, chaos theory, and social science: Hearn, (2002) and Hazy and Tivnan (2003) are good references to related work. A value of OrgSim, for example in understanding the acceptance of new business processes, would be anticipating some of the reasons people would or would not accept to work under these new processes. Such could be anticipated early in simulation to allow time to both pursue culture change approaches and to possibly modify the processes so they would be more acceptable.

CONCLUSIONS

Research in artificial intelligence has produced an array of knowledge representation and problem solving methods that can be used to engineer functionality into organizational simulations. In this chapter, we restricted our focus to a task-based view of AI where tasks might be performed by one or more intelligent agents to perform monitoring, planning, or analysis tasks. Such tasks could be useful in organizational simulation to help design challenging scenarios, provide more efficient simulations by allowing software to replicate human roles, helping in the design and early evaluation of new team processes, and helping in the analysis of emergent behavior. A critical issue, as much cultural as technical, is adopting a culture of software reuse. Organizational simulations, especially if empowered by cognitively capable agents, will be large software development efforts. Hence maximal reuse should be made of AI methods, architectures, and implemented systems.

REFERENCES

Allen, J. (1998), AI Growing Up. AI Magazine, 13-23.

Barr, A., Feigenbaum, E., & Cohen, P. (1989). Handbook of Artificial Intelligence, Vol. 1-4, Cambridge (MA): Addison Wesley.

Berners-Lee, Hendler, J., & Lassila, O. (2001). The Semantic Web, Scientific American, www.sciam.com

Brooks, R. (1991). New Approaches to Robotics. Science (253), September 1991, 1227–1232.

Carley, K., Altman, N., Kaminsky, B., Nave, D., & Yahja, A. (2004). BioWar: A City-Scale Multi-Agent Network Model of Weaponized Biological Attacks. Technical Report (CMU-ISRI-04-101). Pittsburgh, PA: CASOS, Carnegie Mellon University.

Fowler, N., Cross, S., & Owens, C. (1995). Progress in Knowledge-Based Planning, IEEE Expert, 10, (1), 4-9.

Fu, D., Houlette, R. (2002). Putting AI in Entertainment: An AI Authoring Tool for Simulation and Games. IEEE Intelligent Systems. 81-84.

Griss, M. & Pour, G. (2001). Accelerating Development with Agent Components. Computer. 37-43.

Hazy, J. and Tivnan, B. (2003). Simulating agent intelligence as local network dynamics and emergent organizational outcomes. Proceedings of the 2003 Winter Simulation Conference. New York: ACM Press. 1774-1778.

Hayes, C. (1999). Agents in a Nutshell-A Very Brief Introduction. IEEE Transactions on Knowledge and Data Engineering. 127-132.

Hearn, P. (2002). Emerging Behavior of Complex Collaborative Networks, Proceedings of the ThinkCreative Workshop, Brussels Belgium. www.uninova.pt/~thinkcreative/D1eWorkshop.pdf.

Himmelspach, J., Rohl, M, and Uhrmacher, A. (2003). Simulation for testing software agenda – an exploration based on JAMES. Proceedings of the 2003 Winter Simulation Conference. New York: ACM Press. 799-807.

Jones, R. M., Laird, J.E., Nielsen, P.E., Coulter, K.J., Kenny, P. G., & Kiss F. (1999) Automated Intelligent Pilots for Combat Flight Simulation. AI Magazine, 20(1), 27-41.

Kramer, P. (2001). IBM gets smart about artificial intelligence, IBM Think Research, http://researchweb.watson.ibm.com/thinkresearch.

Leake, D. (Ed) (2000). Case-based Reasoning: Experiences. Lessons, and Future Directions 2nd Edition. Cambridge (MA): The MIT Press.

Manley, K. Vacuums sweep into history. USA Today. B03.

McArthur,D., & Klahr, P. (1982). The ROSS Language. Santa Monica (CA): Rand Report.

Menzies, T. (2003). 21st Century AI: Prod, Not Smug. IEEE Intelligent Systems, 18-24.

Mulvehill, A. (2005). Authoring Templates with Tracker. IEEE Intelligent Systems. in press.

Newell, A. (1990). Unified Theories of Cognition. Harvard University Press. Cambridge, Massachusetts.

Newell, A., & Simon, H. (1963). GPS: A program that simulates human thought. In E. Feigenbaum and E. Feldman, Eds., Computers and Thought. New York: McGraw-Hill.

Nilsson, N.(1980). Principles of Artificial Intelligence, Palo Alto (CA): Tioga Publishing Co.

Port, O., Arndt, M., & Carey, J. (2003). Smart Tools, Business Week Online, http://www.businessweek.com/print/bw50/content/mar2003.

Rasmussen, J. (1983). Skills, Rules, and Knowledge: Signals, Signs, and Symbols and Other Distinctions in Human Performance Models. IEEE Transactions on Systems, Man, and Cybernetics (13). 257-266.

Shah, A. and Pritchett, A. (2004). Work Environment Analysis: Environment Centric Multi-Agent Simulation for Design of Socio-technical Systems, Proceedings of the Joint Workshop on Multi-Agent and Multi-Agent-Based Simulation), New York: in press.

Riley, P. and Riley, G. (2003). SPADES – a distributed agent simulation environment with software-in-the-loop execution. Proceedings of the 2003 Winter Simulation Conference. New York: ACM Press. 817-825.

Simon, H. (1996). The Sciences of the Artificial, 3rd Edition, Cambridge (MA): The MIT Press.

Thompson, C. (1998). Characterizing the Agent Grid, Technical Report, Object Services and Consulting, Inc. www.objs.com/agility/tech-reports/9812-grid.html.
Turing, A. (1950). Computing Machinery and Intelligence. Mind. Oxford University Press, 433-460.

Vu, T., Go, J., Kaminka, G., Veloso, M., Browning, B. (2003). MONAD: A Flexible Architecture for Multi-Agent Control. Proceedings of the Adaptive Agents and Mutiagent Systems Conference. New York: ACM Press.

Wildberger, A. (2000). "AI & Simulation: When is smart *smart*?" Simulation. http://www.modelingandsimulation.org/wildberger.html

Winston, P. (1993). Artificial Intelligence, 3rd Edition, Reading (MA): Addison-Wesley.

Zeng, D., & Zhao, J. (2002). Achieving Software Flexibility via Intelligent Workflow Techniques. Proceedings of the 35th Hawaii International Conference on System Sciences. IEEE Computer Society.

CHAPTER 16

SIMULATING HUMANS

IRFAN ESSA AND AARON BOBICK

ABSTRACT

Simulations designed for training or experimentation have (at least) two elements that are critical to their success. The majority of simulation research focuses on content validity – assuring that the simulation is accurate in terms of the domain. For example, ensuring that simulated productivity in a given situation approximates the productivity that would be observed in the real world should the corresponding context arise. However, for organizational simulation or any other system in which human decision making is critical, content validity is not sufficient to provide an adequate simulation experience. For example, in military or emergency training the emotional content is just as important as situational accuracy. A principal goal of such exercises is to give experience to decision makers in evaluating options and making choices in the face of a variety of stressful or emotional factors. To achieve this emotional state requires interaction plausibility with sufficient believability. That is, the interfaces with the systems and the people must reflect the tenor of what the interactions would be in the real world. For organizational simulation this task is particularly challenging because organizational structure implies particular interpersonal interactions, and these are the interactions that must rendered with sufficient fidelity. Specifically, the system must be able to render the output of simulated human characters in a manner that can provoke the appropriate interpersonal and/or emotional response of the user. In this chapter, we describe a variety of recent advancements in computer graphics and animation technologies that are directly relevant to this daunting task. Our exposition will cover all aspects of human modeling and animation. We will discuss prevalent techniques for simulating human behaviors as it supports organizational simulation and describe approaches and tools that are appropriate to do such forms of simulation. We will specifically focus on the rendering of "talking heads" – upper torso simulations of people speaking arbitrary text and also containing a variety of emotional content. Recently there have been numerous developments in image- and video-based rendering that produce photorealistic video sequences of talking characters. These methods, combined with recent advances in computer vision analysis of videos of people, make plausible the prospect of highly effective output for organization simulation.

INTRODUCTION

Organizations are for and about people. Interpersonal interaction and related dynamics form the core of how people (are expected to) behave in their daily lives and therefore such behaviors both at the individual and collective levels are topics of study in many disciplines. The study of the organization, and its simulation to extract inferences as to how the organization "behaves," needs to leverage studies in individual and collective behaviors of its members/participants. Effective simulation of the behaviors of the members of an organization can inform us, at a minimum, about (a) how such behaviors add to the complex dynamics that is an organization, (b) how such behaviors evolve, and (c) how can these be predicted and modeled at the organization level.

It is important to emphasize that behaviors can be (i) local and over a short duration and (ii) global and over extended periods. Local and short-term behaviors are mostly how a person behaves at any point in time, *e.g.*, movements of the person, facial expressions, *etc.* While global and extended behaviors are more goal-oriented *e.g.*, people trying do a specific task or being cooperative or uncooperative within a teamwork situation. There is a vast space in the middle of these two types of behaviors that also needs to be studied and modeled.

The field of computer animation, especially of 3D humans has matured considerably of late due to advancements in (a) in computer graphics technologies, both hardware and software, and (b) in our understanding (facilitated by automatic methods of analysis) of how humans move. The greatest benefactors of these advancements, which are directly obvious, are animations and simulations of people in full-length feature films and in computer games. There is, however, a significant difference in quality of both how the simulated people look like and how they move in games and in movies.

PIXAR has led the way with huge successes like *Toy Story* (1995) and *Toy Story 2* (1999), followed closely by PDI/Dreamworks with Academy Award Winning *Shrek* (2001) and *Shrek 2* (2004). These movies have a great combination of good stories, nice models of the environments with detailed renderings, and very clean motions of the characters. Note that these movies are made with significant manual creativity of some very talented artists. As further progress is made, some automation is showing up, but the higher demand for good animation has left excessive automation off the table for most animation houses, as (and the claim is) it compromises the artistic expression. In addition to such feature-length animation films, animation and simulation of humans is showing up considerably in video films where computer graphic humanoids are added to live footage. Columbia Pictures' *Hollow Man* (2000) and New Line Cinema's *Lord of Rings Trilogy* (2001-03) are some well-known examples from a list of thousands, each requiring many manual and artistic abilities.

On the other hand, simulation of human motion in computer games needs and has benefited from more automated techniques. The athletic motions in EA Sports', *Madden* and *NBA Live* games are generated by modeling and then tracking real athletes in action and scripting their behaviors to respond to the needs of the game dynamics and the user interactivity. Success of games like Rock Star Games' *Grand Theft Auto* and online multi player role-playing games like Lucas Arts' *Star Wars Galaxies* suggests that the graphics, simulations, and the interactive stories are engaging and are here to stay.

In this chapter, leveraging our knowledge of computer graphics as it applies to computer animation, our goal is to outline the state of the art in how humans are modeled and how human simulation can be brought to bear to the task of organizational simulation. On one hand, our goal is to summarize where the technology stands in the field and on the other our goal is to attempt to cast a wide net around what is current to the domain of organizational simulation. As researchers, we remain interested in studying how each of these two goals lead to more questions about how each domain can aid in enhancing human simulation and the related organizational simulation.

OVERVIEW OF METHODS FOR MODELING HUMANS

The first step in any form of human simulation is building models of humans and then attaching to these models various controllers to allow for manipulating the various degrees of freedom of the model. For instance, for 3D animation of a human, we need a 3D model of a person and then controls are added to this model to allow for movement of each of the limbs of the model, synthesizing motions like walking, gesturing, nodding, and to generate facial expressions. Accurate modeling of 3D humans is still a research question in graphics and to-date the two methods are widely used.

- Manual or user-driven, modeling. In this method, an expert modeler creates a model of the subject to be animated. Many modeling packages are available are in wide use for this purpose. The process is cumbersome and requires significant time and effort, with limited scalability.

- Scanning existing models. In this approach, an existing model is scanned to generate a 3D model. Then a modeler cleans up and refines the model for use.

After a model is constructed, it needs to be provided with control knobs to facilitate controlled movement; a process called "rigging", or extracting "armatures." Doing this automatically is yet another research problem. Modelers are usually known to resort to rigging a 3D model manually. An important aspect with modeling is also how to render it correctly so that it looks real. We undertake a brief overview of these approaches as it serves to guide our understanding of humans can be simulated.

Manual and User-defined Modeling of Humans

The predominant method used for manual modeling of humans is the use of well-known packages like Alias's *Maya* software or Discreet's *3dsmax* software (to name a few). Modelers using these types of software "draw" out models of people and then add to the control knobs to manipulate the various limbs. Each modeler has a special way to start the modeling process, but most start with a stick figure or skeleton and then add to the stick figure surfaces to represent limbs, torso, head, *etc.* Modeling software like the above provide some tools to add surfaces to the underlying skeleton and to blend and merge surfaces to form the whole-body shape. The existence of the underlying skeleton also aids with the eventual extraction of control knobs to animate the model.

Modelers often rely on photographs and sometimes even video to get a visual feel of the subject being modeled. After the person is modeled, there is also a need to model things like clothes and hair on the subject. Providing very exact details of the model, for instance the face and how it moves is a very time consuming process. For this reason, more and more data-driven approaches are being studied using data of human models to generate a wider variety of human models. We discuss some of the widely used techniques to aid a modeler next.

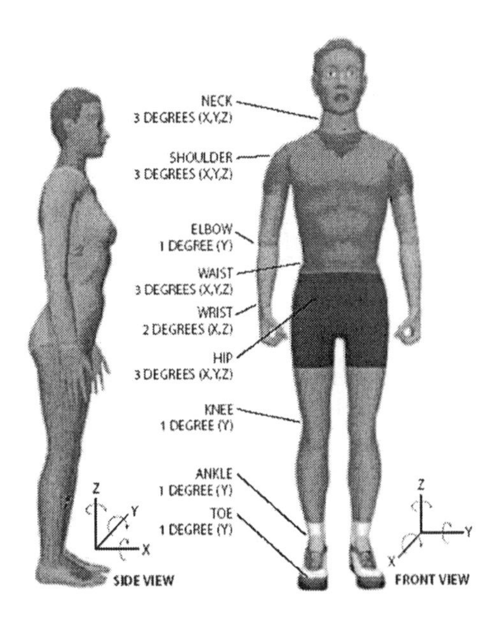

FIGURE 1. 3D models of Man and Woman with various degrees of freedom highlighted. (Image Courtesy of Jessica Hodgins).

Data-Driven Modeling of Humans

Recently scanning techniques are in wide use for modeling of humans. There are many human body scanners that are currently available commercially. Some of the more prevalent techniques rely on illuminating the subject with some sort of a structured light pattern. Then a visual sensor (or a series of sensors) is used to register this pattern from multiple viewpoints to extract geometry. Some of the high-end scanners also extract color and texture properties, which are useful for rendering.

CyberWare's whole body color scanner, CyberScan (see Figure 2) is an example of such a scanner. Much research has gone into how to extract clean geometric models from such scanners. In most instances, the geometric model still requires some manual cleanup before it can be useful. Most commercial scanners have problems scanning reflective materials and or materials with low-contrast. Hair and fine skin features are of course beyond the regular resolution of these devices. One additional problem is that most scanners requires a subject to remain completely still during the scanning procedure. Furthermore, to capture the shape of the perception as they perform simple tasks is out of the question as they are bound to be inside the volume of the scanner.

For this reason, methods for scanning 3D models still remain a topic of much research and development. In the upcoming sections, we briefly highlight some of the new approaches to support extraction of complex human models, especially with the goal of animating such models.

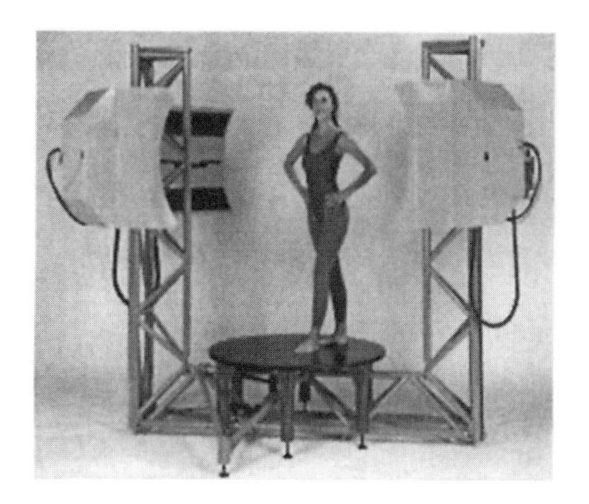

FIGURE 2. CyberWare's Whole body Color 3D Scanner. (right) A 296,772 polygon model of a person.

In addition, it is important to note that while the scanners, like the one described above, provide us with a polygonal model of the human, several important steps remain before the model can be effectively animated. One of these steps is the extraction of a skeletal model that underlies the geometric model, and then use of the skeletal model to define a set of control knobs, sometimes in the instances of skeletalized structure also referred to as the skeletal parameters, to deform the model as it is animated (see Figure 4)

Generating New Human Models from Data.

Recently, there have been new and interesting methods introduced to use a library of scanned or manually built models of humanoids to generate more models. In one specific example of this, work use cyber scans of many subjects to build a space of human body shapes, which in turn is then parameterized, allowing for interactive generation of appropriate models as needed (Allen *et al.* 2003). This study used scans of 250 volunteers in the same pose. Then they fit a canonical model to each one of the subjects in the dataset, which allowed them to generate a parameterization of the body surface of each of the data sets. A very useful aspect of this parameterization is that it also allows for the modeling of variation in body shape and size using a well-known dimensionality reduction and pattern analysis technique called Principle Component Analysis (PCA). Once such a model is learned, it is possible to generate novel individual shapes as shown in Figure 3. A similar approach for modeling and synthesized faces was proposed by Blanz and Vetter (Blanz & Vetter 1999), where CyberWare scans of heads were used to clean up datasets and generated complex face models.

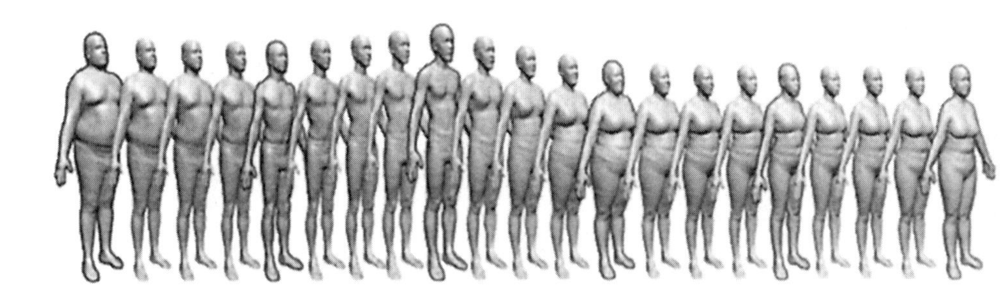

FIGURE 3. A parameterized model of human is generated using scanned data and then models are synthesized by changing few of the parameters. The models in red were synthesized (Allen *et al.* 2003).

Automatic extraction of human body shape from multiple video streams observed from multiple viewpoints has also been studied and serves to provide an effective approach for generating models quickly. In one example of such work Sand *et al.* (Sand *et al.* 2003) where deformable human geometry is acquired from silhouettes, captured from multiple cameras. In this work, the acquired models were also used to synthesis missing motions not originally captured.

Brostow and his colleagues (Brostow *et al.* 2004) have taken this process further by not only extracting the model from multiple viewpoints but also extracting an underlying structure of the subject (see Figure 4). In this approach, a subject is placed in a special environment instrumented with a number of cameras. The number of cameras in this work varied from 4 to 18 depending on the subject and the requirements of details of the model. All the cameras are calibrated to aid in fusing the information from each one with the other. The first step in the analysis is to extract the subject from the background and generate a silhouette. The silhouettes from each view are then intersected with the silhouettes from all the views to generate a visual hull. The better the coverage of cameras better is the visual hull reconstruction. This visual hull is a volumetric representation of the data and is converted to polygonal mesh.

Additional processing is then undertaken to convert the polygonal mesh (in Figure 4 (B)) to allow for representing the articulated structure, underlying the mesh. A spine, defined as a branching axial structure representing the shape and topology of a 3D object's limbs, and capturing the limbs' correspondence and motion over time, is extracted by relying on temporal and spatial consistency of the structure over time. A significant aspect of this approach is that it is purely data-driven, with no a priori information provided as to the number of limbs of the subject. A system based on this approach that is aimed at human modeling, with prior information about the human skeleton can be of much use.

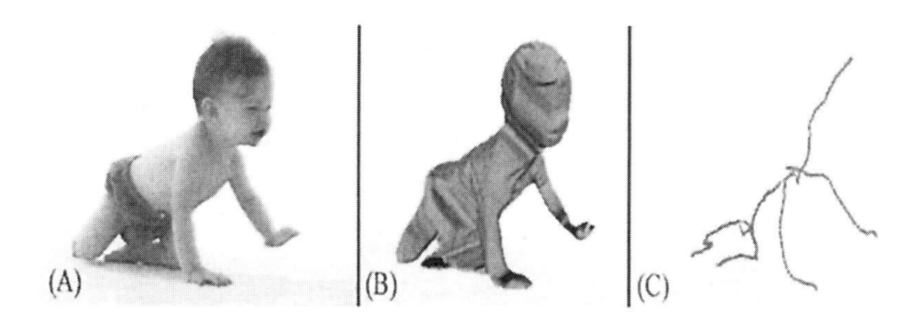

FIGURE 4. Process to Model from Multiple Viewpoints and Extracting Geometry and Underlying Structure. (A) One of the 10 views used, (B) Geometric model, and (C) Underlying structure (Brostow *et al.* 2004).

Modeling of Surface Mesh.

Generating a detailed surface model (i.e., mesh) in 3D and also making it so that it remains consistent with motions is referred to as "skinning" (Mohr & Gleicher, 2003), as the mesh effectively serves as the skin of the skeletal model. In the skinning process, the mesh geometry must be attached to the underlying structure so that as the skeleton deforms, the mesh also deforms appropriately. There are two aspects of skinning, namely authoring and computation. Authoring refers to the modelers' use of tools to define skin geometry as the skeleton moves. Computation on the other hand, is the process by which the deformed mesh geometry is evaluated for display at some skeletal state. For high-end visualization, authoring methods drive the process, as the modeler will specify significant details at smallest of movements of the skin. Computation methods are more used for interactive systems.

Some of the well-known techniques for generating human models involve modeling skin substructure such as muscles and tendons to drive the skin geometry (Wilhelms & Gelder, 1997; Scheepers *et al.*, 1997). Many systems that deform skins by linking their control points to the skeletal motion with custom expressions or scripts are also available. Some examples include free form deformation lattices (Sederberg & Parry, 1986) or wires as geometric primitives (Singh & Fiume, 1998). High-end simulations of characters will require the use a combination of these techniques and different tools are appropriate for different parts of the body. This form of generality in modeling and the related control suggests that the computational aspects for high-end characters can be highly customizable, tightly coupled to user authoring, and therefore, unbounded in terms of effort that could be used towards it.

FIGURE 5. Skeleton used to drive a skinned model of human. (Image Courtesy Jessica Hodgins).

Interactive systems require fast computation and small memory size to represent characters. Meaning that the character computation model is fixed and users must restrict their tool set to author characters in direct support of it. The most common skin computation model in games and interactive systems are referred to as enveloping, smooth skinning, or linear blend skinning. This technique assigns a set of influencing joints and blending weights to each vertex in the character. The skin is computed by transforming each vertex by a weighted combination of the joints' local coordinate frames. While this approach is considered fast, due to being compact in memory, it is significantly difficult to author and has several undesirable deformation artifacts. Nonetheless, this method is in wide use as these humanoids can be used with arbitrary amounts of animation data and can be posed at runtime.

Rendering of Humans

Modeling just the geometry of a subject is of course not sufficient. Accurate rendering of the subject is also needed. While scanners like the CyberScan do capture color values to support reproduction of the model with color, accurate reproduction still requires dealing with reflectance properties of, for example, the skin and how the lighting effects the model.

Significant advances have recently been made in realistic synthesis using global illumination and appearance modeling based on techniques like Photon Mapping (Jensen, 2001). Photon mapping is an extension of ray tracing, a well-known process where all light rays are traced to their eventual destination while trying to simulate the physics of light in a CG world. Photon mapping can simulate caustics (focused light, like shimmering waves at the bottom of a swimming pool), diffuse inter-reflections (*e.g.*, the "bleeding" of colored light from a red wall onto a white floor, giving the floor a reddish tint), and participating media (such as clouds or smoke). Photon mapping makes it possible to efficiently simulate global illumination in complex scenes. The goal of global illumination is to compute all possible light interactions in a given scene, and thus obtain a truly photorealistic image. All combinations of diffuse and specular reflections and transmissions must be accounted for. Effects such as color bleeding and caustics are also included in a global illumination simulation. Global illumination is widely used in computer graphics these days especially on feature films. The computational cost of computing images with global illumination still prevents use in interactive and real-time situations.

There is also some very exciting work on capturing reflectance fields of human faces by Debevec and his colleagues (Debevec *et al.* 2000). In this work, a Light Stage is designed, in which a small spotlight moves around a person's head (or a small collection of objects) illuminating it from all possible directions in a small period of time. A series of video cameras are used to capture the person's face under these changes in lights. From this data, a person's appearance under any complex lighting condition is simulated.

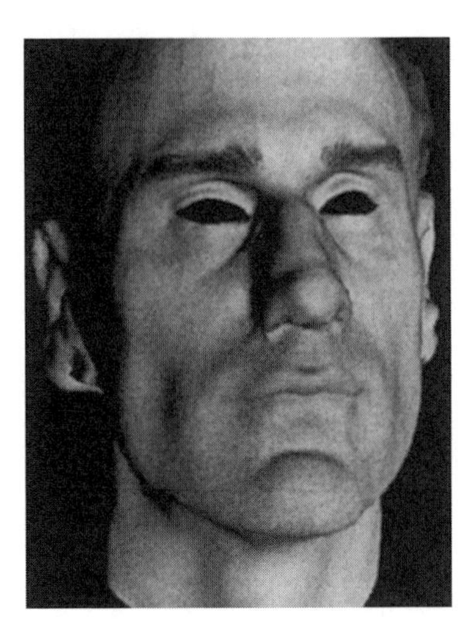

FIGURE 6. Face Model with Skin Texture and Normal Maps (Haro *et al.* 2001).

Haro and his colleagues (Haro *et al.* 2001) have also added to the realism of faces by developing a highly-detailed model of human skin for rendering faces. In this approach, in addition to just showing the texture map of the skin, finer details of skin microstructure are added by incorporating normal maps, which are modeled after real skin. Such texture maps with normal maps are synthesized over the whole face, with different characteristics. The results make the faces look even more real. This process is real-time and can be used for interactive approaches too.

In addition to skin details, there is also some important work dealing with light scattering from human hair fibers by Marschner (Marschner *et al.* 2003).

OVERVIEW OF METHODS FOR ANIMATING HUMANS

Once a geometric model of human is constructed and the material properties (texture, color, *etc.*) are modeled, the next step is making this model move. We provide a very brief overview of some the methods used to simulate such movements and list their advantages and limitations. (Parent 2001) presents a more thorough overview of the field of computer animation. (Lasseter 1987) provides a very interesting overview of how traditional animation has moved to computer animation. A concise description of simulation humans is presented by (Hodgins 1999). As categorized by Hodgins, there are three main approaches to generate realistic human motion simulation. In addition to these three, we add one more approach that is needed for organizational simulation application, i.e., goal-driven animation, which builds on the first three approaches of key framing, data-driven animation, and physical simulation.

Key frame-Driven Animation

The technique of key framing for 3D animation of humanoids is basically inherited from traditional animation. In traditional animation, an artist would manually specify a few critical or key positions of the object along a path that is desired and then interpolate the motion. In computer animation, a computer is used to generate the in-between frames with constraints that allow for smooth motion. Key framing is perhaps the most used approach for animated films. PIXAR claims to rely on key framing as its sole character animation tool.

For 3D computer animation of humans, key framing is a very powerful tool as the computer can be used to aid in both specification of the key poses of the human and then synthesizing all the transitional frames between those poses. The more key frames or poses the animator specifies, the more realistic behavior is visible as the control is more refined on the motion. The problem of course is the more key frames that need to be specified, more the labor for the animator. It is claimed that each character in PIXAR's *Toy Story* (1995) had about 700 controls, which means that an animator had to specify key frames by manipulating that many number of controls. It should also be obvious that defining of the control knobs (*i.e.*, the rigging process) is therefore extremely crucial as how well these are defined effects how good the motion will be. It is for this reason, animators are directly involved with the modeling and rigging process as much as they can.

For animation of the articulated motions of the whole body, the underlying stick figure, or the skeleton is the most significant. Consider the example of how an animator will make a humanoid model walk. The animator will move each joint of each foot to specific locations and then visualize the interpolated data. After getting the feet to move well (and this includes the ankle, knee, and hip joints), the animator will add key frames for each hand and then the head, torso etc. After doing one whole step, he has to repeat the whole process. Luckily, computers support copying and this reduces excessive repetition of the same key frame specification. However, suppose that after a few steps, the humanoid has to turn or step over something. Both of these and variety of other similar motions will require a complete new set of keys and interpolation. Now consider even a more difficult scenario and that is the humanoid is wearing a flowing skirt, which moves with the movement of the character. This requires adding key frames for cloth motion that is coupled with the whole-body motion. While this type of motion simulation was done in the past using key framing, more recently physical simulation methods are used as will be discussed later.

Animation of facial motion leads to yet another set of interesting challenges as now the motion to be simulated is no longer articulated, but non-rigid. The bigger problem with this form of non-rigid motion synthesis is determining an appropriate set of parameterizations allowing for control, such as the skeletal mechanism does for the whole body. For this purpose, computer animators rely on two different forms of representations.

In one approach, a parameterization proposed by psychologists and anthropologists to code (and categorize) facial expressions was used. This was representation called Facial Action Coding System (FACs) is based on seminal work of Ekman and Friesen (Ekman & Friesen 1978). While not completely suitable for computer-based analysis and synthesis of facial expressions as shown by Essa and Pentland (Essa & Pentland 1997), its impact as the underlying key frames for facial animation is hard to deny. Figure 7 shows some of the variations associated with the FACS AUs.

The second approach builds on the FACs representation, but adds another layer to provide control for animation. This approach presented by Waters (Waters 1987) relies on using the underlying muscles of faces as the controller for facial motion, with an animator specifying keys on the muscles to generate motion. Parke and Waters (Parke & Waters 1994) discuss both of these approaches and related techniques and technical details.

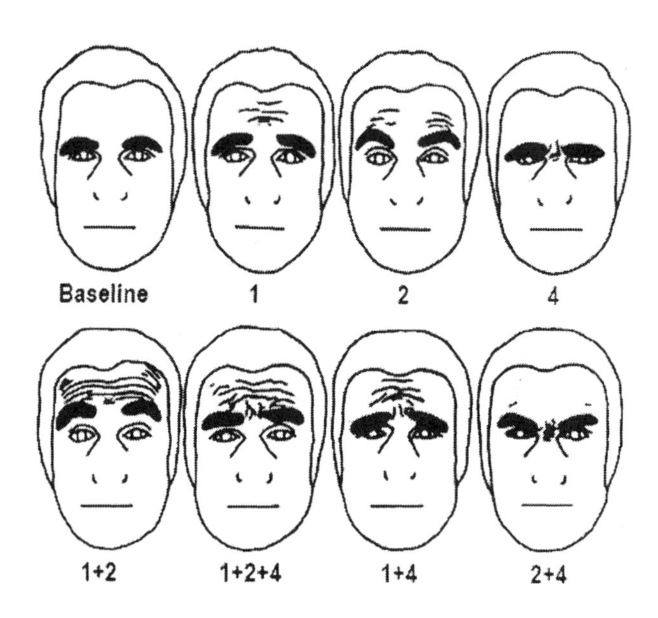

FIGURE 7. Facial Action Coding System, Action Units (AUs) 1,2, 4 and their combinations shown (Ekman & Friesen 1978).

Data-Driven Animation

Recently a new approach has become very common that removes the complexity associated with providing a large number of key frames. Initially called performance-based animation (Williams 1990) this approach is based on the recognition that what, in essence, is needed to animate any motion is to just have someone perform it. This approach is now referred to as motion-capture or data-driven animation. The basic premise still remains and that is to animate a specific motion, have someone perform it. The main issues with this method are (a) how such a performance can be captured, (b) how such performance data can be repurposed and reused, and (c) how such data can be used to generate data not in the database.

Capturing Performances.

The concept of using performance to guide animation is as old as traditional animation. Animators used to trace pictures of characters to inform themselves about how to generate motions. In fact, even in early traditional movies like Disney's *Snow White and the Seven Dwarfs* (1937), animators used to study stock footage of dancers to help them animate the dancing sequences. Legendary innovators like Etienne-Jules Marey and Eadweard Muybridge were using photographs to study human motion -- though not for the purpose of animation -- as early as the mid to late 1800s (Braun 1992). Since those early days, the concept and the technologies that support capturing motion have matured considerably.

There are a wide variety of available sensing infrastructures for capturing facial and whole body motions. These include technologies to track limb movements using magnetic sensors to optical sensors to track passive markers (illuminated by special IR light) or active markers. Figure 8 shows these motion capture systems. These systems are now in wide use, being applied for motion capture in movies, computer games, and also for motion analysis for physical therapy.

Motion capture technologies, while demonstrating huge successes of late, are still limiting. One basic limitation is that the users have to be somehow instrumented (*i.e.*, markers attached or wear a special costume, *etc.*) to aid in tracking. This, while still practical, is a cumbersome process, sometimes directly interfering with the performance. Another limitation is the use of multiple views and special purpose capturing equipment that requires he performance be really done for the purpose of capture.

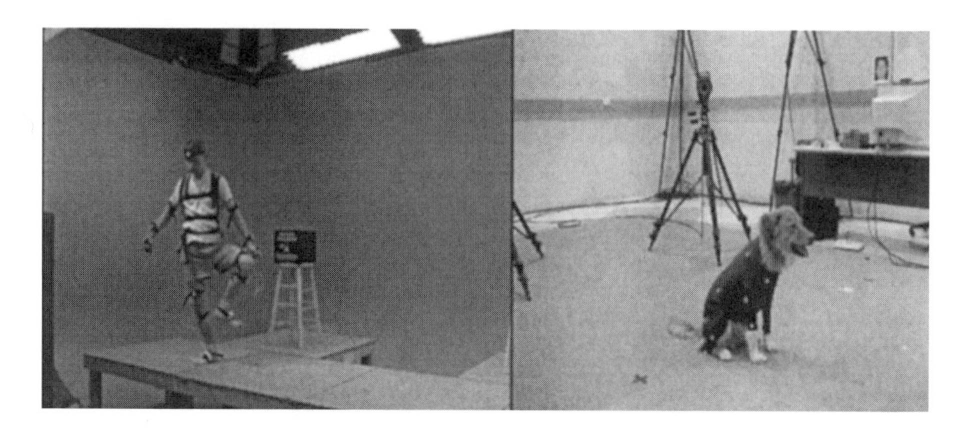

FIGURE 8. (Left) A person wearing a magnetic motion capture system. (Right) a pet wearing a suit with balls painted with reflective material, which is illuminated by an IR light and tracked by the 12 cameras (one seen behind the pet).

These limitations have led to computer vision research for tracking movements in monocular (single view) footage. Success of such efforts opens many interesting doors for motion capture. For example, Cham and Rehg (Cham & Rehg 1999) introduce a multiple-hypothesis tracker to extract the articulated motions of Fred Astaire in the classic *Shall We Dance* (1937) (See Figure 9). Capturing of motion information from all forms of legacy footage can aid in extracting motions, which in turn can be used for effective synthesis of realistic motions, which has also been demonstrated in further work by Rehg and his colleagues (Rehg *et al.* 2003) for articulated motions and by Essa (Essa *et al.* 1996) for head and face movements.

FIGURE 9. Vision-based tracking used to track dancing motions from monocular view in legacy footage (Cham & Rehg 1999).

As discussed, earlier, facial animation, while in principle similar, differs in details from motion capture of whole-body movements. Figure 10 shows a subject wearing markers, which were used to track facial motions, which in turn were registered to the facial mesh. In this work by Reveret and Essa (Reveret & Essa 2001), detailed models of lip motions were generated for synthesis with speech. To build these models, one subject was tracked in a motion capture facility while speaking. A model of how the geometry deformed for various utterance was matched to phonemes, which are segments of words that are the most discriminative in the audio channels. The mesh transformation during various phonemes was then modeled using PCAs to generate a compact representation. Then this compact representation was used to track other subjects speaking with no markers and tracked in video. Figure 11, shows the schematic of this approach. The new tracking provided data that could then be targeted to any model of a face for animation.

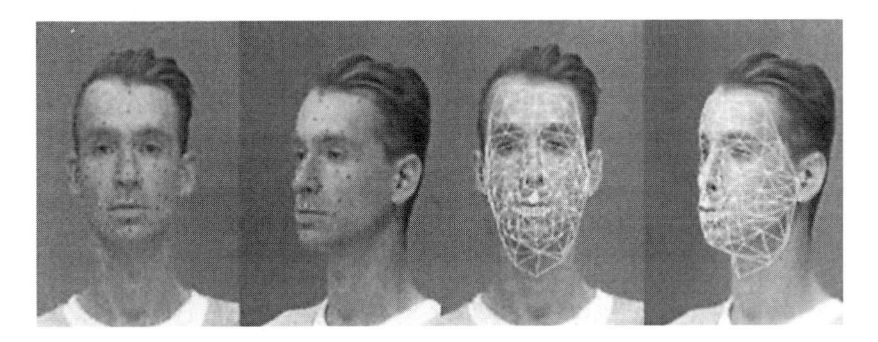

FIGURE 10. A human participant with IR reflective markers on his face and a mesh grid used to model the face. Image from (Reveret & Essa 2001).

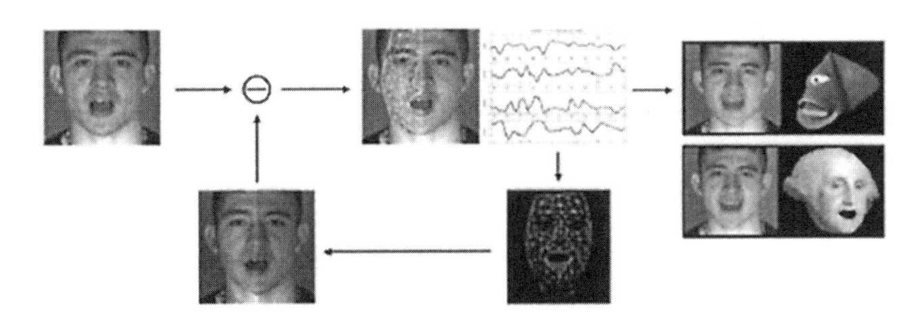

FIGURE 11. Process to measure audio signal for lip motion analysis (Reveret & Essa 2001).

Some new work on video-based analysis has also focused on just using the video information itself to generate animations. The novelty of this approach is simply that no 3D geometric model is constructed, and the video is itself replayed in a controlled manner to generate animations. We discuss this approach in a bit more detail in the next section. In final analysis capturing data of performances has been a huge success for simulation human motion, as perhaps is evident from availabilities of large databases of motions for research and testing (see mocap.cs.cmu.edu) and its application is production and games.

Using Captured Motions.

Once the motion is captured, the work of an animator is not done, as it is highly unlikely that the data captured is perfect for the task at hand. In addition to providing tools for cleanup of motion capture data, there have been many tools developed to aid in using the motion capture data, and also to generate new motions by combining snippets of motions from the data set.

Some of the earliest efforts in processing motion capture data were aimed at efficient processing and editing of such data. These methods included applying signal processing techniques to extract sampling and frequency information (Bunderlin & Williams, 1995), to adding constraints for editing of motions (Gleicher, 1997) to retargeting the motion to models of different shapes and sizes (Gleicher, 1998). Rose and his colleagues (Rose *et al.* 1998) proposed a method to generate motions from the data by interpolation that relied on a higher-level description of motions and actions. The next round of research in this area of motion capture for human simulation took a bit of a different approach. In these methods the emphasis was more on using and reusing existing data then to transform it into different parametric spaces. To this end, following the lead of example-based synthesis for video by Schödl and his collaborators (Schödl *et al.* 2000; Schödl & Essa 2001), a method for splicing together similarly repeating motions has been introduced for simulating humans from motion capture data (Kovar et. al., 2002).

Use of such example-based techniques, which rely on re-sequencing existing datasets with additional variants, has proven to be a strong animation technique for simulating various complex set of motions. Arikan and Forsyth (Arkian & Forsyth 2002) have demonstrated such techniques for motion capture data, while Schödl and Essa (Schödl & Essa 2002) have used this to generate novel animations by re-sequencing video regions with Monte-Carlo Optimization. FIGURE 12, shows four frames from this synthesized video with 2 hamsters. Only one hamster was video taped and then both were composited on the new video plate with the books to generate this simulation.

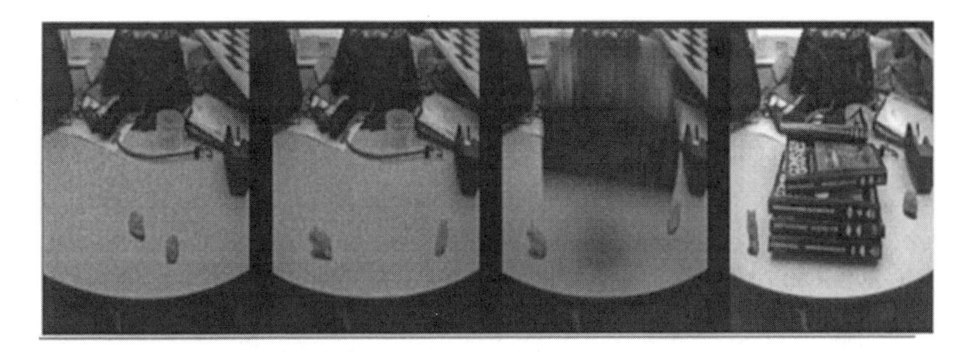

FIGURE 12. Animation by controlled sequence of video sprites from a given data set (Schödl & Essa 2002).

Physics-Driven Animation

While the data-driven methods have significantly enhanced the field of human simulation, there are still a few limitations that prevent its usage in all situations. Consider the task of simulating a runner, running very fast, or a diver, diving off a platform. To motion capture a runner at fast speed would require a very large space with a lot of high-speed sensors. For diving, it would require setting up a performance capture in *situ*. While, both of these are being addressed and developed to enhance motion capture, one problem remains and that is that these types of motions are inherently dynamic, i.e., there is a significant amount for variation in time resulting from the physical forces in the environment. This means that any kind of adaptation or changing the motion requires that correct dynamics of motions be preserved. This is primarily complicated by the fact that as humans we are very sensitive to the "correctness" of motions and we can tell if some principles of physics are being compromised. This has led to work on animating humans using physical simulations.

For physical simulation, in addition to having geometric models, we also need to represent, hopefully simply, the physical attributes of the human body. This means that each limb has to have some mass, and connections between limbs have to be modeled to deal with inertial and damping forces. In addition, contacts and collisions between parts of the body and of the body with the environment (e.g., foot hitting the ground) also need to be represented. Hodgins, working with her team (Hodgins *et al.* 1995; 1997) has perhaps done the most amount of work in this domain. A main contribution of this work was the building of controllers to allow for simulation of those types of characters that were doing physical activities like running, jumping, and diving. The characters in this work would adapt to simple changes in the environment. One problem with this type of animation is that it is not always easy to determine the ideal controller for a given simulation. This has been addressed using numerical machine learning techniques (e.g.,

(Grzeszczuk *et al.* 1998)), though applying it to complicated articulated structures is a challenging problem, addressed to some extent in the literature.

Physical simulation of humans has definite benefits for simulating real physical movements, but the present feeling in the animation community is that for simple motions like walking, data-driven methods are more productive, as the variety is easy to capture and simulate. The domain where motion capture (and for that matter key framing) does not offer a good solution is for highly-physical motions like hair, clothes, and fluids.

It is for this reason that simulation of hair and clothes on human bodies and fluids in an environment are primarily left to physics-driven methods. This is also evident from recent film productions like Sullivan's Hair in Pixar's *Monsters Inc* (2002). While these forms of physical simulation methods to generate motions that support the character animation (these animations are therefore referred to as secondary motions) are widely used in production, most of them are rarely real-time. Some recent work on GPUs for numerical processing is being brought to bear on this to problem to aid in simulation in interactive games. There is of course significant need for simple simulations in games with many computer game engines now supporting simple physical simulations to show motions of objects colliding and bouncing off each other.

Goal-Driven Animation

So far, we have concentrated on techniques for simulating and animating humans, with emphasis on low-level types of movement and also local actions like walking, running, pointing and making expressions. Now let's move to higher-level behaviors which are essential for synthesis of humans in contexts like organizational simulation. For example, we want to simulate a humanoid that can undertake a series of purposive actions as desired by the context of the situation. This leads us to the domain of autonomous and believable graphical agents.

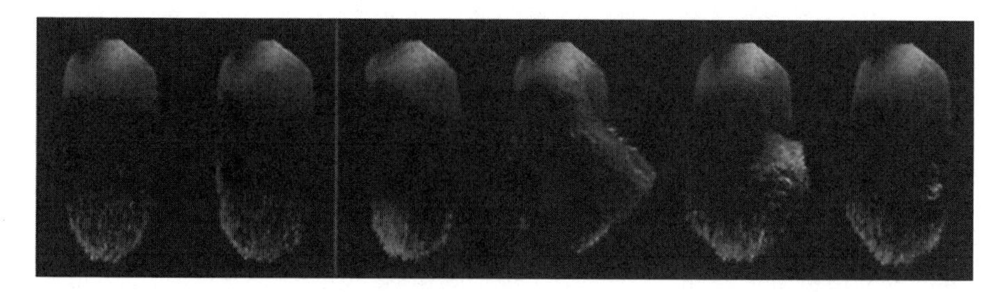

FIGURE 13. Hair Simulation 25 Simulated Hair Filaments Driving 25000 Hairs to Show the Hair Blowing Effect. Project in Professor's Essa's class on Computer Animation by David Cunningham in 2002.

Most work in believable agents (autonomous agents incorporating rich personalities, emotion and social interaction) has focused on AI architectures for solving the action selection problem in the context of personality expression and intentional, affective, and social states. Some of the best work in this area includes (Bates *et al.* 1992; Blumberg 1996; Hayes-Roth *et al.* 1997; Lester & Stone 1997; Cavazza *et al.* 2001; Mateas 2002). Of course, a rich computational model of a human model is useless unless it can be expressed through action, typically the movements of an animated, screen-based agent.

The autonomous agent research has focused on the developing the minds, the intelligence and reactiveness of the agents. The methods described above have served to build bodies that play back and layer animation data, perhaps with a small amount of procedural warping. (This is the typical solution employed in computer games.) While significant work has been done in both communities, there are some gaps between the richness of the character model and the ability to express this model through motion. The graphics community has focused on realistic humanoid animation, solving problems such as foot skate, with no modeling of intentional or affective state, and consequently no attempt to express this state, producing bodies without minds. This work has tended to be either fully procedural, viewing humanoid animation as a complex control theory problem or data-driven, viewing humanoid animation as an interpolation problem over a large set of motion capture data. Neither approach incorporates the "semantic knobs" necessary to produce expressive parameterization of motion, e.g. "walk across the room sadly but with purpose", or "lift the drink to your mouth thoughtfully".

There are however some attempts at building responsive animated characters that are in 3d worlds. For example, in the IMPROV project (Perlin & Goldberg, 1996) characters with complex sets of behaviors were developed to interact with the users in real-time. Most of these characters and their behaviors were handcrafted. However, a simple, yet effective approach for layering their actions was introduced to make these characters appear believable. A narrative was also used to aid in engaging the user with the character. A similar approach was used by Mateas (Mateas 2002; 2002) for interactive stories in form of dramas.

Use of story in an immersive and interactive setting is best demonstrated by MacIntyre and Bolter (MacIntyre & Bolter 2003), in their augmented reality, dramatic experience entitled "Three Angry Men" where a user wearing a head mounted display can view and interact with other characters. The human characters are simply following a script and have limited autonomous behaviors. However, as noted earlier this area is making much progress under the guise of autonomous agent's research.

A notable example of such autonomous creatures is work by Blumberg (1996, 2002) where ethologically inspired models are merged with learning algorithms to generate believable synthetic characters.

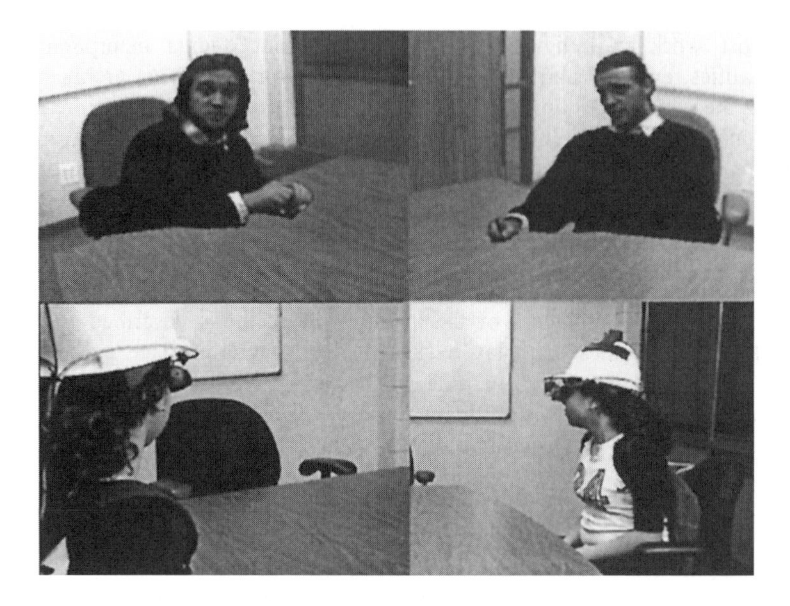

FIGURE 14. User with Head-Mounted Display Interacting with Virtual Characters in "Three Angry Men", (MacIntyre & Bolter, 2003).

CONCLUDING REMARKS

We have presented a brief yet fairly thorough exposition of the issues that are important for synthesizing humans. We have described various approaches that are in wide use for generating human models and then animating them. The descriptions presented in this chapter should not in any way be considered exhaustive, as this field of research, development and practice is quite large.

As also noted, the largest impact to date of human simulation has been in the fields of film production and computer games. While there is some recent work on simulating people for planning and synthesizing for human factors experiments (Badler et al., 2002) and in battlefield situations (Gratch, 2002), the impact of human simulation in organizational context is still unrealized.

However, it is important to also emphasize that although the state of the art in human animation and simulation is quite advanced and has demonstrated significant progress in the last decade or so, this work is far from complete. To achieve effective organizational simulation with human animation, we still need to undertake much work to develop engaging and believable agents. We need humanoids that can act autonomously and work with us effectively so that we can interact with them naturally. While simple interactions are possible at present, we need to undertake much work in building persistent, and socially acceptable agents that appear real and can support extended interactions that are required to simulate complex organizations.

REFERENCES

Allen, B., Curless, B., & Popović, Z., (2003). "The space of human body shapes: reconstruction and parameterization from range scans," ACM Transactions on Graphics (TOG), Vol. 22, No. 3, (Special Issue on Proceeding of ACM SIGGRAPH 2003 Conference, San Diego, CA).

Badler, N., Phillips C. B., & Webber, B. L., (1993). Simulating humans: computer graphics animation and control, Oxford University Press, Inc., New York, NY.

Badler, N. Erignac, C., & Liu, Y., (2002) "Virtual humans for validating maintenance procedures, Comm. of the ACM, Vol. 45, Issue 7, Pg. 56-63.

Bates, J., Loyall, A. B., & Reilly, W. S. (1992). Integrating Reactivity, Goals, and Emotion in a Broad Agent. Proceedings of the Fourteenth Annual Conference of the Cognitive Science Society, Bloomington, Indiana..

Blanz, V., & Vetter, T. (1999). "A Morphable Model for the Synthesis of 3D Faces". In Proceedings of ACM SIGGRAPH 99, ACM Press/Addison-Wesley Publishing Co., New York, A. Rockwood, Ed.,Computer Graphics Proceedings, Annual Conference Series, 187.194.

Blumberg, B. (1996). *Old Tricks, New Dogs: Ethology and Interactive Creatures.* Ph.D. Dissertation. MIT Media Lab.

Blumberg, Downie, Ivanov, Berlin, Johnson, and Tomlinson (2002) "Integrated Learning for Interactive Synthetic Characters." In Proceedings of SIGGRAPH 2002, San Antonio, Texas.

Braun, M., Picturing Time. (1992) University of Chicago Press.

Brostow, G., Essa, I., Steedly, D., & Kwatra, V. (2004) "Novel Skeletal Representation For Articulated Creatures" in Proceedings of European Conference on Computer Vision (ECCV 2004), May 2004, Springer-Verlag Vol III: 66-78.

Bruderlin, A., & Williams, L., (1995) "Motion signal processing," Proceeding of ACM SIGGRAPH. ACM Press.

Cavazza, M., Charles, F., & Mead, S. (2001). Characters in Search of an Author: AI-based Virtual Storytelling. Proceedings of the International Conference on Virtual Storytelling. Avignon, France.

Cham T.-J. & Rehg, J. M. (1999) "Dynamic Feature Ordering for Efficient Registration" Intl. Conf. on Computer Vision, volume 2, pages 1084–1091, Kerkyra, Greece, Sept. 20-27.

Chi, D., Costa, M., Zhao, L, & Badler, N. (2000) "The EMOTE model for Effort and Shape", Proceedings of ACM SIGGRAPH 2000, New Orleans, LA, July 2000, pp. 173-182.

Debevec, P., Hawkins, T., Tchou, C., Duiker, H-P., Sarokin, W., & Sagar, M., (2000) "Acquiring the Reflectance Field of a Human Face" in Proceedings of ACM SIGGRAPH 2000.

Ekman, P. & Friesen W.,(1978).Facial Action Coding System. 577 College Avenue, Palo Alto, California 94306: Consulting Psychologists Press Inc.

Essa, I. & Pentland, A. (1997). Coding, analysis, interpretation, and recognition of facial expressions. IEEE Transaction on Pattern Analysis and Machine Intelligence, 19(7), 757-763.

Essa, I., Basu, S., Darrell, T., & Pentland, A. (1996). "Modeling, tracking and interactive animation of faces and heads using input from video". In Proceedings of Computer Animation Conference 1996, (pp. 68-79). IEEE Computer Society Press.

Gratch, J., Rickel, J., André, E., Badler, N., Cassell, J., & Petajan, E. (2002) Creating Interactive Virtual Humans: Some Assembly Required. IEEE Intelligent Systems July/August 2002, 54-63.

Gleicher, M., (1997) "Motion Editing with Spacetime Constraints." Proceedings of the 1997 Symposium on Interactive 3D Graphics. 1997.

Gleicher, M., (1998) "Retargetting Motion to New Characters" Proceedings of SIGGRAPH 98. In Computer Graphics Annual Conferance Series. 1998.

Grassia, S. (2000). Believable Automatically Synthesized Motion by Knoweldge-Enhanced Motion Transformation. Ph.D. Dissertation. Tech report CMU-CS-00-163, Carnegie Mellon University.

Grzeszczuk, R., & Terzopoulos D., (1995) "Automated learning of muscle-actuated locomotion through control abstraction," R., Proc. ACM SIGGRAPH 1995 Conference, Los Angeles, CA, August, 1995, in *Computer Graphics* Proceedings, Annual Conference Series, 1995, 63-70.

Haro, A., Guenter, B., & Essa, I. (2001) "Real-time, Photo-realistic, Physically Based Rendering of Fine Scale Human Skin Structure." Proceedings 12th Eurographics Workshop on Rendering, London, England, June 2001.

Hayes-Roth, B., van Gent, R. & Huber, D. (1997). "Acting in character". In R. Trappl and P. Petta (Eds.), Creating Personalities for Synthetic Actors. Berlin, New York: Springer.

Hodgins, J. K. (2000) "Animating Human Motion," Scientific American, pp 64—69, March 2000.

Hodgins, J. K., & Nancy S. Pollard (1997) "Adapting Simulated Behaviors for New Characters." in Proceedings of ACM SIGGRAPH 1997. ACM, 1997.

Hodgins, J. K., Wooten, W. L., Brogan, D. C.., & O'Brien, J. F. (1995)

"Animating Human Athletics." in Proceedings of SIGGRAPH 1995. ACM,1995.

Jensen, H. (2001), Realistic Image Synthesis Using Photon Mapping, AK Peter, July 2001.

Kovar, L., Gleicher, M., & Pighin, F. (2002) "Motion Graphs" ACM Transactions on Graphics (TOG), Proceedings of the 29th annual conference on Computer graphics and interactive techniques, 21(3), (SIGGRAPH 2002) July 2002.

Lasseter, J., (1987) "Principles of Traditional Animation Applied to 3D Computer Animation", Computer Graphics, pp. 35-44, 21:4, July 1987.

Lester, J., & Stone, B. (1997). Increasing Believability in Animated Pedagogical Agents. Proceedings of the First International Conference on Autonomous Agents. Marina del Rey, CA, USA, 16-21.

MacIntyre B., & Bolter, J., D., (2003). "Single-Narrative, Multiple Point-of-View Dramatic Experiences in Augmented Reality" In The Journal of Virtual Reality .

Mateas, M. (2002). Interactive Drama, Art and Artificial Intelligence. Ph.D. Dissertation. Tech report CMU-CS-02-206, Carnegie Mellon University.

Mateas, M. & Stern, A. (2002). "A behavior language for story-based believable agents", *IEEE Intelligent Systems,* July/August 2002, 17 (4), 39-47.

Marschner, S., R., Jensen, H. W., Cammarano, M., Worley, S., & Hanrahan, P. (2003): "Light Scattering from Human Hair Fibers". Proceedings of ACM SIGGRAPH 2003.

Mohr, A. & Gleicher, M., (2003) "Building Efficient, Accurate Character Skins from Examples", in ACM Transactions on Graphics (TOG), Volume 22 Issue 3, (Special Issue on Proceeding of ACM SIGGRAPH 2003 Conference, San Diego, CA), July 2003.

Mubridge, E. (1955), The Human Figure in Motion, Dover Publications. From material initially printed in 1887.

Parent, R. (2001), Computer Animation: Algorithms and Techniques, Morgan-Kaufmann, San Francisco, 2001.

Parke F., & Waters, K. (1996). Computer Facial Animation. AK Peters.

Perlin, K. & Goldberg, A. (1996) "Improv: A System for Scripting Interactive Actors in Virtual Worlds" Computer Graphics; Vol. 29 No. 3., 1996.

Rehg, J. M., Morris, D. D., & Kanade, T.(2003) "Ambiguities in Visual Tracking of Articulated Objects Using Two- and Three-Dimensional Models", Int. J. of Robotics Research, 22(6):393-418, June 2003.

Reveret, L. & I. Essa (2001), "Visual Coding and Tracking of Speech Related Facial Motion", In Proceedings of CUES 2001 Workshop, Held in conjuction with

CVPR 2001, Lihue, Hawaii, Dec 2001, also available as Georgia Tech, GVU Center Tech Report No. GIT-GVU-TR-01-16

Rose, C., Cohen, M., & Bodenheimer, B. (1998), ``Verbs and Adverbs: Multidimensional Motion Interpolation'', IEEE Computer Graphics and Applications, v. 18, no. 5, Sept. 1998, pp. 32-40.

Sand, P., McMillan, L., & Popović, J., (2003) "Continuous Capture of Skin Deformation" in ACM Transactions on Graphics (TOG), Volume 22 Issue 3, (Special Issue on Proceeding of ACM SIGGRAPH 2003 Conference, San Diego, CA), July 2003.

Scheepers, F., Parent, R. E., Carlson, W. E., and May, S. F. (1997), "Anatomy based Modeling of the Human Musculature," in Proceedings of ACM SIGGRAPH 1997, pp. 163-172.

Schödl, A. & I. Essa. (2001) "Machine Learning for Video-Based Rendering". In Todd K. Leen, Thomas G. Dietterich, and Volker Tresp, editors, *Advances in Neural Information Processing Systems*, volume 13, pages 1002-1008. MIT Press, USA, 2001.

Schödl, A., Szeliski, R., Salesin, D. H. & Irfan Essa. (2000) "Video textures." Proceedings of ACM SIGGRAPH 2000, pages 489-498, July 2000.

Schödl, A., & Essa, I, (2002) "Controlled Animation of Video Sprites." In Proceedings of the First ACM Symposium on Computer Animation (held in Conjunction with ACM SIGGRAPH 2002), San Antonio, TX, USA, July 2002.

Sederberg, T. W., & Parry, S. R. (1986). "Free-form deformation of solid geometric models." In Proceedings of ACM SIGGRAPH 86, Annual Conference Series, ACM SIGGRAPH, 151-160.

Singh, K., AND Fiume, E. L. (1998). "Wires: A geometric deformation technique" .In Proceedings of ACM SIGGRAPH 98, Annual Conference Series, ACM SIGGRAPH, 405-414.

Tomlinson, and Blumberg B. (2002) "Social Synthetic Characters" Computer Graphics Vol. 26, No. 2. May 2002

Waters, K. (1987) "A muscle model for animating three-dimensional facial expression". ACM SIGGRAPH Conference Proceedings, 21(4):17–23, 1987.

Wilhelms, J., & Gelder, A. V. (1997), "Anatomically Based Modeling," in ACM SIGGRAPH 1997, ACM Press, pp. 173-180.

Williams, L., "Performance-driven facial animation". ACM SIGGRAPH Conference Proceedings, 24(4): 235–242, 1990.

CHAPTER 17

MODELING CROWD BEHAVIOR FOR MILITARY SIMULATION APPLICATIONS

R. BOWEN LOFTIN, MIKEL D. PETTY, RYLAND C. GASKINS III, AND FREDERIC D. MCKENZIE

ABSTRACT

Crowds play a large and increasing role in modern military operations, and often create substantial difficulties for the military forces involved. Military operations in Mogadishu, Bosnia, and, most recently, Iraq exemplify the possibly significant effects crowds may have in warfighting, peacekeeping, and nation-building activities. Yet, in spite of the difficulties and risks imposed by crowds, models of crowds are either absent or present in overly simplified forms in the current simulations used by the military. In the past, crowds were not a significant factor in the types of military scenarios anticipated, so their absence from military simulations was tolerable. However, the threat has changed, and crowds are more important in current and future military actions. Including realistic crowd models in simulations can provide substantial increases in validity and usefulness for training, analysis, experimentation, and acquisition applications.

A requirements analysis method proposed for human behavior modeling in general has been adapted and applied to identify crowd-modeling requirements and relate them to the intended uses of military simulations. A large number of crowd modeling requirements were found; those requirements were expressed in various forms, including crowd behaviors needed, missions affected by crowds, and crowd effects to model. The requirements provide a firm basis for implementing useful crowd models, but because of the requirements' diversity in scope and intent no single crowd model is likely to satisfy all of the requirements.

There are many psychological variables that may influence crowd behavior. Few existing models of crowd behavior have strong underpinnings in psychology. An improved understanding of the contributions of cognitive psychology, including better connection of cognition to behavior, is essential to provide a sound psychological basis for crowd models. Previous psychological research suggests that models of crowd behavior should consider crowd members' past experiences and expectations for future outcomes as well as those cultural differences that can have an important influence on crowd behavior.

Implementing a crowd simulation that is useful with current simulations requires resolving both technical and modeling challenges. The technical challenges include developing a reconfigurable software architecture to allow future advances in crowd cognitive and physical modeling to be more readily integrated and ensuring correlation between the disparate terrain database formats used by the crowd simulation and the other simulations with which the crowd simulation would interoperate. The modeling challenges include basing the crowd simulation on psychologically credible models of how crowds behave in relevant scenarios and in developing physical and visual models of crowd members that are realistic enough to provide utility without being excessively computationally expensive.

INTRODUCTION

This section gives a motivation for the study of crowd modeling, connects crowd modeling to organizational simulation in general, and overviews the crowd modeling study methodology used in the research described in the rest of this chapter.

Crowd Modeling Motivation

Crowds of non-combatants play a large and increasing role in modern military operations, and often create substantial difficulties for the combatant forces involved.

> "In Somalia, U. S. Marines often faced hostile crowds of rock-throwing women and children. In Bosnia, U. S. Army soldiers had to disperse angry mobs of Serb hard-liners near the town of Banja Luka. More recently, Danish, French, and Italian forces attempted to control riots between ethnic Albanians and Serbs in Mitrovice, Albania." (Kenny & Gilpin, 2002)

> "All military operations, large or small, have a crowd control/crowd confusion factor. ... (C)rowds are one of the worst situations you can encounter. There is mass confusion; loss of control and communication with subordinates; potential for shooting innocent civilians, or being shot at by hostiles in the crowd; potential for an incident at the tactical level to influence operations and policy at the strategic level." (Ferguson, 2003)

In spite of the military challenges and risks imposed by crowds, models of crowds are essentially absent from current production military simulations. This omission has been understandable in the context of legacy simulations that were historically focused on large-scale engagements between heavy mechanized forces in primarily non-urban settings. However, in the last decade the threat has changed and future engagements are expected to often involve lighter forces in urban settings. In simulations of such scenarios the absence of crowds and of non-combatants in general is a more serious departure from realism. The absence of models of crowds in military simulation, and the need to include them, has been widely recognized:

> "Military forces are increasingly called upon to support operations other than war in which they come into contact with civilian populations. In some cases, the interaction takes place with crowds of civilians. Unfortunately, the computer generated forces that support virtual training systems do not yet support the simulation of crowds of civilians." (Reece, 2002)

> "Representations are needed for ... (neutrals or civilians) to represent operations other than war and the interactions among these forces." (Pew et al, 1998)

> "With the Army's growing emphasis on low-intensity conflicts and operations other than war, the need to consider the civilians that live in the environment in which our forces will operate has become increasingly important. ... (C)ivilian populations can have a profound affect in a crowded battle space. ... There is, however, little representation of the civilians in today's military simulations." (Fields & Spradlin, 2000)

> "(T)he ability to represent the behavior of crowds is currently lacking in military modeling and simulation technique." (Department of the Air Force, 2003)

Missions are changing for the U.S. military, especially for the Army and the Marine Corps. They are increasingly likely to be conducting operations in urban or built-up areas and interacting with crowds of noncombatants, as illustrated in Figures 1 and 2. Hostile forces may even use noncombatants for cover and concealment. Defeating such adversaries while minimizing civilian casualties, avoiding excess collateral damage, and maintaining popular support is a challenge (Hall & Kennedy, 2000). A critical need exists for realistic training in this area (Greenwald, 2002). Simulation has a significant part to play in developing an acceptable level of expertise in urban and crowd-related operations; and to provide this capability crowd models are needed.

FIGURE 1. U. S. Soldiers Confront a Crowd in Iraq in 2003 (photo: AP/World Wide Photos).

FIGURE 2. A Crowd Obstructs U. S. Operations in Iraq in 2003 (screen capture: CNN). (Usage of this CNN material does not constitute an implied or expressed endorsement by CNN).

Crowd Modeling and Organizational Simulation

Organizations need not be formal in their nature to be of interest to those seeking to simulate organizations and organizational behavior. Just as physicists study both crystalline and amorphous materials, those simulating organizations may choose to examine collections of humans as diverse as a military battle staff or a crowd composed of individuals. As the following discussion will show, crowds may be assembled from those who, by random chance, are in the same place at the same time. They may become part of a crowd due to some catalyzing event (e.g, a vehicular accident on a busy street). Other crowds may form due to the design and execution of an elaborate plan (e.g., a staged riot meant to disrupt a meeting of high-level government officials).

The analysis of any organization relies on psychological (and other) principles that may apply equally well to a crowd as to the employees of a company. Similarly, the simulation of an organization, while it may be applied to specific examples of humans working together, contains elements and techniques that may be useful in modeling a crowd. In this chapter we focus on crowds, but elements of this body of work will find other chapters that report complementary research results.

Crowd Modeling Study Overview and Methodology

To develop an improved understanding of the requirements and issues associated with crowd modeling in military simulation, a three-part study was undertaken, consisting of a requirements analysis, a state-of-the-art survey, and a design study.

A requirements analysis to identify and analyze requirements for crowd modeling in military simulation was conducted. We consulted with M&S (modeling and simulation) users in the military community regarding their current and anticipated needs for crowd modeling in military simulations. We also surveyed published sources calling for crowd modeling. An adaptation of a methodology previously proposed for human behavior modeling requirements analyses (Chandrasekaran & Josephson, 1999) was used for the analysis. That methodology calls for a series of case studies, using task analysis techniques, to establish the relationship between simulation use requirements and cognitive model fidelity/capability requirements. Each case study was focused on a different cognitive modeling topic. The crowd modeling requirements analysis was essentially one of the case studies, focused on crowd modeling, with adaptations to the methodology to specialize it for the crowd-modeling domain.

A survey of past and current psychological research in the area of crowd modeling, as well as existing models and simulations of potential applicability, was performed. We conducted a survey of the current state-of-the-art in crowd modeling, from two perspectives. The first perspective was psychological; here we surveyed the psychological research literature for research relevant to understanding and modeling the behavior of crowds. Special attention was given to research that considered crowd behavior in military scenarios, but other scenarios, including civil unrest and sporting event riots were also considered. Both descriptive (qualitative) and predictive (quantitative) models associated with crowd modeling were studied. Examination of the literature found research of some relevance, though it was not often focused on military scenarios. The second perspective was engineering; here we identified models and simulations with capabilities relevant to crowd modeling that have been or are being implemented as computer systems and assessed their capabilities. Both models of crowd cognitive behavior and crowd physical behavior were of interest.

Concepts leading to the design of a simulation of crowd behavior that can operate in a distributed simulation system were studied, and aspects of the design considered central and/or potentially problematic were experimentally tested. We developed concepts of a design for a crowd simulation. The design study included consideration of details of the crowd simulation's input and output, software architecture, essential algorithms, and data assumptions. We designed multiple notional crowd simulation software architectures to address the evolving simulation and scenario concepts. We reviewed existing software and simulations with capabilities related to crowd modeling. Aspects of the evolving design were implemented and tested experimentally to confirm (or refute) the design concepts.

CROWD SIMULATION REQUIREMENTS ANALYSIS

This section details an analysis conducted to identify and organize the requirements for crowd modeling capabilities in military simulation. First, the methodology used for the requirements analysis is explained. Following that, the requirements identified are listed and discussed. Finally, summary findings of the analysis are given.

Requirements Analysis Introduction

Chandrasekaran and Josephson explain guiding principles and propose a method for performing requirements analysis in the domain of human behavior modeling (Chandrasekaran & Josephson, 1999). These ideas are directly relevant to the process of establishing requirements for crowd modeling in military simulation. Their two main points are:

- The intended uses or purpose of a simulation should determine its fidelity requirements; arbitrarily pursuing more fidelity, or as much fidelity as possible, without having a basis in requirements is unjustifiable.

- A "strategy" (i.e., a method) based on purpose-driven task analysis is a recommended means for "empirically investigating" and determining the requirements, both fidelity and capability, for cognitive models.

They describe the recommended requirements analysis strategy and argue that it should be applied in a set of studies to determine the fidelity requirements for human behavior modeling:

> "... case studies should be performed to analyze the actual requirements for cognitive models in real-world examples of military uses of simulation." (Chandrasekharan, 1999)

Though the need for crowd modeling in military simulation has been recognized, it is not clear how much fidelity the models must have to be effective. From a training perspective, psychological research is necessary to assess its effect on training and human performance and to identify its important instructional aspects. Possible variables for consideration in such research include crowd behavior fidelity, skill levels necessary for optimal training performance validity, and cognitive models most representative of crowd behavior in military scenarios. Further analysis of the simulation could also yield useful research information in the area of situation awareness and tactical decision-making under stress, training technology, and human behavioral representation techniques. From an experimental perspective, the effect that different levels of crowd behavior fidelity have on scenario outcome is of interest. Similar considerations apply to analysis and acquisition applications.

There is a widespread tendency to assume that more fidelity in models is better for a simulation. For example, in the case of crowd modeling in an individual level simulation, it might be assumed that simulated crowd members must have realistic bodies, emotional states, and distinct personalities, and complex behavior repertoires, based on sophisticated cognitive models. However, assumptions like this may be an oversimplification if the purposes of the simulation are not considered, as Chandrasekharan and Josephson assert:

> "We argue that the pursuit of high fidelity cognitive models, unfettered by detailed considerations of what we want the models for, is so unfocused as to be almost useless for practical purposes. No cognitive model can be detailed enough, in every respect, to serve for every purpose we might conceivably use it for, and no practical purpose is such that an unbounded degree of fidelity is significantly useful for its practical achievement. We argue that what a cognitive model needs to contain is vitally affected by what kinds of questions one wishes to answer, i.e., the goals of the simulation." (Chandrasekharan, 1999)

There are possible counter-arguments to their first point, i.e., that pursuing fidelity beyond requirements is unjustifiable. For example, additional fidelity beyond that strictly required by the initial requirements may allow a cognitive model to support later emerging requirements that were initially unanticipated. Nevertheless, the argument that fidelity should be tied to requirements is convincing. As for their second point, that determining human behavior modeling requirements requires methodical analysis, we see the requirements analysis reported here and the behavior fidelity experiment later proposed as future work as examples of such studies.

Requirements Analysis Process

We first analyze the method of Chandrasekharan and Josephson and then explain how it was adapted for the crowd modeling requirements analysis. In the list that follows, the italicized items are the steps of the method as given by Chandrasekharan and Josephson (Chandrasekharan, 1999); they are followed by commentary regarding the steps:

1. *Conduct a set of case studies wherein concrete examples of military uses of the simulation are studied empirically.* Note that "uses" of simulation are studied. The basic idea of the method is that simulation requirements follow from simulation uses. However, the general term "simulation use" has several possible interpretations. It could arguably refer to broad categories of simulation applications, e.g., training or analysis; narrow subapplications within those broad applications, e.g., command staff training or doctrine development; generic use cases, e.g., training a company commander for situations where the opposing side adopts a certain tactic (this example is from (Chandrasekharan, 1999)); or specific use events, e.g., the Unified Endeavor 03-1 exercise or the Millennium Challenge 02 experiment.

2. *For each case, analyze the demands on simulated entities and characterize these demands relative to the purposes for which the simulation is conducted.* "Simulated entities" does not necessarily mean automated or computer controlled; there may have been a need for entity behaviors that was met via role-player control. "The purposes for which a simulation was conducted" suggests that the answer to the question in step 1 is specific use event.

3. *Map out the tasks to be performed, both by the simulation user and by simulated agents, in terms of goals and subgoals.* To understand this step, it is necessary to first clarify what is meant by "tasks" and "goals and subgoals." Considering tasks first, Chandrasekharan and Josephson explain that there are two types of tasks in a simulation use, the tasks of the simulation users and the tasks of the simulated entities[1]. The tasks for simulation users are interpreted as those actions they may perform while using the simulation, such as constructing a scenario or commanding a force. The tasks for simulated entities are interpreted as their run-time behaviors. Run-time behaviors of simulated entities include both

[1] Chandrasekharan and Josephson (Chandrasekharan, 1999) use the term "agents" to refer to the simulated people and groups of people that are performing cognitive tasks in the simulation. We avoid that term because it also has a separate and widely used meaning as a software organization technique (Weiss, 1999; Wooldridge, 2002). Although simulated agents may be implemented using software agents, the two are distinct concepts. Software agents can be used to implement processes other than simulated agents and simulated agents need not be implemented using software agents. To reduce the potential for confusion, we use the term "entities" instead.

scenario-specific actions (e.g., capture Basra) and domain-generic behaviors (e.g., conduct hasty attack). Generic behaviors are more likely to be relevant to understanding general requirements for behavior modeling. As for goals, they give an example that makes clear that here the "goals" are the users' intended purposes for the simulation (e.g., training), and not the objectives of the simulated entities' behaviors in the simulation (e.g., capture the bridge).[2] Note that in the previous step the question was the "demands" on simulated entities, whereas in this step, the concern is the "tasks" to be performed. This is understood to mean that the demands on the simulated entities are expressed in terms of tasks for them to perform, and that those tasks are interpreted as run-time behaviors, which must be performed at some sufficient level of fidelity. Hence, the set of demands on simulated entities imply a repertoire of run-time behaviors they must be capable of exhibiting, and some indication of the level of fidelity those behaviors should have. Those behaviors and their associated fidelity levels together express the requirements the method is designed to elicit.

4. *Develop concepts and vocabulary for describing these task structures and for characterizing the similarities and differences among the different uses of simulation.* In this case, "different uses" may make more sense in terms of application areas or levels of warfare. A behavior such as "conduct deliberate attack" would have different requirements and implementations for different areas and/or different levels. As for "concepts and vocabulary", we will rely on the existing concepts and terms for application areas and levels of warfare to identify simulation uses[3] and on behavior names to identify "task structures". Of course, unlike military behaviors where behavior names are doctrinally defined and widely understood, crowd behaviors are not so clear. Some terminology for crowds can be drawn from the psychological literature, but there is some variance in those sources. The need to develop vocabulary exists.

5. *Seek generalizations that aid description and analysis and help to predict the characteristics of further examples.* The "generalizations" may be findings that certain types of behaviors, at certain fidelity levels, are needed for simulation uses in certain application areas and/or at certain levels of warfare. To apply this method to crowd modeling, it would be

---

[2] Even though Chandrasekharan and Josephson say "goal", we will generally use the terms "purpose" or "intended uses", rather than "goal", to denote the intended uses for a simulation. We do so because the latter term has a secondary connotation of entity behavior goal that the other terms do not have and is not relevant here.

[3] This is detailed later in this chapter.

useful to identify simulation uses, either use events or use categories, where crowd modeling was present or was needed, and to produce generalizations by examining those uses.

To repeat, the fundamental idea of the Chandrasekharan and Josephson method is that a simulation's requirements follow from its intended uses or purpose. That idea is evident in their summarization of the method:

> "The kind of case study we are proposing is aimed as identifying the requirements on agent behaviors, whether or not the agents are to be humans or CGFs. In each study, analysis would identify which entities need to be modeled for achieving the simulation goals, what kinds of information each entity needs, what kinds of behavior and information each entity produces, and what kinds and degrees of fidelity are needed." (Chandrasekaran & Josephson, 1999)

We distill the Chandrasekharan and Josephson method still further into these logical steps (here the symbol "→" means "determines"):

- Purpose → Entity types required
- Purpose and entity type → Cognitive tasks (Behaviors) required
- Purpose and Cognitive task (Behavior) → Fidelity required

Clearly, the starting point for the method, and for an analysis that uses it, is the simulation's purpose. But there are multiple ways to categorize and characterize simulation purpose. Again, the possible types of simulation use are:

- Broad application areas
- Narrow subapplication areas
- Generic use cases
- Specific use events

These different types of simulation use are related in this way: broad application areas may be partitioned into narrow subapplication areas; narrow subapplication areas may be defined as sets of generic use cases; and generic use cases may be instantiated and executed as specific use events. (For brevity, the four types will be referred to without the adjectives, i.e., as application areas, subapplication areas, use cases, and use events.)

Two questions arise. First, which of these simulation use types is most appropriate for requirements analysis for human behavior modeling, in general? Second, which of these simulation use types is most appropriate for requirements analysis of crowd modeling in military simulation, in particular? The answer to the first question seems to be specific use events, according to Chandrasekharan and

Josephson; for example, they write, "... study real instances of uses of simulation ..." (Chandrasekharan, 1999). However, as already noted, examples of the method are given using the generic use case type. In any case, the first question need not be answered here, as this study is concerned with the second question.

For the purposes of analyzing the requirements for crowd modeling in military simulation, simulation use categories are defined (for brevity, "use categories") using two orthogonal dimensions: application area and warfare level.[4] Using application area to define simulation use type is relatively intuitive. Warfare level was used as well because, as will be seen, it was found to make an important difference in the requirements for crowd modeling.[5] The possible combinations of application area and warfare level were used to define the use categories considered in the requirements analysis. For each of the use categories, crowd-modeling requirements generally applicable to that use category were identified. In addition, for some of the use categories, narrower use types (subapplication areas, use cases, or use events) were considered and their special crowd modeling requirements detailed as components or examples of their use categories. The crowd modeling requirements were defined primarily in terms of the crowd behaviors needed for each use category, though other requirements, such as crowd size and behavior fidelity, were also considered.

Use categories based on application area and warfare level are clearly broader than the use events or use cases that seem to be called for in the Chandrasekharan and Josephson method as the basis for requirements analysis. Conversely, in another respect this analysis of crowd modeling requirements is narrower than the case studies they describe. In their method, the case studies are intended to identify the full range of human behavior modeling requirements needed for a simulation use type. They advocate a broad analysis of a narrow simulation use type (a use case or use event). Our analysis of crowd modeling requirements starts from broad simulation uses (the use categories) but the analysis is narrow; the entity types and their control are given (crowds of non-combatants, controlled by software rather than human operators), rather than being determined during the analysis. The more specific focus on crowds balances the more general coverage of simulation use categories. The goal is to determine "what kinds of information each entity needs, what kinds of behavior and information each entity produces, and what kinds and degrees of fidelity are needed" for crowds in each use category.

---

[4] The application areas used were training, analysis, experimentation, and acquisition. The warfare levels used were tactical, operational, and strategic. The application areas and warfare levels are defined in some detail in the next section.

[5] We initially intended to use level of resolution of simulated entity as the second dimension for defining simulation use categories. However, the simulation users consulted during the requirements analysis process preferred to express their requirements using level of warfare, so we adjusted our scheme. There is a rough mapping from level of warfare to level of resolution of simulated entity. At considerable risk of oversimplification, that mapping (using ground units for illustrative purposes) is tactical, individual to company; operational, company to corps; strategic, corps and above.

With this in mind, we add an initial logical step to the method summary given before. In that additional step, the simulation purpose for this analysis is determined from application area and warfare level. We call a combination of application area and warfare level a use category; each use category specifies a purpose, or more precisely, a class of simulation purposes. From those purposes, entity types, cognitive tasks, and fidelity can be determined. However, in this analysis, attention is restricted to crowd entities. With the additional step and the restriction to crowd entities, the logical steps of the method become:

- Use category (application area and warfare level) → Purpose
- Purpose → Entity types required (crowd entities assumed)
- Purpose and entity type (crowd entities assumed) → Cognitive tasks (behaviors) required
- Purpose and Cognitive task (behavior) → Fidelity required

This approach is, we believe, an adaptation of the Chandrasekharan and Josephson method. Consider the points of the method in turn:

1. *Conduct a set of case studies.* This analysis of crowd modeling requirements is a single case study consistent in intent and closely related in method to those called for.
2. *Analyze the demands on simulated entities relative to simulation purpose.* The demands on crowds are analyzed for each of the use categories. The use categories represent the different classes of purposes typical for those use categories.
3. *Map out the tasks to be performed.* The tasks to be performed, given as run-time behaviors, are identified through reference to simulation users and research literature.
4. *Develop concepts and vocabulary.* For the most part the terminology used by simulation users and researchers, who identify the run-time behaviors with descriptive names, are applied. Inconsistent or overlapping behavior names are clarified.
5. *Seek generalizations.* Patterns of consistent required behaviors or fidelity levels for application areas or warfare levels are identified.

Application Areas and Warfare Levels

Military simulations are used for a variety of purposes, and those purposes may be grouped or categorized in a variety of ways. For this analysis, those purposes are grouped into four application areas: training, analysis, experimentation, and acquisition.[6] Each is defined in turn.

Training. Training simulations, in general, are intended to induce learning in human trainees who interact with or participate in the simulation. The trainees interact with or participate in the simulation, which provides an instructive experience. Flight simulators and command staff exercise drivers are well-known examples of training simulation; the former can teach aircraft control psychomotor skills via an immersive experience, while the latter can teach cognitive and decision-making skills by providing a realistic battlefield context.

Analysis. Analysis is the use of simulation to answer questions about some aspect of the system or scenario being simulated. Military analysis simulations are often used to assess the effectiveness of new weapons systems, test new force structures, or develop doctrine. For example, simulation was used to conduct experimental trials testing the design of a new naval surface combatant (Ewen, Dion, Flynn, & Miller, 2000). In analysis, applications simulation is used in a carefully controlled way, with run-to-run parameter differences restricted to the factors under question (e.g., different weapons performance levels). Repeatability, determinism, and the ability to isolate the cause of an observed effect are desirable characteristics of analysis simulation.

Experimentation. Experimentation is similar to analysis, in that the simulation is being used to answer questions, but in experimentation the questions are more open-ended and exploratory and "insights", rather than specific answers, are often the goal (Ceranowicz et al, 1999). In experimentation, strict control of run-to-run outcome differences is less important than exploring a space of possible outcomes. The objective of such experimentation is "not to evaluate system effectiveness, but rather, to provide an environment and tools that will allow operators and analysts to discover new insights" (Ceranowicz et al, 1999). Large simulation-based experiments have been conducted by U.S. military commands, notably including the recent Millennium Challenge 02 experiment (Ceranowicz et al, 1999).

---

[6] Chandrasekharan and Josephson identify these simulation "purposes", i.e., application areas, in our terminology: "training, mission rehearsal, doctrine evaluation, and acquisition decisions" (Chandrasekharan, 1999). We map their applications into those used here as follows: "training" and "mission rehearsal" are training; "doctrine evaluation" is analysis; and "acquisition decisions" is acquisition. They did not indentify any purposes that map to experimentation. Another categorization of simulation purposes found, for example, in the OneSAF Operational Requirements Document has these application areas: Research, Development, and Acquisition (RDA), Advanced Concepts and Requirements (ACR), and Training, Exercise, and Military Operations (TEMO). We map these application areas into those used here as follows: RDA is analysis and acquisition; ACR is experimentation; and TEMO is training.

Acquisition. Increasingly, simulation is being used to support acquisition.[7] Here the use of simulation can be broadly categorized into two general subapplication areas, colloquially described as "building the right thing" and "building the thing right" (Castro et al, 2002). In the case of "building the right thing", simulations are used to compare or evaluate proposed or notional systems, combat or otherwise. The projected performance characteristics of those systems are stipulated and inserted into the simulation, so that the effects the given capabilities might have on mission outcomes can be assessed in an analysis-like process. In the case of "building the thing right", simulation is used to support the system design and engineering process, at levels from individual components to integrated systems-of-systems. High-fidelity engineering simulations can be used to determine if a design will operate as intended (Castro et al, 2002).

Military simulations model operations at different levels of warfare, from tactical engagement simulations that model individual combatants to strategic theater simulations that model aggregate military units. Military simulation users are accustomed to specifying requirements for simulations in general by reference to those levels. Doctrinally, warfare is divided into three levels: strategic, operational, and tactical. Military operations at the different levels differ in important ways, including scale of forces involved, time duration, geographic scope, support and logistical requirements, and consequences. Although the boundaries between the levels can be fuzzy, and events may cut across the boundaries, the distinction is nevertheless useful; the levels are "doctrinal perspectives that clarify the links between strategic objectives and tactical actions" (Department of Defense, 1995). In an official doctrine manual the warfare levels are defined as follows:

> "Strategic level of war – The level of war at which a nation, often as a member of a group of nations, determines national or multinational (alliance or coalition) security objectives and guidance, and develops and uses national resources to accomplish these objectives. Activities at this level establish national and multinational military objectives; sequence initiatives; define limits and assess risks for the use of military and other instruments of national power; develop global plans or theater war plans to achieve these objectives; and provide military forces and other capabilities in accordance with strategic plans." (Department of Defense, 2001)

> "Operational level of war – The level of war at which campaigns and major operations are planned, conducted, and sustained to accomplish strategic objectives within theaters or operational areas. Activities at this level link tactics and strategy by establishing operational objectives needed to accomplish the strategic objectives, sequencing events to achieve the operational objectives, initiating

---

[7] This application area is often called "simulation based acquisition" (SBA). Here the simulation-based aspect of the application is assumed.

actions, and applying resources to bring about and sustain these events. These activities imply a broader dimension of time or space than do tactics; they ensure the logistic and administrative support of tactical forces, and provide the means by which tactical successes are exploited to achieve strategic objectives." (Department of Defense, 2001)

"Tactical level of war – The level of war at which battles and engagements are planned and executed to accomplish military objectives assigned to tactical units or task forces. Activities at this level focus on the ordered arrangement and maneuver of combat elements in relation to each other and to the enemy to achieve combat objectives." (Department of Defense, 2001)

For each use category, i.e., combination of application area and warfare level, different aspects of crowds and crowd behavior will be needed or emphasized. For example, in a immersive virtual environment used to train individual combatants at the tactical level, simulated crowds of non-combatants could make target acquisition and selection more realistic by providing visual clutter and make mission decision making more challenging by adding the need to minimize non-combatant casualties. In an operational level simulation used for course of action analysis, the presence of crowds of refugees could make maneuver planning more accurate by slowing unit movement along road networks and through urban areas. Crowds can be important in any of the use categories, but the nature and degree of their importance, the behaviors that the crowd should be able to perform, and the fidelity needed in the crowd models, can vary by use category. The examples illustrate this; in a simulation of individual combatants, the actions and appearance of individual crowd members are important, while in an aggregate level simulation the outcome may depend on the cumulative effect of many crowd members.

Crowd Modeling Requirements Examples

The crowd modeling requirements found were diverse in both content and form. There was a good consensus that crowds were needed in military simulation. For example:

"The potential of this capability could provide a major benefit for many applications." (Ferguson, 2003)

However, there was less agreement on what the specific requirements were, and even less on how the requirements were expressed. The requirements, both as given by the users and drawn from published sources, were stated in several different ways. The first was in terms of needed crowd behaviors, e.g., "take hostile action against combatants", sometimes with a description of needed fidelity level. The second was in terms of military mission types for which crowds were generally needed, e.g., "urban warfare". The third was in terms of the effects that a crowd might have on a scenario that needed to be modeled, e.g., "road congestion."

To illustrate the variety and types of crowd modeling requirements found, the requirements for a single use category are listed. The entries remain true to the sources' responses and freely mix these types of requirements, fully aware that they are different types of requirements, or at least different ways of expressing requirements.[8]

Crowd modeling requirements for the tactical training use category include:

- Tactical constraints imposed by crowds, including restrictions on movement ("don't move over civilians") and free use of firepower ("avoid collateral damage") (Sokolowski, 2003; *One Semi-Automated Forces*, 2001). At the tactical level more detail and fidelity in crowd behavior is needed than at the operational level; in virtual simulations, detail and fidelity in crowd member appearance is also needed (Sokolowski, 2003).
- Needed crowd size: ~100 persons (Sokolowski, 2003).
- Needed crowd size: ~50-100 persons in a confined urban combat area (Ferguson, 2003).
- Needed crowd size: ~5-10 persons for a small arms or aircrew trainer (Ferguson, 2003).
- Needed crowd size: "Somalia" (Bailey, 2003).
- Urban warfare (Sokolowski, 2003; Ferguson, 2003; Bailey, 2003).
- Noncombatant evacuation operations (NEO) (Sokolowski, 2003).
- Military operations other than war (MOOTW) (Sokolowski, 2003) (Ferguson, 2003).
- Homeland security operations (Sokolowski, 2003); including crowd response to events such as natural disaster, terrorism, industrial accidents, disease epidemics (Ferguson, 2003).
- Crowd behavior in response to non-lethal weapons used for crowd control (Sokolowski, 2003; Bailey, 2003; Kenny et al, 2001).
- Needed crowd behaviors: move randomly in scenario area, approach battle, flee battle, take hostile action against combatants (e.g., throw

[8] It is possible that deeper analysis of the military mission types or desired crowd effects mentioned in the requirements would lead to a more uniform set of requirements consisting of crowd behaviors needed to support the mission types or produce the desired crowd effects. This is suggested by the Chandrasekharan and Josephson method (Chandrasekharan, 1999).

rocks), assist combatants (e.g., medical aid), conduct negotiations, ask for assistance, react to threats, react to directions concerning movement, riot to acquire food (Sokolowski, 2003).

- Needed crowd behaviors: mass action to force combatant withdrawal, looting, provide human shield cover for combatants, mass action to overwhelm checkpoint security (Ferguson, 2003).
- Needed crowd behaviors: react to gunfire, react to military police, react to barriers, drive vehicles, employ crude or improvised weapons, take hostile action against other crowd members, be repelled, threaten checkpoint, attack barrier, burn objects (Bailey, 2003).
- Trafficability problems and road network congestion, especially at bridges and other choke points (Ferguson, 2003).
- Displaced personnel requiring humanitarian assistance; such persons consume logistical supplies, requiring re-supply of food and water, and affect logistics of combat operations (Ferguson, 2003; *One Semi-Automated Forces*, 2001).
- Movement of non-combatant people and vehicles to provide "clutter" to obscure military movements for sensor platform operations (*One Semi-Automated Forces*, 2001; D. F. Swaney, electronic mail, 2002) and for intelligence processes such as fusion, correlation, and targeting (Ferguson, 2003).
- Crowd hysteria and confusion caused by misinformation or partial situational awareness (Ferguson, 2003).
- Population evacuation or flight in Homeland Security scenarios (Creech & Petty, 1996; Ferguson, 2003).
- Crowd control security operations for major events, e.g., Olympic games (D. F. Swaney, electronic mail, 2002).
- General crowd control operations (Bruzzone & Signorile, 1999; Miller, 2002).
- Movement behaviors: stay at a point, move to a point, follow an entity, move with neighbors (flocking), and stay within or beyond a distance from a point or entity (Reece, 2002).
- Individual actions: perform mission, maintain personal safety, satisfy curiosity, and confront antagonists (Reece, 2002).
- Crowd actions: respond to crowd actions (Reece, 2002).

- Needed crowd behaviors: seek goal, flocking, safe wandering, following, goal change, group control (Musse, Babski, Capin, & Thalmann, 1998).
- Protection of civilian populations (*One Semi-Automated Forces*, 2001).
- Civilian transport by combat entities (*One Semi-Automated Forces*, 2001)
- Crowds in stability and support operations: "displaced civilians, refugees, mass migration, rioters, disaster victims" (*One Semi-Automated Forces*, 2001).

Similar lists, partially but far from completely overlapping the tactical training list, were also developed for the other use categories: operational training, strategic training, tactical analysis, operational analysis, strategic analysis, tactical experimentation, operational experimentation, strategic experimentation, tactical acquisition, operational acquisition, and strategic acquisition. Those lists are not given here, but are available (Gaskins, McKenzie, & Petty, 2003). Interestingly, no requirements for crowd modeling were identified for any of the strategic use categories.

Crowd Modeling Requirements Comments

In addition to specific requirements, some sources asserted a general need for crowd modeling (Miller, Battaglia, & Phillips, 2002; Reece, 2002; Pew et al, 1998; Fields & Spradlin, 2000; Department of the Air Force, 2003) or training in dealing with crowds in various contexts (Kenny & Gilpin, 2002). Some sources observed that in simulation-supported training contexts the lack of crowd modeling capabilities might have influenced the training objectives; i.e., training objectives were established that did not require crowds because that capability was known to be not available (Sokolowski, 2003; Ferguson, 2003). This is an undesirable situation. Some sources emphasized the logistical effects that crowds of displaced persons could have on friendly forces. For example:

> "(Crowds of non-combatants) will consume logistic supplies, etc., driving the training audience to look at re-supply of food and water, and assess the logistical impact on combat operations. The U.S. will rarely, if ever, allow a displaced personnel problem to devolve into a humanitarian crisis. Operating forces will be challenged to maintain their own momentum and handle refugees." (Ferguson, 2003)

The need for an explicit psychological basis for crowd behavior modeling was not clearly established by the sources. Some sources discounted the need for psychological modeling for crowd simulation, while others saw sought greater fidelity based on psychological models as increasing simulation utility:

> "Lack of science causes some lack of value in analytical results, prevents us from access to root causes of effects, makes crowds 'black boxes.' " (Bailey, 2003)

To exemplify this difference, these seemingly contradictory excerpts are from the same source:

> "Due to the limited training objectives at the operational level, I do not think that a psychological basis would be required." (Ferguson, 2003)

> "Currently there is little 'behavioral' aspect to the objects and entities, as they exist in the current models. Behavioral object states such as passivity, hostility, aggressiveness, confusion, could all contribute to an (entity's) behavior and the reliability of the information output by the simulation." (Ferguson, 2003)

In any case, the sources were more interested in the capabilities than the implementation method, and psychological modeling would be part of the latter.

Many of the sources blurred the distinction between crowds and population, asking for behaviors, mission types, and effects associated with the general population, as opposed to a crowd. For example, general road congestion and logistical supply consumption are both more likely to be population effects than crowd effects. Both crowd and population behaviors can be understood in this requirement:

> "(Homeland Security) training objectives driven by natural disaster, terrorism, large industrial accidents, disease or epidemic control would be greatly enhanced by crowd modeling." (Ferguson, 2003)

Others sources were sensitive to the difference. The following excerpt, for example, shows considerable insight into the issue of how and when a group of previously independent persons from the population could become a 'crowd':

> "I look for (a model) that does 'normal' civilian activities and scales in response vice a crowd or a mob to start. To me there is a difference. ... The people moving through the NY streets doing Christmas shopping are not a crowd to me. The "street" is crowded but the people are not a crowd, as they have no ulterior group purpose. Crowds can be formed by external stimulation: traffic accident, bomb blast, street vendor/performer, etc. To me the difference between a crowd and just normal traffic is density and an interest in a common theme. However, (they are) still relatively

> mindless and benign. Mobs share (a) crowd's density but differ in a direction/purpose and leadership. Mobs have emergent leaders - crowds don't. The leaders may be plants but members are civilians. In all three cases above I think we are talking about civilians transiting up a scale of complexity. I consider this different than what we experienced in Seattle or Mogadishu. There, military and law enforcement activities were up against a planning, fore thinking enemy with premeditated direction and actions, which is a little different problem set." (A. Cerri, electronic mail, 2002)

An important difference between crowds and population is scale. Crowds usually occupy a reasonably bounded area for a limited time, and their actions are somewhat focused, e.g., "storm the food warehouse". By comparison, a population may occupy a large geographical region more or less indefinitely, and their actions are often unfocused, such as "commute" or "evacuate the city." From a modeling point of view, the distinction between crowd and population is worth preserving, because modeling crowd-specific behaviors, such as providing human shield cover for combatants, appear to require rather different modeling techniques than modeling general population effects, such as large-scale citizen evacuation, e.g., (Creech & Petty, 1996).

Requirements Analysis Findings

The sources identified a wide variety of crowd modeling requirements, expressing those requirements in a variety of ways, including crowd behaviors, military missions, and crowd effects. We elaborated the distinctions between application areas and warfare levels and associated the requirements with them to make apparent the different requirements for crowd modeling in simulations in each of the use categories. Table 1 counts the requirements found in each use category - for a complete list of the requirements, see Gaskins et al, 2003.

The primary finding of this requirements analysis, which follows from the number and variety of requirements found, is that a single crowd behavior model is unlikely to satisfy all of them. This finding has two implications. First, research and development efforts aimed at developing a crowd model should be explicitly oriented towards its intended uses (i.e., application area and warfare level) and the requirements associated with those uses; of course, this is just what had been previously asserted for human behavior models in general (Chandrasekharan, 1999). Second, because multiple crowd models are likely to be needed to meet all the crowd modeling requirements in military simulation, a technical capability to compare the effects, fidelity, and utility of different crowd models would therefore be useful.

Application Area

Warfare Level: Use Categories	Training	Analysis	Experimentation	Acquisition
Tactical	Sources 12 Reqs 27	Sources 7 Reqs 16	Sources 6 Reqs 16	Sources 5 Reqs 11
Operational	Sources 7 Reqs 16	Sources 5 Reqs 9	Sources 5 Reqs 10	Sources 3 Reqs 5
Strategic	Sources 0 Reqs 0	Sources 0 Reqs 0	Sources 0 Reqs 0	Sources 0 Reqs 0

TABLE 1. Counts of Crowd Modeling Requirements Found in the Requirements Analysis (Petty, 2003).

A secondary finding is that crowd modeling requirements seem to be most important at the tactical level, of some importance at the operational level, and not important at the strategic level of warfare. Indeed, as noted, no requirements were expressed for crowd modeling at the strategic level. Possible interpretations of the latter outcome are that crowds are simply not significant at that level of warfare, that the sources consulted were focused on the tactical and operational levels, and that there are non-combat factors at the strategic level that need to be modeled (e.g., political, ethnic, and religious alliances) but that these are not considered by the experts to be "crowd phenomena" (W. B. Rouse, personal communication, 2004). In any case, if there are crowd-modeling requirements at the strategic level, it is likely that they are rather different from the tactical and operational requirements, and this supports the primary finding.

PSYCHOLOGICAL AND COMPUTATIONAL MODELS OF CROWD BEHAVIOR

In this section, a survey of the state of the art in psychological and computational models of crowd behavior is presented. Motivating issues of crowd behavior are discussed and the methodology used for the survey is explained. Next, the psychological research literature on crowd behavior is covered. Some computational models of crowd behavior are reviewed as well. The difficulty and importance of achieving realism in such models is discussed. Finally, summary findings of the survey are listed.

Survey Introduction

To meet the requirements and provide the expected benefits of crowd modeling identified earlier, the crowd models should have a basis in the psychological characteristics of crowd behavior in military situations. Relevant research from both the cognitive and the social domains of psychology should guide the development of models of crowd behavior in military scenarios.

In an attempt to provide that psychological basis, the crowd research literature was surveyed. The surveyed work draws largely from cognitive psychology, social psychology, sport psychology, sociology, police, and military research literature. Information on non-combatant crowd behavior during military operations was of primary interest. Of secondary interest, but much more plentiful, was information regarding riots and sport fan behavior. Relevance to the survey was evaluated on the following criteria: credible psychological basis, closeness to implementation as a computational model, relevance to military scenarios, level of detail, validation, and quantitative data about crowd behavior.

Psychological Research on Crowd Behavior

Looking first at the psychological research literature regarding crowd behavior, it is possible to broadly classify it into two categories: analyses of types of crowds and phases of crowd behavior, and identification of variables affecting crowd behavior.

Types of crowds and phases of crowd behavior

Varwell identified four main crowd types: aggressive crowd, an escaping crowd, an acquisitive crowd and an expressive crowd (Varwell, 1978):

- *Aggressive.* Aggressive clearly describes the sort of situation where a crowd is intent upon destruction of some sort or another.
- *Escapist.* A crowd may be intent upon escape. These circumstances may arise when a serious fire or explosion occurs in a dance hall or in a busy hotel, when what was originally a passive crowd becomes frightened.
- *Acquisitive.* A crowd may become acquisitive and begin looting for various reasons
- *Expressive.* The purpose of the crowd may be expressive - it may be primarily concerned with the expression of feelings or emotions.

An important point is that crowds can exist for any combination of these reasons. Further, a crowd can change its type due to the unfolding situation (Moore, 1990). Virtual crowds can be varied on many pertinent dimensions. These include composition, ethnicity, size, noise, cause, location, weapons, verbal abuse, projectiles, and the use of women, children and elderly as human shields. The paradox of a "crowd psychology" makes this an even more difficult task considering people's individuality, judgment, and critical thinking are subsumed by the group, while the breakdown of social structure simultaneously results in more individualistic acts of asocial behavior (Dupuy, 1990).

Recent conclusions by Kenny et al. (2001) have pointed to the difficulties in examining crowds due to the diverse nature of participants not only within the crowd but also across time, cultures, and individual motives. There is clearly a lack of unanimity on motivation or crippled individual cognition as was once advocated by some of the earliest investigators of crowd behaviors (Le Bon, 1895; Park, 1904; Park, 1930; Blumer, 1939). Kenny et al. (2001) indicate the need for more research to address the unknowns of crowd behavior as well as to test new theories against existing ones recognizing discrepancies. Ideally, such research would allow for models to be built with probabilities of specific crowd behavior. Further basic research is needed to address the psychological variables that have been related to crowd behavior in order to establish such probabilities of occurrence in a specific scenario within various cultures of eminent interest.

Kenny et al. (2001) identify three specific phases of a gathering: assembling, gathering, and dispersal. The assembly stage consists of the process and motivation behind the initial collection of people. For example, the crowd may assemble for a planned event or people may collect without warning. The assembly stage contains variables that determine the motivation behind this collective activity (Kenny et al, 2001).

Once the crowd has assembled it enters the gathering stage. During this stage a crowd begins to engage in collective behaviors. These behaviors can range from peaceful actions such as singing or cheering to violent behaviors and the use of weapons. One example of this type of behavior was seen in the Los Angeles race riots following the Rodney King verdict in 1992. A more violent example is the conflict between American troops and thousands of heavily armed Somalis in Mogadishu in 1993. Most recently, the protesters of the Seattle World Trade Organization meeting in 1999 involving a large mix of activists and anarchists, from around the world, who were intent upon disrupting the launch of the new organization.

Eventually the crowd will discontinue its collective behaviors and disperse. Crowd dispersal is the final stage of a crowd's evolution and this process may be either forced or routine. The type of dispersal that occurs often depends on several factors present during this stage (Kenny et al, 2001).

Kenny's model provides peacekeeping forces with a starting point for understanding crowd behavior, however, this theory lacks a complete examination of crowd dynamics and the variables that cause violent crowd behavior. Several researchers have adopted the flash point philosophy to help further uncover this dimension of crowd dynamics (Kenny et al, 2001; Waddington, 1987). According to these researchers, a "flash point" is the point at which seemingly docile crowds suddenly become violent in response to a trigger or stimulus. The violence may be a response to any number of factors.

Crowds are not always unintentional. Greer describes crowds in Bosnia that were not spontaneous formations of people with complaints or accidental collections of people caught in a military engagement (Greer, 2000). Rather, they were the result of a planned response to the United Nations stabilization force operation. Greer calls this type of crowd a "rent-a-crowd" as many of the people in the crowd were paid by opposition leaders to demonstrate against or attack the stabilization force. The crowd organizers members knew that the forces would not knowingly hurt civilians or unarmed protesters. This crowd used rocks and sticks as their weapons and later Molotov cocktails in attempts to set vehicles on fire. The crowd consisted of women and children as well as men of every age. Instigators in the crowd consisted of military-aged men who also engaged in hand-to-hand combat with the troops. The crowds of Serbs usually numbered several hundred against a single platoon of about twenty soldiers. The crowds generally fought for short intense periods followed by retreat and periods of rest and eating where they awaited further instruction from the instigators. The Serbs had maintained control of the crowd by signaling them to assemble with air raid sirens and local radio stations to provide instructions. These tactics resulted in crowds of 500-800 to assemble within thirty minutes of the first siren. The crowds sang patriotic Serb songs to raise the intensity of the demonstrations. These tactics allowed for excellent control of the Serb crowds.

Successful crowd control methods used in Bosnia include movement utilizing speed, mobility and height; using multiple platoons approaching from different directions to break up the crowd mass and cause confusion; exhausting the crowd by forcing them to chase vehicles and attempt to overcome soldiers behind well built barricades. Some less successful methods include using grenades, tear gas, or negotiation (Greer, 2002).

Variables Affecting Crowd Behavior

In the literature, a wide variety of situational and psychological variables have been identified that can affect the behavior of crowds in the scenarios of interest. Kenney's three-phase model of crowd behavior provides a useful framework for organizing those variables; the relevant psychological variables that have been identified in the crowd context can be related to each of these phases. Table 2 summarizes those variables by phase, and the following paragraphs provided some details.

Factors that contribute to aggressive crowd behavior have been studied by social psychologists such as McPhail (1991) and Horowitz (2001). There is no unanimity on the particulars, but in general the common factors that tend to contribute include: presence of weapons, authoritarian government, lining up behind a barricade, drawing lines between "us" and "them", dramatizing issues

Assembling	Gathering	Dispersal
• Territoriality (Weller, 1985) • Presence of agitator (Meyer, 1978) (Feinberg, 1988) • Presence of police (Aveni, 1977) • Curiosity (Berk, 1974) • Political events (Sullivan, 1977) • Time stress (Berk, 1974) • Media violence (Weller, 1985) • Presence of media (Silverman, 2002) • Making a statement (Fogelson, 1971) • Proximity to speaker (Meyer, 1978) • Racial event (Weller, 1985) • Religious (Duggan, 1990) • Frustration (Berk, 1974) (Dollard, 1947) • Potential for reward (Berk, 1974) • Communication (Sullivan, 1977) (McPhail, 1973) • Association with known others (Aveni, 1977) (McPhail, 1985) • Boredom (Berk, 1974) • Time of day (Stark, 1972) (Stark, 1974) • Homogenity (Feinberg, 1988) (Derlega, 2002) • Convenience (Aveni, 1977) • Comfort (Aveni, 1977) • Self efficacy (Bandura, 1977)	• Presence of other like-minded (Berk, 1974) (Blumer, 1936) (Klapp, 1972) • Proximity to like minded (Berk, 1974) • Weapon availability (Berkowitz, 1964) • Opposition (Canetti, 1981) • Knowledge visibility (Berk, 1974) • Fatigue (Berk, 1974) • Conformity (Weller, 1985) • Ignorance of others' reactions/motives (Berk, 1974) • Age (Russell, 1998) • Physically Aggressive (Russell, 1998) • Gender (Russell, 1995) • Single (Russell, 1995) • Underemployment (Russell, 1995) • Commitment to cause (Derlega, 2002)	• Fatigue (Berk, 1974) • Control of media (Veno, 1992) • Disappointment (Duggan, 1990) • Perception of own strength/self-efficacy (Bandura, 1977)

TABLE 2. Summary of Psychological Variables and Research that Affect Crowd Behavior (Gaskins, 2004).

(e.g., in a speech) and making victims, large spatially concentrated crowds, and presence of television camera and crew (Silverman, Johns, O'Brien, Weaver, & Cornwell, 2002).

There are relatively few studies of crowd aggression that address personality measures as independent variables (Forward & Williams, 1970; Meier, Mennenga, & Stoltz, 1941; Ransford, 1968). Consequently, explanations regarding the sorts of people involved tend to rely almost exclusively on a mix of speculation and generalizations from the social-experimental literature (Russell, 1995). Confirming evidence of participants in crowd violence or would-be participants is sparse not only for this reason but also for methodological differences in observation. There is ongoing debate as to the psychological nature of crowds as well as the very existence of such an entity or rather it should be called a gathering of small clusters.

Russell and Arms (1998) showed that would-be rioters tend to be younger and physically aggressive. They also tend to be male and single (Russell, 1995). The evidence that they are variously described as underemployed, disaffected or marginally employed and economically disadvantaged males has also been consistently shown in a wealth of studies by European sociologists investigating football hooliganism (Adang, 1992; Bakker, Whiting, & van der Brug, 1990; Murphy, Williams, & Dunning, 1990; Pilz, 1989; Roversi, 1991; Van der Brug, 1992; Van Limbergen, Colaers, & Walgrave, 1989; Zani & Kirchler, 1991). Haddock and Polsby (1994) note similar characteristics of Los Angeles rioters. Not only did they conclude that they were largely poor, underemployed, and victims of racism but also unique in their jubilant moods rather than fury as they ran with their new cameras and VCR's. "It is hard to accept that these rioters were protesting the jury system, the state of race relations in Southern California or anything else - they were, in fact, having a party".

The Russell (1995) study found that those admitting to strong inclinations to assault, those attracted to violence also exhibited psychopathic, or antisocial personality tendencies. In addition to a proclivity for aggression, psychopathic inclinations has been variously seen to include: a lack of empathy, guilt, remorse, and fear of punishment, an extreme degree of selfishness, impulsiveness, and irresponsibility, a callous disregard for the feelings and welfare of others, and weak inhibitory controls (Williamson, Hare, & Wong, 1987; Hare, 1994). This in addition to the strong tendency of antisocial youths to perceive hostile intent in others (Sarason, 1978) and the result is that public disorders become an even greater likelihood.

Russell (1995) also found a cognitive component in his study. Subjects' attraction to the fights was positively related to their estimates of the percentage of other fans that were in attendance for the same reason, to see a fight. This effect has been described as the false consensus effect (Ross, Greene, & House, 1977). In other words, if the spectators feel that other like-minded spectators are also there to see a fight then there is a general acceptance for aggression. These spectators are emboldened by the thought that others will not only approve but also applaud and cheer them on in joining a fight.

In a study of the Watts riots by Ransford (1968) locus of control was linked to violent behavior of crowds. In an interview of 312 male heads of households conducted while buildings were still smoldering, Ransford asked, "Would you be willing to use violence to get Negro rights?" Those with a belief in an external locus of control, or alternatively, men who felt powerless in their circumstances were more willing to resort to violence as a solution (Russell, 1995). Age was not a factor. However, in another study by Forward and Williams (1970) of a much younger sample, aged 12-18 years, and using a different measure of locus of control the opposite was found. These younger black males with favorable attitudes toward violence instead had an internal locus of control orientation.

Meier et al. (1941) examined the personality factor of extroversion/introversion. They showed subjects a scenario describing an escalating riot. Those who indicated that they would likely take part in the mob action were predominantly males who exhibited tendencies toward extroversion and lower intelligence.

Territoriality is one of the more basic areas of ethnological research, or the study of instinctive behavior. Humans are very sensitive to relatively minor aspects of the home territory and "tenaciously defend minor boundaries, such as those created by housing clusters, street layouts, garages and gardens, with great tenacity" (Whythe, 1961). The need for territory for the raising of infants links aggression with courtship (Weller, 1985). Weller concludes that the pressure to assert dominance amongst males has a sexual basis and helps to explain the prevalence of violent behavior among young men.

Weller (1985) also attributes aggressive behavior to the media. The frequent exposure of experimental groups to violent incidents on film and television has led to long lasting effects (Bandura, 1965, 1973; Roth, 1979). Aggressiveness has been positively correlated in 19 year olds having viewed violent scenes on television 10 years earlier (Eron, Huesmann, Lefkowitz, & Walder, 1972). The actual enactment of violence seems to induce further violence in children and adults (Buss, 1966; Loew, 1967; Nelsen, 1969) that Weller (1985) attributes to the behavior of crowds, mobs, and riots.

Empirical evidence on an individual, who believes his identity is unknown, such as in a large crowd, indicates that he is more likely to behave aggressively (Rehm, Steinleiter, & Lilli, 1987; Watson, 1973). A more recent study by Silke (2003) found that anonymity also increases the range of a crowd's violent behavior. This effect has also been demonstrated in cross-cultural surveys. Evidence here indicates that warriors who use body or face paint are more likely to kill, mutilate and torture captured prisoners than warriors who do not use such masking (Watson, 1973). Similarly, a study of violence in Northern Ireland between 1994 and 1996 demonstrates that the use of disguises by attackers was significantly more associated with aggression at the scene of the crime and with more punitive treatment of the victims. Disguised attackers also showed a wider range of aggressive behavior (Silke, 2003). Psychologist have found that in large cities where potentially lethal emergencies, accidents, thefts, or personal attacks occur more frequently, people exhibit bystander apathy whereby they watch but seldom help in what has become known as the bystander effect (Latane & Darley, 1968).

Terror Management Theory is a psychological theory of how people cope with their awareness of the inevitability of death, and how a core fear of human mortality and vulnerability leads to a need for self-esteem, faith in a cultural worldview, and hostility toward those who hold different cultural worldviews (Pyszczynski, Solomon, & Greenberg, 2003). The findings of these and other basic psychological research studies can readily be applied to a model used to determine simulated crowd behavior of foreign cultures with known ethnic customs and norms.

Incorporating a substantial body of research from anthropology and from organizational, developmental and cognitive psychology, a "Cultural Lens" model has been developed that can capture cultural differences in reasoning, judgment, and authority structure (Klein & Steele-Johnson, 2002). This model allows for someone to see the world as if through another's eyes and to understand and evaluate options as others might. This "decentering" allows the outsider to anticipate actions, judge accurately, and intervene effectively. A "decentering" training program has been developed, implemented, and evaluated for use in understanding knowledge structures in terrorist groups. It can be utilized in a simulated model of crowd behavior in foreign settings to develop profiles of goals, patterns, and weaknesses of new potential terrorist groups (Klein & Steele-Johnson, 2002).

Crowd management and control models based on both military Command and Control and Management Information System concepts have been utilized and validated by police chiefs and urban administrators to develop procedures for massive demonstrations (Alghamdi, 1992). These models are based on military and civilian law enforcement experiences with pre-planned managed events, "spontaneous" demonstrations, terrorism, riots, natural disasters, and even peaceful special events and demonstrations requiring extreme management concentration. The model is separated into two phases, crowd management and crowd control. Both phases support the Command Center control teams. The crowd management phase pre-supposes prior notification of an event. Crowd management incorporates management information concepts that apply the management of technological and human resources in a dynamic environment. It utilizes important modeling and simulation considerations of support activities such as planning, data base design, scenario generation, event modeling, and training.

The crowd control phase within the model is precipitated by an unanticipated incident similar to those needed for urban combat operations training exercises. The emphasis is on the essential Command and Control coordination concepts that permit complex operations to function in a coordinated way. Here, Command and Control has two central functions: situation modeling and resource allocation. Situation modeling requires the fusion of data from all sources. Resource allocation is a decision making process for the marshalling and direction of equipment and manpower under a unified command. Systems that support this process include event evaluation systems, decision support systems and crowd control planning and monitoring systems.

A model of human crowd behavior focusing on group inter-relationship and collision detection variables proposed by Musse and Thalmann (1997) treats the individuals as autonomous virtual humans that react in presence of other individuals and change their own parameters accordingly. The model's focus is on individual and group parameters and distributed group behaviors to determine the global effect generated by local rules. The group parameters include defining the goals as specific positions that each group must reach, number of autonomous virtual humans in the group, and the level of dominance from each group. The individual parameters are a list of goals and individual interests for these goals, an emotional status, the level of relationship with the other groups, and the level of dominance following the group trend. This rather simplified model may assist in the development of a simulation of crowd behavior by focusing on the group inter-relationships.

Johnson and Feinberg (1977) describe a model of the internal dynamics of crowd processes based on the idea that crowds strive to reach a consensus for behavior through a process of interaction or milling. They report that individual crowd members seek support from a sample of crowd members. The successful ones attempt to influence the crowd toward a particular course of action. Consensus is achieved with a sufficiently reduced variability of opinions in the crowd. Independent variables examined include the initial distribution of opinions, crowd suggestibility, and the average position of individuals seeking to influence the crowd. Findings indicate that most of the simulated crowds reach consensus within a given time limit; rapidity and extremity of consensus are considerably affected by interaction among the manipulated variables.

This finding is supported by recent research (Derlega, Cukur, Kuang, & Forsyth, 2002) in a test of inter-group relations and social identity theory. Here, subjects were asked how they would respond to a conflict, either with another individual, between their group and another group or between their country and another country. Participants responded more negatively to inter-group and international conflicts than to interpersonal conflicts. Self-construal, or the feeling that one is separate and bounded from others (Markus & Kitayama, 1991), moderated this effect. Collectivist cultures emphasize the development of an interdependent construal of self, and individualistic cultures emphasize the development of an independent construal of self (Singelis, 2000). Controlling for country of origin, subjects who were high in interdependence endorsed threat more and acceptance of the other's demands less in an international conflict versus an interpersonal conflict. Those low in interdependence differed less in their endorsement of conflict resolution strategies in an international versus an interpersonal conflict. These findings would support the development of a model of crowd behavior in different cultures once the level of agreement is established.

The presence of outside agitators is another variable examined in a crowd behavior model offered by Feinberg and Johnson (1988) to determine the success of an extremist leading a group into "radical" action by a gathering of moderates not otherwise disposed to militancy. Findings indicate the variables most related to agitator success are crowd suggestibility, the action-choice advocated by the agitator, and the probability of movement within a physical space. Thus, the agitator is likely to be successful only in the specific and infrequent circumstances of a small gathering in a highly ambiguous situation in which the crowd members are not suspicious of the outsider.

A third body of research suggests a strong link between environmental variables and crowd violence. For example, Stott and Reicher (1998) believe that in some cases the presence of a perceived environmental threat, such as a police force, may cause a crowd to display hostile behavior toward that threat. Crowds may also become more aggressive in the presence of an aggressive or violent activity such as a violent sporting event (Miller, 1993). Finally Feinberg and Johnson (1988) found that an ambiguous environment is one factor that causes a crowd to become more readily accepting of violent behavior.

Several researchers believe that violent crowd behavior is not the result of predisposition or emergent group norms but is instead due to the complex interaction between individual and group goals (McPhail, 1991). This interaction determines the crowd's type or mission; a fourth factor that may lead to violent crowd behavior. These researchers explain that each crowd member is cognitively capable of setting the terms of his or her cooperation with the group's goals. Therefore, crowd behavior is determined by the extent to which a consensus is reached between the rational calculation of the individual members and those of the group (McPhail, 1991).

Violence occurs in crowds when individual members, called agitators, successfully influence other members and direct the goals of the crowd toward violent behavior (Feinberg & Johnson, 1988). The extent to which these agitators influence individual crowd members is dependent on several variables. For example, according to Feinberg and Johnson, members of small crowds are more likely to be influenced as well as members that are not suspicious of the agitators. An earlier study by these two researchers found that the diversity of opinions and suggestibility of the individual members also affected the likelihood that these members would become coaxed into violent behavior by crowd agitators (Johnson & Feinberg, 1977).

Crowd research has uncovered many potential factors that can cause crowds to reach a flash point. However, the research lacks an overall structure and coordination of these variables within a psychological framework. A well-organized and comprehensive model of the factors that cause crowd violence is needed because it may enable peacekeepers to answer some important questions. First, which of these flash points are the most common among crowds? Second, which, if any, of these variables are consistent across all crowds and in all situations? Third, at which stage of a crowd's evolution does each of these potential flash points influence crowd behavior? Providing peacekeepers with answers to these questions will greatly contribute to the creation and implementation of training programs designed to help eliminate this violence.

Psychological and Computational Models of Crowd Behavior

A model of crowd behavior created by Jager, Popping, and van de Sande, (2001) was initially developed to explore hypotheses about what might trigger riot behavior in crowded conditions. The model represents crowd behavior as internal states, with perceptual based knowledge of its environment and a repertoire of available behaviors. The model involves multiple agents, each of which belongs to one of two parties and whose interaction with agents of their own and the other party depended on the level of an agents' aggression and the number of agents surrounding that agent at a particular time. The model demonstrated emergent behavior similar to the rioting behavior seen in actual crowd situations.

Jager makes two main assumptions about individuals in crowd conditions. The first is that the desire to fight can change within an individual, depending on the circumstances in which those individuals find themselves and what their previous experience has been. The second assumption of the model is that different people may be more or less provoked in any given situation. The model utilizes a relatively simple framework that allows for possible variation in simulation. The first model behavior is clustering which varies in size and may be composed of agents from different parties. The second model behavior is global agent disposition representing varying levels of aggression. Each of these model behaviors will influence the others. For instance, decreasing the disposition will result in agents moving towards members of their own party if their own party outnumbers the other party, yet when disposition is high, party members move toward the party with the largest number of members. Similarly, decreasing vision and number of agents slows down clustering because of the increase in time required to cluster. Keeping the variables in mind makes possible more accurate models of military scenarios in foreign urban settings of varying ethnic composition.

Crowd simulation is becoming increasingly common in special effects for movies and games. Examples of such efforts include the crowd of digital passengers waving farewell on the Titanic or the army androids of soldiers in the recent Star Wars episodes. Further, sports games often include large numbers in spectator crowds with very limited and simple behaviors. These special effects lack the element of real time necessary for computer-generated crowds in virtual reality educational or training systems (Ulicny & Thalmann, 2002). Real time applications pose the challenge of handling interactions among crowd agents as well as the limited computational resources available.

The science of crowd simulators has been conceived from a number of more basic fields of study. Reynolds (1987) first studied distributed behavior modeling for simulating the aggregate motion of a flock of birds. He revolutionized the animation of flocks of animals, in particular birds, or "boids", by utilizing theories from particle systems relating each individual bird to a particle. Individual and collective actions were studied in what McPhail, Powers, & Tucker (1992) prefer to call temporary gatherings. A technique utilizing a combination of particle systems and transition networks to model human crowds in the visualization of human spaces employed by Bouvier and Guilloteau (1996) was the basis of later work in agent dynamics. Brogan and Hodgins (1997) addressed the issue of significant dynamics in simulating group behaviors. Their work simulated groups of creatures traveling in close proximity, whose motions are guided by dynamical laws. The focus to this type of animation is on collision avoidance.

Virtual humans have been represented through dynamically generated impostors in the work of Aubel and Thalmann (2000). Mobile cellular agents moving in an automatic fashion were used by Still (2000) in the simulation and analysis of crowd evacuations. Tecchia et al. (2002) employed image-based methods for real-time rendering of animated crowds in virtual cities. O'Sullivan et al. (2002) presented the Adaptive Level of Detail for Human Animation (ALOHA) model of crowd and group simulation. The ALOHA model incorporates levels of detail for not only geometry and motion, but also includes a complexity gradient for natural behavior both conversational and social. Few models (Musse & Thalmann, 2001; Ulicny & Thalmann, 2001) have attempted to examine more general crowd behaviors, integrating several sub-components such as collision-avoidance, path-planning, higher-level behaviors, interactions, or giving up. In Musse and Thalmann's (2001) ViCrowd system a virtual crowd can be created where the individuals have variable levels of autonomy including scripted, rule-based, and guided interactively by the user (Musse, Garat, & Thalmann, 1999).

Ulicny and Thalmann (2001) advocate a real-time crowd simulation with an emphasis on individuals in contrast to groups in a multi-agent system. Levels of variety range from zero variety where for a given tasks there is only a single solution; to level one where it is possible to make a choice from a finite number of solutions; and level two, where it is able to use solutions chosen from an infinite number of possible solutions. Seo, Yahia-Cherif, Goto, and Magnenat-Thalmann (2002) exemplify such a model whereby a level one system presents a crowd that is composed of multiple humans selected from a pre-defined set composed of sets of exchangeable parts such as heads, bodies, and textures. A level two system described in the same study displays a potentially infinite number of unique humans generated by a parameterized anthropometric model generating humans with different morphologies. Ulicny and Thalmann (2002) point out that such higher levels of variety can be unnecessary when perfect visualizations are distracting and simpler uniform visualizations could help emphasize problems.

The model of the world varied by Ulicny and Thalmann (2002) distinguishes between dynamic and static objects and environments. They define dynamic objects as those that may change place during the scenario such as fire or gas clouds. Boulic et al. (1997) model agents with more complex visualizations. Theirs are able to perform certain low-level actions such as pre-recorded body animation sequences of gestures and changes of posture or walking to a specified location with different gaits. Varying facial animation sequences were illustrated by Goto et al. (2001) looking at specified places or playing 3D localized sounds. These lower-level actions can be combined for higher-level behaviors in the Ulicny and Thalmann model (2002).

Ulicny and Thalmann's (2002) model of the world recognizes the importance of internal psychological or physiological states of agents. These include memory, fear, mobility, or level of injuries in simpler states. Higher-level complex behaviors include wandering, fleeing, or following a path. Agents follow a set of rules for determining the appropriate behavior as a result of the static or dynamic environment. Further, interactions between the agents and the dynamic objects simulate the reciprocal role that each plays, e.g., the effects of the agent on the fire or the effects of the fire on the agent.

Behavior models of Ulicny and Thalmann (2002) focus on the perception of the agents in their surrounding environment. This includes reaction to changes, to other agents, and to the real humans interacting in the virtual world. Behaviors, which can be mixed as needed, range in level from simple scripted behaviors to high-level autonomous behaviors such as wandering, fleeing, neutralizing the threat, or requesting and providing help.

Ulicny and Thalmann's model is a good example of efficiently managing variety. The levels of variety for multiple agents are perhaps one of the most advanced in the concept of crowd simulation. Their current work focuses on enhancing the levels of variety and incorporating motion models able to synthesize variations of a given motion (Lim & Thalmann, 2002). Further research is called for to enhance the behavior models by including new behaviors based on sociological observations from gatherings in real world settings.

Realism of Crowd Cognition and Behavior Models

Some agent modeling work is of potential relevance to crowd modeling. Increasing the realism of agent cognition and crowd behavior models is a important concern (Silverman et al, 2002). Such realism contributes to transfer of training for simulators of war-gaming and operations rehearsal. Including realism in the simulated crowd agents contributes to improved ability to explore alternative strategies and tactics when playing against them and higher levels of skill attainment for the human trainees (Pew et al, 1998; Sloman & Logan, 1999). Silverman identifies several variables necessary for inclusion in simulators to allow for more realistic agents. Agent behaviors need to change as a function of cultural values, emotions, level of fatigue and stress, time pressures, physiological pressures, and the group effectively overcoming an opposing group. All, of which may lead to limits of rationality. He notes, that in most available combat simulators the agents conduct operations endlessly without tiring or making mistakes of judgment. The agents, unrealistically, are uniform and predictable in following systematic doctrines in the defeat of their opponents.

The believability issue has been noted in the graphics and animated agent literature noting especially the lack of deeper reasoning ability of agents (Laird & Van Lent, 2001; Bates, 1994; Elliot, 1992). Silverman emphasizes the importance and need for a focus on planning, judging, and choosing types of behavior that are necessary of embodied agents. They note, perhaps most importantly, that "the human behavior literature is fragmented and it is difficult for agent developers to find and integrate published models of deeper behavior." The approach taken by Silverman investigates the duality of mind-body interaction and the impact of environment and physiology on stress and in turn, of stress on rationality. Their objective is to accurately model a crowd and predict when one may become violent while another remains peaceful.

The ability of humans to detect even slightly unnatural human behavior is a particular challenge in the simulation of crowds of humans. Realistic gestures and interactions between the individuals in the crowd are important cues for realism (O'Sullivan et al, 2002). Their model emphasizes the importance of not only verbal communication but also non-verbal expression as a means of communication. They point out that many of the existing methods look great at a distance but upon closer observation it is obvious that the agents are behaving in a cyclical manner and not interacting naturally with each other. One reason, according to Sullivan et al. (2002), is that most of the human crowd simulations tend to follow the flocking approach presented by Reynolds (1987), whereby the collision avoidance focus yields a crowd that is too sparse. A natural crowd consists of people conversing with each other, touching each other either intentionally or unintentionally. People who know each other in the crowd may be walking alongside or holding hands. Further, there are innate differences in how people react to being touched by a stranger in a crowd. Following behavioral rules of psychology makes it possible to achieve a "more chaotic, less military look to the crowds."

The O'Sullivan et al. model proposes an Embodied Conversational Agent with the same properties as humans in face-to-face conversation especially the ability to produce and to respond to both verbal and non-verbal expression. A nonverbal behavior generation toolkit called Behavior Expression Animation Toolkit (BEAT) has been developed to address this multimodal generation and interpretation of what is being said in the conversational content and how turns are managed in conversational management (Cassell, Vihjalmsson, & Bickmore, 2001), allowing for more realistic simulations of crowds.

Future work on the O'Sullivan et al. model calls for the integration of an intelligent agent based role passing technique into the ALOHA framework. Role passing involves the layering of roles on top of basic agents. This makes it possible for the agents to engage in collective motivations that recognizes psychological drives and attributes such as personality traits. Level of Detail Artificial Intelligence (LODAI) allows for incorporating these higher order behaviors in the more visible agents while omitting them for the less important agents (MacNamee, Dobbyn, Cunningham, & O'Sullivan, 2002).

Early work on embodied conversational agents (Cassell, Sullivan, Prevost, & Churchill, 2000) and animated pedagogical agents (Johnson, Rickel, & Lester,

2000) has laid the groundwork for face-to-face dialogues with users. An example of such a possibility is known as Steve (Rickel & Johnson, "Animated agents," 1999; Rickel & Johnson, "Virtual humans," 1999), and is particularly relevant. Steve is a collaborative agent used in 3D virtual worlds who can act as a member of a crowd, a teammate, or an instructor.

Steve has already been applied to Army peacekeeping scenarios. Steve's behavior is not scripted. Rather Steve consists of a set of general, domain-independent capabilities operating over a declarative representation of domain tasks (Rickel et al, 2002). According to Rickel, virtual worlds like the peacekeeping example introduce new requirements (Rickel et al, 2002). To create a more engaging and emotional experience for users, agents must have realistic bodies, emotions, and distinct personalities. To support more flexible interaction with users, agents must use sophisticated natural language capabilities. Finally to provide more realistic perceptual capabilities and limitations for dynamic virtual worlds, agents such as Steve need a human like model of perception. These new requirements have recently been tested in Mission Rehearsal Exercise (Swartout, 2001) implementing them in a peacekeeping scenario. The scenario uses three Steve agents (a sergeant, a medic, and a mother) who interact with the user, playing the role of lieutenant. The scenario also uses modified versions of Boston Dynamics' PeopleShop for additional scripted virtual humans in the crowd of locals and four squads of soldiers. The lieutenant's decisions influence the way the situation unfolds, culminating in a glowing news story praising the user's actions or a scathing news story exposing decision flaws and describing their sad consequences. The military is well aware of the challenges associated with training soldiers to handle difficult and stressful dilemmas in foreign cultures. This type of immersed training in realistic worlds allows soldiers valuable personalized experiences.

Survey Findings

There are many variables that may influence crowd behavior. Few existing models of crowd behavior have strong underpinnings in psychology. An area of psychological research that seems particularly relevant to crowd behavior and lacking in much of the existing research is cognitive psychology. Studying social cognition factors such as schemata or strategies for decision-making, judgments, and reasoning could enhance understanding human behavior in abnormal situations such as military operations.

The question of how direct is the relationship between cognition and behavior is not an easy one to answer. There has been little research that includes behavioral dependent measures, so opportunities to examine the relationship are relatively few. One challenge to understanding the cognition-behavior relationship is that we may expect too many and too varied behaviors to be related to any given cognition. Improving measurement of cognition and behaviors is essential for determining probabilities of behavioral consequences. At issue is the level of situational specificity with which each is measured. Attitudes and behaviors may be inconsistent given a particular cultural climate such as degree of interdependence or religious conviction. This suggests that the consistency will be highest when one examines behaviors that are prototypically related to particular cognitions, but that cognition-behavior consistency will be lower when one examines behaviors that are less centrally related to the cognitions in question.

Overall, attitudes that matter to a person such as those that are based on personal experience, held with confidence or that have implications for one's future show a stronger relationship to behavior than those that matter little (Kelman, 1974). The implication for developing models and simulations of crowd behavior is that we must consider the past experiences and the implications for one's future outcomes rather than focusing on cognitions that develop from mild curiosity or fleeting interest.

In addition to the psychological factors, situational factors may also influence the cognition-behavior relationship by making certain cognitions more salient as guides for behavior. For instance, social norms can be strong situational determinants of behavior that overwhelm seemingly relevant attitudes (Bentler & Speckart, 1981). Models of crowd behavior need to consider the presence of an audience or when one's attention is directed outward toward the situation which is when behavior is most likely to be strongly affected by self-presentational concerns (Snyder & Swann, 1978). Therefore, to predict and model what cognitions will cohere with behavior one must understand what factors such as self-presentational concerns, prior beliefs, attributions, expectations, or other influences in a situation are salient.

Future research on crowd behavior during military operations should address these more challenging areas of cognitive psychology. The people who exist within an urban area during a military operation are the only thinking component of an operational area and have the capacity to significantly modify operations (Medby & Glenn, 2002). Furthermore, the cultural differences must also be acknowledged in future research. We are seeing an increasing number of unarmed combatants being used as human shields to protect gunmen as in Mogadishu. Such conviction is very different from the cognitive processes to which American soldiers are accustomed and must also be included in future modeling and simulation endeavors.

CROWD SIMULATION DESIGN STUDY

In this section a design study of a crowd simulation federate is reported. The design study was based on a set of implementation experiments; those experiments and their results are described.

Design Study Introduction

In order to learn about salient challenges and possible solutions in the implementation of a simulation of crowd behavior, we conducted a design study of such a simulation. The design study began with certain premises:

- The crowd simulation would be implemented as a federate in a distributed simulation, and so would need to operate with an interoperability protocol such as the High Level Architecture (HLA), and other federates in the federation would be responsible for the non-crowd entities.
- The crowd entities would be individual human characters, rather than aggregations.
- The behavior of the crowd entities would be controlled by behavior models, rather than by a human operator.
- The federate could be used in a virtual simulation environment, so both the behavior and the visual appearance of the crowd entities was of interest.

Design Process and Goals

The design process started with an investigative software development process to initially create an exploratory prototype aimed at elaborating requirements of the Crowd Behavior application programming interface (API). The goals of developing this prototype were twofold. Firstly, we sought to mitigate risk by investigating areas of the design that contained either much complexity or involved reuse of software which may be difficult to integrate. Specifically, this involved the reuse of game AI solutions from the entertainment industry and integrating them with traditional military distributed simulation solutions. Those risk areas involved software integration and terrain correlation. Secondly, growing familiarity with the capabilities of the commercial-off-the-shelf/government-off-the-shelf (COTS/GOTS) software and the theoretical necessities of crowd behaviors and their effect on military outcomes, allowed iterative refinement of the requirements of the Crowd Behavior API. The exploratory prototype and refined requirements could then be used to implement a functional prototype designed to support the Crowd Behavior API. This prototype could be expanded as additional capabilities are needed.

This process was, in effect, the beginnings of a spiral development effort tailored to the goals and objectives of crowd modeling for military simulation. Upon conclusion of the design study the following design and implementation goals were achieved:

- A prototype was implemented that demonstrated the integration of game AI software within a military-oriented distributed simulation architecture, such as HLA. This prototype resulted in a federation that combined crowd-modeling capabilities from entertainment industry software with semi-automated forces modeling capability of the military simulation JSAF.
- An initial draft of the Crowd Behavior API was achieved to a level of detail specification that could be fully implemented later.
- Crowd models with differing fidelity were created as it was important to determine the effects of the different fidelity crowds on the outcome of a military mission.
- The team also developed and correlated terrain between the game technology and the military simulation. This process was evaluated and documented.

Crowd Simulation Architecture

This section describes design considerations and ideas for the crowd federate architecture.

Architecture Background

Crowd behavior models in the literature, ranging from purely physical representations to purely cognitive representations and combinations of both, have been successful to varying degrees in the domain for which they were intended. Motivations range from improving graphical rendering performance to providing realistic behavior representations with emotions and other drivers. Crowd models include particle systems, flocking systems, or behavioral systems, with the difference being an increasing level of interaction among participants in the crowd and with the environment in general. Along with increased interaction comes increased attention to modeling the social and emotional interactions among crowd members.

A recurring theme throughout the literature is the necessity to model behavior either at the individual level and allow crowd behavior to emerge as a result of many goal-oriented individuals within the same location or the behavior of the crowd may be completely controlled by a crowd entity. Again, there is a range of efforts between these two extremes. Recent findings (Kenny et al, 2001) indicate that there is empirical evidence that crowds are not wholly individuals nor wholly a single entity but rather a congregation of individuals and small groups from which crowd behavior emerges out of a process of assembling, temporarily interacting, and then dispersing. This suggests that there is a hierarchy of behavior needed at the individual and group levels and an awareness and affect of the evolution of the crowd over time. Such complexity could only arise from behavioral models that include aspects of cognition. Nevertheless, a variety of models do exist. Where realism and human decision-making based on simulation stimuli are important, models that focus on physical visual representation of crowds are important. Where motivations and consequences take precedence over visual realism, robust cognitive models are necessary. A framework that allows the integration of diverse models at the cognitive level with differing physical models would serve to provide useful crowd models to fit a variety of needs. In recognition of this variety, the Crowd Behavior API must be flexible enough to allow such mappings of differing philosophy.

We used COTS tools such as AI.implant, Maya, and simulation engines from the gaming and entertainment industry to provide much of the physical and visual realistic representation. AI.implant is an autonomous character tool that calculates and updates the position and orientation of each entity, chooses the correct set of animation (locomotion) cycles, and enables the correct simulation logic. Maya is a graphical design and animation tool that provides the workspace for AI.implant and produces the animation clips that a game engine would use to visualize the simulation environment.

Military alternatives at the physical layer of crowd representation include the military simulation DISAF (Reece, 2002). Recently, DISAF began including a flocking-based representation of crowd behavior allowing the incorporation of standing-around behavior and flocking to waypoints. In addition, crowd individuals have traits such as curiosity, fear, motivation, and hostility. The mental states achieved in individuals stimulate crowd behavior resulting in stay or flee.

More robust models of crowd cognitive behavior are being developed (Silverman et al, 2002; Cornwell, Silverman, O'Brien, & Johns, 2002; Musse et al, 1998; Musse & Thalmann, 2001; Allbeck et al, 2002). Figure 3 shows Silverman's model, called PMFserv. These models include crowd behaviors such as dispersing, gathering, swirling, clustering, flocking, safe wandering, following, goal changing, attraction to a location, repulsion from a location, group splitting, and space adaptability. Crowd evolution is also tracked so that one can determine whether the crowd is advancing, gathering, retreating, or dispersing. These models include environmental knowledge, group beliefs, intentions or goals, and also desires. Silverman builds a cognitive model based on Markov chains and a BDI (Belief, Desires, Intensions) model (Silverman et al, 2002).

Musse and Thalmann (2001), in particular, discuss three ways of controlling crowd behavior: 1) scripts, 2) rules with events and reactions, and 3) real-time external control. An entity hierarchy is provided where the smallest unit is the virtual human agent. Next up in the hierarchy are groups that are composed of agents and then crowds that are composed of groups. "Crowd behavior corresponds to a set of actions applied according to entities' intentions, beliefs, knowledge and perception" (Musse & Thalmann, 2001). Different levels of realism are explored to support simple to complex crowd behaviors. Simple crowd

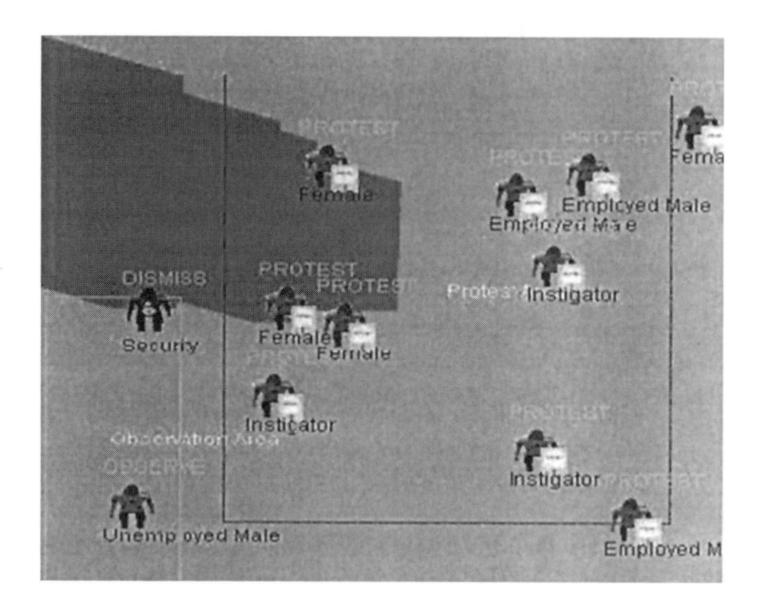

FIGURE 3. PMFserv, An Existing Model of Crowd Behavior (Silverman, 2002; Cornwell, 2002).

behaviors would approximate crowd behavior based on flocking systems. Additionally, crowds may have a predominant emotion (sad, calm, regular, happy, or explosive, which is extremely happy). These emotions may affect the posture and walk characteristics of individuals.

Our desire is to benefit from the variety of models by creating an architecture with an API specifically to support crowd behaviors in distributed simulations. This framework would allow the integration of robust cognitive models with models existing at the physical layer.

Architecture Approach

The Crowd Behavior API facilitates integration, communication, and interoperability of crowd cognitive models with crowd physical models. HLA compliance allows another level of interoperability that may be exploited. Physical models that are uniquely military oriented are immediately available with the use of the HLA. This combination of the Crowd Behavior API and HLA compliance allows for the development of a crowd federate architecture that may draw from the military as well as the entertainment industry to utilize the best of both worlds in carrying out realistic simulations of crowds in military scenarios.

The approach used in the design for the crowd federate is to start with a generalized API for an architecture that supports the reuse of existing cognitive models as well as existing physical models. The API is intended to facilitate control of the physical model by the cognitive model as well as event and state feedback from the physical model to the cognitive model. An example portion of the API is shown in Figure 4. The API is also intended to provide access to the entity and interaction data exchanged with other simulations with which the crowd federate may be operating. The details of the API syntax are not as important as the basic idea suggested by the figure; the API provides a set of operations through which the cognitive and physical models communicate. Figure 5 illustrates a generic crowd federate architecture and shows the Crowd Behavior API as a layer between the cognitive model and the physical model. This API facilitates the exchange of information between the two models tempered by the reconfigurable mapping data. The Mapping Data module correlates terms for specific actions and events in the Cognitive Model with similar actions and events that also exist in the Physical Model but are named differently. The Run-Time Infrastructure (RTI) is part of the High Level Architecture (HLA), a communication infrastructure for linking distributed simulations via a network. The RTI Interface module connects the crowd federate to other simulations via the HLA RTI. Information about the crowd members is exchanged across the RTI with other federates such as a semi-automated forces (SAF) simulation, which controls the combatants in the simulation.

```
ActionName <= CreateAction(ActionType, Parameters[]);
CharID = CreatCharacter(Attrs[], Actions[], GroupID[]);
GroupType = CreateGroupType(Attrs[], ActionNames[]);
GroupID = CreateGroup(GroupType, CharIDs[]);

Boolean = AddAttr(CharID/GroupID/GroupType, Attrs[]);
Boolean = SetAttrValue(CharID/GroupID/GroupType, Attrs[], Values[]);
Boolean = DelAttr(CharID/GroupID/GroupType, Attrs[]);
Boolean = AddAction(CharID/GroupID/GroupType, ActionNames[]);
Boolean = DelAction(CharID/GroupID/GroupType, ActionNames[]);

Boolean = JoinGroup(CharID, GroupID);
Boolean = LeaveGroup(CharID, GroupID);
```

FIGURE 4. Sample Crowd Federate API Specifications.

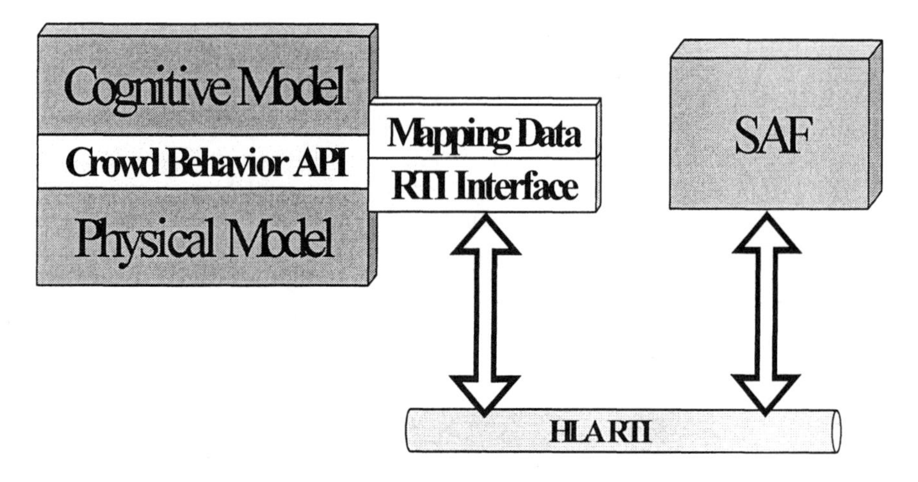

FIGURE 5. Generic Crowd Federate Architecture.

In addition to refinement of the API through examination of background and current material, we conducted small experiments such as modifying force-identifiers of remote entities to emulate individuals in a crowd turning hostile. Other experiments include developing crowds using two levels of fidelity, controlling of entities using game API's in Maya and in the AI.implant software developers kit (SDK), and evaluating the development of a crowd simulation using simulation game engines. All these experiments are intended to flesh out the requirements for a Crowd Behavior API.

The initial exploratory prototype implemented is shown in Figure 6. Note that the Crowd Behavior API, although implemented, was not integrated into the prototype. However, the prototype along with the various experiments aided in clarification of requirements for and refinements of the Crowd Behavior API. The Script module in the figure is provided in place of a Cognitive Model and directs the physical actions of the crowd members using fixed pre-determined commands.

The physical model used for this prototype was the AI.implant SDK while the SAF used was JSAF (Joint Semi-Automated Forces, a military simulation widely used for combat analysis and experimentation). In the prototype, characters in the crowd were registered with the HLA RTI and subscribed to by JSAF. The RTI objects representing the crowd characters were displayed on the JSAF plan view display (PVD) as remote entities. The prototype provided only one-way entity flow as illustrated in Figure 7. The JSAF PVD is a graphical user interface that includes a two-dimensional map view of the terrain showing entity velocities and status. The generation of remote representations of JSAF military entities within AI.implant will be a follow-on activity.

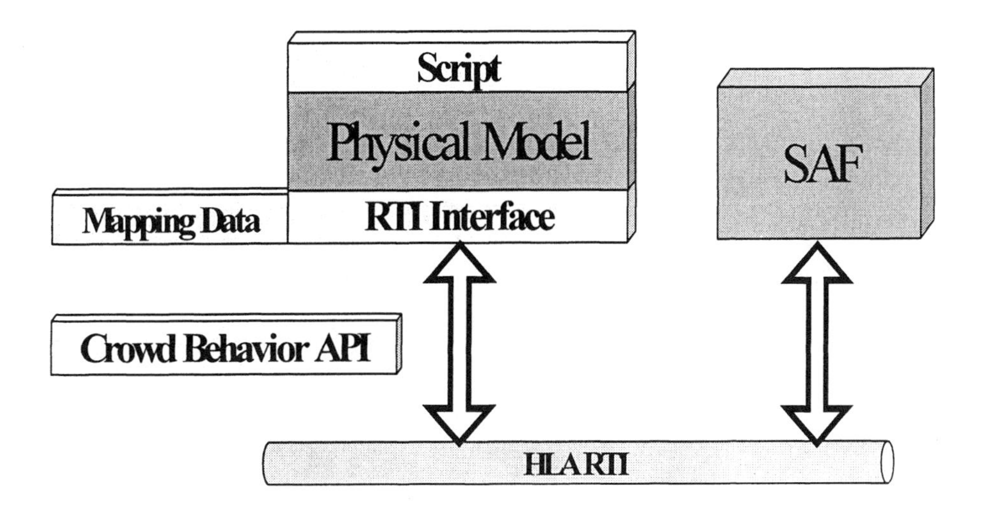

FIGURE 6. Implemented Exploratory Prototype Crowd Federate.

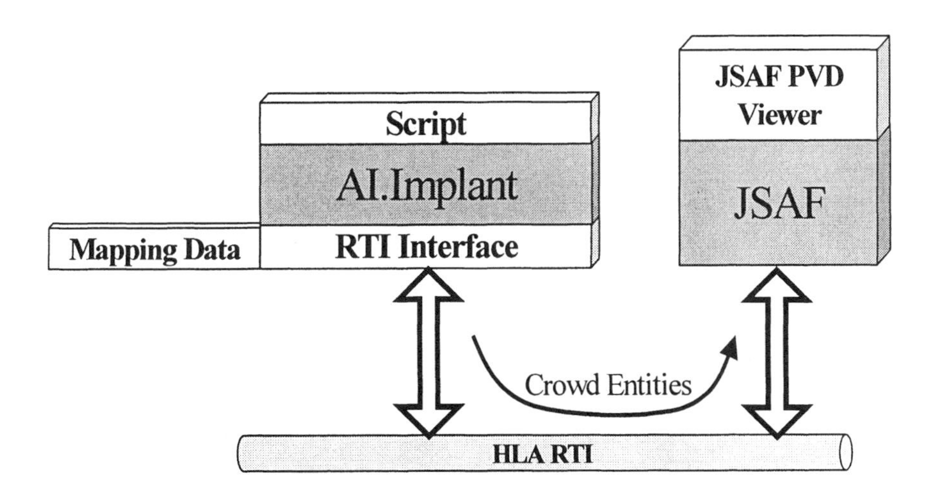

FIGURE 7. Running Crowd Federate Prototype.

Architecture Design Details

Details of the crowd federate are most apparent in the Crowd Behavior API. In Figure 8, general designs for the cognitive model and the physical model are shown. These designs are intended to represent functionality that may be mapped to any given specific cognitive or physical model. This is important since the Crowd Behavior API will support such functionality and, therefore, specifics of the API contains the framework necessary to convey functionality and information given the necessary set of mappings.

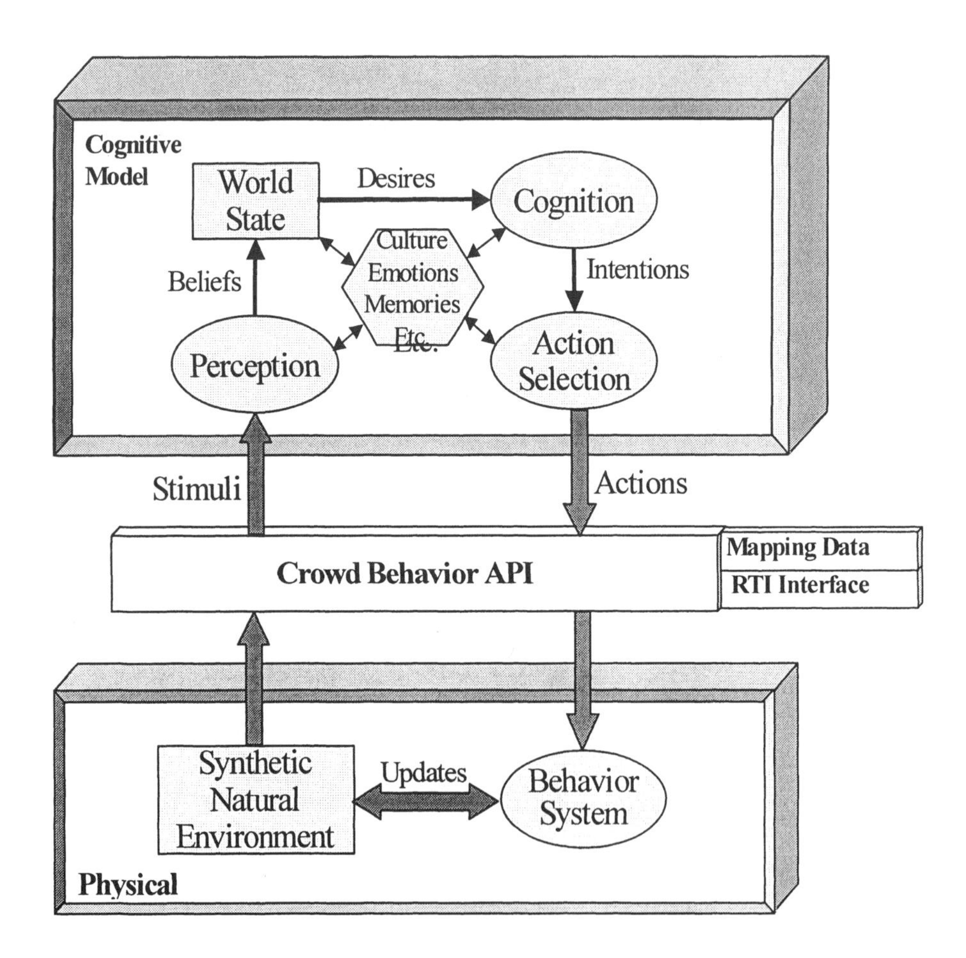

FIGURE 8. General Cognitive and Physical Model Designs.

Shown in the design of the general cognitive model are the two main interfaces with the physical model and thus the API. These are the components that include perception and action selection. The API contains mechanisms to provide sensory data to the cognitive model so that this data may be perceived in the required manner. The API must also contain mechanisms to provide command and control to the cognitive layer over the physical layer. The converse is true with regard to the physical layer. Sensor stimuli must be provided to the cognitive layer and actions provided by the cognitive layer must have a means to be executed. All this must occur via the Crowd Behavior API.

The specifications of the interface have been designed and refined based upon these general cognitive and physical designs derived from the crowd modeling literature, the implemented prototype, and the experiments conducted. Specific design criteria for the interface include the flexibility of the API to handle behavior transference from an individual influencing group behaviors or a group influencing an individual's behavior. Also of interest are individual or group behaviors that are influenced by triggers or persistent phenomena. It is believed but not proven that such capability is needed, as these are unanswered questions in the psychology community. Whether or not these capabilities are needed in military simulations are questions we may be able to answer in surveying the military simulations community. Our intent is to provide the flexibility in the API to support the changing psychology of crowd behavior.

Human Character Animation

Characters and groups of characters are defined by a set of attributes and set of actions to be carried out by these groups and the characters in the groups. These distinctions allow individual characters to carry out individualized actions while sharing actions designated to be group activities with members of its group. Figure 9 shows representative attributes and actions.

Crowd Command and Control

The ability to perceive and affect the characters and their environment is provided through the execution of a predetermined set of crowd behavior actions that may be mapped to similar actions in the cognitive and physical models. It is believed that such a selection may be possible since this API is focused on the behaviors of crowds and there are reasonable sets of activities that are realistic for crowds to perform. Further, mappings of semantically exact actions, although desirable, may not be a necessary condition in that semantic differences should be small enough to be insignificant in the randomness of what it is to be human and unconstrained within the context of civilian life. This is, of course, based on the axiom that civilians are significantly more non-deterministic than military units, and therefore

Attributes Category:
External states => position, orientation, speed ...
Internal states => alive, anger, hungry ...

Actions Category:
Observation: see, hear, feel ...
Movement: stay, walk, run, flock with, follow path ...
Interaction: throw rock, shoot, talk ...
State change (callback): set attribute value

FIGURE 9. Example Attributes and Actions Categories.

there is a wider range of believable behavior for crowds and the individuals within them.

Distributed Simulation Interface

Mappings for exchange of information with other federates depend upon the ability of the Federation Object Model (FOM) to support the necessary information. Initially, the Real-time Platform Reference Federation Object Model (RPR FOM) has been used to support interoperability with selected semi-automated forces (SAF) systems. It is believed that there will be a set of attributes and interactions that will not be supported by the RPR FOM nor any of the SAF systems. Such a set would remain part of the simulation object model (SOM) of the crowd federate unless and until the cognitive model portion of the crowd federate were to be executed on a separate node from the physical model portion of the crowd federate to in fact create a distributed crowd federate. In such a case, the RPR FOM would need to be augmented to support transfer of the extraneous set of data. The HLA interface therefore is a mapping dependent upon the FOM used.

Implementation Experiments and Results

Six implementation experiments were conducted to learn about potential problems and solutions in implementing a crowd federate. All of the experiments were designed to evaluate the ability of the Crowd Behavior API to support various configurations that might be necessary. Those experiments were:

- Character modeling
- Crowd fidelity modeling
- Terrain development processes for game technology interoperability

- Standalone CrowdFed
- CrowdFed federation (includes testing JSAF support for crowds)
- Game view CrowdFed

In the following sections details for the first three experiments are individually discussed while the last three experiments are discussed in common given their similarity.

Implementation Experiment: Character Modeling

Objective. The objective in character modeling was to develop character models at varying levels of fidelity that could support close views of crowd participants as well as distant views of the whole crowd. Additionally, low-resolution models were needed to support large crowd animations.

Method. Initial development of a low-resolution human figure visual model tested performance within the AI.implant and Maya behavioral environment. Walking, climbing and interpolation motion animations were developed and converted to animation clips that could be called by AI.implant during a simulation. These incorporated horizontal movement, e.g., during a walk cycle animation clip, the figure walks from point A to point B; an animation methodology that utilizes Maya's ability to offset translation synchronously according to the speed of the animation cycle. AI.implant however does not support this functionality so it was found that animations had to loop on the spot. A running 'on the spot' clip was developed and tested accordingly.

Next, a high-resolution human figure model was used to create models with varying polygonal geometry complexity for real-time performance. Material properties were assigned to the textured figure for improved realism. To control the geometry and to prepare for animation, the figure was rigged with an inverse kinematics skeleton. The geometry was subsequently bound to the skeleton so that it would deform with the bones when these were animated. This included deformations for bulging muscles/twisting upper body and constrained movements for elbow, shoulder and knee joints.

Results. Several character models were developed at varying levels of resolution. Polygon counts of the models range from over 10,000 down to 440 polygons. As expected the high polygon count characters could only support two to three characters in real-time rendering. Approximately fifty low polygon count characters could be easily supported in real-time. An example high-resolution model obtained from reference material is shown in Figure 10 on the left while a low-resolution model developed for this experiment is illustrated on the right.

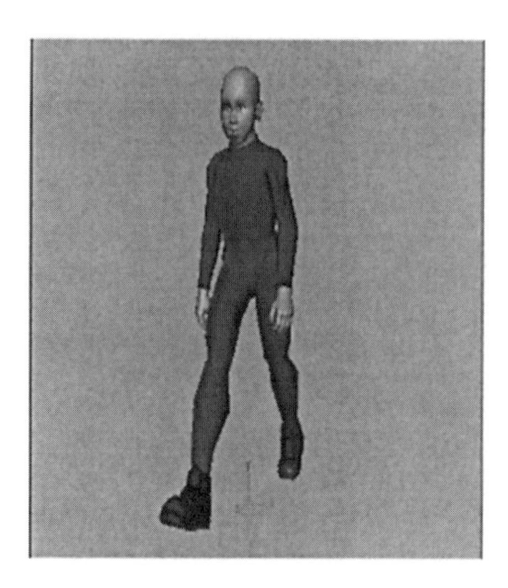

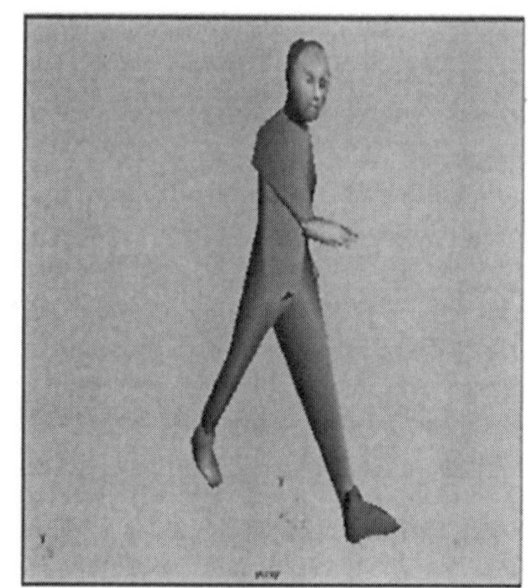

FIGURE 10. Higher- and Lower-Resolution Human Character Visual Models.

Conclusions. High-resolution models are difficult to work with in real-time rendering. However, for non real-time purposes, these high-resolution models can be rendered off-line and played back in real-time. It should be noted that these difficulties arose using the AI.implant plug-in within the Maya environment. Using the AI.implant SDK, the numbers of characters that can be supported are much larger since rendering is processed separately and by a more optimized rendering engine.

<u>Implementation Experiment: Crowd Fidelity Modeling</u>

Objective. The objective was to examine two levels of crowd fidelity using AI.implant – one where the crowd is wandering around and the other where they have purposeful movement.

Method. Crowd fidelity modeling includes the development of scenario characters with varying resolution (polygon count), applying animation characteristics to the characters which includes providing them with skeletons and specific movement clips, and incorporating them into the Maya environment where they are given AI.implant behaviors. The type of AI.implant behaviors they are given creates the level of crowd fidelity. For this experiment, AI.implant crowds were given the two levels of fidelity. The Quantico Combat Town terrain database, illustrated in Figure 11, was used as a virtual venue for the experiment.

Results. Two levels of behavior fidelity were studied. At fidelity level 1 (lower fidelity), people were simply wandering around. Figure 12 shows an image at low fidelity; in it, autonomous characters randomly are moving around on the terrain

FIGURE 11. Quantico Combat Town Urban Terrain Digital Representation.

FIGURE 12. Lower Fidelity Crowd Behavior (Relative to Figure 13).

and avoiding buildings automatically. The AI.implant behaviors used were "Wander Around" and "Avoid Barrier".

In contrast, Figure 13 shows an image at fidelity level 2 (higher fidelity); in it, the autonomous characters are flocking, wandering, following paths, and seeking. Several of them form groups and flock with each other. Some follow different paths as individuals. One character seeks an object in a building; when he gets there, he stops. The others characters just wander around as before. The AI.implant behaviors used are "Wander Around," "Avoid Barrier," "Flock With," "Follow Path," "Seek To," and "Avoid Obstacle." Without the avoid obstacle behavior, characters would bump into each other and various features of the environment. Barriers are used for static features of the environment; therefore, the avoid barrier behavior keeps the characters within the bounds of the Quantico town and stops the characters from walking through the walls of buildings.

Conclusions. We were able to successfully create two levels of crowd fidelity even with the use of limited animation clips and standard behavior mechanisms. More complex behavior mechanisms using a decision tree approach is a capability available in AI.implant that would provide significantly more realistic behaviors.

FIGURE 13. Higher Fidelity Crowd Behavior (Relative to Figure 12).

Implementation Experiment: Terrain Development for Game Interoperability

Objective. The objective here is to document the process involved with the development and correlation of terrain databases used in military simulations with those employed in game simulations. This process was documented during the creation of an urban environment for the Somalia compact terrain database (CTDB) terrain. CTDB is the terrain format used by JSAF.

Method. In order to integrate buildings into the Somalia CTDB dataset we used the DART software tool from TERREX. By using DART we were able to re-use existing CTDB data. DART was used to import the CTDB dataset into TERRAVISTA, another product from TERREX. Once in TERRAVISTA we were able to generate a 3D terrain database in a more common format. The common-format database then can be populated with buildings and geo-referenced satellite imaging. An example of the satellite imaging used is shown in Figure 14 while the satellite imaging overlaid on Mogadishu, Somalia flat terrain with associated buildings is shown in Figure 15.

FIGURE 14. Satellite Image of a Portion of Mogadishu Somalia. (Courtesy of Space Imaging).

FIGURE 15. Mogadishu Digital Terrain Representation with Imagery and Non-Geo-Referenced Buildings.

The buildings were extracted from a video game called Black Hawk Down from Novalogic. Two pieces of software were used for obtaining the models in a format that could be transferred into the CTDB database. The first software tool was used to extract the compressed data format used by this gaming company. The second software tool was used to convert the extracted format into a file format that then can be imported into Multigen Creator that allowed the georeferenced combination of buildings, satellite imaging, and terrain. TERRAVISTA was then used to convert this final combined database back to the CTDB terrain database format.

Results. Two CTDB datasets of Somalia are available. The original that contains no features and one created with the satellite imaging and Mogadishu buildings. Additionally, terrain datasets correlated to these were created for Maya.

Conclusions. Terrain generation and correlation of needed geographical regions is not a simple task. Legacy databases should be reused where appropriate. However, the process is tenable, as has been demonstrated.

Implementation Experiments: Standalone CrowdFed, Federation, and Game View

Objective. Standalone CrowdFed encompasses the ability to simulate the different fidelity crowds outside of the Maya environment using the AI.implant SDK. Next, integrating an HLA component with the standalone version would allow interoperability with other HLA compliant federates such as JSAF. The idea would be to have civilian entities move amongst combating entities and supposedly not be harmed. These entities can then be made to become combative at which time they may be attacked by the federated fighting units. Note that crowd behavior observed in a federation would be restricted to the allowable human activities outlined within the particular FOM used for the federation. To visualize high fidelity behaviors of crowd members that would exceed the capabilities of a standard FOM such as the Real-time Platform Reference (RPR) FOM, a visualization component would need to be added to the crowd federate by way of a rendering engine. This introduces the concept of the game view CrowdFed experiment. The objective is to view CrowdFed entities along with JSAF entities in real-time using the AI.implant SDK along with a rendering engine such as the RenderWare game engine.

Method. Our first experiment creating simple objects and using simple behaviors was a success. A sphere with a bounding volume was created on flat terrain and moved around using the AI.implant wandering behavior. Follow-on experiments recreated the low and high fidelity crowd models with only the AI.implant SDK. Crowds would wander around using information from the AI.implant "marked up" terrain used earlier within Maya. "Marked up" terrain contains added AI.implant-specific boundaries and movement areas that are not visible when rendered. This standalone configuration does not contain a viewer but when instantiated in a CrowdFed federation would create objects that would be viewed on participating federates with viewers such as the JSAF plan view display (PVD).

Once the HLA component was integrated, civilian entities in the crowd federate were created and placed in the vicinity of military entities (M1A1 platoon). The civilian entities appear in JSAF as remote entities. The military entities were created within JSAF so they are local entities. The rules of engagement on these military entities were set to weapons free. Remote civilians, with ForceIDs of neutral and friendly, were not shot at by the military entities. The neutral civilians were also placed within the context of a battle with military entities that were fighting (shooting each other). Again the civilian entities were not shot at. As soon as the ForceIDs of some of the civilians were changed to enemy, those civilians with changed ForceIDs were shot. The other civilians with ForceIDs of neutral were untargeted.

The events described were observed using the JSAF PVD. Although, undergoing implementation at the time of this writing, the architecture of the game view capability is shown in Figure 16. The architecture shows the AI.implant SDK

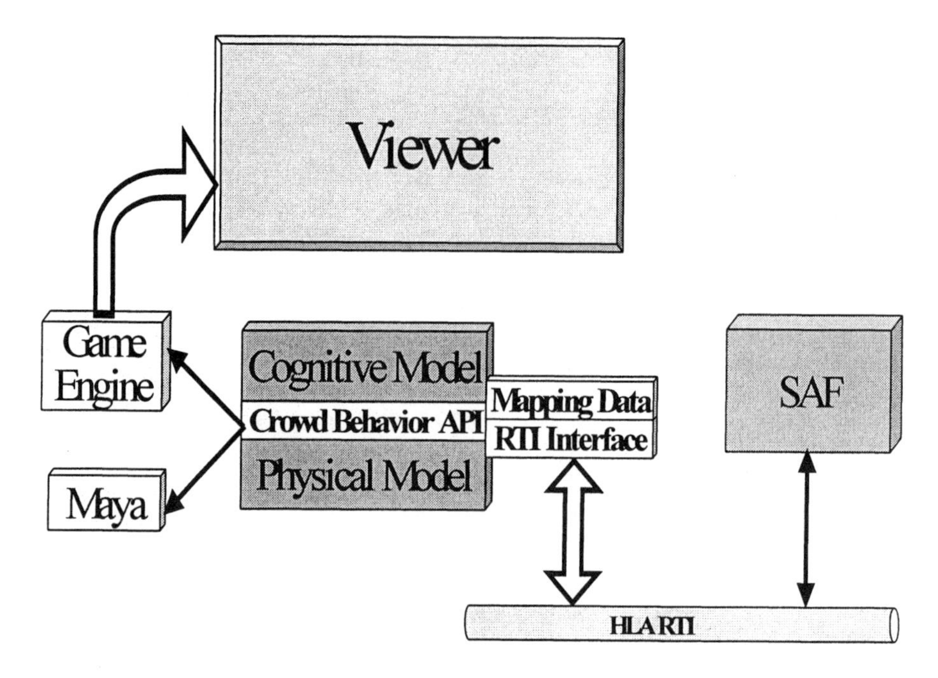

FIGURE 16. Game View CrowdFed Architecture.

accessing graphical models generated using Maya for use as entities in the scenario. Then AI.implant provides state information to the Game Engine that renders the scenario on a chosen platform/viewer (PC, Xbox, etc.). This visualization would provide valuable insight into what the crowd is doing and sensing beyond the allowable capabilities of the RPR FOM so that the scenario can be more easily validated.

Results. Crowds were successfully created and executed outside of Maya using the AI.implant SDK. The HLA-compliant CrowdFed non-combatant civilian entities viewed on JSAF changed their ForceIDs from neutral to enemy; they were then attacked by friendly forces in JSAF.

Conclusions. The AI.implant SDK is an advantageous method to use given the speed and flexibility achieved. CrowdFed civilian crowd members can be viewed on JSAF and can change their ForceIDs to emulate neutral crowd members turning hostile. Individual members in the crowd can turn hostile while others remain neutral. Friendly forces can then target the hostiles but not the neutrals within the crowd. However, it is unclear whether the neutral crowd members can obscure line of sight (i.e., provide cover and concealment). Allowing for the detection of JSAF entities in AI.implant is task yet to be undertaken. However, it will be addressed in follow-on efforts.

Design Study Findings

We were able to mitigate COTS/GOTS integration risk as well as terrain correlation risk by successfully implementing the crowd federate exploratory prototype. Additionally, an enhanced understanding and refinement of the Crowd Behavior API requirements was achieved. Many of the issues discussed above are resolvable and will be addressed in follow-on efforts.

Also of interest is the significant level of flexibility found in COTS entertainment industry tools such as AI.implant. With these tools, civilian/crowd behavior is easily created in a believable and high fidelity manner, where as, military behaviors are not easily available in realistic fashion in these tools. There is a realization that military simulations model military behaviors well, while tools from the entertainment industry model other behaviors well. The combination of the two seems to bring the best of both worlds together in the creation of a federate for representing civilian crowds in military scenarios.

CONCLUSIONS

This section summarizes the findings of this report and recommends further related research and development.

Summary of Findings

An adaptation of a previously proposed requirements analysis method was applied to identify crowd-modeling requirements and relate them to uses of military simulation. A large number of crowd-modeling requirements were found; these requirements included needed crowd behaviors, military missions affected by crowds, and crowd effects to model. The identified requirements should provide a good basis for implementing a useful crowd simulation. The primary finding of the requirements analysis was that because of the requirements' diversity in scope and intent, no single crowd behavior model is likely to satisfy all of the requirements. Therefore, research and development efforts aimed at developing a crowd model should be explicitly oriented towards its intended uses and the requirements associated with those uses.

There are many variables that may influence crowd behavior. Few existing models of crowd behavior have strong underpinnings in psychology. An area of psychological research that seems particularly relevant to crowd behavior and lacking in much of the existing research is cognitive psychology. Improving measurement of cognition and behaviors is essential for determining probabilities of behavioral consequences. At issue is the level of situational specificity with which each is measured. Research suggests that models of crowd members' behavior should consider past experiences and implications for future outcomes. In addition to the psychological factors, situational factors may also influence the cognition-behavior relationship, such as the presence of an audience or the affect of attention on self-presentational concerns. Future research on crowd behavior during military operations should address these more challenging areas of cognitive psychology. Furthermore, cultural differences must also be modeled because American soldiers are increasingly faced with behaviors, such as the use of human shields, with which they are unfamiliar.

Module integration risk and terrain correlation risk in an eventual crowd federate were both mitigated by successfully implementing the crowd federate exploratory prototype. The implementation experiments identified possible problems and solutions in federate development. An understanding of how to implement a Crowd Behavior API was achieved. Also of interest was the significant level of flexibility found in COTS entertainment industry tools, such as AI.implant. With these tools, civilian/crowd behavior can be created in a believable and high fidelity manner with reduced difficulty. There is a realization that military simulations model military behaviors well, while tools from the entertainment industry model other behaviors well. The combination of the two seems to bring the best of both worlds together in the creation of a federate for representing civilian crowds in military scenarios.

Research and Development Requirements

As a result of the study documented in this chapter, we have formulated several research questions regarding crowd modeling, which are listed here:

- *Crowd model types.* What types of crowd models have or can be developed, and which types are best suited for different applications and levels of resolution in military simulation?
- *Crowd simulation architecture.* Is there a simulation software architecture that could serve as a reconfigurable support structure for developing and testing multiple crowd models?
- *Behavior fidelity requirements.* How large a repertoire of crowd behaviors, and how much fidelity in those behaviors, is needed for different intended uses?
- *Cognitive model requirements.* Are cognitive models of the reasoning and emotional states of individual crowd members needed to generate realistic crowd behavior, and how much detail must those models have?
- *Psychological model applicability.* Can current psychological models of crowd behavior, which are sometimes descriptive and qualitative, be adapted to serve as the basis for quantitative computational models of crowd behavior suitable for implementation as software in the context of a simulation?

- *Crowds as civilians.* Is there a difference between crowd behaviors and more general civilian/non-combatant behaviors, and if so, which of the crowd/civilian behaviors are required in military simulation?
- *Control vs. emergence.* Should the behaviors of crowds, which are made up of individuals, be modeled using methods that depend on representing the state or controlling the behavior of the crowd at an overall level, or instead using methods that model only individual behavior and allow crowd behaviors to emerge from the individuals' cumulative actions?

Follow-on research to actually develop and test a crowd federate, based on the lessons learned in this study, is recommended. The recommended research would focus on modeling crowd members in the context of simulations that model military combatants and other scenario participants. To some extent, the research should focus on the training and experimentation applications, though this is primarily a reflection of degree of validation; with additional validation effort the capabilities we recommend developing could be applied to the more demanding analysis and acquisition applications as well.

The recommended research and development program has two objectives. The first objective is to design, implement, and test a multi-layered reconfigurable software architecture for a crowd federate, i.e., an HLA-compliant simulation that models crowd behavior. The crowd federate architecture would be in the form of a usable crowd federate that models crowd behavior in the context of a real-time, individual combatant-level simulation federation and includes initial models of crowd member cognition and of crowd member physical actions. The second objective is to use the crowd federate and federation to conduct two experiments. Those experiments will investigate crowd behavior fidelity requirements and assess the reconfigurability of the crowd federate architecture.

Expected Benefits

The benefits that are expected to accrue from a crowd modeling capability in military simulation are implicit in the motivation for this study. As noted earlier, in spite of the military difficulties and risks imposed by crowds, models of crowds are essentially absent from current production military simulations. Consequently the military forces and commands of the U.S. and coalition allies are incompletely prepared for dealing with crowds. Adding a crowd modeling capability would address this shortcoming and allow the military, which already uses simulation for many training tasks, to prepare for crowd operations as well. The inclusion of realistic crowd models into military simulation can provide substantial increases in realism and usefulness for training, analysis, experimentation, and acquisition applications.

ACKNOWLEDGEMENTS

The research upon which this chapter is based was sponsored by the Defense Modeling and Simulation Office and supervised by that office and the Air Force Research Laboratory. That support is gratefully acknowledged. The authors also thank the experts who provided input during the requirements analysis process. Old Dominion University graduate students Taimur Khan and Qingwen Xu contributed to the study.

REFERENCES

Adang, O. M. J. (1992). *Crowds, riots and the police: An observational study of collective violence.* Siena, Italy: Presented at the meeting of the International Society for Research on Aggression in September 1992.

Alghamdi, A. A. (1992). *Crowd control model with a management information system approach for command and control*, Dissertation #0075. Washington, DC: George Washington University.

Allbeck, J., Kipper, K., Adams, C., Schuler, W., Zoubanova, E., Badler, N., Palmer, M., & Joshi, A. (2002). ACUMEN: Amplifying control and understanding of multiple entities. In the proceedings of the First International Joint Conference on Autonomous Agents and Multiagent Systems. Bologna, Italy, July 15-29.

Aubel, A., & Thalmann, D. Thalmann (2000). Real-time display of virtual humans: Level of details and impostors. *IEEE Transaction on Circuits and Systems for Video Technology, 10*, 201-217.

Bailey, M. P. (2003). *Questionnaire response*, June 18.

Bakker, F., Whiting, H. T. A., & van der Brug, H. H. (1990). *Sport Psychology.* Chichester, England : John Wiley & Sons.

Bandura, A. (1965). Influence of model's reinforcement contingencies on the acquisition of imitative responses. *Journal of Personality and Social Psychology, 1*, 589-595.

Bandura, A. (1973). *Aggression: a social learning analysis.* Englewood Cliffs, NJ: Prentice-Hall.

Bates, J. (1994). The role of emotion in believable agents. *Communications of the ACM*, Special issue on Agents.

Bentler, P. M., & Speckart, G. (1981). Attitudes cause behaviors. A structural equation analysis. *Journal of Personality and Social Psychology, 40*, 226-238.

Blumer, H. G. (1939). *Collective behavior.* In R. E. Park, (Ed.) (1939). *Principles of Sociology.* Pagination from A. M. Lee (Ed.) (1946). *Principles of Sociology.* (pp. 219-288). New York: Barnes & Noble.

Bouvier, E., & Guilloteau, P. (1996). Crowd simulation in immerse space management. In the proceedings of the Eurographics Workshop on Virtual Environments and Scientific Visualization '96, 104-110.

Brogan, D., & Hodgins, J. (1997). Group behaviors for systems with significant dynamics. *Autonomous Robots, 4*, 137-153.

Bruzzone, A. G., & Signorile, R. (1999). Crowd Control Simulation in Java Based Environment. In the proceedings of the 1999 Western MultiConference, San Francisco, CA, January 17-20.

Buss, A. H. (1966). Instrumentality of aggression, feedback, and frustration as determinants of physical aggression. *Journal of Personality and Social Psychology*.

Cassell, J., Sullivan, J, Prevost, S., & Churchill, E. (Eds.) (2000). *Embodied Conversational Agents*. Cambridge: MIT Press.

Cassell, J., Vilhjalmsson, H., & Bickmore, T. (2001). BEAT: Behavior expression animation toolkit. In the proceedings of SIGGRAPH 2001, 477-486.

Castro, P. E., et al. (2002). *Modeling and Simulation in Manufacturing and Defense Acquisition: Pathways to Success*, National Research Council. Washington, DC: National Academy Press.

Ceranowicz, A., Torpey, M., Helfinstine, B., Bakeman, D., McCarthy, J., Messerschmidt, L., McGarry, S., & Moore, S. (1999). J9901, Federation Development for Joint Experimentation. In the proceedings of the Fall 1999 Simulation Interoperability Workshop, Orlando, FL, September 12-17.

Chandrasekaran, B., & Josephson, J. R. (1999). Cognitive Modeling for Simulation Goals: A Research Strategy for Computer-Generated Forces. In the proceedings of the Eighth Conference on Computer Generated Forces and Behavioral Representation, Orlando, FL, May 11-13, 117-126.

Cornwell, J. B., Silverman, B. G., O'Brien, K., & Johns, M. (2002). A Demonstration of the PMF-Extraction Approach: Modeling the Effects of Sound on Crowd Behavior. In the proceedings of the Eleventh Conference on Computer-Generated Forces and Behavior Representation, Orlando, FL, May 7-9, 107-113.

Creech, R. C., & Petty, M. D. (1996). A Model of Large-Scale Citizen Evacuation for Emergency Management Simulation. In the proceedings of the Sixth Conference on Computer Generated Forces and Behavioral Representation, Orlando, FL, July 23-25, 67-78.

Department of the Air Force (2003). *Program Research & Development Announcement NR 03-01-HE*. Retrieved from http://www.eps.gov /spg/USAF/AFMC/AFRLWRS/PRDA-03-01-HE/listing.html.

Derlega, V. J., Cukur, C. S., Kuang, J. C. Y., & Forsyth, D. R. (2002). Interdependent construal of self and the endorsement of conflict resolution strategies in interpersonal, intergroup, and international disputes. *Journal of Cross-Cultural Psychology, 33*(6), 610-625.

Department of Defense (1995). *Doctrine for Joint Operations*. Joint Publications 3-0, February 1.

Department of Defense (2001). *Department of Defense Dictionary of Military and Associated Terms*. Joint Publications 1-02, April 12.

Dupuy, J. P. (1990). *Panic: From the myth to the concept, Report 068857006.* Arcueil, France : Etablissement Technique Central De' Larmement, Centre de Recherches et D'Etudes D'Arcueil.

Elliot, C. (1992) *The affective reasoner: A process model of emotions in a multi-agent system*, Ph.D. Dissertation. Evanston, IL: Northwestern University.

Eron, L. D., Huesmann, L. R., Lefkowitz, M., & Walder, L. O. (1972). Does television violence cause aggression? *American Psychologist, 27*, 253-263.

Ewen, D., Dion, D. P., Flynn, T. F., & Miller, D. M. (2000). Computer Generated Forces Applications to a Simulation Based Acquisition Smart Product Model for SC-21. In the proceedings of the Ninth Conference on Computer Generated Forces and Behavioral Representation, Orlando, FL, May 16-18, 353-361.

Feinberg, W. E., & Johnson, N. R. (1988). Outside agitators and crowds: Results from a Computer Simulation Model. *Social Forces, 67*, 398-423.

Ferguson, M. (2003). *Questionnaire response*, May 8.

Fields, M. A., & Spradlin, G. (2000). Modeling Civilian Crowds in a Battlefield Simulation. In the proceedings of the Ninth Conference on Computer Generated Forces and Behavioral Representation, Orlando, FL, May 16-18, 451-458.

Forward, J. R., & Williams, J. R. (1970). Internal-external control and black militancy. *Journal of Social Issues, 26*, 75-92.

Gaskins, R. C., McKenzie, F. D, and Petty, M. D. (2003). Crowd Modeling in Military Simulations: Requirements Analysis, Survey, and Design Study. Technical Report, Virginia Modeling Analysis and Simulation Center, Old Dominion University, April 30.

Gaskins, R. C., Boone, C. M., Verna, T. M., Bliss, J. P., and Petty, M. D., "Psychological Research for Crowd Modeling," Proceedings of the 2004 Conference on Behavioral Representation in Modeling and Simulation, Arlington, VA, May 17-20 2004, pp. 401-402.

Greenwald, T. W. (2002). *An analysis of auditory cues for inclusion in a close quarters battle room clearing operation*, M.S. Thesis. Monterey, CA: Naval Postgraduate School.

Greer, J. K. (2000). The urban area during stability missions case study: Bosnia-Herzegovina, Part 2. In R. W. Glenn (Ed.), *Capital preservation: Preparing for urban operations in the twenty-first century.* Pittsburgh PA: RAND.

Haddock, D. D., & Polsby, D. D. (1994). Understanding riots. *Cato Journal, 14*(1), 1-13.

Hall, M. T., & Kennedy, M. T. (2000). The urban area during support missions case study: Mogadishu. In R. W. Glenn (Ed.), *Capital preservation: Preparing for urban operations in the twenty-first century.* Pittsburgh, PA: RAND.

Hare, R. D. (1994). Predators: The distrubing world of the psychopaths among us. *Psychology Today*, January/February 1994, 54-56, 58, 60-63.

Horowitz, D. (2001). *The deadly ethnic riot.* Berkeley, CA: University of California Press.

Jager, W., Popping, R., and van de Sande, H. (2001). Clustering and fighting in two party crowds: Simulating the approach-avoidance conflict. *Journal of Artificial Societies and Social Simulation, 4*(3). Retrieved from http://www.soc.surrey.ac.uk/JASSS/4/3/7.html.

Johnson, N. R., & Feinberg, W. E. (1977). A computer simulation of the emergence of consensus in crowds." *American Sociological Review, 42*(6), 505-521.

Johnson, W. L., Rickel, J. W., and Lester, J. C. (2000). Animated pedagogical agents: Face-to-face interaction in interactive learning environments. *International Journal of Artificial Intelligence in Education, 11*, 47-78.

Kelman, H. C. (1974). Attitudes are alive and well and gainfully employed in the sphere of action. *American Psychologist, 29*, 310-324.

Kenny, J. M., McPhail, C., Farrer, D. N., Odenthal, D., Heal, S., Taylor, J., Ijames, S., & Waddington, P. (2001). *Crowd Behavior, Crowd Control, and the Use of Non-Lethal Weapons*, Technical Report. University Park, PA: Penn State Applied Research Laboratory.

Kenny, J. M., & Gilpin, W. L. (2002). Insertion of Crowd Behavior Models into the INIWIC Course. In the proceedings of the 2002 Interservice/Industry Training, Simulation, and Education Conference, Orlando, FL, December 2-5, 1064-1074.

Klein, H. A., & Steele-Johnson, D. (2002). *Training cultural decentering*, Technical Report. U.S. Army Research Institute.

Laird, J., & Van Lent, M. (2001). Human-level AI's killer application, interactive computer games. *Artifical Intelligence Magazine, 22*(2), 15-25.

Latane, B., & Darley, J. M. (1968). Group inhibition of bystander intervention in emergencies. *Journal of Personality and Social Psychology, 10*(3), 215-221.

Le Bon, G. (1960) *Psychologie de Foules* (TRANSLATOR, Trans.). New York : Viking (Original work published 1895. Paris, France : Alcon).

Lim, L., & Thalmann, D. (2002). Construction of animation models out of captured data. In the proceedings of the IEEE International Conference on Multimedia and Expo '02.

Loew, C. A. (1967). Acquisition of a hostile attitude and its relationship to aggressive behavior. *Journal of Personality and Social Psychology.*

MacNamee, B., Dobbyn, S., Cunningham, P., & O'Sullivan, C. (2002). Men behaving appropriately: Integrating the role passing technique into the ALOHA system. In the proceedings of the AISB '02 Symposium: Animating Expressive Characters for Social Interactions.

Markus, H. R., & Kitayama, S. (1991). Culture and the self: Implications for cognition, emotion, and motivation. *Psychological Review, 98*, 224-253.

McPhail, C. (1991). *The myth of the madding crowd.* New York: De Gruyter.

McPhail, C., Powers, W. T., Tucker, C. W. (1992). Simulating individual and collective actions in temporary gatherings. *Social Science Computer Review, 10* (1), 1-28.

Medby, J. J., Glenn, R. W. (2002). Street Smart: Intelligence preparation of the battlefield for urban operation. Santa Monica: Rand.

Meier, N. C., Mennenga, G. H., Stoltz, H. J. (1941). An experimental approach to the study of mob behavior. *Journal of Abnormal and Social Psychology, 36*, 506-524.

Miller, W. (2002). Dismounted Infantry Takes the Virtual High Ground. *Military Training Technology, 7*(8).

Miller, T. E., Battaglia, D. A., & Phillips, J. K. (2002). Cognitive Train Challenges in Operations Other Than War. In the proceedings of the 2002 Interservice/Industry Training, Simulation, and Education Conference, Orlando, FL, December 2-5, 417-427.

Moore, T. (1990). Keep it cool! *Police, 22*(11), 32-33.

Murphy, P. J., Williams, J. M., Dunning, E. G. (1990*). Football on trial: Spectator violence and development in the football world.* London: Routledge.

Musse, S. R., & Thalmann, D. (1997). A model of human crowd behavior: Group inter-relationship and collision detection analysis. In the proceedings of the Eurographics CAS 97 Workshop on Computer Animation and Simulation, Budapest Hungary.

Musse, S. R., Babski, C., Capin, T., & Thalmann, D. (1998). Crowd modelling in collaborative virtual environments. In the proceedings of the ACM Symposium on Virtual Reality Software and Technology, Taipei, Taiwan, November, 115-123.

Musse, S. R., Garat, F., & Thalmann, D. (1999). Guiding and interacting with virtual crowds in real-time. In the proceedings of the Eurographics Workshop on Computer Animation and Simulation '99, 23-34.

Musse, S. R., & Thalmann, D. (2001). Hierarchical model for real time simulation of virtual human crowds. *IEEE Transactions on Visualization and Computer Graphics, 7*(2), 152-164.

Nelsen, E.A. (1969). Social reinforcement for expression vs. expression of aggression. *Merrill-Palmer Quarterly of Behavior and Development, 15*, 259-278.

One Semi-Automated Forces (OneSAF) Operational Requirements Document (ORD) Version 1.1. Retrieved February 27, 2001, from http://www.onesaf.org/public1saf.html.

O'Sullivan, C., Cassell, J., Vilhjalmsson, H., Dobbyn, S., Peters, C., Leeson, W., Giang, T., & Dingliana, J. (2002). Crowd and group simulation with levels of detail for geometry, motion and behavior. In the proceedings of the Third Irish Workshop on Computer Graphics, 15-20.

Park, R.E., (1904). *Masse und Publikum* (Charlotte Elsner, Trans.). Chicago: University of Chicago Press. (Original work. Bern: Lack and Grunau).

Park, R. E. (1930). Collective Behavior. In *Encyclopedia of the Social Sciences* (Vol. 3, pp. 631-633). New York: MacMillan.

Pew, R. W., et al. (1998). *Modeling Human and Organizational Behavior: Application to Military Simulation*. Washington, DC: National Research Council, National Academy Press.

Petty, M. D., McKenzie, F. D., and Gaskins, R. C., "Requirements, Psychological Models, and Design Issues in Crowd Modeling for Military Simulation," Proceedings of the Huntsville Simulation Conference 2003, Huntsville, AL, October 29-31 2003, pp. 237-245.

Pilz, G. A. (1989). Social factors influencing sport and violence: On the "problem" of football fans in West Germany. *Concilium—International Review of Theology, 5*, 32-43.

Pyszczynski, T. A., Solomon, S., & Greenberg, J. (2003). In the wake of 911: The psychology of terror in the 21st Century (in the press). Washington, DC: American Psychological Association.

Ransford, H. E. (1968). Isolation, powerlessness, and violence: A study of attitudes and participation in the Watts riot. *American Journal of Sociology, 73*, 581-591.

Reece, D. A. (2002). Crowd Modeling in DISAF. In the proceedings of the Eleventh Conference on Computer-Generated Forces and Behavior Representation, Orlando FL, May 7-9, 87-95.

Rehm, J., Steinleiter, M., & Lilli, W. (1987). Wearing uniforms and aggression: A field experiment. *European Journal of Social Psychology, 17*, 357-360.

Reynolds, C. W. (1987). Flocks, herds, and schools: A distributed behavioral model. In the proceedings of SIGGRAPH '87, 25-34.

Rickel, J. W., & Johnson, W. L. (1999). Animated agents for procedural training in virtual reality: Perception, cognition, and motor control. *Applied Artificial Intelligence, 13*(4-5), 343-382.

Rickel, J. W., & Johnson, W. L. (1999). Virtual humans for team training in virtual reality. In *Proceedings of the 9th International Conference of Artificial Intelligence in Education* (578-585). Amsterdam, Holland: IOS Press.

Rickel, J. W., Marsella, S., Gratch, J. Hill, R., Traum, D., & Swartout, W. (2002). Toward a new generation of virtual humans for interactive experiences. *IEEE Intelligent Systems,* July/August 2002, 32-38.

Ross, L, Greene, D., & House, P. (1977). The "false consensus effect": An egocentric bias in social perception and attribution processes. *Journal of Experimental Social Psychology, 13,* 279-301.

Roth, M. (1979). Pornography and Society: A psychiatric view. Goodman Lecture, BBC Network, 3 January.

Roversi, A. (1991). Football violence in Italy. *International Review for Sociology of Sport, 26,* 311-332.

Russell, G. W. (1995). Personalities in the crowd: Those who would escalate a sports riot. *Aggressive Behavior, 21*, 91-100.

Russell, G. W., Arms, R. L. (1998). Toward a social psychological profile of would-be rioters. *Aggressive Behavior, 24*, 219-226.

Sarason, I. G. (1978). A cognitive social learning approach to juvenile delinquency. In R. D. Hare, D. Schalling D. (Eds.), *Psychopathic Behavior: Approaches to Research* (299-317). New York: John Wiley and Sons.

Seo, H., Yahia-Cherif, L., Goto, T., & Magnenat-Thalmann, N. (2002). Genesis: Generation of e-population based on statistical information. In the proceedings of Computer Animation 2002.

Silke, A. (2003). De-individuation, anonymity and violence: Findings from Northern Ireland. *Journal of Social Psychology*, in press.

Silverman, B. G., Johns, M., O'Brien, K., Weaver, R., & Cornwell, J. B. (2002). Constructing Virtual Asymmetric Opponents from Data and Models in the Literature: Case of Crowd Rioting. In the proceedings of the Eleventh Conference on Computer-Generated Forces and Behavior Representation, Orlando FL, May 7-9, 97-106.

Singelis, T. M. (2000). Some thoughts on the future of cross-cultural social psychology. *Journal of Cross-Cultural Psychology, 31*, 76-91.

Sloman, A. Logan, B. (1999). Building cognitively rich agents using the SIM_AGENT toolkit. *Communications of the ACM, 42* (3), 71-77.

Snyder, M. & Swann, W. B. (1978). When actions reflect attitudes: The politics of impression management. *Journal of Personality and Social Psychology, 34*, 1034-1042.

Sokolowski, J. A. (2003). *Questionnaire response*, April 22.

Still, G.K. (2000). Crowd dynamics. Ph.D. Thesis, Warwick University.

Swartout, W. (2001). Toward the holodeck: Integrating graphics, sound, character, and story. In *Proceedings of the 5th International Conference of Autonomous Agents* (409-416). New York: ACM Press.

Ulicny, B., & Thalmann, D. (2001). Crowd simulation for interactive virtual environments and vr training systems. In the proceedings of the Eurographics Workshop on Animation and Simulation, 163-170.

Ulicny, B., & Thalmann, D. (2002). Towards interactive real-time crowd behavior simulation. *Computer Graphics, 21* (4) 767-775.

Van der Brug, H. H. (1992). *Football hooliganism in The Netherlands*. Unpublished manuscript. Amsterdam, The Netherlands: Department of Political and Social Sciences, University of Amsterdam.

Van Limbergen, K., Colaers, C., & Walgrave, L. (1989). The societal and psycho-social background of football hooliganism. *Current Psychology: Research and Reviews, 8*, 4-14.

Varwell, D. W. P. (1978). *Police and public*. Plymouth: MacDonald and Evans.

Watson, R. I. (1973). Investigation into de-individuation using a cross-cultural survey technique. *Journal of Personality and Social Psychology, 25*, 342-345.

Weiss G. (1999). *Multiagent systems: A modern approach to distributed artificial intelligence*. Cambridge, MA: MIT Press.

Weller, M. P. I. (1985). Crowds, mobs and riots. *Medical Science Law, 25*(4), 295-303.

Whythe, W. H. (1961). *The Organization Man*. Harmondsworth, England: Penguin Books.

Williamson, S., Hare, R.D., Wong, S. (1987). Criminal psychopaths and their victims. *Canadian Journal of Behavioural Science, 19*, 454-462.

Wooldridge, M. (2002). *An Introduction to Multiagent Systems*. Chichester, England: John Wiley & Sons.

Zani, B., Kirchler, E. (1991). When violence overshadows the spirit of sporting competition: Italian football fans and their clubs. *Journal of Community and Applied Social Psychology, 1,* 5-21.

CHAPTER 18

APPLICATION OF IMMERSIVE TECHNOLOGY FOR NEXT-GENERATION SIMULATION

RICHARD D. LINDHEIM AND JAMES H. KORRIS

ABSTRACT

In response to shifting training needs in the post Cold War environment, the U.S. military is reviewing many closely-held assumptions in simulation that may, over time, trade attrition-based for behavior-based models. This trend could prove useful in developing the tools for Organizational Simulation. A US Army research facility, the Institute for Creative Technologies at the University of Southern California, combines research in immersive technologies including Artificial Intelligence, graphics and sound with entertainment industry know-how in a "hybrid approach" to meet this challenge. An analysis of the Joint Fires & Effects Trainer System project is offered as a case study.

INTRODUCTION

Like a venerable but ageing warrior resting on his laurels, some would argue that simulation technologies have moved to a flattening region of the "S-Curve." Traditional simulation technologies may be, in fact, well developed and in many ways mature. But this ignores a nearly ubiquitous characteristic of the "state of the art."

Current simulations have a limited scope. Often, the limitations arise from the current tool set in the area of human behavior modeling. At a fundamental level, current human behavior modeling technologies do not fully support current and future needs for a broad range of simulation applications. Neither the technology, nor the analytic understanding for expanding the breadth of simulation is available.

But recent developments in artificial intelligence, combined with a better understanding of human interaction and decision-making in crisis circumstances, are enabling a new generation of highly complex, immersive and realistic simulation capabilities to emerge. The addition of multi-faceted human behaviors and the effects of personalities upon decision-making will provide essential elements to be integrated into any OrgSim architecture.

MODELING HUMAN BEHAVIOR

For a number of reasons, current simulations tend to ignore human behavior in crisis situations. Technology limitations, especially in the world of artificial intelligence, forced unacceptable constraints upon simulation systems. Avatars were clumsy with primitive recognition and understanding of speech. Certainly, emotion was well beyond their capabilities (Gratch & Rickel, 2000).

Integrating personalities and human behavior into simulations has almost always been met with resistance by the military. When the Institute for Creative Technologies (ICT) began and proposed integrating emotion and character into simulations, there was vocal opposition from many Army officers, including General (Ret) Paul Gorman, one of the most far-sighted and visionary Army advisors. General Gorman's position was simple. If the enemy is close enough to see and identify personality, then you have lost the battle. In conventional warfare today the enemy is usually an icon on a map, a speck in the sky, or a blip on a computer screen.

Paul Gorman and others repeatedly asked: "What does character and emotion bring to a simulation? Why are they needed? How can the demonstrability of improving the training of the warfighter be measured?"

Underlying these criticisms were two factors. The first factor is a general lack of understanding or appreciation about the importance of emotion. The military position generally espoused is that emotion has little or no role in training simulations. People are supposed to perform their tasks, regardless of stress, sleep depravation, and other factors. Human reactions are often dismissed and "buried under the carpet" known as the "Fog of War."

Further, military training tends to be doctrine oriented. Hence, the military has an aversion to simulations where things do not proceed as scripted or encompass specific pedagogical learning objectives. While it can be argued that simulation is dedicated to learning, and learning from mistakes is one of the most powerful ways to accomplish this task, military people are uncomfortable with simulations that do not go perfectly. The simulation is stopped and the situation is reset. This happens on simulations of varying scales. The most recent example was the huge Millennium Challenge simulation. Because it was so complex and correspondingly expensive, the "reset" received unfavorable press coverage.

The second factor is that personality conflicts and interpersonal communication failures can clearly create "turbulence" in conventional battlefield simulations. Human personality conflicts dramatically increase the complexity of such simulations. This complexity, when combined with the limited understanding of complex interactions, could easily make the task beyond the capability of current technology. Therefore, it is more convenient to label them as "unnecessary."

Finally, the U.S. military experience since the end of the Cold War, in what is described as the "Contemporary Operational Environment", has emphasized far more human interaction than in the past. Relief operations, peace-keeping and security in many instances rely on tight person-to-person contact and split-second judgments from soldiers. This is a far cry from the attrition-based, force-on-force simulations of the past. For training and modeling alike, behavior-based simulation, with rich computational cognitive and emotional models, will ultimately fill this need.

Fortunately, there is a shortcut to solving these problems. It involves drawing upon the expertise of a group totally different in culture – film and stage actors.

LESSONS FROM STAGECRAFT

The question, simply put, follows: is it necessary to understand the complex reasoning process behind a behavior, or is the emulation of such a behavior sufficient? Actors don't have to understand the critical decision-making process of the brain. What they understand is motivation and response, and they use these tools to simplify the problem.

Actors study the character they are going to play. They learn the demonstrated behavior of that character. They know, for example, that if threatened, the character will react with rage. If afraid, the studied character will seek companionship. They don't understand the workings of the brain that lead to these behaviors. They don't need to. They just understand the stimulus and reaction.

Applying the actor's technique to simulation significantly scales back the magnitude of the challenge. Building such character profiles into a simulation would enable the incorporation of human interaction within current simulations.

A HYBRID APPROACH

Up to now, most conventional military simulations have focused upon performance characteristics. The pedagogical goals of the simulation have been clearly defined. There is a Red Team to be defeated by the Blue Team. The Red Team may have various levels of capability. In firearm training systems the opposition is "dumb." They are shooting gallery targets without any intelligence of strategy. In complex simulations like those held at the National Training Center, the Red Team is supremely qualified and almost always defeats the Blue Team.

Regardless, virtually all of these simulations are skill trainers of some sort – driving, flying, shooting - it is simpler to deal exclusively with engineering and computer parameters of these acts. Setting system requirements is also familiar to government contractors. Requests for proposals outline the requirements for the system in distinct, tangible form. It is much harder to quantify personal interaction and human behavior as part of these requirements. As a result, military simulations are generally tied to implicit engineering requirements, rather than end-state goals.

The Institute for Creative Technologies has taken a different approach. As a result of its interaction with the entertainment industry, ICT quickly learned that the military and scientific communities approach requirements completely differently from the entertainment industry.

As noted, the traditional military approach to simulation is to define the capabilities desired. This is usually done in terms of area of terrain to be modeled and the level of realism required, as well as other specific, numeric parameters. While there is value to this approach, it means that the level of realism in the system is determined by the lowest common denominator. This is usually an equation defining the amount of terrain to be modeled, the speed of the processor, storage capabilities of the computer system, and acceptable cost.

For example, let's assume that the requirements call for 20 square kilometers of terrain at 1 meter resolution. The result will be that a blade of grass at the fringes of the terrain database will be rendered at the same resolution as the city. The rolling terrain at the edge of the simulation may look fine. That same resolution of the city will look inferior.

The design assumes that every piece of the terrain map is of equal significance. Consequently, many resources go to modeling all 400,000,000 square meters of the database in equivalent detail.

The entertainment industry (and the entertainment software business in particular) knows that all terrain is not "created equal." The interactive experience design will inevitably emphasize different parts of the terrain map. If the simulation's focus is on urban warfare, for example, the level of the detail in the city will be much higher than it is the hinterlands. By serving the "end state," (i.e., an urban warfare experience) much of the effort will go into modeling the city.

But just as significantly, the experience design will ensure that you get to the city. Techniques commonly employed in the computer game industry ensure that the user "rides on a rail." When artfully employed, the user is not even aware of this manipulation (Morie, 2002).

There is another important consideration that results from this approach. The military simulation almost always looks for a single technological solution to the task. The typical research and engineering approach to any problem is to seek the single best solution to a problem. The scientific method reinforces this. What does science suggest? Explore the problem (need) and form a hypothesis to explain (solve) it. It is singular. One problem. One solution. If the hypothesis can be proven, the problem is solved. If not, the hypothesis is discarded or modified. But, again, it is focused upon one solution. That is not the best approach to solving the complex problems in the multi-disciplinary, multi-modal world that exists today.

The entertainment industry, as improbable as this might sound, has found an alternative and, in many circumstances, better way. The approach is diametrically different. After all, movies can be considered passive simulation; while you cannot interact, the Hollywood masters create a world and immerse you into it. They focus upon the end-state.

They ask the question: What do we want? What is the goal of this simulation? What immersive experience do they want the viewer to have? What story point do they want to communicate? To do this they combine resources to achieve the desired end result. They readily mix computer-generated material with live action footage (using plywood cutouts and models). The results can be seen in almost every new high-action motion picture, from "The Matrix" to "Spiderman" to "Lord of the Rings."

The ICT has adapted a hybrid approach in developing its simulations (Lindheim & Swartout, 2001). While it retains the parameters and system requirements doctrine, it does not rely upon a sole solution to the problem. The requirements for the simulation are parsed into several different technologies, depending upon the task needs. They are then integrated to produce the desired results. Approaching the simulation task in this manner has allowed the ICT to create high resolution, impactful simulations rapidly with enormous savings in computer "horsepower."

THE "WILLING COLLABORATOR"

Entertainment industry professionals use their knowledge and expertise in understanding an audience and creating emotional involvement to "lead" the viewer. Like a magician, who uses distraction as a tool to keep the audience from discovering the trick, entertainment industry people learn how to "drive" the simulation.

One can readily evaluate the effectiveness of their work on a DVD. Look at one of those hugely expensive action-adventure films. Any one will do. It could be "Gladiator" or "X-Men." Turn off the sound; that is a key factor in controlling the simulation. And look, analytically, at the edges of a fantastic special effects shot. Almost assuredly you will notice that the simulation is flawed, obviously painted, and not realistic at all. You don't notice these things when watching the movie, because you are being emotionally manipulated by the simulation.

While the level of realism is variable within the frame, you don't notice it. You accept the entire simulation as having the highest level of resolution shown, not the lowest common denominator. That is the value of the approach of entertainment industry professionals. They have learned that there are two critical factors in accepting a simulation as being realistic.

One, they know that the mind is a willing collaborator. Scott Trowbridge, an executive at Universal Studios Theme Parks, when interviewed about the new state-of-the-art Spiderman ride that combines 3-D photography, a motion platform on a moving track, stage props and 4-D effects such as water sprays and heat, was quick to state that the reason the ride was so immersive was because it was designed to maximally utilize this capacity of human perception. If not, Trowbridge pointed out, the participants would be aware of the technical devices (film, props, effects) that create the simulation. Thus, the impact would be minimized. But, because the mind is a willing collaborator, the effect of compositing these various simulation forms is multiplied. The result is an extremely immersive, effective entertainment experience.

The question that this posed to the ICT researchers was: "Could this technology be utilized for learning, where real pedagogical goals were the results, rather than simple entertainment?" What all creative people involved in the entertainment industry know is that if you engage the participant, he or she will forgive the less realistic aspects of the scene (simulation).

Two, they understand the critical importance of suspension of disbelief. The most effective words to suspend disbelief are "Once Upon a Time." Another version is "A long time ago in a galaxy far, far away." If you accept these statements, the film producer can take great latitude in his/her presentation of reality. And, it will be accepted.

SETTING LIMITS

A simple example: A number of years ago, the authors worked on a fantasy television series called "Knight Rider." The core concept of "Knight Rider" was a futuristic car that, among other things, could talk (Barris & Fetherston, 1996). The television series was a huge hit because the audience wanted to believe that such a car could be built. They suspended disbelief.

So long as the producers didn't violate the audience's willingness to suspend disbelief, the audience was absorbed in the show. The fact that it took a dozen stunt cars to perform the tasks shown was irrelevant (the studio needed a dozen cars, because half of them were in the shop for repairs from the crashes, jumps and other fantastic activities attempted in the show). But, in one show, the producers went too far. They showed KITT, the car, driving on water. The audience response was immediate and highly critical. "The car can't do that!" they wrote. The producers quickly retreated and never repeated the mistake.

Any Organization Simulation (OrgSim) needs to recognize the parameters of accepted action that the system will tolerate. Initially, this would probably be a trial-and-error discovery system. The important factor is that the Orgsim is designed to be aware and track such variations. By doing so, it would compile the acceptable (and unacceptable) behaviors. Over time, the simulation would become stronger and more accurate, providing guidelines for Orgsim construction.

In the case of the "Knight Rider" TV series, Universal Studios decided to put one of the cars at its theme park. They figured that children would like to climb into the driver's seat. It was a modest investment with no expectation of great success. To provide the "voice" of KITT, they hid a performer in a blind, where he could see the person in the car and speak. The Studio quickly found out it was wrong in its assessment. Not only was the attraction a huge hit, adults were elbowing aside their children to climb into the driver seat to "talk" with KITT. They were ANXIOUS to suspend disbelief, and they accepted the fantasy as reality.

The KITT experience also demonstrates the ability to effectively use simulation as an analytic tool. Some of the characteristics of that fantasy car were eventually incorporated in Detroit's best, such as voice-activated systems, information assistance systems, variable ride characteristics, computer diagnostic systems, and GPS guidance. Unfortunately, cars still can't crash through walls and jump over barriers without damage.

POTENTIAL FOR ANALYSIS

Most training simulations can be equally effective as analytical simulation tools. In the simulation environment, prototype devices and concepts can be made real. Potential new products can be evaluated and their usage determined before they are actually built. Training programs for these new devices and systems can similarly be developed by using simulation as an analytic tool. These simulations will be more effective, however, if they incorporate human interaction into the system. As the entertainment industry demonstrates, the emotional context of a system is crucial. Such inclusion, however, vastly increases the complexity of the task for the analytic community. Combining "soft" factors, such as emotion, with hard facts about performance places more dependence upon the analysis capability of the individual; effectiveness becomes more than simple measurement.

Similarly, simulation technology to explore the human-machine interface is a natural expansion of today's efforts. Robots exist today in myriad forms. While Robbie the Robot from "Forbidden Planet" and "Lost in Space" and Data from Star Trek are still fantasy, robots exist in a multitude of forms, from Roomba, the robotic vacuum cleaner (Roomba, 2004) to UAV's and smart sensors being developed and employed by the military. In the Air Force there are serious conversations about the need for a human pilot on reconnaissance and certain other kinds of missions. This discussion dates back to the initiation of the space program and was heavily mentioned in Tom Wolf's book (and subsequent movie) "The Right Stuff." It has become a major subject of interest with the development of UAV's for targeting and surveillance activities. And NASA, of course, relies heavily on robotics for space exploration. No one need look further than the Mars rovers.

Communication between these devices and humans is currently very primitive. At best, they report information in digital form and accept similar commands. This sustains an awkward interface between man and machine. Men are used to communicating through gesture, speech, facial expression. With the exception of developing speech recognition software, robots do not respond to such communication avenues. While speech recognition is developing, the computer speech recognition and generation systems take no account of anxiety, humor, anger, or other human emotions. Try screaming at the voicemail assistant. No recognition. No response.

Recently, an L.A. Times reporter wrote a column about his frustration trying to install a new wireless system in his home. His anger and frustration in dealing with the voice activated "help system" was humorous, because it was so relatable. He kept seeking a human, who would understand both his problem and his frustration. The black humor of the column was successful, because almost all of us have encountered similar experiences.

THE "GOOD BORG"

An exciting new arena for simulation technology centers about the commercial development of the Massive Multiplayer Game. In a short period of time, hundreds of thousands of young people have paid monthly fees to become subscribers and participants in these Internet "adventures."

What is particularly fascinating about this activity is the phenomena of behavior that has emerged among the game players. With no formal leadership training or organization, the players have learned that they are most effective if they play as teams. The team concept has emerged as the dominant form of behavior; an individual cannot succeed in these adventure games, they must become a member of a team. The teams themselves are virtual and highly diverse. The players may never meet one another physically. Their knowledge may extend only to the behavior and skill of their teammates (Gehorsam, 2002).

There is another aspect to this phenomenon, which could be called "the Good Borg." For the uninitiated, the Borg are a warrior race in the Star Trek universe. Created with Cold War sensibility, the Borg are a fearsome enemy that absorb all races and cultures they contact. Their greeting is, "You will be assimilated." In other words, Communism. But, there is a particular aspect of the Borg that goes beyond political systems. Once assimilated, all knowledge of the captive's culture becomes part of the "collective" to be shared by all. Thus, as the Borg conquer races in other worlds, their knowledge base grows. And each individual Borg member can instantly access all knowledge of the Collective.

In the Internet world of Massive Multiplayer Games a similar phenomenon is occurring. Through the team selection process, individual members reach out to others in the "collective world" of the game. They instantly expand their knowledge base via Chat and other Internet clients in order to become more effective players.

If applied in a systematic methodology and with specific learning objectives, the application of these characteristics of team behavior and shared knowledge present exciting possibilities for learning.

In all of these applications, combining the use of artificial intelligence, computer game resources, and entertainment industry expertise in human emotion and involvement, the ultimate value of such simulations can only be realized with the application or organizational simulation. This can take two forms.

First, organizational simulation can be applied across simulation technologies. For example: Massive Multiplayer Game technology can be integrated with military simulation and crisis leadership training tools. Second, organizational simulation can be used in coordinated multi-echelon applications of the same simulation. For example, training of a platoon can be extended to company and division on one level and squad on another level. Obviously, the two forms may also be combined in more complex organizational structures.

TOP-DOWN VS. BOTTOM-UP

Effectively, these two methods amount to a "top-down" (the massively multiplayer case) and "bottom-up" (collaborative echelon case) approach. In the massively multiplayer example, human players in large measure, supply the intelligence of the system. Non-player characters (NPC's), generated by the games' artificial intelligence, typically play a limited role with narrowly scripted player interactions. When human players need a break from the virtual action, their character dozes or simply disappears.

From the outset, the game is intended to model an entire community. As a practical matter, today's popular massively multiplayer games limit players to a small subset of the total game community in servers apportioned geographically. Thus, while there is an illusion of existing in a universe of a very large number of human players, at any given time, the actual on-line interaction may be with a group in the hundreds or low thousands. A truly "massive" community is still in the future, although rapid progress is being made.

For prospective Organizational Simulation applications, the massively-multiplayer simulation is, essentially, the organization itself without bricks and mortar, at least until NPC's can be developed to take on meaningful roles. While there is a value in creating a forum for interaction and experimentation with new organizational designs and processes, the methodology, with large numbers of connected role-players, is logistically challenging. Also, the ability to compress time and replay alternative "what-if" scenarios, central features of many kinds of simulation, is not realistic for massively-multiplayer products. That being said, when it comes to the task of preparing and conditioning an organization for change, the massively multiplayer approach has great promise.

BOTTOM-UP AT ICT

At ICT, the "bottom-up" approach arose as a natural outgrowth of small maneuver unit command simulators developed for the US Army. "Full Spectrum Warrior 1.0", a light infantry squad simulation and "Full Spectrum Command 1.0", a light infantry company simulation were both greeted with genuine enthusiasm by end users. A typical early question asked, "How may we connect these to simulate a larger force?"

Effectively, there was a natural affinity in the user base for larger, even theater-size simulation. The key distinction in this case was that, in contrast to many past Army efforts, the journey did not begin with a plan to simulate thousands of troops and vehicles, but with an attempt to model -- *with immersive realism* -- the job of a sergeant leading eight soldiers, and a captain leading one-hundred thirty troops. With the current development of a new echelon simulator through a partnering agreement with the Singapore Armed Forces, the platoon leader simulation "Full Spectrum Leader 1.0," interest in aggregating larger simulated forces has only grown.

The lieutenant's "first-person" perspective in "Full Spectrum Leader 1.0" includes his responsibility for thirty troops along with a more granular decision-making requirement. While the captain works in the broader strokes of Company command, it falls to the lieutenant to parse the captain's commands and adapt to circumstances. For example, in "Leader," the lieutenant has a side-arm and, in some circumstances, will join the actual fighting.

"Leader" is shaping up as a "Petrie dish" to explore the role of Platoon leaders as forward observers to call for fires and effects: effectively, the eyes and ears of the Company commander. This will further enrich the echelon interaction in these simulations.

The user engagement and immersive realism these products provide could only have been possible with a "bottom-up" development approach supported by a considerable body of practical entertainment software experience and a number of research themes.

LESSONS FROM THE MARKETPLACE

As in many domains, market forces and the attendant capitalist Darwinism has done much to focus the minds and sharpen the skills of commercial game developers. As with the makers of filmed entertainment, the market place is a simple and reliable means of keeping score. Game development is a stunningly competitive field, both in content creation and distribution. Most developers scratch out a living while hoping for the royalties from a hit. Hits, as in motion pictures, television, and recorded music are a rare occurrence. Survival is keyed to capturing and holding the mind of the audience; successful game developers always seem to have their audience in mind.

When this skill set is applied to a military simulation challenge, the results are compelling game play. But it is the research effort, in the vein of ongoing work at the ICT that holds the ultimate promise of propelling the second "S-Curve" of Organizational simulation (Swartout, 2001).

VIRTUAL HUMANS

One of the most challenging areas of research concerns virtual humans. Whether in the example of Massively Multiplayer game NPC's or the entities that populate a single or multi-player game, there are inevitably some number of human-like intelligences that are central to the effectiveness and, ultimately, the accuracy of the simulation.

Virtual humans can manifest in a wide range of forms. Consider *Eliza*, the first "chatterbot". In 1966, with only about 200 lines of computer code, Joseph Wiezenbaum was able to model the "active listening" human interaction with a 1960's style "talking therapy" analyst. Two important considerations made this simulation so effective and so strikingly economical. First, it was text-based: all interaction was between a user typing at a terminal and a computer mainframe application. This eliminated need for facial or whole-body animation, speech recognition and speech synthesis. Second, it was highly domain-specific. The Rogerian style of analysis was centrally Socratic: It responded to almost all user input with questions meant to elicit descriptions of feelings, with a limited database of significant words (e.g. mother, obscenities etc.) mapped to special responses.

Every vehicle, soldier, enemy, civilian, coalition member and relief worker found in a simulation is, on some level, a virtual human. An aircraft, for example, does not pilot itself across a sky. There is a human intelligence flying it, or in the case of unmanned platforms, determining its mission. If there is more than one aircraft flying in a formation, task organization and teaming processes, along with the more obvious obstacle avoidance and path planning capabilities are needed. Human factors (e.g. fatigue) and cognitive modeling (e.g. a fight or flight emotional model) may also need to be included. If players need to communicate with the pilots, speech recognition, natural language understanding and speech synthesis will be required as well. All this technology is for an aircraft that may be only a distant blip on the horizon.

What makes this particularly challenging is that, as far as people are concerned, humans are the most universally recognized living thing on this planet. It is extremely difficult to fool humans when it comes to humans. Needless to say, the greater the distance, the easier the illusion. When "synthespians," computer-generated human avatars, first appeared in motion pictures, they typically were found in large crowds in the background. It is no accident that the first fully computer-generated foreground character, Stuart Little (of the eponymous motion picture series), was a talking mouse.

Advances in virtual human technology will come from a combination of Artificial Intelligence research and Moore's Law, as solutions ultimately will likely be computationally intensive.

BEYOND PHOTO-REALISM

A more obvious aspect of the immersive simulation experience is the graphical techniques employed to create the visual field. For years, the term "photo-realism" defined the quest; current work now challenges the notion of what looks real.

There are the obvious considerations of resolution and frame rate; under ideal circumstances, individual pixels should not be visible and the rate of frame generation should exceed a human's ability to detect changes. Early motion pictures displayed 16 frames per second. The highly visible flicker gave rise to the vernacular moniker "flicks." The introduction of sound saw the rate move to 24, but it wasn't until the development of the double shutter that exposed each projected frame twice, with interleaved dark intervals providing an effective 48 frame per second rate, that human persistence of vision created the illusion of visual continuity (Grossman, 2000).

NTSC television of the type broadcast in North America runs at 30 frames per second composed of 60 half-fields that create the familiar scanning line artifacts of an interlaced scanned image ("line twitter") (Blackburn, 2001). LCD and DLP progressive scan computer displays generally run at 60 frames per second (CRT's are usually higher). Typically, commercial game developers seek frame rates exceeding forty frames per second. But resolution and frame rate alone are not determinative of realistic appearance.

Steady progress in colorimetry has done much to solve the problem of mapping a continuous, analog information-space (an aspect of both the physical world and photo-chemical processes) to the digital domain. Currently, there is a considerable body of work focusing on image-based lighting and high dynamic range imaging.

In the physical world, we routinely experience wide swings in lighting dynamic range. Stepping out of a movie theater at mid-day, after our eyes have become adjusted to the dark, we are apt to be momentarily blinded, squinting until our iris and fovea have fully adjusted. Entering a darkened room from bright sunlight requires a similar adjustment. Photographic imaging, limited to a dynamic range of roughly 1,000 to 1 falls far short of the experience of the real world. You are not apt to be blinded by a projected image of the mid-day sun, for example.

High dynamic range imaging aims to recreate the full dynamic range of lighting in the physical world (Larson, 1998). New advances in graphical rendering hardware have a floating point mathematical capability to enable a range substantially in excess of the 255 gradations of intensity older equipment featured (Macedonia, 2002). The problem remains, however, of how to display it.

There is some current experimentation with display hardware that can reproduce an extraordinary range of illumination: pitch black and bright enough to make a subject squint (Stuerlinger & Seetzen, 2003). The natural question arises: "Is it better to simulate extreme changes in illumination or for users to experience it?" In other words, is it better to display a view that briefly "whites out" when a character steps outside or to turn the display up so high that the viewer has to shield his eyes?

Like many questions in simulation, we won't be able to search for answers until we have working prototypes and can evaluate alternatives. For years, questions about simulation immersiveness were taken as an article of faith. Most people would agree that the more engaged they are in an experience, the more they are apt to pay attention and remember. We are just beginning to have the tools to evaluate which factors are the most significant in creating that immersion.

A final emerging area in immersive graphical technology is that of image-based lighting (Debevec, 2002). In other words, how do you illuminate an extrinsic or wholly synthetic object in a scene so that it looks like it belongs there? Current methods involve capturing the lighting in a scene and using the resulting photonic database to illuminate the new object. This is particularly challenging when the reflectance properties of the object are irregular, like the diffuse and specular properties of human skin. These techniques extend to the illumination of scenes and objects under varied and novel conditions (i.e. time of day and weather).

SOUND'S GOOD....

Sound has its own complications, not the least of which are our own expectations. On average, an adult in an industrialized society today will have experienced tens of thousands of hours of filmed entertainment content hearing a great many things that do not specifically correspond to the physical world (Edgerton & Mardsden, 2002). People are often startled to hear the "pop" of gunfire, having trouble reconciling it with the complex report of a movie handgun. What most of the audience does not realize is the sound may be composed of shotguns and jet aircraft.

In developing immersive simulation, the big issues in sound are resolution and directionality. Resolution is relatively a simpler business, as there are excellent capture and reproduction tools available for the faithful recording and reproduction of sound.

Directionality is an extremely challenging area given the characteristics of an environment and the location of a listener. A simulation of a bathroom shower, for example, would require the modeling of a large number of sound reflections to play convincingly in a relatively anechoic room. Headphones solve some problems, but introduce others. As long as the listener doesn't move, it may be possible to model a great deal of an environment's characteristics. With movement, there is the added challenge of tracking the subject's head and recomputing the sound environment on the fly.

With multiple participants, it is possible to create unique sound fields for each listener using beam-forming technology. As these listeners move, however, the beam-forming calculus changes, resulting in an immense computational load. These challenges are vastly more complex than the familiar directional sound field in motion picture and home theaters. In those cases, a pre-rendered sound field is reproduced in a relatively limited playback "sweet spot." Computationally, it is analogous to the difference between watching a pre-rendered video and participating in a realistic interactive game.

CASE STUDY: JOINT FIRES & EFFECTS TRAINER

A recent example of Organizational Simulation involving immersive technologies is the Joint Fires & Effects Trainer at the US Army's Artillery training facility at Fort Sill, Oklahoma. The project was conceived to demonstrate a new kind of training in a force structure, command and technology environment that not only does not exist; it had, up to that point, never been fully described.

In the Fall of 2002, it was decided that the Artillery School at Fort Sill, under the command of Major General Michael Maples, would take the bold step of demonstrating training for what was described as a "Universal Observer." In the future, it was imagined, any US soldier, sailor, airman or marine could call for a wide range of fires and effects from the joint, interagency, or multinational domain.

Now this is quite a tall order. The motivation for the Universal Observer had arisen from the advent of Maneuver Warfare as the new accepted means for the US military to conduct its campaigns. It was vividly demonstrated in Operation Iraqi Freedom, as the US Maneuver Warfare recalled the German Blitzkrieg in its speed and effectiveness. With non-contiguous forces, there was no fixed front and the enemy was constantly kept off-balance (Swanson, 2003).

With this new approach came some surprises. Because of the fluidity of the battlespace, rear-guard logistical personnel could find themselves suddenly and seriously in harm's way. This, coupled with the wily elusiveness of an often asymmetric enemy, signaled the need, in the view of many, to prepare all who wore a uniform to be able to call for indirect (usually, non line of sight or beyond line of sight) fires and effects either to press an advantage or to mount a defense.

At a high level, this seems like a worthy, even obvious goal. There are, however, two practical problems: (1) the technology and hence the technological proficiency necessary to call many kinds of fires and effects is significant and (2) military organizations are fundamentally hierarchical in nature. The ability, for example, for an Army soldier to, at his sole discretion, tap into Air Force munitions, would today cut across significant organizational boundaries.

ICT, by virtue of its concept development and visualization work throughout its four-year history, has grappled with significant questions of military transformation. Early on, its combination of academic research, coupled with entertainment industry creative energy, attracted many in the military who sought fresh perspectives on its hard problems.

The ICT creative team, working with Fort Sill subject matter experts, defined a solution that involved three domains. These domains were likely to be a part of operations and, consequently training in a future, transformed military environment. Specifically, an Open Terrain Module, Urban Terrain Module and Fires & Effects Command Module were chosen.

From the outset, every effort was made to create immersive simulation environments. In the Open Terrain case, ICT built a thirty-six foot wide curved screen with a 120-degree field of view. In front of this, a Humvee mock-up on a motion platform with a laser target designator (Long Range Advanced Scout Surveillance System) carried an Army scout crew of three.

In the urban setting, an Air Force Enlisted Terminal Air Controller worked along side a Special Forces team with a twist: the spotter and sniper were virtual humans, providing background information and even challenging the trainee's decisions. In the command module, the three trainees were in different cities, sharing the immersive simulation by means of a distributed learning system.

In describing this system, itself a significant departure from the current way of doing business, ICT portrayed familiar characters in an unconventional context. In fact, it was the command module that was most innovative, with a small, mobile Battalion-level command group that monitored, as opposed to approved, decisions operators on the ground were making. There was a significant reliance on an "under the hood" system that performed a number of critical functions, including situational awareness and effects resource optimization, which seemed likely candidates for automation. The command group was seen doing work that required essential human qualities of insight, judgment and heuristic thinking -- qualities that, in essence, could not reasonably be off-loaded to a computational system.

This exercise in Organizational Simulation was unusual in that nothing that was demonstrated was interactive. The entire demonstration was scripted. The trainees were all characters performed by professional actors. All of the video was pre-rendered. Even the virtual humans were human actors composited into a synthetic scene.

Nonetheless, this "simulation of a simulation" proved effective in one essential aspect: it provided senior leadership with a way to envision the way in which the military might one day fight and described the means to train for it. In the strikingly short period of development, under six months, the effort simulated an organizational alternative with an effectiveness that could not have been equaled with publications, presentations or briefings.

Moreover, it worked because it tried hard to be honest about how such a system might function. There were mistakes, errors in judgment, and failures of technology. Ultimately, human invention and resourcefulness were the measure of the system's viability.

CONCLUSION

Consider the entertainment industry's overarching question: How can one most effectively and efficiently deliver an effect and engage the viewer? Immersive technologies are ultimately merely tools in the simulation designer's arsenal. If we maintain focus on the purposes and audience for a simulation...if we avoid blind allegiance to any specific method, approach or technology...if we heed the lessons of the filmed entertainment and computer game industries...we are apt to create a new generation of high-impact, high-value Organizational Simulations.

REFERENCES

Barris, George & Fetherston, David (1996) Barris TV & Movie Cars, Motorbooks International

Blackburn, Doug (2001) "Video Noise: Interlaced Video Explained", Home Theater Sound

Debevec, Paul (2002) "Image-Based Lighting", IEEE Computer Graphics and Applications

Edgerton, Gary R. & Marsden, Michael, T. (2002) "The Teacher-Scholar In Film And Television", Journal of Popular Film and Television.

Gehorsam, Robert (2002) "The Coming Revolution in Massively Multiuser Persistent Worlds", IEEE Computer

Gratch, John "Modeling the Interplay Between Emotion and Decision-Making" (2000), 9th Conference on Computer Generated Forces and Behavioral Representation, Orlando, FL.

Gratch, Jonathan & Rickel, Jeff et al (2002) "Creating Interactive Virtual Humans: Some Assembly Required", IEEE Intelligent Systems.

Grossman, Stephen, (2000) "Frame Rate Technique Delivers Flicker-Free Motion-Picture Performance", Electronic Design

Larson, Gregory Ward (1998) "Overcoming Gamut and Dynamic Range Limitations in Digital Images", Proceedings of the Sixth Color Imaging Conference

Lindheim, Richard & Swartout, William "Forging a New Simulation Technology at the ICT" (2001), IEEE Computer

Macedonia, Michael (2002) "The Computer Graphics Wars Heat Up", IEEE Computer

Morie, Jacquelyn F. "Coercive Narratives, Motivation and Role Playing in Virtual Worlds" (2002) World Multiconference on Systemics, Cybernetics and Informatics Industrial Systems and Engineering II.

Roomba Robotic Floorvac (2004) http://www.roombavac.com

Swanson, Bill (2003) "Boyd's Tactics and Operation Iraqi Freedom", Tester, dcmilitary.com

Swartout, William & Hill, Randall et al (2001) "Toward the Holodeck: Integrating Graphics, Sound, Character and Story", Proceedings of the Fifth International Conference on Autonomous Agents

Wolfgang Stuerlinger & Seetzen, Helge et al "High Dynamic Range Display System", SIGGRAPH 2003

CHAPTER 19

FROM VIZ-SIM TO VR TO GAMES
How We Built a Hit Game-Based Simulation

MICHAEL ZYDA, ALEX MAYBERRY, JESSE MCCREE, AND MARGARET DAVIS

ABSTRACT

Program managers want games for their next training simulator or combat-modeling system. Corporations want their messaging put forward in game form. These desires are sharpened by the enormously successful career of the *America's Army* game, the first "serious" large-scale game ever produced. In this chapter, we discuss why people want their next-generation simulation to look like a game and where they got that idea. We then describe the development of *America's Army* to elucidate what is required for such an effort. *America's Army's* can be studied as an example of the challenges we will encounter as we go forward with game-based simulation for training and combat modeling.

PLATE 1. Soldiers Spill from a Stryker in the *America's Army* Online Game

INTRODUCTION

Why do so many people want games for their next training simulator? For one thing, games boast intuitive interfaces, which is one reason kids the world over spend hours playing games. The average *America's Army* fan spends something like sixty hours in the game, counting those who completed the basic-combat training, and it is only one of the top-five online games: their cumulative hours must be staggering. Ask any parent of an avid online gamer—the number of kids hooked and time spent is scandalous. Games and their interfaces have become second nature to youth.

As new games appear, they are adapted to instantly. Game interfaces are as standardized as automobile dashboards—drive one, drive them all—and in any case, setup functions allow for preferences. Because next-to-no training time is needed to joyride the latest game, attention is riveted to the story and challenges to be traversed.

Games are also attractive for their immersive qualities. As a rule of thumb, there is more immersion in a typical game than in a typical training simulator. Teenagers often enter a game world before dinnertime, after which it is difficult to prise them out to eat: need more be said? The same is rarely true of training simulators. If the training world were to achieve this level of immersion, they would have to invest heavily, as the game world does, in story and design. Training developers spend little on story and even less on design; most time and money goes to technology. Conversely, technology gets perfunctory treatment from game makers, who use entertainment tricks to convey story rather than worry about the real modeling of the displayed system.

So there are strong reasons to move our training simulations to a game basis. But there are problems.

One of the larger problems is the generation gap. Games mean "frivolous wastes of time" to the older generation, so it is hard to convince them to buy off on such training systems or even the term "game-based simulation." Eventually this resistance will fade, but at present it is our biggest impediment. Meanwhile, we know we have to move on. When we hear stories about nine-month learning curves for the latest combat-modeling system, we can't but think of the five minutes it takes to drive a game. As a community, we want our systems to offer training in five minutes. We want our systems as immersive as games. We want them entertaining, so that work is play and people don't leave. In short, we want our training systems so immersive that soldiers forget to eat.

Where Did We Start?

If we go back to the mid-1980s, when we launched the field now known as virtual reality, the motivation was to make 3D virtual environments available to everyone who could afford a workstation.

At that time, all we had were very expensive (multi-million-dollar) visual-simulation systems. In the NPSNET project (Macedonia, 1994; Singhal & Zyda, 1999), we deemed ourselves successful when we had over a hundred organizations ask for tapes of the NPSNET source so they could adapt it to their training needs. We simplified lives by giving away the source codes to NPSNET I through IV. We enabled anyone with a $60K workstation to play in SIMNET and DIS simulations or extend that code for their own purposes.

So how do we get back to such a notion for games? Again, remember that games are mainstream entertainment—and big money. Games look way better than the old-style virtual worlds and visual-simulation systems we used to build. With games, we harness the creativity of artists and designers, rather than engineering acumen, to get our training simulators built.

Why Did We Start Thinking About Games?

The 1997 National Research Council report entitled "Modeling and Simulation – Linking Entertainment and Defense" (Zyda & Sheehan, 1997) states that games and interactive entertainment—not defense research expenditures—have become the main drivers for networked virtual environments. To keep up with developments in modeling and simulation, that report indicated, DoD ought to examine networked entertainment for ideas, technologies, and capabilities. We thought a lot about this insight when forming the MOVES Institute[1] as a center for research in modeling, virtual environments and simulation, and game-based simulation became a focus.

PLATE 2. A Black Hawk helicopter as modeled in *America's Army*

[1] The Modeling, Virtual Environments & Simulation Institute at the Naval Postgraduate School, Monterey, California.

What Does Game Development Cost?

So if we make games, what's the bill? In Table 1, we see a notional cost for *America's Army*. *America's Army* was built as an entertaining vehicle for strategic communication (Davis, 2003; YerbaBuena, 2004; Zyda, 2003a&b). We start by discussing a notional/approximate cost for that development. With luck, our training simulator will be less expensive.

The first row lists notional game-engine costs. Game engines licensing for use in one game runs from $300K to $1.5M. ("Game engine," by the way, is a poor term. It ought to be "game engine and authoring-tool set," as that is what you expect with your license.)

We want to get our game out in twenty-four months, so for the moment let's banish the notion of developing our own engine and toolset. Let's assume the lowest cost, $300K, is the figure to use notionally for the price of a game engine. Then there is software maintenance on that engine, usually about 33% of the cost of the engine, so add another $100K per year. Let's bear in mind that the engine is good for about three years (until the next generation comes out), so in year four we see both the purchase of the next-generation engine and the software-maintenance fee for the old engine. And when we build on that licensed engine, we can't send the source code for our training simulation to anyone not licensed. So having chosen to license a commercial game engine to save time, we are stuck paying licensing forever.

The moral: if we are really to follow the path towards game-based simulation, DoD needs an open-source game engine yesterday. DoD also needs to consider open sourcing the painstakingly developed art within its games, so departments don't throw scarce resources at reinventing 3D soldiers, weapons, and training bases.

Development costs are the next line in the table. In the first year, we are building a lab (comprising computers and servers for the development team) and getting software tools installed. We are growing from zero staff towards, say, twenty-six. So in year one, we will spend about $2M on the development team and setup. Year two has us spending $2.5M for our team of twenty-six, plus management and admin costs. At twenty-four months, the game debuts on the Internet. In the case of *America's Army*, there were then four single-player levels and six multi-player levels (the complete release history through version 2.0.0a is

Typical Costs	Year 1	Year 2	Year 3	Year 4
Game Engine	$300K	$100K	$100K	$400K
Dev Costs	$2.0M	$2.5M	$2.5M	$2.5M
Operation Costs	$1.5M	$1.5M	$1.5M	$1.5M
Total	$3.8M	$4.1M	$4.1M	$4.4M

TABLE 1. Typical entertainment game costs (loosely based on *America's Army* costs).

presented later in this chapter). Year three, we are adding new content for additional online releases and again spend some $2.5M. We ought to be spending *more* as we start the second version of the game. We ought to bubble up in cost by something like $1M to $2M at the start of the third year. For this chapter, however, we will eke by with a spartan staff and not show such a bubble. Year four is again $2.5M, and so on.

Operations costs begin near the start of the project, as we fund servers to host the game, a marketing firm to build booths for E3, and travel costs associated with promotion. If we are building a training system, we don't really a incur substantial publicity cost, but we cannot get around server costs. So building a game with as complex an agenda as *America's Army's* (say, infantry-based combat in a small-terrain box) is on the order of $2M to $3M per year. Add in bigger pieces of terrain, HLA networking, and costs go up.

The Tough Issue Is Team Building and Maintenance

What is the biggest challenge in building games? If you're coming at this from the visual-simulation or virtual-reality world, it's team building—which suddenly becomes a whole new proposition.

If we were building a visual simulation in the mid-1990s, we might hire twenty-six programmers—and if one of those programmers had taken an art class in college, we would consider ourselves good to go. What we would end up with was a well-engineered training simulation with displays that sport "engineer art." Engineer art is not immersive. Nor is it engaging. It inspires the outsider to utter the developer's most dreaded words: "my kid's video game looks better than that and it only cost $50." The ignorant public will also point out that Game X's AI seems superior, the scoring system is way more thought out, and the networking is better. These comments are industry standards—your mileage may vary.

Team building for game development is different. In a team of twenty-six, we will have, say, four game programmers (perhaps two with CS degrees and two self taught, who can do scripts but maybe not C++). The remaining twenty-two will be level designers and artists. The formal education of the designers and artists is of practically no interest. What is important is their demo reel showing past work, whether in school, game companies, or on their own. Of highest important is the recommendation of persons you already hired and trust. Because many first-rate artists and designers lack degrees, traditional hiring procedures beat the wrong bushes and come up empty. Human-resource departments and program managers should not be expected to build effective game teams; insiders build these teams.

Getting your team to function pipeline-fashion is the job of the executive producer (EP) or creative director. He may be thirty—maybe younger—but he is the father figure for the group. Under the executive producer are a lead programmer, lead artist, and lead designer (for the story and presentation of the game). The EP's job is to make sure his team masters the selected game engine and tool suite and maintains an efficient resource-management system, and that this cross-cultural, interdisciplinary group behaves well enough and long enough that a game pops out after twenty-four months of concerted effort. Whiners are culled. In the game-development community, exactly how to create this teamwork is widely understood.

Back to our goal of building training systems with such a team: we begin to perceive an incipient cultural challenge; namely, we will have to ensure that the game people and training people get along. Put military officers in charge of the project, and we have an extra dimension of fun and understanding. One group shows up at 11am in t-shirts and flip-flops. The other group comes in at 6am in uniform—but leaves at 5pm, while the gamers toil till midnight. This makes for a prickly cultural interface and requires patience and understanding. You can help things along by supplying the right management and keeping the program manager away from the development team.

AMERICA'S ARMY DEVELOPMENT PIPELINE

To suggest the development process, we sketch the production of *America's Army (AA)*. We then cover *AA* as a case history of what can be done in a given time through that process. In the industry, a game like *AA* is called a first-person shooter (FPS). This genre assumes that the game is rendered in real-time and the point of view is that of the player looking through the eyes of his character. To develop an FPS, skilled individuals are needed in some key positions.

Positions and Duties

Programmer: Programmers are the technical glue that holds everything together. They maintain the game engine, merge code updates, add features and tools, ensure hardware compatibility, identify and fix bugs, and integrate all content into the one package that users install on their machines. They interact with all other team members to weave strands of content into a final product. Without programmers, creating a game would be impossible.

Level Designer: Level designers provide the biggest tangible piece of the game. Their job is to design and construct worlds in which the player can interact. They create terrain and buildings, place objects and sounds, add special effects, and, like stage managers, array each environment for its particular use. Level designers maintain frequent contact with everyone on the team.

Artist: Artists are responsible for the look and feel of the game. They create the surface, or "texture," of every wall, ceiling and floor, as well as flora, fauna, and faces. Artists typically develop the user interface and game icons and provide the artwork for special effects such as explosions, fire, water, smoke, muzzle flashes, lightning, etc. Generally speaking, if it can be seen in the game, an artist had something to do with it.

3D Modeler: While artists give you the skin, 3D modelers construct the bones. They create the frameworks for the artifacts that populate the game environment, from furniture to fire hydrants, phone poles to forearms. Without 3D modelers, game environments would be nothing but static, empty shells. Some 3D modelers develop specialties. Two that pertain to *America's Army* are as follows:

Character Modeling: A character modeler must have a highly developed sense of body proportion and structure to create realistic figures. They must also have a good sense of bipedal locomotion for realistic animation. Character modelers typically work with increased polygons, which adds extra complexity to their craft. Generally speaking, they will work primarily on characterization tasks throughout the course of development.

Weapons Modeling: The weapons modeler determines how each weapon will be animated. Since weapons will typically be the largest element on a user's screen, the weapons modeler works in minute detail (for which he has a large budget of polygons) to ensure verisimilitude. He will typically work on weaponry alone, with little time for any other modeling work.

PLATE 3. Weapons, such as these M-9 pistols, receive fine detail

Sound Engineer: The sound engineer creates, mixes, and imports into the game engine all the sounds the player hears. From bullets to footsteps to crickets, the sound engineer provides the hundreds (perhaps thousands) of effects that make an environment sound alive.

Project Leaders: Games can be highly complex, and the FPS is one of the more difficult genres to work in. Every good development team includes a number of project leaders, including a producer/director, lead designer, lead artist, and lead programmer. Depending on the size of team and the complexity of project, a small support staff will also be necessary.

Core Game Components

The following diagrams illustrate the core components of a typical FPS game. Figure 1 depicts a hierarchy of these components, while subsequent diagrams break down positions and interdependencies. Note that there is a deeper interdependency that cannot readily be depicted.

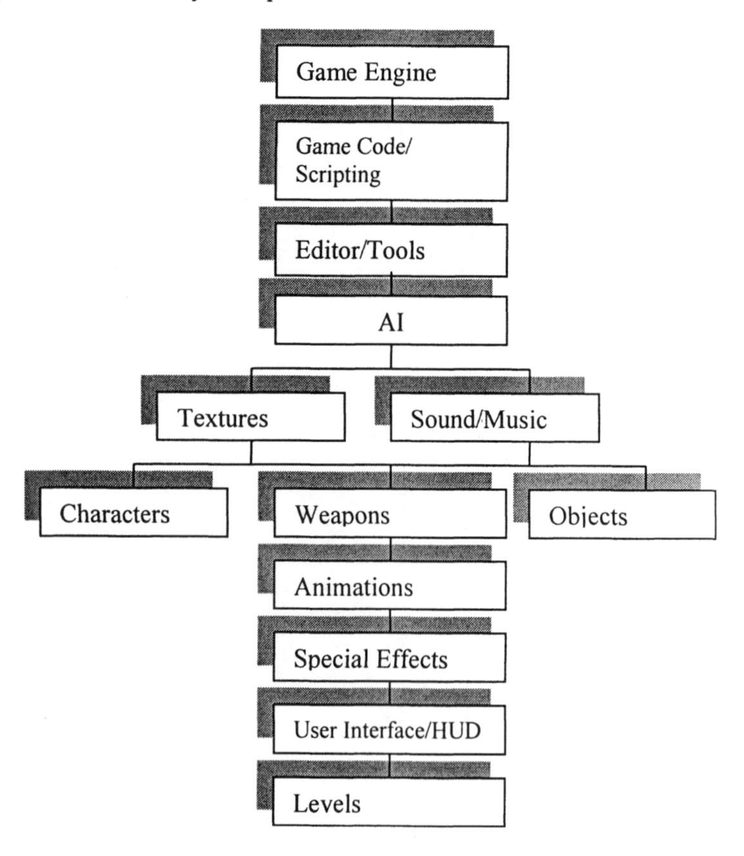

FIGURE 1. General Hierarchy of Core Components

At the foundation of every game is the game engine (Figure 2). Every element of the game will depend on this low-level piece, and it is the task of the programming department to ensure the game engine can support the final product. It is extremely important that this complex and crucial element be maintained and organized properly. If the game engine fails, the project fails with it.

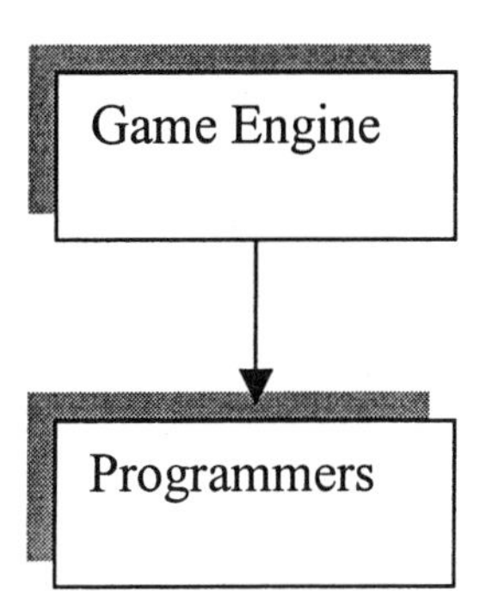

FIGURE 2. Game Engine

Programmers write game code and scripts to produce the game's peculiar atmosphere and identity. Written on top of the game engine, this code incorporates all assets into a coherent interactive experience. Programming and every other department work together in a give-and-take manner to successfully integrate the pieces. The game code and scripting set the scope of the overall design and provide the functionality that distinguishes your game from all other games based on the same engine.

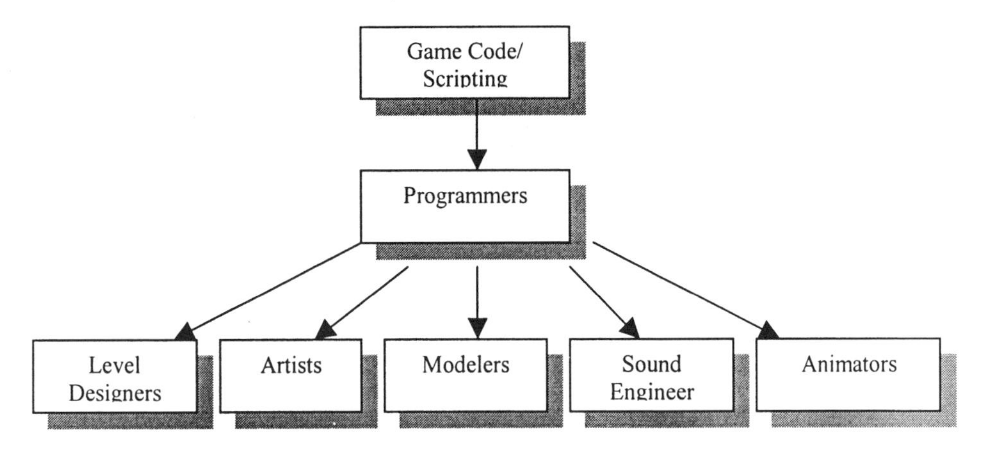

FIGURE 3. Game Code/Scripting

To facilitate the use of game code, the programming department provides the team with a game editor and tools for importing assets. Although these tools can be time consuming to create and maintain, ultimately they save countless man-hours and prevent bottlenecks by providing an assembly line for developing and integrating content. As the game evolves, so must the tools that support the team.

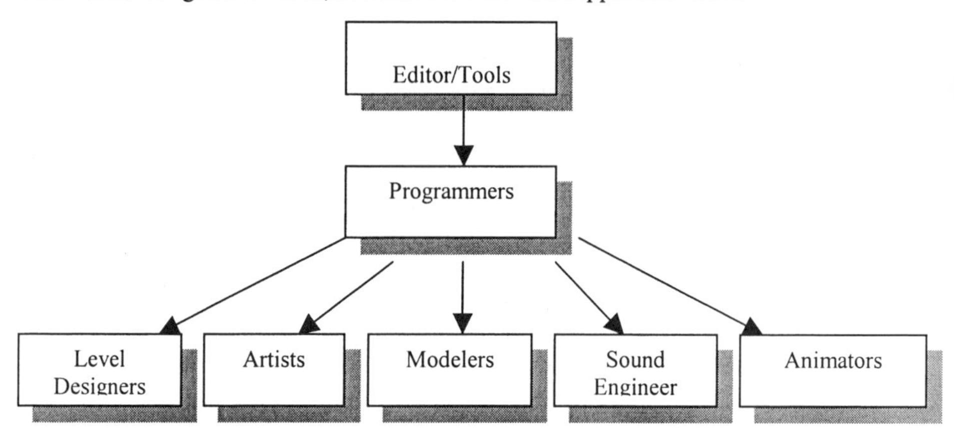

FIGURE 4. Editors/Tools

Artificial Intelligence (AI) is similar to game code, but more specialized and complex. To create AI, programmers work directly with each department of the team. For example, computer-controlled characters need an environment to run around in, so the programmers work with level design to ensure their proper setup. The art and modeling teams provide character models to attach the AI to, and the animator and sound engineer breathe life into these characters through movement and sound. Only when all these elements come together is AI fully functional in the game. It is typically a long process and requires one or more dedicated programmers through the course of the development cycle.

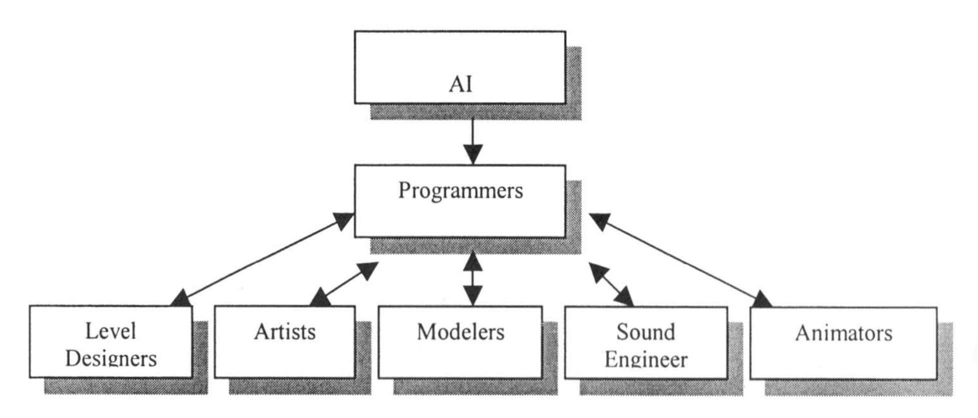

FIGURE 5. Artificial Intelligence

The artists provide texture maps to the level-design team, who then place them on the walls, ceilings, and floors of their game environments. Texture maps are used on all game objects (such as furniture, characters, weapons, etc.), in the user-interface screens, and for all in-game icons. For 3D objects, texture maps must be painted so that they wrap precisely around the model in a custom fit, while environments require that textures be painted according to a mathematical paradigm. Texture maps are essentially the visual matter of the game; without them, the characters, weapons and environments they cover would be invisible.

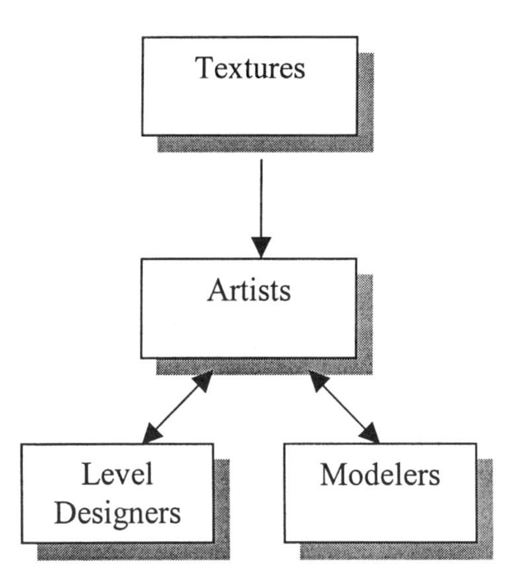

FIGURE 6. Textures

The sound engineer creates all sound and music files in the game. Background noises are distributed to the level designers for implementation. Sounds for the user interface and other effects go directly to the programming staff. Those that need to be synchronized with the movement of weapons and characters, the engineer collaborates with the animation team. These elements, along with the models they are associated with, are then given to the programmers, who import them into the code and ensure their unified functioning.

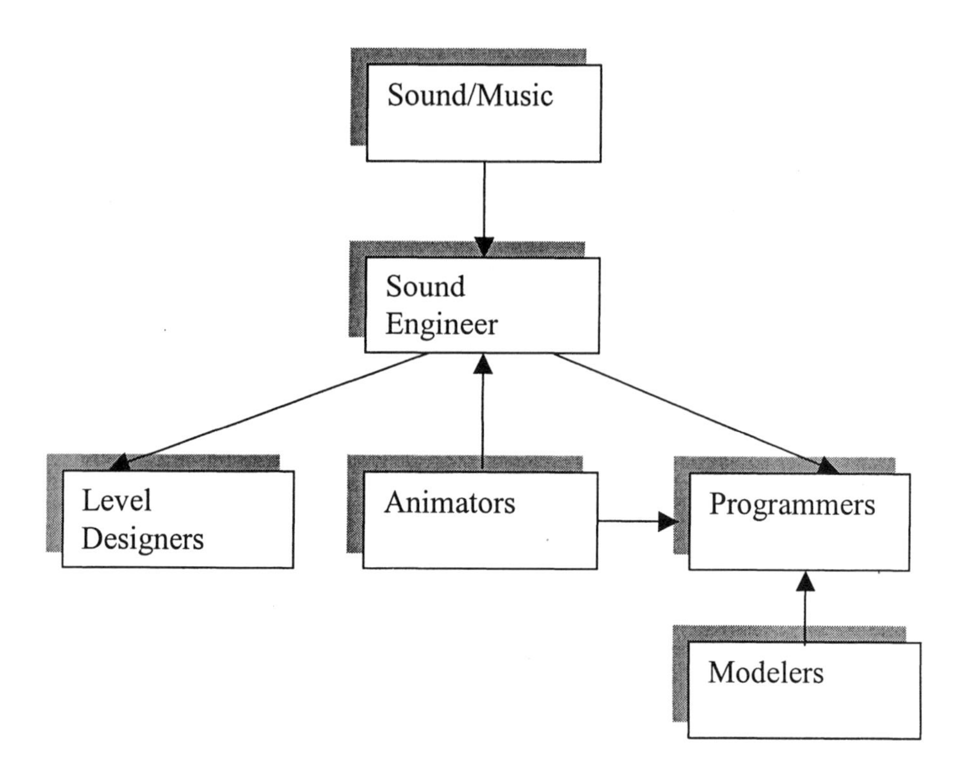

FIGURE 7. Sound/Music

Character models are created by a specialist 3D modeler then texture-painted by an artist (or the modeler, if he has the skill). When finished, the painted character is passed to an animator. Motion-capture data is applied to the object and hand tweaked. The completed object and animation data are sent to the programming team, who integrate it with the game code and attach any available AI functionality. The sound engineer then creates and synchronizes sounds for the character for addition by the programming team. Finally, the level-design team places the functional character into the game environments.

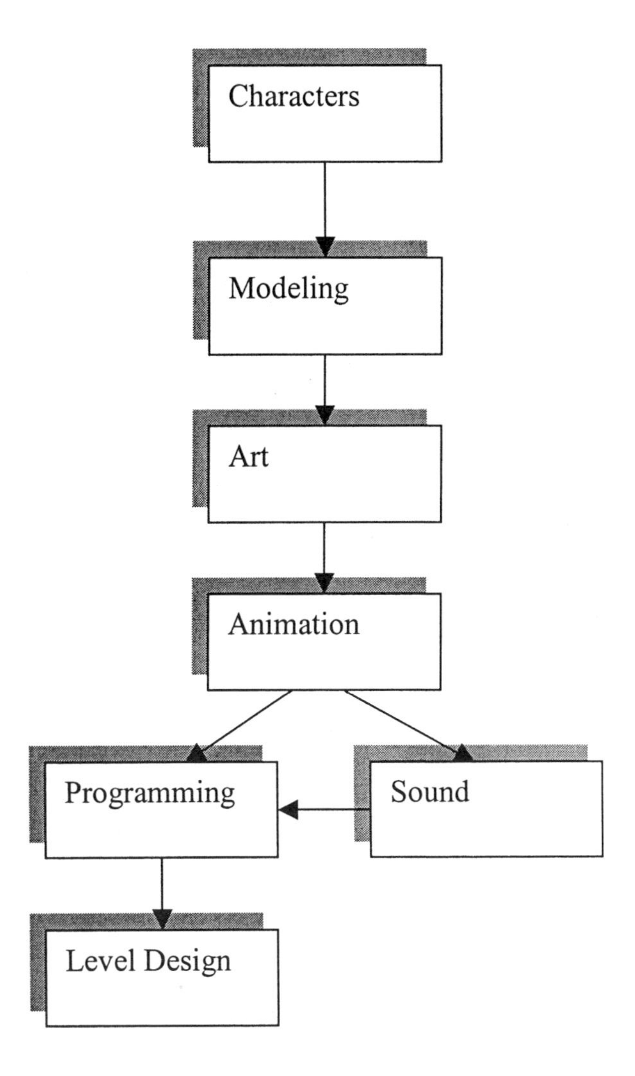

FIGURE 8. Characters

A specialized 3D modeler creates weapons models. Once the model has been crafted, he or an artist paints a texture for it. It is then handed to an animator, who sets up the model, animates it, and sends the resulting data to programming, for incorporation into the game code. The sound engineer provides sounds for the weapon and the artists create special effects. Programmers then integrate these elements and write game code that defines the weapon's functionality. The level designers add the finished weapon to the game's environments.

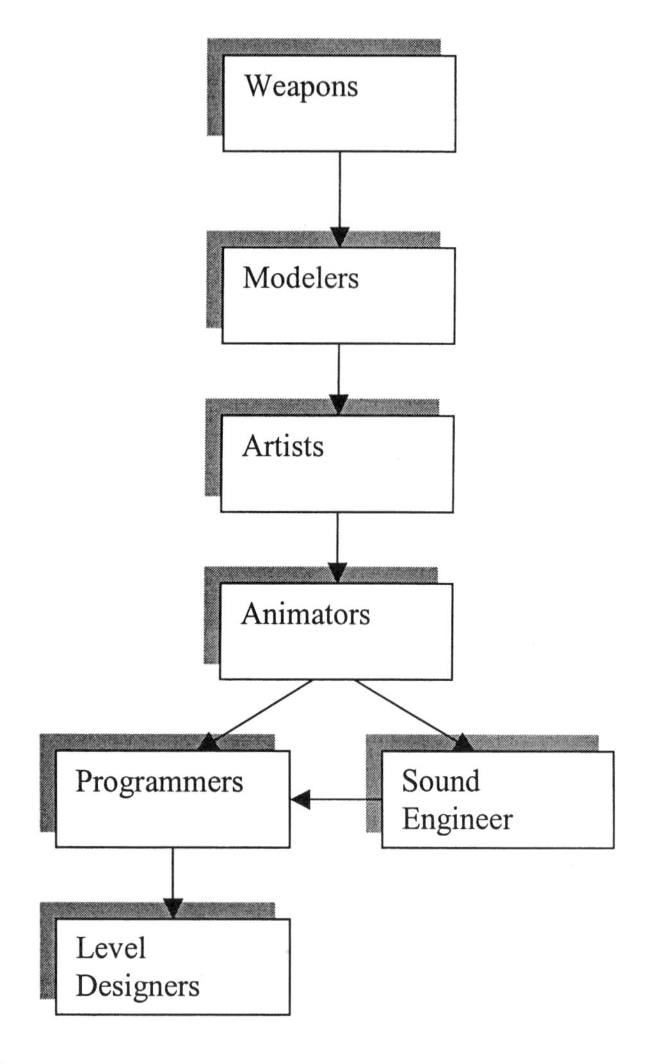

FIGURE 9. Weapons

3D Modelers create the world objects that are placed in the game environments: for example, light fixtures, vehicles, trees, grass, bushes, fences, and rocks. Once a world object has been created, an artist paints its texture map. The finished object is imported into the game code and placed in the game environments by level designers.

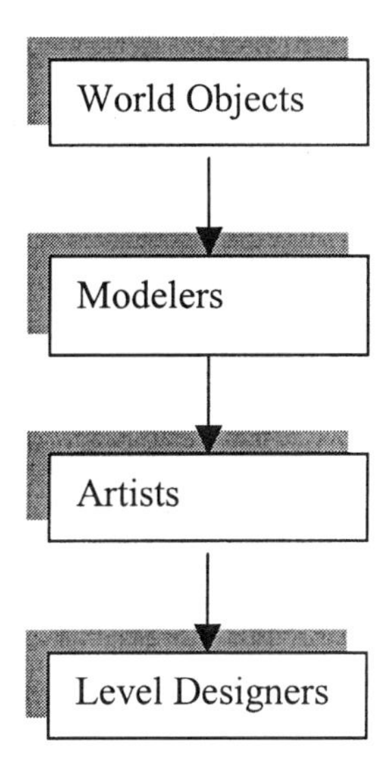

FIGURE 10. World Objects

The animator determines the entire range of motion for all moving elements in the game. If AI is to be implemented, character behaviors are examined to determine what animations are necessary. Once this has been decided, the animator directs a motion-capture session, in which an actor performs specified movements (usually these services must be contracted out, at high cost). The animator takes the modeler's objects, applies motion-capture information (adjusting as needed), and distributes the assets to programming and sound engineering. The sound engineer supplies audio to the programmers and level design adds the finished components to the environments.

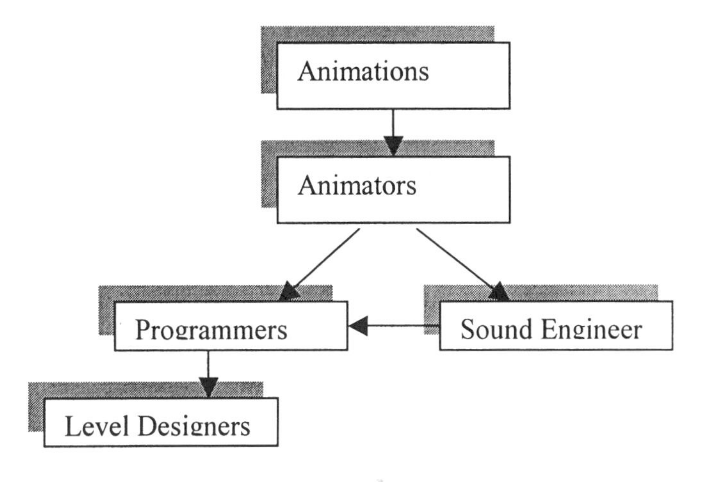

FIGURE 11. Animations

Special effects are an often-overlooked element that can be applied to virtually every aspect of the game, adding polish and interest. Clouds that pan across the sky, muzzle flashes, tracer fire, and water dripping from a leaky pipe are just a few of the effects that can make a game environment feel alive. These effects are usually created by the art team, who relay them either to level design for integration into the environments or to programming for placement directly into code. Typically, special effects are added towards the end of the project, when all other assets have been completed.

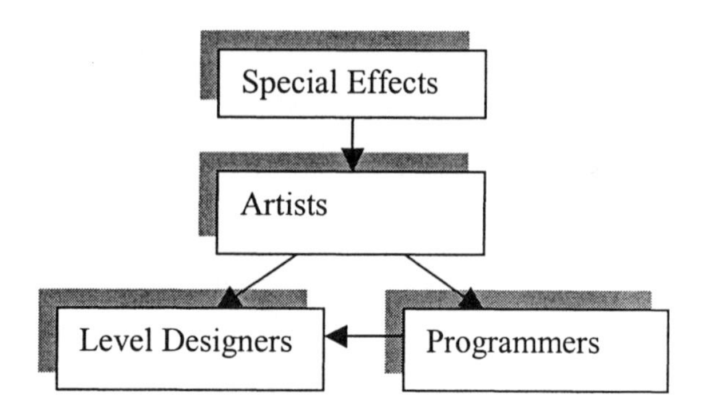

FIGURE 12. Special Effects

For the user to understand and play the game, a user interface and icons must be designed and implemented. These assets are typically created by the art department, who distribute them to the programming team for writing into code. Because the interface has to be updated as new features appear, it is important that it be robust and dynamic enough to grow as the game evolves.

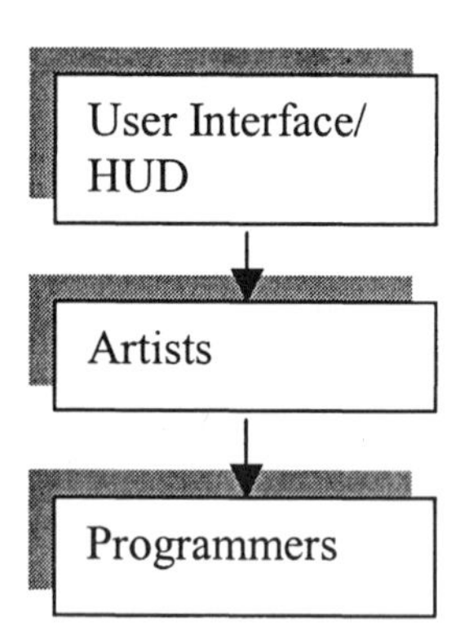

FIGURE 13. User Interface/Heads-Up Display

Level designers create the game environment. Like the dish on which a fine meal is served, it is here that all the components come together and the user enjoys the final presentation. With this in mind, the level designers work closely with the team to ensure that each component works as planned. Just as the programming department is the hub for all technical elements of the game, level design is the hub for all content. If the level-design team misses the target, the entire game will suffer.

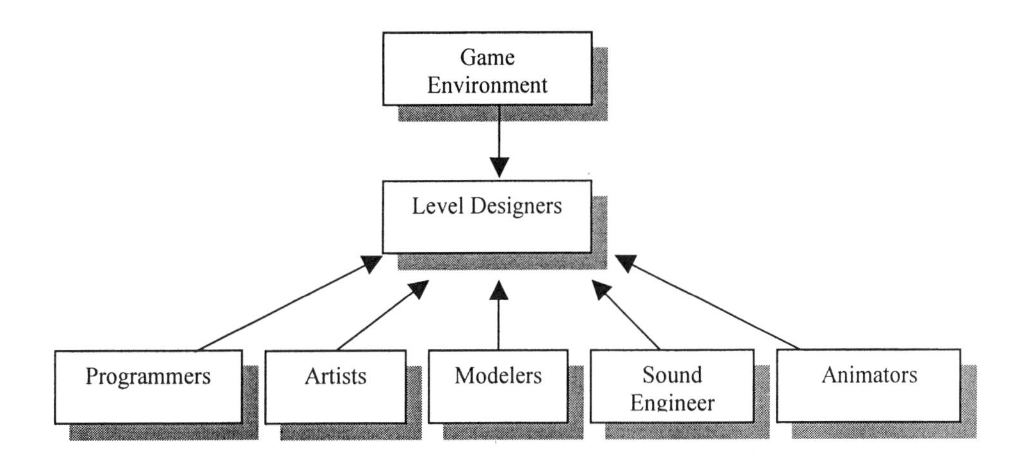

FIGURE 14. Game Environment

SUMMARY

There are many pieces to a game like *America's Army*. Identifying them is half the battle, because it leads to a solid plan of action, which starts with good game design and project leaders who can communicate the design to the team. Scheduling – deciding which pieces are constructed when and by whom – helps the project meets its goals (yes, we're talking Microsoft Project). As illustrated above, there are many interdependencies among the components of an FPS game, and many risks: if one element fails, the ripples are felt throughout the enterprise. But with planning, good staffing and coordination, the development team can overcome these risks and produce a well-constructed, high-quality game.

AMERICA'S ARMY – A CASE HISTORY

To show how much can be accomplished within three years, the following section describes the *America's Army* development from inception to the 2.0.0a release of December 21, 2003. We will enumerate the contents of each release discuss the concerns associated with it, and approximate the time spent. This close look at *AA's* development reveals issues that come up in developing large-scale games.

Some of the difficulties we encountered will doubtless crop up in your project. We hope our experience will prove useful.

America's Army Pre-Release – August 2001

In August 2001, the *AA* project was seriously understaffed and unable to prosecute proper development. Major obstacles to success included the following issues:

Improper Team Balance and Organization. At this juncture, the team was not well structured to develop a first-person shooter. No one had experience in creating and shipping an actual product, and the team structure was inefficient and inadequate to the task. An overabundance of designers was coupled with a severe dearth of art support. We had a character animator, but no character modeler, and no one on sound. Overall, the team lacked cohesion and leadership. Solution: We hired three game-industry veterans as team leaders to rebalance the team, hired a character modeler, and acquired a sound engineer.

Lack of Design and Common Vision. The absence of a thorough design document fragmented the team's vision and precipitated confusion between the development team and the customer (i.e., the US Army). Without a proper design, it was difficult to guide the team, schedule tasks, and track progress. Solution: We focused on the overall mission statement, which was to develop a game with appeal similar to that of *CounterStrike*. To this successful model, we added heavy emphasis on realism and Army values and training.

Technical Issues. The game engine licensed for *America's Army* was still in development; in fact, during the entire course of development, the technology was constantly in flux. Many systems were not in place or inadequate for the game's needs, and completion of the engine was not anticipated until after the scheduled release of *AA*. Due to the development team's inexperience, the game's database structure was vastly inefficient and lacked consideration for distribution. Many of the game's assets were un-optimized or beyond the technical specifications of the game engine. Many of the steps and tasks necessary for success went unaddressed.

Solution: The engineering team wrote a number of new systems from scratch (approx 150,000 lines of code for the initial release of the game). We reorganized the art database and created a standard structure for all file formats and a team-wide methodology for database organization. Game assets were optimized to run well under the game engine. We cut a number of elements that were outside the engine's specifications. Task-management software was implemented to organize and track progress.

Version 1.0 Release –July 4, 2002

The first version of *America's Army* was released on July 4, 2002. With the game a runaway success, the Army and development team were unprepared for the sheer volume of players that flocked to the game. Game servers were massively overloaded, and the need for a professional quality-assurance team became apparent as the public discovered critical bugs that detracted from the experience and even prevented some players from running the game. Several features had been delayed at launch so that the July 4th deadline could be met. Because of this, the Army labeled the initial launch of the game the "recon" version. Most players understood it was really a beta version. Issues that the development team dealt with during this phase are as follows:

Server Overload. Initially, the Army stood up only 140 servers for the launch of the game. The average server could accommodate 24 players. With the game downloaded over 500,000 times that weekend, the servers were swamped and many players had to wait days to play. Additionally, the game used an authentication server that validated players' having completed basic training (required for multi-playing) before allowing them onto a game server: this authentication server, too, was overwhelmed, making it even more difficult for players to enter. Because so many players had never played the game at once, many dormant errors emerged. Solution: The Army quickly stood up additional game servers and authentication servers. The development team went to work on addressing the most critical errors and applying server-side fixes.

No Server-Browser/Community-Server Support. At the release of version 1.0, the in-game server browser was not finished. As a stopgap, *Gamespy Arcade* was included with the download and was required to find and join game servers. There was no mechanism by which users could set up their own servers or use other server-browser software to find game servers. This shortcoming exacerbated the problem of server overload and irritated players by forcing unwanted software on them. Solution: The development team completed the in-game server browser, as well as packages for setting up user servers and user-created browser software.

Game-Play Bottleneck. The initial release of the game required that all players complete the single-player training courses (rifle-range, obstacle, weapons-familiarization, and tactical). Once these courses were finished, players had to go online and participate in a multiplayer training exercise before additional scenarios could be played. Additionally, until a user had played online and was part of a winning team in the MOUT McKenna training level, he could not proceed to other missions. While this seemed a good idea, in practice it created additional bottlenecks and yet another barrier to entry. To make matters worse, the game did not adequately describe the requirements for participation in further missions, so people were confused about what they were supposed to do. Solution: We did away with the online-training requirement and changed the game so that only completing the single-player levels was necessary.

Training-Level Bugs. Both the rifle range and obstacle course suffered critical bugs. In the case of the rifle range, players discovered an exploit that allowed them to cheat and bypass qualification. In the obstacle course, a logical error in the script prevented many players from finishing and proceeding with the game. Solution: The development team immediately fixed these problems.

Multi-Player Bugs. A number of critical bugs in the multiplayer portion of the game were discovered after initial release, ranging from graphical glitches to serious flaws in game play that marred the overall experience. In the case of the collapsed-tunnel mission, a logical flaw in the objective system miscounted victories and losses. In many cases, a victorious team was credited as having lost. This frustration led most users to avoid the mission. Solution: Identified the most severe problems and began working on fixes.

Version 1.0.1 Release –July 12, 2002

Version 1.0.1 of *America's Army* was released on July 12, 2002. As implied by the version number, this was a minor release, consisting primarily of a patch for the worst problems of version 1.0. The main issues addressed were as follows:

- Corrected client and server-flooding issues. This fix stabilized servers that were overloaded by network traffic.
- Fixed training level bugs. These fixes addressed the most critical issues involving the rifle range and obstacle course.
- Added community game-server support. This allowed the use of alternative server browsers for finding game servers.
- Added a dedicated server executable, allowing players to stand up their own game servers.

Version 1.1.1 Release –August 1, 2002

On August 1, 2002 the development team released version 1.1.1, the "marksmanship pack." This release added the Army's sniper schools and the M24 and M82 rifle positions to the game, features originally scheduled for the initial July 4th release, but fallen behind schedule. Eligibility to play the marksmanship levels was based on scores from the original rifle-range training level. A player who shot 36 out of 40 targets in the final test could try to qualify as a sniper. Only those players who passed the marksmanship training levels could take a sniper position in online play.

While finalizing this version, an unfortunate database error was discovered: the authentication server was logging only pass/fail results for the rifle range. Once a player was determined to have passed the course (with a score of 23 or above), the authentication server did not bother to record subsequent attempts, so that players who had met the basic qualifications could not return to the rifle range and try for better scores so they could move on to sniper school. In the end, we reset the rifle-range scores for all players to force the necessary changes to the authentication system. Many players who had already qualified for sniper (an extremely difficult feat) found they were obliged to qualify again. This naturally had a very negative impact on the player community.

To make matters worse, *AA* opened the sniper role only after other team positions were filled, meaning there were only a few sniper positions available at any time. With the release of the marksmanship pack, *everyone* wanted to be a sniper. Virtual fratricide broke out as people killed team members just to steal their sniper rifles. Needless to say, we did not anticipate this abuse and had to brainstorm ways to curb it.

During this release we also did away with the MOUT McKenna online-training requirement. Ironically, this caused an outcry from those who had gone through the painful launch experience and saw completion of the training as a badge of honor. Many felt that since they went through MOUT McKenna, others should too. Regardless, it was necessary to remove the requirement to free up server bandwidth.

Other changes in this release included:

- Added idle-player kick. In the initial release, it was discovered that with the limited server space, many players neglected to log off when they weren't playing (to preserve their slot). This infuriated players who couldn't get in and annoyed those in the game who saw a team member just standing there. A fix was added to time idle players and kick them off after a certain period. Occasionally players who were not idle would get the boot, requiring additional fixing in subsequent releases.
- An in-game server browser was finally added. While offering only the most rudimentary functionality, it at least appeased players and removed the necessity of using *Gamespy Arcade*.
- Added MILES grenades to MILES missions. MILES is a laser-tag system the Army uses for training. At the Army's request, a number of *AA* missions were based on MILES scenarios (the irony of simulating a simulation was not lost on the development team). With release of version 1.1.1, the Army wanted to add a MILES-equipped grenade to these missions. Opinions concerning this addition by the community were mixed.
- The development team was asked to change the tracers of enemy fire from amber to green.

Version 1.2.0 Release –August 22, 2002

Released on August 22, 2002, version 1.2.0, the airborne/ranger pack, introduced airborne and ranger schools to the game. While the airborne school came with two training levels that depicted an abridged version of the Army's actual training, the ranger school offered no training levels at all. The original design called for ranger training to take place online with other players, but after the debacle of the MOUT McKenna training scenario, adding another round of multiplayer training requirements was determined not worth the risk. Instead, the ranger-training levels were converted to standard online scenarios. The disadvantage was that there was nothing players had to do to qualify for these maps. In the end, we required that all other training be completed before ranger maps could be played. While this was a workable compromise, it clashed with existing paradigms in the game.

Other problems encountered with this release revolved around the airborne portion of the game. The technology used for *AA* was not ideal for simulating flight, and the artists had to depend heavily on tricks to create the illusion of parachuting. While this worked well in the single-player training missions, where the experience could easily be constrained, multiplayer missions posed hurdles and challenges that were never fully resolved. Parachuting introduced a host of bugs, not to mention heavy demands on the processor. While ultimately the team this feature adequately, associated problems plagued the entire production cycle, including parachutes not opening (and players falling to death), parachutes deploying inside planes, parachutes stuck on the body after landing, players stuck together or stuck on other objects, players unable to move after landing, and a host of related technical issues. Although this was only a small feature in the game, it consumed a great many man-hours.

Additional highlights for version 1.2.0 included to following.

New Voice-Overs for Radio Commands, Shouts and Whispers: During development, team members and Naval Postgraduate School students were often used as voice actors for the game. While this saved the cost of hiring professionals, it meant that creating good voice-overs (VO) was a struggle. A particularly good reader might be a military officer, stationed at NPS for only a short time, or an original reader might no longer care to participate. When this happened, a new VO candidate had to be located and the entire voice-over sequence recreated. Moreover, voice files tend to be quite large, and the continual changes frequently increased the download size of subsequent releases. This aspect of development proved frustrating, an ever-changing facet of the game.

Adjusted Team-Balance System: In multiplayer games, it is customary to include team balancing. If one team heavily outnumbers the other, the system will shuffle players to achieve equity. Also, if one team consistently beats another by large margins, the system will exchange players to make the teams equitably matched. While this sounds good in theory, it can create problems. Players may not understand the computer's arbitrarily changing the conditions of the game, and the system itself may respond to very specific contexts only. Without a professional QA department, many of the flaws in the auto-balancing system aren't discovered until after a new version of the game is released and feedback is received from irritated players. In the case of *America's Army*, this feature was adjusted several times before it was deemed acceptable. In all likelihood, it was never truly perfected and there are still players who are not satisfied with it.

Adjusted Vote-Kick Feature: The vote-kick system was created so that players themselves could enforce the rules of the server. If an unruly player were causing havoc, a player could call for a vote to kick him off the server. While this is a common tool in multiplayer games, we didn't foresee the ways in which it might be abused. We found that many players were being tossed for reasons outside the scope of the system. Like the team balancing system, it was necessary to adjust vote-kick numerous times. It's difficult for a computer to identify and regulate human behavior, so a perfect solution to game pests was never truly achieved.

Adjusted Weapon Distribution: In the original version, players were not allowed to select their weapons, but instead chose what role they wanted and were given the accompanying weapons, based on the actual structure of Army infantry units. The weapon-distribution system regulated how the various weapons were dispersed among players. The problem was that most players maintained a personal-weapon preference and wanted to find out what to do to obtain the favored weapon; but the system relied on mathematical voodoo that did not always provide consistent results. The result was great confusion among the players and constant modification by the development team.

Three New Multi-Player Maps: Version 1.2.0 added three new multiplayer missions to the game: the FLS assault, the swamp raid, and the mountain ambush. Because we had few testers at this point (as well as an internal network that did not allow us to test maps with a full contingent of players), a host of new problems appeared with these levels. The most dramatic involved the mountain-ambush level. It was found that if someone changed teams and then left the server after the mission began, the round immediately ended. With players entering and leaving servers frequently, this level was in effect unplayable and was temporarily removed from server rotation.

Version 1.2.1 Release –August 24, 2002

On August 24, 2002, only two days after the release of version 1.2.0, a patch was created to deal with the critical errors introduced in the previous release. Specifically, several fixes were made to new missions and to the team-structure system to make the mountain-ambush level playable.

Map Pack Release –October 3, 2002

On October 3, 2002 the development team released a map pack including two new missions for the game: JRTC Farm and Weapons Cache. These two maps had been finished for some time, but were delayed by request of the Army so that they could be used for strategic-marketing purposes: before releasing them to the public through standard distribution channels, these missions were first available through Army recruiters only. After a time of exclusivity, the missions were added to the next release. Although this practice seemed straightforward, it actually caused the development team several distribution problems. Patches were created with every new release so that players had only to download the new rather than retrieve the full version again. With the map pack however, our engineers now had to account for two different versions of the game (one with the new missions, one without) and apply the patch accordingly. Since this map pack fell outside the scope of the team's normal distribution methodology, extra engineering was required to ensure that all players would be able to update the game seamlessly for the next release.

Version 1.3.0 Release –October 10, 2002

Released on October 10, 2002, this version of *AA* added a host of new features, bug fixes, and adjustments. Since the game's initial release, the development team had been scrambling to finish uncompleted features. With version 1.3, they were finally able to consider the initial launch finished and begin focusing on new features and adjustments based on user feedback. While this release offered only one new multiplayer level (the mountain-pass arctic mission), great effort was put into improving the game overall. Some of the changes made in this release are as follows.

Combat Effectiveness Meter (CEM): Because *America's Army* attempted to portray a realistic combat system, there were a number of factors that could affect a player's accuracy and effectiveness while engaging the enemy, including posture (standing, crouching or kneeling), movement (e.g., running versus walking), use of weapons' iron sights, scopes, and bipod supports, and proximity to team leaders. While this allowed for a system more closely resembling the experience of real combat, the calculations were done behind the scenes, and players often were confused about the variance of weapon accuracy in the game. In version 1.3, a meter was added to the player's screen, resembling the equalizer bar on a stereo system: the higher the bar, the more effective the player in combat. As the player moved (for example, changed posture and speed), the bar rose or fell to reflect the effectiveness of the player's actions. This feature brought the inner workings of the combat system to the fore, allowing better understanding of how to be effective and what might cause poor performance.

Honor System: For some time, the Army had been looking for the development team to provide players with a comparative statistic showing accomplishment within the game. Version 1.3 answered this desire by adding an honor system. The honor system attached a persistent score (between 1 and 100) to every player. By tracking points scored against points lost, players could build their honor score and wear it as a badge for all to see.

Inevitably, however, many players wanted the score to reflect actual ability, rather than simple time invested in the game. Moreover, the honor system created a distinction between official and unofficial game servers, because only experience racked up on official servers was counted towards honor gain (to prevent exploitation of the system). This caused players to avoid unofficial servers and play on Army-sponsored servers only, hampering the growth of the game community. Over the course of the project, there were also several bugs and situations that could cause honor scores to be lost or reset, precipitating an outcry. While the development team made many alterations to the honor system, its full potential was never achieved.

Auto Weapon Lowering: In early releases, it was discovered that occasionally the player's weapon would penetrate level geometry and give away his position. In response, a system was modified so that when a player was too close to an object, his weapon automatically lowered to avoid it. While this solved one problem, players found that their weapons did not always return to proper position when needed. These glitches were addressed in subsequent releases of the game.

"Hit the Dirt" Feature: The new version of the game gave players the ability to perform a combat dive while running, quickly hitting the ground. While the feature was well received, it was eventually scaled back because players were sometimes stuck in level geometry after performing the maneuver. While scaling back solved the problem, many players were disappointed by the changes.

Night Vision for Spectator Mode: In *America's Army*, once a player is killed he is out of the action and may watch the game from a number of spectator cameras or by viewing a particular team member. In night missions, spectators often couldn't see the action in the low lighting. To compensate, night vision was provided to spectators and camera points.

Adjusted Server Browser: More detailed player and game info was added to the server browser so that players could better select the game servers they wanted to participate on. More options were also provided to sort the data received in the server browser.

Adjusted M249 Fire Model: In previous versions, it was discovered that many players had learned to tap the fire key of the M249 to turn it into a powerful, long-range weapon. This was at odds with the weapon's real performance, so adjustments were made to add variance to the burst-fire capabilities of the weapon.

Adjusted Weapon Accuracy System: We made adjustments so that all weapons fired with increased realism in shot patterns and bullet spread.

Adjusted Prone Movements: Movement in the prone position was adjusted to provide better performance over terrain and more flexibility when performing certain actions.

Adjusted Footstep Volume: It was discovered in previous versions that footsteps were too hard to hear. The volume was turned up to give better immersion.

Adjusted Sniper Rifle Accuracy: Adjustments were made to the sniper-rifle accuracy system, so that shots fired always hit the exact spot where the crosshair was targeted, but decreased combat effectiveness was translated to the player through greater wavering in the weapon's scope.

Adjusted Grenades: It seemed that the development team would forever be adjusting and balancing the way grenades were depicted in the game. We discovered that a realistic grenade does not necessarily equal a fun experience, leading to constant rebalancing and enhancing of the feature. In version 1.3, the following changes were made to the grenade system:

- Auto Grenade Notification: Many players were dying from grenades because they were unaware that they had been thrown. The development team added a feature whereby throwing a grenade triggered an audible warning to other players in the area. To reward stealth, the warning could be overridden if players moved slow in lobbing a grenade.
- Auto Weapon Switch Upon Grenade Throw: Many players were dying after throwing a grenade because they couldn't raise their weapons in time afterwards to defend themselves. We added automatic switching back to the primary weapon after a throw. Realizing that some players might dislike the feature, we included a menu option for disabling.
- Grenade Spin: In previous versions, grenades did not observe physics and traveled in a frozen position. For better realism, spin was applied.
- Dive on Grenades: The ability to dive on grenades was added, thus letting players save buddies from harm. Unfortunately, because of game perspective, it was difficult to judge exactly where to land. It turned to be out rare for anyone to exploit this ability; the feature was mostly ignored.
- Grenade Physics by Material Type: Changes were made so that grenades would react differently depending on the type of surface they encountered. Like the grenade spin, this increased apparent realism.
- Adjusted Variance of Fuse Length: Originally, all grenades possessed the same length fuse. We became aware that players had learned exactly how long they could hold a live grenade before throwing it, pulling off precision attacks that would not be possible in the real world. To compensate, the development team varied the fuse length, making accurate judgment impossible. From version 1.3 on, if players held on to live grenades, they risked blowing themselves up.
- Adjusted Auto Roll Distance: We adjusted how far grenades could be rolled.

Version 1.4.0 Release –November 25, 2002

Released on November 25, 2002, version 1.4 of *America's Army* was a minor release that offered one new mission (River Basin), and a handful of new features and bug fixes.

New Scoring System: A new scoring system de-emphasized killing the enemy and rewarded acting as a team and completing objectives. While hard-core gamers did not immediately embrace the system, many players found they were able to achieve higher scores without necessarily using violence. Ultimately, this created a more balanced experienced while simultaneously improving the marketing message the Army sought to express.

"Report In" Feature: Based on user feedback, players' ability to hit a single key and report their location was added. This well received featured required the development team to make substantial adjustments to, and testing of, every level in the game.

Binoculars: Team leaders were provided with binoculars to better scout positions and coordinate with team members.

Movement With Iron Sights: In previous versions, if the player was using the iron sights of a weapon, any movement would drop him to the normal weapon perspective. With version 1.4, players could move (albeit very slowly), while looking through the sights.

News for Login Screen: A news section was added to the login screen so that the Army could make general announcements about the game.

Adjusted Automatic Weapon Fire System: Adjustments were made so that if a player switched from standing to crouching while firing an automatic weapon, the weapon would continue firing during the posture change. Players had brought this need to the attention of the development team.

Fixed Multiple Login Exploit: It was discovered that players were using multiple machines to login to different game servers under the same account. By playing simultaneous games with one account, players were building their honor score at an unacceptable rate. To address the issue, the development team caused the authentication servers to check for multiple logins and kick offenders from the server.

Another Grenade Adjustment: To increase grenade realism, a change was made so that if the player pressed the fire button while selecting a grenade, the grenade was made available with the pin already pulled and ready to throw.

Version 1.5.0 Release –December 23, 2002

On December 23, 2002, the development team released version 1.5. Around this time, the game had come under fire by a Miami attorney on a crusade against violence in video games. As the funders of AA, the U.S. government proved an irresistible target.. The development team was required to make several modifications to counter the negative press generated by this man, including the elimination of the word "sniper" from the game (which involved major changes to several levels and weapon systems, as well as new voice-overs for the marksmanship schools). Parental controls were added so that parents could monitor language, weapon usage, and mission types, and limit displays of blood. These changes were designed to differentiate *AA* from most commercial games by letting parents control content.

In addition to parental controls, other changes in this release included:

Weapons-Cache Special-Edition Map: One of the most popular levels in the game was the weapons-cache mission. Many fans pointed out flaws in the map, as well as desired improvements. Based on this feedback, a new version of the mission was created, effectively doubling its scope. These changes were applauded, and the mission remains one of the most popular to date. By implementing improvements per popular demand, the team was able to foster goodwill and to assure the community of their voice in the game's evolution.

New Enemy Voices: With the help of the Defense Language Institute, the development team created a fictive enemy language, based on a combination of natural languages. Voice-overs of foreign students were recorded to create realistic shouts and enemy radio commands while ensuring that no speakers of an actual foreign language would be depicted as enemies of the United States. As a bonus, because the enemy language had its roots in reality, players found they could learn and understand the commands issued by opposing forces.

Optional Reason for Vote-Kick System: Previous versions revealed that the vote-kick system was inadequate because players were often in the dark as to why a player had called to ban another player. An optional reason was added so that when a player called a vote, the others could see why.

Army Star to Player Listing: The development team added to the scoreboard the ability to show whether a player was an active member of the US Army (subject to verification). When a verified soldier played in the game (and there were many of them), an Army star appeared next to his name. This allowed the community to know when they were interacting with actual soldiers and strengthened camaraderie between military and civilian players.

ROE Penalty Adjustments: Whenever a player injured or killed a team member or performed specific detrimental actions in the game, he suffered a penalty to his score for violating the "rules of engagement" or ROE. While this was an effective way to enforce Army values, the development team often found it necessary to tweak the system to ensure proper play balance.

Server-Browser Adjustments: Adjustments were made so that the server browser distinguished between leased servers and official servers. How many LAN servers could be displayed at a time was also increased.

Version 1.6.0 Release –March 16, 2003

Version 1.6 of *America's Army* was released on March 16, 2003. This release took considerably longer to complete than previous versions due to an update of the game's core technology: Epic Games, who created the software *AA* was built on, had released a major update to the game engine. The development team had to merge the updated technology with the game's current code base. After the months of work had been put into the game, there were vast differences between the code base and that update. The merger took about six weeks of programming, as well as a number of weeks to adjust content to work with new features. While painful, it was a requirement if *America's Army* were to keep its cutting edge.

Although only one new mission accompanied this release, the radio-tower level was the largest map the development team had created. This mission pushed technological limits, and frequent adjustments were made to reach a smooth playing experience. Flaws in the authentication and loading systems were discovered, and it was found that low-end machines were taking so much time to load the level that the authentication server would time them out and drop them from the server. A number of band-aids were applied before this version could be released.

Other changes in version 1.6 included:

Projectile Penetration: Previously, any time a bullet struck an object, the bullet was blocked and considered spent. Version 1.6 introduced penetration, by which bullets passed through penetrable objects and continued with diminished velocity and force (depending on the material hit) as well as condign entry and exit effects. This yielded a dynamic change in game play, because objects that had previously served as cover could no longer be depended on.

Projectile Ricochets: The tendency for bullets to ricochet when fired from certain angles was introduced for more realistic ballistics and added tension.

Bullet Decals on Static and Dynamic Objects: The technology update allowed bullets to leave marks on static and moving objects. While this increased realism, it also increased processor overhead. To avid sluggishness in low-power machines, settings were added to control how many bullet marks could be displayed at once.

New Sound Effects: New sounds were added for ricochets, as well as for footsteps on concrete and carpet.

New Texture-Detail Options: An array of new settings in the menu system enabled players to adjust texture detail to suit the power of their machines.

Password-Entry Window to Server Browser: To allow users to set up private servers and control access to them easily, a new window was added to the server browser for passwords.

Spam Control for Messaging System: It had been discovered that players were flooding the in-game messaging system, effectively ruining communication during play. To compensate, the engineering team controlled how many messages could be sent by a player in a given time.

New Desert Camouflage: During this development period, we learned that the Army had changed its desert-uniform camouflage. Desert uniforms in the game were changed accordingly.

New Loading Screen: A new loading screen was added to indicate when the game engine was tied up with loading new content into memory.

Fatigue Element to Jumping Abilities: Many players were demonstrating a tendency to jump up and down in the game, a term known to gamers as "bunny hopping." Since soldiers are typically weighted down with equipment, such action was not in keeping with the degree of realism we were attempting to portray. Fatigue was therefore added so that repeated jumping caused the player's character to tire and be unable to continue.

Grenade Aiming: Players found that, because of the perspective in the game, aiming a grenade accurately was extremely difficult, requiring a great deal of guesswork. To make the system more intuitive, the player's onscreen hands were changed so that the gap between the forefinger and thumb of the lead hand was positioned over the center of the screen, enabling the player to use it as a guide.

Improved Weapon-Jam System: The algorithm for weapons jamming was altered to reflect the jam rate of real-world counterparts.

Version 1.7.0 Release –April 21, 2003

Version 1.7 of the game was released April 21, 2003. Most of the development team was tied up with preparations for the Electronic Entertainment Expo (E3) in Los Angeles the following month. Since there was no time for a proper update, the only addition to the game was a new single map, a special-edition version of the popular Bridge Crossing. Although the team did not plan to increment the version number with this release, the Army requested it be labeled version 1.7. The internal version of the code was in heavy flux, so the previous version was rebuilt as 1.7.0. Unfortunately, a few bugs crept into the code packaged in this new version, while an improper assumption was made that the only change to the code was the version number; it was thus released without thorough testing. The result was a sub-par release that inflicted several critical bugs upon the community. Once again, the development team felt the pain of an inadequate testing solution.

Electronic Entertainment Expo – May 2003

The E3 show in Los Angeles is about showing the world what new things are in store for your players. The tendency is to shove as much into the game as possible and somehow make it all work through smoke and mirrors. While the goal of the show *per se* can be met this way, afterwards developers find themselves with a roster of features and systems that are incomplete, in need of optimization and reworking. The *AA* development team spent the rest of the summer trying to deliver on promises made at the show.

Version 1.9.0 Release –August 8, 2003

On August 8, 2003, version 1.9 was released, the biggest update yet. It was a difficult period for the development team, as there were more features needing work than time to work on them. Although the plan had been to prepare a comprehensive release addressing all E3 fallout, it became apparent that the load would have to span multiple updates. The team's loss of two employees during this time jeopardized the schedule further. Examining the various features in progress, it was eventually determined that version 1.9 would focus on introducing medics: thus this version was labeled the "combat-medic pack." New features were as follows:

New Damage Model: To create combat medics for *America's Army*, a new damage model for the game was designed. In previous versions, all bullets inflicted a specified amount of damage on striking a player. This system was changed for version 1.9 so that the player initially suffered a percentage of damage, while the remaining portion was doled out over time in the form of blood loss. If a combat medic reached a wounded player in time, the bleeding could be staunched and further damage avoided. The system worked well by supporting the concept of medics without making it seem they had magical healing powers, but it was a dramatic change that players had to get accustomed to.

New Character Models: Because version 1.9 was released more than a year after the initial launch of *America's Army*, it was deemed acceptable to raise system requirements for the game. Most conspicuously, the game's character models had never satisfied the team. A decision was made to raise the bar and replace all characters with new, highly detailed versions. While the result was a dramatic improvement, it entailed a colossal amount of work for the artists.

New Interface: The original menu system for the game had been created at the last minute, just before the initial launch in July 2001, and its design was inadequate for an ever-evolving product. Aesthetically, it was unpleasing; operationally, unintuitive. For version 1.9, an entirely new interface was designed, with great thought put into navigability, expandability, and tie-ins to the game's official website. While the result was an extraordinary improvement that gave users the impression that *AA* was a whole new game, the work required to pull it off was incredibly tedious and time-consuming. There were so many pieces to the new menu system, with such a vast array of interdependencies, that the development team worked on it till, literally, the last minute. Of necessity, many smaller elements of the interface went unfinished, and polishing of the system would be completed over the next several releases.

New Theme Song: Originally, *America's Army* had no music. To open the game and augment the new look and feel, a distinctive, patriotic theme song was commissioned for the franchise. The development team did not create this work. Nevertheless, it involved many iterations and frustrating changes before the score was finally approved.

Detail Textures: Capitalizing on a previously unused feature of the engine, new artwork was created so that when a player got close to any surface, a high-resolution texture was swapped with the normal, lower resolution texture usually seen from a distance. This allowed for a high degree of realism when studying world geometry up close, but kept system overhead manageable.

Combat Medic: To become a combat medic, players had to complete a four-level training sequence involving three classroom lectures and a field-training exercise. These levels were heavily scripted and presented actual first-aid techniques and quizzes. Much research went into making a realistic course, including consultation with medical professionals. Once qualified in the combat-medic course, players were able to treat injured comrades.

Player Shadows: Detailed player shadows were finally added to the game. This feature became available in the previous code merge and technology update, but required extensive engineering to work properly.

Lip-Sync and Facial Animation: In previous versions, facial expressions of characters were fixed. By licensing a middleware package developed for Unreal technology, the development team was able to add facial animations with speech synchronized to mouth movement. This capability, combined with the improved character models, boosted character realism tremendously.

Punkbuster: For a year, the development team tried to combat multiplayer cheating, but simply didn't have the time and expertise to squelch the growing number of hacks that were becoming available for *America's Army*. The job was finally contracted to a commercial anti-cheating firm, who added Punkbuster service to the game. The several weeks it took to port cheat protection to the Unreal technology were well worth it: the feature was a huge success with the player community, effectively stymieing those who wished to ruin the game for others.

ROQ Video Support: Support was added for ROQ-format video-clip playback within the game engine, expanding the team's ability to add supplemental content and offering another means of providing education about the Army.

New Scoreboard, Team-Selection, and Class-Selection Interface: In keeping with the new look and feel of the menu system, a new scoreboard and team- and class-selection interface was created. Unfortunately, there were so many elements involved with the new menu that it wasn't discovered till the last moment that we had failed to redesign these particular portions of it. Realizing the game could not be released without completing these elements, the development team spent the final days of the production cycle working feverishly to finish them.

New Server Admin Commands: An array of new commands was created so that those running their own servers could easily monitor, organize, and customize the game experience.

Demo Recording: We added a feature enabling players to record and view game-play sequences within the game engine.

Multiple Bug Fixes: A great many longstanding bugs were finally addressed in this version.

Version 2.0.0 Release - November 6, 2003

As a follow-up to version 1.9, the development team released the 2.0 special-forces pack on November 6, 2003, completing another segment of the features that had been originally planned for that spring, as well as tying a number of loose ends from the previous release. Many players viewed Version 2.0 as the development team's finest release ever. The changes included:

"Special Forces" Role: After the successful completion of three training segments, players were qualified to play four new multiplayer missions as green berets. The special-forces (SF) role introduced new character models to the game, as well as the ability to use and customize an assortment of new weapons.

"Indigenous Forces" Role: We made it possible for players who did not pass SF training qualifications to play the new missions in the role of indigenous soldier. This ensured that the missions were available to all players while reinforcing the point that a major duty of SF units is to train and fight alongside indigenous forces in foreign countries.

New Weapons: The following new weapons were added:

- SOPMOD M4 carbine (SF weapon)
- SPR (SF special-purpose rifle)
- Thermite grenade (SF weapon)
- VSS Vintorez (enemy weapon)
- AKS-74U (enemy weapon)
- RPG-7 (enemy weapon)
- M9 pistol (snipers only)

Weapon Modifications: The SOPMOD M4 allowed a number of weapon customizations by the player. A new interface section was added by which players could view their weapon and add and remove interchangeable parts, configuring as desired. This was a major feature in version 2.0, and proved to be one of the most appealing aspects of play as an SF soldier. The customizable elements are as follows:

- ACOG 4x scope
- ACOG reflex sight
- M68 Aimpoint sight
- M203A1 grenade launcher
- M583A1 flare launcher

- Harris bipod
- M4QD suppressor
- Iron sight
- Heat shield

3D Iron Sights: Additional changes to weapons came in the form of true 3D iron sights. In previous versions, the iron sights for all weapons were depicted using 2D overlays. The new method involved three-dimensional geometry for more accurate portrayal.

In-Game IRC Chat Client: A new page was added to the interface to provide an in-game internet-relay chat (IRC) client, enabling players to speak with other users who were not necessarily playing at the time. This new tool further supported the community.

Andromeda Server Browser: Although for some time the game had employed licensed and proven server-browser technology, the Army contracted a third party to develop a new browser specifically for the game. In development several months, the product finally made it into the game in version 2.0. This technology never quite lived up to its design and proved a source of difficulty to the developers, and ultimately a major point of contention between the development team and the Army.

Interface Modifications: Continuing the work begun in version 1.9, the team made several adjustments to the new interface. These included:

- New progress bar for the server browser
- New mission-deployment page
- New in-game icon key
- New loading/connecting-message text boxes
- New glossary page
- Various detail settings on the video-options page
- Tour icons for the server browser
- Three new weapon-camouflage skins (desert, forest, arctic) were added to the weapon-modification page
- Resized server browser page (for better screen fit)
- Page and resolution sizing
- Ultimate Arena tournament server functionality for the server browser page

- An updated support page

New Weapon Animation System: To accommodate the weapon-modification feature, a new method was developed for efficient display of third-person weapon animations.

New Authentication System: During this period, a third party took over the task of running the authentication system. Because of contract issues, this required the development of new authentication technology. Since the authentication system was part of the game's technological foundation, a vast amount of work was required to make the transition to the new company. Even so, the transition was rough and there was an extended period when authentication was unavailable. Additionally, it was not possible to transfer the full player database from the previous third-party company to the new provider. Because of this, account information for an excessive number of players was irretrievably lost. The most frustrating aspect of this changeover was that many elements were out of the control of the developers, and though the development team had not supported the decision to change, the burden of making it work fell on their shoulders.

Version 2.0.0a Release – 21 December 2003

Originally unscheduled, this release reflected the Army's wish to provide an update over the Christmas holiday. Despite the detrimental impact on the schedule then underway, the developers effected the following changes:

New Multiplayer SF Mission: The mission "SF Sandstorm" was created.

Resolved Punkbuster Issues: Several operational issues with the Punkbuster anti-cheating system were addressed.

Distribution Partner and Version Tracking: A new system was created to enable the Army to improve version tracking and assess distribution efficiency.

Interface Adjustments: Several lingering issues with the game's new interface were addressed, including:

- Overlapping problems with the training menu
- Unnecessary authentication messages
- Need for new authentication messages
- Changing the new-account in-game URL
- Changing the default in-game IRC server
- Updating game credits

- Updating the support menu
- Adding server browser adjustments
- Adjusting the news page

Summation: March 8, 2004

Version 2.0.0a was the last release of *America's Army* developed by the MOVES Institute. In March 2004, the Army chose to take control of development and move the project off the Naval Postgraduate School campus. Although the MOVES Institute created one of the world's most popular video games for the US Army, differences between MOVES and Army management saw the game's production take a different turn. For many on the project, the whirlwind development cycle had taken an emotional and physical toll over the years. In the circumstances, a lesser team would have found it to impossible to deliver a game of such high caliber as *America's Army*, illustrating that the importance of selecting a team more for attitude and work ethic than seniority cannot be overstated.

LESSONS LEARNED

We obviously learned a lot from this endeavor, but three lessons are particularly salient:

- Pick the best team you can and support them. We accommodated our development team's creature comforts by supplying videogames and sofas for relaxation and (of vital importance) an industrial-strength, well-stocked canteen, and encouraged collaboration by offering a dim, cubicle-free workspace (allowing each to see what the other was working on and thereby to keep hold of the big picture). We assigned them a secretary for hated administrative chores and shielded them from direct contact with the client. Result: they stuck together and worked like madmen.
- Talk to your clients till you hammer out what they want, and have them sign off on it. If they choose to deviate, tell them in writing what alterations will cost in time, money, and the abandonment of agreed-on features.
- Don't just build a game; build the infrastructure for a game community. Our fan website proved of incalculable worth. Well beyond providing a forum for suggestions and bug reports, the *AA* site enabled far-flung individuals, alone at their computers, to become a tight-knit virtual brotherhood that circled the globe. The community displayed an intense regard for our development team; they were thrilled when a developer signed on to play, and the news spread like wildfire. The fans' pumped-up energy and

immediate appropriation of the game was a source of refreshment and inspiration throughout our time on the project.

CONCLUSION

We began this chapter under the premise that future training simulations and combat-modeling systems need to look and feel like games to be embraced by soldiers. We then showed how to organize a full game-development team, like *America's Army's*. We embarked on a history of *AA's* various releases and the problems and solutions involved. As an exercise in development, *America's Army* represents a huge success; we can look at the vexation level of its various setbacks as the least one can expect in such an undertaking, a lower bound on the difficulties developers can encounter. That going forward with game-based simulation in a governmental or corporate environment will always produce stresses and issues should be well understood. Nevertheless, with eyes wide open and heads stuffed with guidance, knowledge, and peer sympathy, let us stride confidently into the game-based future of training simulation.

ACKNOWLEDGEMENTS

The authors salute the development team, pictured in the Yerba Buena guide (YerbaBuena, 2004), for their incredible efforts in producing one of the top-five played online games—the first game ever produced fully inside a research institute (the MOVES Institute) or based on a university campus (the Naval Postgraduate School). We wish to acknowledge Michael Capps as the original executive producer of *America's Army,* from May 2000 through the 1.0 release in July 2002. Michael did a spectacular job getting this project off the ground, and we think fondly of his time with the project. John Falby's role in making all contracting, hiring, purchasing, and operations happen flawlessly and expeditiously from May 2000 to May 2004 is gratefully acknowledged. Thanks to Rosemary Minns, who as team mom kept admin far away from the development team and guaranteed the flow of sugar snacks so necessary for the game's proper development.

REFERENCES

Davis, M., Shilling, R., Mayberry, A., McCree, J., Bossant, P., Dossett, S., Buhl, C. , Chang, C., Champlin, E., Wiglesworth, T., & Zyda, M., (2003). Researching *America's Army*. in B. Laurel, (Ed.), Design Research: Methods and Perspectives, (pp. 268-275). Boston: MIT Press.

Macedonia, M., Zyda, M., Pratt, D., Barham, P. & Zeswitz, S. (1994). NPSNET: A Network Software Architecture for Large Scale Virtual Environments, Presence, Vol. 3, No. 4 pp.265-287.

Singhal, S. & Zyda, M. (1999). Networked Virtual Environments - Design and Implementation, New York: ACM Press/Addison-Wesley

Yerba Buena Art Center (2004) America's Army PC Game - Vision and Realization. Margaret Davis, (Ed.) Monterey: MOVES Institute & US Army.

Zyda, M., Hiles, J., Mayberry, A., Capps, M., Osborn, B., Shilling, R., Robaszewski, M., & Davis, M. (2003). Entertainment R&D for Defense, IEEE CG&A, January/February, pp.28-36.

Zyda, M., Mayberry, A., Wardynski, C., Shilling, R., & Davis, M. (2003). The MOVES Institute's *America's Army: Operations* Game. Proceedings of the ACM SIGGRAPH 2003 Symposium on Interactive 3D Graphics, pp. 217-218, color plate pp. 252.

Zyda, M., & Sheehan, J. (Eds.).(1997). Modeling and Simulation: Linking Entertainment & Defense. Washington, D.C.: National Academy Press.

CHAPTER 20

DISTRIBUTED SIMULATION AND THE HIGH LEVEL ARCHITECTURE

RICHARD M. FUJIMOTO

ABSTRACT

An overview of technologies concerned with distributing the execution of simulation programs across multiple processors is presented. Here, particular emphasis is placed on discrete event simulations. The High Level Architecture (HLA) developed by the Department of Defense in the United States is first described to provide a concrete example of a contemporary approach to distributed simulation. The remainder of this chapter is focused on time management, a central issue concerning the synchronization of computations on different processors. Time management algorithms broadly fall into two categories, termed conservative and optimistic synchronization. A survey of both conservative and optimistic algorithms is presented focusing on fundamental principles and mechanisms. Finally, time management in the HLA is discussed as a means to illustrate how this standard supports both approaches to synchronization.

INTRODUCTION

Here, the term *distributed simulation* refers to distributing the execution of a single "run" of a simulation program across multiple processors. This encompasses several different dimensions. One dimension concerns the motivation for distributing the execution. One paradigm, often referred to as *parallel simulation*, concerns the execution of the simulation on a tightly coupled computer system, e.g., a supercomputer or a shared memory multiprocessor. Here, the principal reason for distributing the execution is to reduce the length of time to execute the simulation. In principal, by distributing the execution of a computation across N processors, one can complete the computation up to N times faster than if it were executed on a single processor. Another reason for distributing the execution in this fashion is to enable larger simulations to be executed than could be executed on a single computer. When confined to a single computer system, there may not be enough memory to perform the simulation. Distributing the execution across multiple machines allows the memory of many computer systems to be utilized.

A second, increasingly important motivation for distributed simulation concerns the desire to integrate several different simulators into a single simulation environment. One example where this paradigm is frequently used is in military training. Tank simulators, flight simulators, computer generated forces, and a variety of other models may be used to create a distributed virtual environment into which personnel are embedded to train for hypothetical scenarios and situations. This paradigm is particularly relevant to the simulation of organizations. Such a simulation might include models for multiple levels of management and workers, supply chains, customers, and external factors such as critical world events or economic trends.

Another emerging area of increasing importance is infrastructure simulations where simulators of different subsystems in a modern society are combined to explore dependencies among subsystems. For example, simulations of transportation systems may be combined with simulations of electrical power distribution systems, computer and communication infrastructures, and economic models to assess the economic impact of natural or human-caused disasters. In both these domains (military and infrastructure simulations) it is far more economical to link existing simulators to create distributed simulation environments than to create new models within the context of a single tool or piece of software. The High Level Architecture (HLA) developed by the U.S. Department of Defense defines an approach to integrate, or *federate*, separate, autonomous simulators into a single, distributed simulation system.

Another dimension that differentiates distributed simulation paradigms is the geographical extent over which the simulation executes. Often distributed simulations are executed over broad geographic areas. This is particularly useful when personnel and/or resources (e.g., databases or specialized facilities) are included in the distributed simulation exercise. Distributed execution eliminates the need for these personnel and resources to be physically co-located, representing an enormous cost savings. Distributed simulations operating over the Internet have created an enormous market for the electronic gaming industry, and could be applicable to organizational simulations where information and/or individuals participating in the exercise are geographically distributed. At the opposite extreme, high performance simulations may execute on multiprocessor computers confined to a single cabinet or machine room. Close proximity is necessary to reduce the delay for inter-processor communications that might otherwise severely degrade performance. These high performance simulations often require much communication between processors, making geographically distributed execution impractial. Historically, the term distributed simulation has often been used to refer to geographically distributed simulations, while parallel simulation traditionally referred to simulations executed on a tightly coupled parallel computer. However, with new computing paradigms such as clusters of workstations and grid computing, this distinction has become less clear, so we use the single term distributed simulation here to refer to all categories of distributed execution.

Two widely-used architectures for distributed simulation are the *client-server* and the *peer-to-peer* approaches. As its name implies, the client-server approach involves executing the distributed simulation on one or more server computers (which may be several computers connected by a local area network) to which clients (e.g., users) can "log in" from remote sites. The bulk of the simulation computation is executed on the server machines. This approach is typically used in distributed simulations used for multi-player gaming. Centralized management of the simulation computation greatly simplifies management of the distributed simulation system, and facilitates monitoring of the system, e.g., to detect cheating. On the other hand, peer-to-peer systems have no such servers, and the simulation is distributed across many machines, perhaps interconnected by a wide area network. The peer-to-peer approach is often used in distributed simulations used for defense.

The remainder of this chapter is organized as follows. First, an historical view of distributed simulation technology and how it has evolved over the last twenty to thirty years is briefly presented. The High Level Architecture is presented to introduce aspects of a contemporary approach to distributed simulation. The remainder of this chapter focuses on the synchronization problem, and time management algorithms that have been developed to address this issue. A more detailed, comprehensive treatment of these topics is presented in (Fujimoto, 2000). Much of this tutorial utilizes material presented in (Fujimoto, 2003)

HISTORICAL PERSPECTIVE

Distributed simulation technology has developed largely independently in at least three separate communities: the high performance computing community, the defense community, and the Internet/gaming industry. Each of these is briefly discussed next.

Distributed simulation in the high performance computing community originated in the late 1970's and early 1980's, focusing on synchronization algorithms (now referred to as time management). Synchronization algorithms were designed, for the most part, in order for the distributed execution to produce exactly the same results as a sequential execution of the simulation program, except (hopefully) more quickly. Initial algorithms utilized what is now referred to as a conservative paradigm, meaning blocking mechanisms were used to ensure no synchronization errors (out of order event processing) occurred. Initial algorithms date back to the late 1970's with seminal work by Chandy and Misra (1978) and Bryant (1977), among others, who are credited with first formulating the synchronization problem and developing the first solutions. These algorithms are among a class of algorithms that are today referred to as conservative synchronization techniques. In the early 1980's seminal work by Jefferson and Sowizral developed the Time Warp algorithm (Jefferson, 1985). Time Warp is important because it defined fundamental constructs widely used in a class of algorithms termed optimistic synchronization. Conservative and optimistic synchronization techniques form the core of a large body of work concerning parallel discrete event simulation, and much of the subsequent work in the field is based on this initial research.

The defense community's work in distributed simulation systems date back to the SIMNET (SIMulator NETworking) project, which can be viewed as a forerunner of organizational simulations that are the subject of this book. While the high performance computing community was largely concerned with reducing execution time, the defense community was concerned with integrating separate training simulations in order to facilitate interoperability and software reuse. The SIMNET project (1983 to 1990) demonstrated the viability of using distributed simulations to create virtual worlds for training military personnel for engagements (Miller and Thorpe, 1995). This led to the development of a set of standards for interconnecting simulators known as the Distributed Interactive Simulation (DIS) standards (IEEE Std 1278.1-1995, 1995). The 1990's also saw the development of the Aggregate Level Simulation Protocol (ALSP) that applied the SIMNET concept of interoperability and model reuse to wargame simulations (Wilson & Weatherly, 1994). ALSP and DIS have since been replaced by the High Level Architecture whose scope spans the broad range of defense simulations, including simulations for training, analysis, and test and evaluation of equipment and components.

A third track of research and development efforts arose from the Internet and computer gaming industry. Some of the work in this area can be traced back to a role-playing game called Dungeons and Dragons and textual fantasy computer games such as Adventure developed in the 1970's. These soon gave way to Multi-User Dungeon (MUD) games in the 1980's. Important additions such as sophisticated computer graphics helped create the video game industry that is flourishing today. Distributed, multi-user gaming is sometimes characterized as the "killer application" where distributed simulation technology may have the greatest economic and social impact.

THE HIGH LEVEL ARCHITECTURE

The High Level Architecture (HLA) was developed in the mid 1990's. It is intended to promote reuse and interoperation of simulations. The HLA effort was based on the premise that no one simulation could satisfy all uses and applications for the Defense community. The intent of the HLA is to provide a structure that supports reuse of different simulations, ultimately reducing the cost and time required to create a synthetic environment for a new purpose. An introduction to the HLA is presented in by Kuhl, Weatherly and Dahmann (1999).

Though developed in the context of defense applications, the HLA was intended to have applicability across a broad range of simulation application areas, including education and training, analysis, engineering and even entertainment, at a variety of levels of resolution. These widely differing application areas indicate the variety of requirements that were considered in the development and evolution of the HLA. For example, HLA is used to provide interoperability with military combat simulations for the crowd simulation described in Chapter 17 (Loftin et al., 2005).

The HLA does not prescribe a specific implementation, nor does it mandate the use of any particular set of software or programming language. It was envisioned that as new technological advances become available, new and different implementations would be possible within the framework of the HLA.

An HLA federation consists of a collection of interacting simulations, termed federates. A federate may be a computer simulation, a manned simulator, a supporting utility (such as a viewer or data collector), or an interface to a live player or instrumented facility. All object representation stays within the federates. The HLA imposes no constraints on what is represented in the federates nor how it is represented, but it does require that all federates incorporate specified capabilities to allow the objects in the simulation to interact with objects in other simulations through the exchange of data.

Data exchange and a variety of other services are realized by software called the Runtime Infrastructure (RTI). The RTI is, in effect, a distributed operating system for the federation. The RTI provides a general set of services that support the simulations in carrying out these federate-to-federate interactions and federation management support functions. All interactions among the federates go through the RTI.

The RTI software itself and the algorithms and protocols that it uses are not defined by the HLA standard. Rather, it is the *interface* to the RTI services that are standardized. The HLA runtime interface specification provides a standard way for federates to interact with the RTI, to invoke the RTI services to support runtime interactions among federates and to respond to requests from the RTI. This interface is implementation independent and is independent of the specific object models and data exchange requirements of any federation.

The HLA is formally defined by three components: the HLA rules, the Object Model Template (OMT), and. the interface specification. Each of these is briefly described next.

HLA Rules

The HLA rules summarize the key principles behind the HLA (IEEE Std 1516-2000, 2000). The rules are divided into two groups: federation and federate rules. Federations are required to define a Federation Object Model (FOM) specified in the Object Model Template (OMT) format. The FOM characterizes the information (objects) that are visible by more than one federate. During the execution of the federation, all object representation must reside within the federates (not the RTI). Only one federate may update the attribute(s) of any instance of an object at any given time. This federate is termed the owner of the attribute, and ownership may transfer from one federate to another during the execution of the federation via the ownership management services defined in the Interface Specification. All information exchanges among the federates take place via the RTI using the services defined in the HLA interface specification.

Additional rules apply to individual federates. Under the HLA, each federate must document their public information in a Simulation Object Model (SOM) using the OMT. Based on the information included in their SOM, federates must import and export information, transfer object attribute ownership, update attributes and utilize the time management services of the RTI when managing local time.

Object Models

HLA object models are descriptions of the essential sharable elements of the federation in 'object' terms. The HLA is directed towards interoperability; hence in the HLA, object models are intended to focus on descriptions of the critical aspects of simulations and federations that are shared across a federation. The HLA puts no constraints on the content of the object models. The HLA does require that each federate and federation document its object model using a standard object model template (IEEE Std 1516.2-2000, 2000). These templates are intended to be the means for open information sharing across the community to facilitate reuse of simulations.

As mentioned earlier, the HLA specifies two types of object models: the HLA Federation Object Model (FOM) and the HLA Simulation Object Model (SOM). The HLA FOM describes the set of objects, attributes and interactions that are shared across a federation. The HLA SOM describes the simulation (federate) in terms of the types of objects, attributes and interactions it can offer to future federations. The SOM is distinct from internal design information; rather it provides information on the capabilities of a simulation to exchange information as part of a federation. The SOM is essentially a contract by the simulation defining the types of information it can make available in future federations. The availability of the SOM facilitates the assessment of the appropriateness of the federate for participation in a federation.

While the HLA does not define the contents of a SOM or FOM, it does require that a common documentation approach be used. Both the HLA FOM and SOM are documented using a standard form called the HLA Object Model Template (OMT).

The Interface Specification

The HLA interface specification describes the runtime services provided to the federates by the RTI, and by the federates to the RTI (IEEE Std 1516.3-2000, 2000). There are six classes of services. *Federation management* services offer basic functions required to create and operate a federation. *Declaration management* services support efficient management of data exchange through the information provided by federates defining the data they will provide and will require during a federation execution. *Object management* services provide creation, deletion, identification and other services at the object level. *Ownership management* services supports the dynamic transfer of ownership of object/attributes during an execution. *Time management* services support synchronization of runtime simulation data exchange. Finally, *data distribution management* services support the efficient routing of data among federates during the course of a federation execution. The HLA interface specification defines the way these services are accessed, both functionally and in an application program's interface (API).

TIME MANAGEMENT

Time management is concerned with ensuring that the execution of the distributed simulation is properly synchronized. This is particularly important for simulations used for analysis (as opposed to training where errors that are not perceptible to humans participating in the exercise may be acceptable). Time management not only ensures that events are processed in a correct order, but also helps to ensure that repeated executions of a simulation with the same inputs produce exactly the same results. Currently, time management techniques such as those described here are typically not used in training simulations, where incorrect event orderings and non-repeatable simulation executions can often be tolerated.

Time management algorithms usually assume the simulation consists of a collection of *logical processes* (LPs) that communicate by exchanging time-stamped messages or events. In the context of the HLA, each federate can be viewed as a single LP. The goal of the synchronization mechanism is to ensure that each LP processes events in timestamp order. This requirement is referred to as the *local causality constraint.* Ignoring events containing exactly the same time stamp, it can be shown that if each LP adheres to the local causality constraint, execution of the simulation program on a parallel computer will produce exactly the same results as an execution on a sequential computer where all events are processed in time stamp order. This property also helps to ensure that the execution of the simulation is repeatable; one need only ensure the computation associated with each event is repeatable.

Synchronization is particularly interesting for the case of discrete event simulations. In this case, each LP can be viewed as a sequential discrete event simulator. This means each LP maintains local state information corresponding to the entities it is simulating and a list of time stamped events that have been scheduled for this LP, but have not yet been processed. This *pending event list* includes local events that the LP has scheduled for itself as well as events that have been scheduled for this LP by other LPs. The main processing loop of the LP repeatedly removes the smallest time stamped event from the pending event list and processes it. Thus, the computation performed by an LP can be viewed as a sequence of event computations. Processing an event means zero or more state variables within the LP may be modified, and the LP may schedule additional events for itself or other LPs. Each LP maintains a simulation time clock that indicates the time stamp of the most recent event processed by the LP. Any event scheduled by an LP must have a time stamp at least as large as the LP's simulation time clock when the event was scheduled.

The time management algorithm must ensure that each LP processes events in time stamp order. This is non-trivial because each LP does not a priori know what events will later be received from other LPs. For example, suppose the next unprocessed event stored in the pending event list has time stamp 10. Can the LP process this event? How does the LP know it will not later receive an event from another LP with time stamp less than 10? This question captures the essence of the synchronization problem.

Much research has been completed to attack this problem. Time management algorithms can be classified as being either *conservative* or *optimistic*. Briefly, conservative algorithms take precautions to avoid the possibility of processing events out of time stamp order, i.e., the execution mechanism avoids synchronization errors. In the aforementioned example where the next unprocessed event has a time stamp of 10, the LP must first ensure it will not later receive any additional events with time stamp less than 10 before it can process this event. On the other hand, optimistic algorithms use a detection and recovery approach. Events are allowed to be processed out of time stamp order, however, a separate mechanism is provided to recover from such errors. Each of these are described next.

Conservative Time Management

The first synchronization algorithms were based on conservative approaches. The principal task of any conservative protocol is to determine when it is "safe" to process an event. An event is said to be safe when one can guarantee no event containing a smaller time stamp will be later received by this LP. Conservative approaches do not allow an LP to process an event until it has been guaranteed to be safe.

At the heart of most conservative synchronization algorithms is the computation for each LP of a Lower Bound on the Time Stamp (LBTS) of future messages that may later be received by that LP. This allows the mechanism to determine which events are safe to process. For example, if the synchronization algorithm has determined that the LBTS value for an LP is 12, then all events with time stamp less than 12 are safe, and may be processed. Conversely, all events with time stamp larger than 12 cannot be safely processed. Whether or not events with time stamp equal to 12 can be safely processed depends on specifics of the algorithm, and the rules concerning the order that events with the same time stamp (called simultaneous events) are processed. Processing of simultaneous events is a complex subject matter that is beyond the scope of the current discussion, but is discussed in detail in Jha and Bagrodia (2000). The discussion here assumes that each event has a unique time stamp. It is straightforward to introduce tie breaking fields in the time stamp to ensure uniqueness (Mehl, 1992).

The algorithms described in Bryant (1977) and Chandy and Misra (1978) were among the first synchronization algorithms that were developed. These algorithms assumed each LP sends messages with non-decreasing time stamps, and the communication network ensures that messages are received in the same order that they were sent. This guarantees that messages on each incoming link of an LP arrive in timestamp order. This implies that the timestamp of the last message received on a link is a lower bound on the timestamp of any subsequent message that will later be received on that link. Thus, the LBTS value for an LP is simply the minimum among the LBTS values of its incoming links.

A principal problem that arises with this approach is the LPs may deadlock, where a circular cycle of processes are blocked, each waiting for the next process in the cycle to proceed. The principal idea used in these early synchronization algorithms is to use *null* messages to avoid deadlock. A null message with timestamp T_{null} sent from LP_A to LP_B is a promise by LP_A that it will not later send a message to LP_B carrying a timestamp smaller than T_{null}. Null messages do not correspond to any activity in the simulated system; they are defined purely for avoiding deadlock situations. Processes send null messages on each outgoing link after processing each event. A null message provides the receiver with additional information that may be used to determine that other events are safe to process. It can be shown that this algorithm avoids deadlock (Chandy and Misra, 1978).

How does a process determine the timestamps of the null messages it sends? The clock value of each incoming link provides a lower bound on the timestamp of the next event that will be removed from that link's buffer. When coupled with knowledge of the simulation performed by the process, this bound can be used to determine a lower bound on the timestamp of the next *outgoing* message on each output link. For example, if a queue server has a minimum service time of T, then the timestamp of any future departure event must be at least T units of simulated time larger than any arrival event that will be received in the future. Whenever a process finishes processing a null or non-null message, it sends a new null message on each outgoing link. The receiver of the null message can then compute new bounds on its outgoing links, send this information on to its neighbors, and so on.

The null message algorithm introduced a key property utilized by virtually all conservative synchronization algorithms: *lookahead*. If an LP is at simulation time T, and it can guarantee that any message it will send in the future will have a time stamp of at least T+L regardless of what messages it may later receive, the LP is said to have a lookahead of L. As we just saw, lookahead is used to generate the time stamps of null messages. One constraint of the null message algorithm is it requires that no cycle among LPs exist containing zero lookahead, i.e., it is impossible for a sequence of messages to traverse the cycle, with each message scheduling a new message with the same time stamp.

The main drawback with the null message algorithm is it may generate an excessive number of null messages. The principal problem is the algorithm uses only the current simulation time of each LP and lookahead to predict the minimum time stamp of messages it could generate in the future. To solve this problem, we observe that the key piece of information that is required is the time stamp of the next unprocessed event within each LP. If a set of LPs at simulation time 100 could collectively recognize that this event has time stamp 200, all of the LPs could immediately advance from simulation time 100 to time 200 without exchanging many null messages. Thus, the time of the next event across the entire simulation provides critical information that avoids inefficiencies in the null message algorithm. This idea is exploited in more advanced synchronization algorithms.

One early approach to solving these problems is another algorithm that allows the computation to deadlock, but then detects and breaks it (Chandy & Misra, 1981). The deadlock can be broken by observing that the message(s) containing the smallest timestamp is (are) always safe to process. Alternatively, one may use a distributed computation to compute lower bound information (not unlike the distributed computation using null messages described above) to enlarge the set of safe messages.

Many other approaches have been developed. Some protocols use a synchronous execution where the computation cycles between (i) determining which events are "safe"' to process, and (ii) processing those events. It is clear that the key step is determining the events that are safe to process each cycle. Each LP must determine a lower bound on the time stamp (LBTS) of messages it might later receive from other LPs. This can be determined from a snapshot of the distributed computation as the minimum among:

- The simulation time of the next event within each LP if the LP is blocked, or the current time of the LP if it is not blocked, plus the LP's lookahead
- The time stamp of any transient messages, i.e., any message that has been sent but has not yet been received at its destination.

A barrier synchronization can be used to obtain the snapshot. Transient messages can be "flushed" out of the system in order to account for their time stamps. If first-in-first-out communication channels are used, null messages can be sent through the channels to flush the channels, though as noted earlier, this may result in many null messages. Alternatively, each LP can maintain a counter of the number of messages it has sent, and the number if has received. When the sum of the send and receive counters across all of the LPs are the same, and each LP has reached the barrier point, it is guaranteed that there are no more transient messages in the system. In practice, summing the counters can be combined with the computation for computing the global minimum value (Mattern, 1993).

To determine which events are safe, the *distance between LPs* is sometimes used (Ayani, 1989; Lubachevsky, 1989; Cai & Turner, 1990). This "distance" is the minimum amount of simulation time that must elapse for an event in one LP to directly or indirectly affect another LP, and can be used by an LP to determine bounds on the timestamp of future events it might receive from other LPs. This assumes it is known which LPs send messages to which other LPs. Other techniques focus on maximizing exploitation of lookahead, e.g., see Meyer and Bagrodia (1999) and Xiao, Unger and colleagues (1999). Full elaboration of these and other technique is beyond the scope of the present discussion.

Another thread of research in synchronization algorithms concerns relaxing ordering constraints in order to improve performance. Some approaches amount to simply ignoring out of order event processing (Sokol & Stucky, 1990; Rao, Thondugulam et al., 1998). Use of time intervals, rather than precise time stamps, to encode uncertainty of temporal information in order to improve the performance of time management algorithms have also been proposed (Fujimoto, 1999; Beraldi & Nigro, 2000). Use of causal order rather than time stamp order for distributed simulation applications has also been studied (Lee, Cai et al., 2001).

Optimistic Time Management

In contrast to conservative approaches that avoid violations of the local causality constraint, optimistic methods allow violations to occur, but are able to detect and recover from them. Optimistic approaches offer two important advantages over conservative techniques. First, they can exploit greater degrees of parallelism. If two events *might* affect each other, but the computations are such that they actually do not, optimistic mechanisms can process the events concurrently, while conservative methods must sequentialize execution. Second, conservative mechanism generally rely on application specific information (e.g., distance between objects) in order to determine which events are safe to process. While optimistic mechanisms can execute more efficiently if they exploit such information, they are less reliant on such information for correct execution. This allows the synchronization mechanism to be more transparent to the application program than conservative approaches, simplifying software development. On the other hand, optimistic methods may require more overhead computations than conservative approaches, leading to certain performance degradations.

The Time Warp mechanism (Jefferson, 1985) is the most well known optimistic method. When an LP receives an event with timestamp smaller than one or more events it has already processed, it rolls back and reprocesses those events in timestamp order. Rolling back an event involves restoring the state of the LP to that which existed prior to processing the event (checkpoints are taken for this purpose), and "unsending" messages sent by the rolled back events. An elegant mechanism called anti-messages is provided to "unsend" messages.

An anti-message is a duplicate copy of a previously sent message. Whenever an anti-message and its matching (positive) message are both stored in the same queue, the two are deleted (annihilated). To "unsend" a message, a process need only send the corresponding anti-message. If the matching positive message has already been processed, the receiver process is rolled back, possibly producing additional anti-messages. Using this recursive procedure all effects of the erroneous message will eventually be erased.

Two problems remain to be solved before the above approach can be viewed as a viable synchronization mechanism. First, certain computations, e.g., I/O operations, cannot be rolled back. Second, the computation will continually consume more and more memory resources because a history (e.g., checkpoints) must be retained, even if no rollbacks occur; some mechanism is required to reclaim the memory used for this history information. Both problems are solved by *global virtual time* (*GVT*). GVT is a lower bound on the timestamp of any future rollback. GVT is computed by observing that rollbacks are caused by messages arriving "in the past." Therefore, the smallest timestamp among unprocessed and partially processed messages gives a value for GVT. Once GVT has been computed, I/O operations occurring at simulated times older than GVT can be committed, and storage older than GVT (except one state vector for each LP) can be reclaimed.

GVT computations are essentially the same as LBTS computations used in conservative algorithms. This is because rollbacks result from receiving a message or anti-message in the LP's past. Thus, GVT amounts to computing a lower bound on the time stamp of future messages (or anti-messages) that may later be received.

Several algorithms for computing GVT (LBTS) have been developed, e.g., see Samadi (1985) and Mattern (1993), among others. Asynchronous algorithms compute GVT "in background" while the simulation computation is proceeding, introducing the difficulty that different processes must report their local minimum at different points in time. A second problem is one must account for transient messages in the computation, i.e., messages that have been sent but not yet received. Mattern describes an elegant solution to these problems using consistent cuts of the computation and message counters, discussed earlier (Mattern, 1993).

A pure Time Warp system can suffer from overly optimistic execution, i.e., some LPs may advance too far ahead of others leading to excessive memory utilization and long rollbacks. Many other optimistic algorithms have been proposed to address these problems. Most attempt to limit the amount of optimism. An early technique involves using a sliding window of simulated time (Sokol & Stucky, 1990). The window is defined as [GVT, GVT+W] where W is a user defined parameter. Only events with time stamp within this interval are eligible for processing. Another approach delays message sends until it is guaranteed that the send will not be later rolled back, i.e., until GVT advances to the simulation time at which the event was scheduled. This eliminates the need for anti-messages and avoids cascaded rollbacks, i.e., a rollback resulting in the generation of additional rollbacks (Dickens & Reynolds, 1990; Steinman, 1992). An approach that also uses a local rollback mechanism to avoid anti-messages involves a concept called lookback (somewhat analogous to lookahead in conservative synchronization protocols) is described by Chen and Szymanski (2002, 2003). A technique called direct cancellation is sometimes used to rapidly cancel incorrect messages, thereby helping to reduce overly optimistic execution (Fujimoto, 1989; Zhang & Tropper, 2001).

Another problem with optimistic synchronization concerns the amount of memory that may be required to store history information. Several techniques have been developed to address this problem. For example, one can roll back computations to reclaim memory resources (Jefferson, 1990; Lin & Preiss, 1991). State saving can be performed infrequently rather than after each event (Lin, Preiss et al., 1993; Palaniswamy & Wilsey, 1993). The memory used by some state vectors can be reclaimed even though their time stamp is larger than GVT (Preiss & Loucks, 1995).

Early approaches to controlling Time Warp execution employed user-defined parameters that had to be tuned to optimize performance. Later work has focused on adaptive approaches where the simulation executive automatically monitors the execution and adjusts control parameters to maximize performance. Examples of such adaptive control mechanisms are described in (Ferscha, 1995; Das & Fujimoto, 1997), among others.

Practical implementation of optimistic algorithms requires that one must be able to roll back all operations, or be able to postpone them until GVT advances past the simulation time of the operation. Care must be taken to ensure operations such as memory allocation and deallocation are handled properly, e.g., one must be able to roll back these operations. Also, one must be able to roll back execution errors. This can be problematic in certain situations, e.g., if an optimistic execution causes portions of the internal state of the Time Warp executive to be overwritten (Nicol & Liu, 1997).

Another approach to optimistic execution involves the use of reverse computation techniques rather than rollback (Carothers, Perumalla et al., 1999). Undoing an event computation is accomplished by executing the inverse computation, e.g., to undo incrementing a state variable, the variable is decremented. The advantage of this technique is it avoids state saving, which may be both time consuming and require a large amount of memory. Carothers, Perumalla and Fujimoto (1999) describe a reverse compiler that automatically generates inverse computations.

Synchronization is a well-studied area of research in the distributed simulation field. There is no clear consensus concerning whether optimistic or conservative synchronization perform better; indeed, the optimal approach usually depends on the application. In general, if the application has good lookahead characteristics and programming the application to exploit this lookahead is not overly burdensome, conservative approaches are the method of choice. Indeed, much research has been devoted to improving the lookahead of simulation applications, e.g., see (Deelman, Bagrodia et al., 2001). Otherwise, optimistic synchronization offers greater promise. Disadvantages of optimistic synchronization include the potentially large amount of memory that may be required, and the complexity of optimistic simulation executives. Techniques to reduce memory utilization further aggravate the complexity issue.

Time Management in the HLA

The HLA provides a set of services to support time management. A principal consideration in defining these services was the observation that different federates may use different local time management mechanisms and have different requirements for message ordering and delay. Two major categories emerged. One class of simulations were designed to created virtual environments for training and test and evaluation (e.g., hardware-in-the-loop) applications. The execution of these simulations is paced by wallclock time, and synchronization algorithms to guarantee time stamp ordering of events are typically not used. Achieving low, predictable delays to transmit messages are important. A second class of simulations are those that require synchronization algorithms, in part to ensure proper ordering of events, and in part as a means to ensure that executions are repeatable, i.e., multiple executions of the same simulation with the same inputs yield exactly the same results. These simulations may use event stepped or time stepped execution mechanisms locally. It was envisioned that some federates may be executing on a parallel processor, and may be using conservative or optimistic synchronization mechanisms within their federate. The HLA time management services were designed to accommodate this wide variety of applications.

There are two principal elements of the HLA time management services: message ordering, and time advance mechanisms. The HLA supports two types of ordering: receive ordered communication, and time stamp order. With receive ordered communication, no guarantees are provided by the RTI concerning the order that messages are delivered to a federate; they are essentially delivered in the order that they are received. This minimizes the latency to transmit messages through the RTI, and is the ordering typically used for real-time training exercises and test and evaluation applications. With time stamp ordering, each message is assigned by the sender a time stamp, and messages are delivered to the federate in time stamp order. In some situations the RTI may need to buffer the message in order to guarantee that it will not later receive a message with a smaller time stamp before delivering it to the federate. Thus, the latency for transmitting messages may be larger when using time stamp ordering. Time stamp order is normally used for analysis applications which are often not paced by wallclock time where correct ordering of events and repeatable execution are important.

The HLA time advance mechanisms are realized by a set of services for advancing simulation (or logical) time. A protocol is defined where federates request a time advance, and the RTI issues a Time Advance Grant when the request can be honored. The RTI ensures that a federate is not advanced to simulation time T, until it can guarantee that no time stamp ordered messages with later arrive with time stamp less than T.

Both time stamp ordering and the time advance mechanisms rely on computation of a lower bound on the time stamp (LBTS) of messages that will later arrive for a federate. To compute LBTS values, federates provide the following information: its current simulation time, a single lookahead value for the federate (L), and guarantees concerning the generation of future events. Regarding the latter, when a federate invokes the Time Advance Request (T) service to request its simulation time be advanced to T, it makes an unconditional guarantee that no messages will later be sent with time stamp less than T+L. This service is typically used by time stepped federates. As noted earlier, use of only unconditional guarantees leads to the lookahead creep problem. To address this issue, federates also provide conditional guarantees. Specifically, when a federate invokes the service Next Event Request (T), it conditionally guarantees that no future messages will be sent with time stamp less than T, *provided the federate does not receive additional messages with time stamp less than T.* This service is typically used by event driven federates, where T is specified as the time of the next local event within the federate.

The HLA time management services define additional services to support optimistic execution. Optimistic execution requires that the federate must be able to process events even though messages with a smaller time stamp may later arrive. For this purpose, the Flush Queue service is defined that delivers all available time stamp ordered messages to the federate. In addition, some mechanism is required to implement anti-messages in Time Warp. This is accomplished through the Retract service. When a federate invokes Retract, it cancels a previously sent message. If the message has already been delivered, the retraction request is forwarded to the receiving federate, who must then cancel the original event. Finally, the Flush Queue service includes specification of time stamp information and advances a federates simulation time much like the Next Event Request service. This is used to advance Global Virtual Time for the federation. It should be noted that it is the federate's responsibility to implement its own rollback, e.g., using a state saving/restoration or a reverse execution mechanism.

While the HLA was originally developed to combine *different* simulators, other work has explored using HLA as an approach to parallelize sequential simulations. The central idea is to use HLA to federate a simulation with itself. Early work using this approach, though not in the context of HLA, is described by Nicol and Heidelberger (1996) for queueing network simulations. Recent work using HLA to parallelize a commercial air traffic control simulation is described in Bodoh and Wieland (2003). This concept has also been applied to parallelizing existing sequential simulators of communication networks (Bononi, D'Angelo et al., 2003; Perumalla, Park et al., 2003). Other work, also aimed at simulating communication networks, parallelizes sequential simulations using a fixed point computation paradigm (Szymanski, Liu et al., 2003). Self-federated HLA-based distributed simulations for supply chain analysis is described in (Turner, Cai et al., 2000).

CONCLUSIONS

Beginning with research and development efforts in the 1970's, research in distributed simulation systems has matured over the years. Much of the early research in this area was motivated purely by performance considerations. As processor speeds have continued to increase at an exponential pace, performance alone has become less of a motivating factor in recent years. For many problems such as simulation of large-scale networks such as the Internet, performance remains a principal motivating objective, however, much interest in this technology today stems from the promises of cost savings resulting from model reuse. Standards such as IEEE 1516 for the High Level Architecture demonstrate the widespread interest in use of distributed simulation technology for this purpose.

What is the future for the technology? It is interesting to speculate. One potential path is to focus on applications. High performance computing remains a niche market that targets a handful of important, computation intensive applications. For broader impacts in society, one must look to the entertainment and gaming industry, where distributed simulation technology has seen the most widespread deployment, and impact in society.

Another view is to observe that software is often driven by advances in hardware technology, and look to emerging computing platforms to define the direction the technology will turn. In this light, ubiquitous computing stands out as an emerging area where distributed simulation may be headed. For example, execution of distributed simulations on handheld computers necessitates examination of power consumption because battery life is a major constraint in such systems.

Interoperable distributed simulations are a key enabling technology for organizational simulations. Techniques, tools, and standards must be developed to automate the development of interoperable simulations. Semantic technologies may aid in this process by providing convenient means to represent and manipulate descriptions of models. Convenient means to easily configure and execute distributed simulations with minimal concern for issues such as resource allocation and execution over heterogeneous computing platforms are required. Approaches exploiting grid computing and web services are emerging, and may help to alleviate these concerns in distributed simulation systems of the future.

ACKNOWLEDGMENTS

The author gratefully acknowledges support for distributed simulation research from the National Science Foundation under grants EIA-0219976 and ECS-0225447.

REFERENCES

Ayani, R. (1989). A Parallel Simulation Scheme Based on the Distance Between Objects. *Proceedings of the SCS Multiconference on Distributed Simulation*, Society for Computer Simulation. 21, 113-118.

Beraldi, R. and L. Nigro (2000). Exploiting Temporal Uncertainty in Time Warp Simulations. *Proceedings of the 4th Workshop on Distributed Simulation and Real-Time Applications*, 39-46.

Bodoh, D. J. and F. Wieland (2003). Performance Experiments with the High Level Architecture and the Total Airport and Airspace Model (TAAM). *Proceedings of the 17th Workshop on Parallel and Distributed Simulation*, 31-39.

Bononi, L., G. D'Angelo, et al. (2003). HLA-Based Adaptive Distributed Simulation of Wireless Mobile Systems. *Proceedings of the 17th Workshop on Parallel and Distributed Simulation*, 40-49.

Bryant, R. E. (1977). Simulation of packet communications architecture computer systems. *MIT-LCS-TR-188*.

Cai, W. and S. J. Turner (1990). An Algorithm for Distributed Discrete-Event Simulation -- the "Carrier Null Message" Approach. *Proceedings of the SCS Multiconference on Distributed Simulation*, SCS Simulation Series. 22, 3-8.

Carothers, C. D., K. Perumalla, et al. (1999). "Efficient Optimistic Parallel Simulation Using Reverse Computation." *ACM Transactions on Modeling and Computer Simulation* 9 (3), 224-253.

Chandy, K. M. and J. Misra (1978). "Distributed Simulation: A Case Study in Design and Verification of Distributed Programs." *IEEE Transactions on Software Engineering* SE-5 (5), 440-452.

Chandy, K. M. and J. Misra (1981). "Asynchronous Distributed Simulation via a Sequence of Parallel Computations." *Communications of the ACM* 24 (4), 198-205.

Chen, G. and B. K. Szymanski (2002). Lookback: A New Way of Exploiting Parallelism in Discrete Event Simulation. *Proceedings of the 16th Workshop on Parallel and Distributed Simulation*, 153-162.

Chen, G. and B. K. Szymanski (2003). Four Types of Lookback. *Proceedings of the 17th Workshop on Parallel and Distributed Simulation*, 3-10.

Das, S. R. and R. M. Fujimoto (1997). "Adaptive Memory Management and Optimism Control in Time Warp." *ACM Transactions on Modeling and Computer Simulation* 7(2), 239-271.

Deelman, E., R. Bagrodia, et al. (2001). Improving Lookahead in Parallel Discrete Event Simulations of Large-Scale Applications using Compiler Analysis. *Proceedings of the 15th Workshop on Parallel and Distributed Simulation*, 5-13.

Dickens, P. M. and J. Reynolds, P. F. (1990). SRADS With Local Rollback. *Proceedings of the SCS Multiconference on Distributed Simulation.* 22, 161-164.

Ferscha, A. (1995). Probabilistic Adaptive Direct Optimism Control iin Time Warp. *Proceedings of the 9th Workshop on Parallel and Distributed Simulation*, 120-129.

Fujimoto, R. M. (1989). "Time Warp on a Shared Memory Multiprocessor." *Transactions of the Society for Computer Simulation* 6 (3), 211-239.

Fujimoto, R. M. (1999). Exploiting Temporal Uncertainty in Parallel and Distributed Simulations. *Proceedings of the 13th Workshop on Parallel and Distributed Simulation*, 46-53.

Fujimoto, R. M. (2000). *Parallel and Distributed Simulation Systems*, Wiley Interscience.

Fujimoto, R. M. (2003). Distributed Simulation Systems. *Proceedings of the Winter Simulation Conference*, 124-134.

IEEE Std 1278.1-1995 (1995). *IEEE Standard for Distributed Interactive Simulation -- Application Protocols*. New York, NY, Institute of Electrical and Electronics Engineers, Inc.

IEEE Std 1516-2000 (2000). *IEEE Standard for Modeling and Simulation (M&S) High Level Architecture (HLA) -- Framework and Rules*. New York, NY, Institute of Electrical and Electronics Engineers, Inc.

IEEE Std 1516.2-2000 (2000). *IEEE Standard for Modeling and Simulation (M&S) High Level Architecture (HLA) -- Object Model Template (OMT) Specification*. New York, NY, Institute of Electrical and Electronics Engineers, Inc.

IEEE Std 1516.3-2000 (2000). *IEEE Standard for Modeling and Simulation (M&S) High Level Architecture (HLA) -- Interface Specification*. New York, NY, Institute of Electrical and Electronics Engineers, Inc.

Jefferson, D. (1985). "Virtual Time." *ACM Transactions on Programming Languages and Systems* 7 (3), 404-425.

Jefferson, D. R. (1990). Virtual Time II: Storage Management in distributed Simulation. *Proceedings of the Ninth Annual ACM Symposium on Principles of Distributed Computing*, 75-89.

Jha, V. and R. Bagrodia (2000). "Simultaneous Events and Lookahead in Simulation Protocols." *ACM Transactions on Modeling and Computer Simulation* 10 (3), 241-267.

Kuhl, F., R. Weatherly, et al. (1999). *Creating Computer Simulation Systems: An Introduction to the High Level Architecture for Simulation*, Prentice Hall.

Lee, B.-S., W. Cai, et al. (2001). A Causality Based Time Management Mechanism for Federated Simulations. *Proceedings of the 15th Workshop on Parallel and Distributed Simulation*, 83-90.

Lin, Y.-B. and B. R. Preiss (1991). "Optimal Memory Management for Time Warp Parallel Simulation." *ACM Transactions on Modeling and Computer Simulation* 1 (4), 283-307.

Lin, Y.-B., B. R. Preiss, et al. (1993). Selecting the Checkpoint Interval in Time Warp Simulations. *Proceedings of the 7th Workshop on Parallel and Distributed Simulation*, 3-10.

Loftin, R.B., D. Petty, R.C. Gaskins III and F.D. McKenzie (2005). Modeling crowd behavior for military simulation applications. In W.B. Rouse and K.R. Boff, Eds., *Organizational Simulation: From Modeling and Simulation to Games and Entertainment.* New York: Wiley.

Lubachevsky, B. D. (1989). "Efficient Distributed Event-Driven Simulations of Multiple-Loop Networks." *Communications of the ACM* 32 (1), 111-123.

Mattern, F. (1993). "Efficient Algorithms for Distributed Snapshots and Global Virtual Time Approximation." *Journal of Parallel and Distributed Computing* 18 (4), 423-434.

Mehl, H. (1992). A Deterministic Tie-Breaking Scheme for Sequential and Distributed Simulation. *Proceedings of the Workshop on Parallel and Distributed Simulation*, Society for Computer Simulation. 24, 199-200.

Meyer, R. A. and R. L. Bagrodia (1999). Path Lookahead: A Data Flow View of PDES Models. *Proceedings of the 13th Workshop on Parallel and Distributed Simulation*, 12-19.

Miller, D. C. and J. A. Thorpe (1995). "SIMNET: The Advent of Simulator Networking." *Proceedings of the IEEE* 83 (8), 1114-1123.

Nicol, D. and P. Heidelberger (1996). "Parallel Execution for Serial Simulators." *ACM Transactions on Modeling and Computer Simulation* 6 (3), 210-242.

Nicol, D. M. and X. Liu (1997). The Dark Side of Risk. *Proceedings of the 11th Workshop on Parallel and Distributed Simulation*, 188-195.

Palaniswamy, A. C. and P. A. Wilsey (1993). An Analytical Comparison of Periodic Checkpointing and Incremental State Saving. *Proceedings of the 7th Workshop on Parallel and Distributed Simulation*, 127-134.

Perumalla, K. S., A. Park, et al. (2003). Scalable RTI-Based Parallel Simulation of Networks. *Proceedings of the 17th Workshop on Parallel and Distributed Simulation*, 97-104.

Preiss, B. R. and W. M. Loucks (1995). Memory Management Techniques for Time Warp on a Distributed Memory Machine. *Proceedings of the 9th Workshop on Parallel and Distributed Simulation*, 30-39.

Rao, D. M., N. V. Thondugulam, et al. (1998). Unsynchronized Parallel Discrete Event Simulation. *Proceedings of the Winter Simulation Conference*, 1563-1570.

Samadi, B. (1985). Distributed Simulation, Algorithms and Performance Analysis. *Computer Science Department*. Los Angeles, California, University of California, Los Angeles.

Sokol, L. M. and B. K. Stucky (1990). MTW: Experimental Results for a Constrained Optimistic Scheduling Paradigm. *Proceedings of the SCS Multiconference on Distributed Simulation*. 22, 169-173.

Steinman, J. S. (1992). "SPEEDES: A Multiple-Synchronization Environment for Parallel Discrete Event Simulation." *International Journal on Computer Simulation*, 251-286.

Szymanski, B. K., Y. Liu, et al. (2003). Parallel Network Simulation under Distributed Genesis. *Proceedings of the 17th Workshop on Parallel and Distributed Simulation*, 61-68.

Turner, S. J., W. T. Cai, et al. (2000). Adapting a Supply-Chain Simulation for HLA. *Proceedings of the 4th IEEE Workshop on Distributed Simulation and Real-Time Applications*, 71-78.

Wilson, A. L. and R. M. Weatherly (1994). The Aggregate Level Simulation Protocol: An Evolving System. *Proceedings of the 1994 Winter Simulation Conference*, 781-787.

Xiao, Z., B. Unger, et al. (1999). Scheduling Critical Channels in Conservative Parallel Simulation. *Proceedings of the 13th Workshop on Parallel and Distributed Simulation*, 20-28.

Zhang, J. L. and C. Tropper (2001). The Dependence List in Time Warp. *Proceedings of the 15th Workshop on Parallel and Distributed Simulation*, 35-45.

CHAPTER 21

HARNESSING THE HIVE

Innovation as a Distributed Function in the Online Game Community

J.C. HERZ

ABSTRACT

This chapter analyzes innovation as a distributed function of computer game culture: specifically, how developers of computer games leverage the time, talent, and enthusiasm of their customers to co-evolve core products and services. The resulting many-to-many knowledge ecology yields broadly applicable insights about how distributed innovation can be modeled.

INTRODUCTION

In the industrial-age organization, there is an explicit model of how innovation happens: the bright boys and girls in the R&D Department experiment in their sandboxes and create prototypes. These prototypes, whether cars or carbonated beverages, are pitched to "the suits" in product management, sales and marketing, and then fed into the product pipeline, where they are refined, focus-grouped and eventually manufactured. Although the motivational posters trill that good ideas can come from anywhere, innovation is organizationally structured as a compartmentalized function. It's not a very well tuned function in large organizations where the risk profile is relatively conservative (hence countless apocryphal tales of fantastic R&D prototypes left to mold on the shelf – until a competitor decides to do something similar). But it is relatively easy to model.

As noted by John Seely Brown and Paul Duguid, innovation is a social function that hinges on implicit knowledge (Brown & Duguid, 2000). As organizations become increasingly networked and distributed, R&D becomes an increasingly networked and distributed function as well as an increasingly social function, within and between organizations and their markets. In this paradigm, the innovative genius lies not just in the cultivation of in-house creativity, but also in the construction of systems that leverage the million monkeys theorem

In 2004, the bleeding edge of massively networked innovation is computer games, a pocket of the software industry whose lurid aesthetics mask transformational advances in technology and business practice. Online games, played by millions of young adults worldwide, are the most highly leveraged kind of networked application, one that harnesses next-generation technology to basic patterns of human behavior: competition, collaboration, the tendency to cluster, and the universal appetite for peer acknowledgement. In other words, the forces that hone games, and gamers, have more to do with the raw social physics of networks than with the particularities of computer game code.

It's useful, therefore, to look at the social ecology of computer game networks for an understanding of how innovation plays out on distributed networks, and how the emergence of intellectual property on distributed networks may be understood in social terms. Through this lens, four lessons about innovation as a networked, distributed function pull into focus:

- R&D estuaries: leveraging community-driven design
- Constructive ecologies: artifacts & social currency
- Beyond collaboration: group-to-group interaction
- Persistence and accretion

R&D ESTUARIES: LEVERAGING COMMUNITY-DRIVEN DESIGN

The development cycle for a computer game, circa 2004, is 18 months, from the generation of the design specification to the release of the product. (Production typically involves 12 to 20 people, with costs ranging from $5 to 7 million , or double both factors for a persistent online world) But for many games, and particularly the stronger-selling PC titles, that process begins before the "official" development period and extends afterwards, with a continuous stream of two-way feedback between the developers and players.

To a large extent, the players co-create the environment once the game launches; their satisfaction with the game hinges on their interactions with each other. They collectively author the human dynamics of the world, and the player-created objects within it – and they can leave if it doesn't suit them. The experience belongs to the players as much as to the developer. So it's in the developers' interest to keep players in the loop as the game takes shape and to leverage their experience. This is not just a marketing ploy ("Make them feel valued and they'll evangelize the product to their friends"), although it does also generate good will. It is part of the core design process on the bleeding edge of networked simulation.

Within existing technologies in well-established genres, the player base is even more actively involved in the design and evolution of computer games. "We put a lot of time and attention into making sure that there were clear and easy hooks for the fans who wanted to be involved in programming work to be able to add and integrate their own work into the game system," says Ray Muzyka, founder and co-CEO of Bioware, a Canadian game developer whose Dungeons & Dragons-inspired *Neverwinter Nights* gives players an unprecedented level of access to the underlying architecture of a fantasy role-playing game. "There's a very modular structure, where you can be a little bit involved in making things, or you can be really involved. Your own technical understanding can either take you into the depths of the code base or you can be very high-level, if you want to just use the easy-to-use Neverwinter tool set."

First-person shooters (FPS) such as *Quake 3 Arena* and *Unreal Tournament* are built on engines that have evolved for years, passed between programming teams and a population of gamers that customizes and often improves the game, just as its sequel is being planned. Player innovations are thus incorporated into the next iteration of the product. A salient example of this phenomenon is in-game artificial intelligence (AI), one of the great engineering hurdles in any game. In first-person shooters, there is a marked difference between real and computer-generated opponents: Human opponents are invariably smarter, less predictable, and more challenging to play against.

AI, however, like all engineering challenges, can be a beneficiary of the million-monkeys syndrome: Put a million gamers into a room with an open, extensible game engine, and sooner or later, one of them will come up with the first-person shooter equivalent of Hamlet. In the case of Id Software's *Quake II*, it was a plug-in called the ReaperBot, a fiendishly clever and intelligent AI opponent written by a die-hard gamer named Steven Polge (who was subsequently employed by Id's main rival, Epic Games, to write AI for Epic's *Unreal* engine). Polge's Reaperbot was far-and-away the best *Quake* opponent anyone (inside or outside Id Software) had ever seen, and the plug-in rapidly disseminated within the million-strong player population, who quickly began hacking away at its bugs. Needless to say, these improvements in game AI were incorporated into the core technology of first-person shooters, a benefit to players and developers alike.

The point here is not that *Quake* has great AI, but how that AI came to be. *Quake's* architecture, the very nature of the product, enables distributed innovation to happen outside the developer's walls. In essence, the player population is transformed from mere consumers into active, vested participants in the development and evolution of the game. Of course, not all players roll up their sleeves and write plug-ins. But even if only 1 percent contributes to the innovation in the product, even if they are making only minor, incremental improvements or subtle tweaks, that's 10,000 (unpaid) people in research and development.

CONSTRUCTIVE ECOLOGIES: ARTIFACTS & SOCIAL CURRENCY

Most of the players who tinker with games aren't programmers. They don't have to be, because the editing and customization tools in today's games require no formal programming skill whatsoever. (Shades of high-level Web services.) Levels of combat games can be constructed in a couple of hours by anyone familiar with basic game play. Real-time strategy games offer similar capabilities. New maps, with custom constellations of opposing forces, can be generated with a graphical user interface. Objects, including custom avatars or "skins," can be constructed with photos wrapped around templates, sculpted with simplified 3D modeling tools, or bitmapped. Beyond the R&D dynamics discussed above, these crafting capabilities foster a constructive ecosystem around the making, enhancement and swapping of functional objects.

Unlike most online communities, games' constructive ecosystems are fueled by an innate human desire to make things, rather than talk about them. Consequently, the dynamics are radically different from the community interaction that occurs around text documents. Social currency accrues not to virtuoso talkers, but to people who make things that other people like to play with – Triumph of the Neato. Because the system runs on functional objects, rather than clever comments, there is greater real and perceived value in players' contributions. Downloading someone's level or map is more than a conversation. It's an acquisition. Similarly, player-creators are validated in a more meaningful way: People are using what they've made, not just agreeing with it. Use, not imitation, is the sincerest form of flattery.

In a commercial context, this tool-based, user-driven activity has several important functions. It extends the life of the game, which both enhances the value of the product at no incremental cost and increases sales: The longer people play the game, the longer they talk about it -- effectively marketing it to their friends and acquaintances. Will Wright, author of Maxis' best-selling *Sim City* series, compares the spread of a product in this fashion to a virus: "Double the contagious period," he says, "and the size of the epidemic goes up by an order of magnitude. If I can get people to play for twice as long, I sell ten times as many copies." Wright's formula bears out on the bottom line. His latest game, *The Sims*, has spawned four expansion packs (developed in response to the creations of its own R&D estuary of fans) and racked up nearly half a billion dollars in retail sales since its 2000 release.

The Sims, which scales Wright's *Sim City* down to the neighborhood level, is noteworthy because it illustrates the level of engagement a game can achieve when its designers incorporate crafting into the culture of the game. Four months before *The Sims* shipped, its developers released tools that allowed players to create custom objects for the game's virtual environment: architecture, props, and custom characters. These tools were rapidly disseminated among *Sim City* players, who began creating custom content immediately. In the months leading up to the game's release, a network of player-run Web sites sprang up to showcase and exchange "handcrafted" Sims objects and custom characters.

By the time the game was released, there were 50 Sims fan sites, 40 artists pumping content into the pipeline, and 50,000 people collecting that content. One quarter million boxes flew off the shelves in the first week. A year later, there were dozens of people programming tools for Sims content creators, 150 independent content creators, half a million collectors, and millions of players reading 200 fan sites in 14 languages. While most of these sites are labors of love, a few are profitable as well.

At this point, more than 90 percent of *The Sims*' content is produced by the player population, which has achieved an overwhelming amount of collective expertise in all things Sim. The player population feeds on itself, in a completely bottom-up, distributed, self-organizing way; none of these people are on the Maxis payroll. So, if these people aren't being paid by game developers (in fact it's the reverse), why do they invest hundreds or thousands of hours whittling 3D models and maps?

Among hardcore gamers, there is an element of competition, and wanting to be noticed on a global scale. But for the casual gamers who furnish *The Sims*' virtual dollhouse, and for much of the level-swapping and map-making community, the practice of creating levels and skins and custom objects is a kind of 21st century folk art - a form of self-expression for the benefit of themselves and their immediate community. It sounds odd to put *StarCraft* maps and Sims dinettes in the same category as fiddle music and quilting, but socially they're congruent. Hence the appeal of sites like "Mall of The Sims," a showcase for items such as the following:

The Mermaid's Cave Rug Decorating Pack

Now you can turn your favorite floor tiles into area rugs...or bathmats, or welcome mats, throw rugs, or anything else you can dream up! Just place floor tiles as usual, in any size and shape desired, indoors or out. Then place these colorful Rug Edgings around the outside, "pulling" away from the "rug" as you go. Brought to you by Hairfish of The Mermaid's Cave, Store G16. (Level 2) here at MOTS.

Some people like to make virtual rug-edgings, and there's a 125K craft tool that lets them do that – almost 4000 downloads at last count. Unlike the R&D being done in the MOD community, this doesn't have anything to do with making the most unprecedented kick-ass Formula One game experience that blows people away. It's a form of social expression not unlike swapping MP3 playlists (mix tapes, in the previous generation) and recipes. Games, which are object-oriented at every level of the experience, provide a substrate for personal construction projects, which are all too rare in the current landscape of corporate capitalism.

In some sense, this mass-market digital crafts fair is an anthropological throwback. It's more like scrimshaw than Web-surfing. And yet, if you look forward to a network where every object is live – the much-trumpeted world of Web services –that experience will be closer to *The Sims* than to the current generation of client-server browsers. *The Sims*' objects are not self-contained executable programs, but they're not static data either. They function in prescribed ways, interact semi-autonomously, and exhibit behaviors within a dynamic framework. New objects contain behaviors that reconfigure the local environment. *The Sims* don't know how to play soccer, for instance, but if a soccer ball (a software object containing all the rules for playing soccer) is dropped into their midst, they will form teams and start playing soccer. Player-created plug-ins and MODs intersect with game engines in a similar fashion.

BEYOND COLLABORATION: GROUP-TO-GROUP INTERACTION

In computer game culture, status is easily established, readily compared and (perhaps most importantly, for the core demographic) quantifiable. Every game ends with a winner and losers. Tournament players are ranked. Player-created content is not only reviewed, but downloaded and therefore measurably popular. The author of a game modification may have an internally driven sense of accomplishment, but he also knows that 18,431 people are playing his song; for a 19-year-old, that's a big deal, particularly when fan sites start pointing to his home page. He gets a few laudatory e-mails from strangers. His friends think he's cool and ask him for map-making advice. A level designer he's never met, but whose work he admires, asks if he'd be interested in teaming up on a *Half-Life* MOD.

It is this web of relationships between players – competitive, cooperative, and collegial – that sustains the computer game industry, no less than the latest 3D engine, facial animation algorithm, or high-speed graphics card. Game code disseminates and thrives because it is an excellent substrate for human interaction, not because it is technologically impressive. Behind every successful computer game is a surge of interpersonal dynamics, both on an individual level and on a group level; games elicit and enable the most basic kinds of human pack behavior.

These group dynamics are best represented by the vast network of self-organized combat clans that vie for dominance on the Internet. No game company ever told players to form clans; they emerged in the mid-90's and have persisted for years. There are thousands of them. The smallest have five members; the largest number in the hundreds and have developed their own politics, hierarchies, and systems of governance. They are essentially tribal: Each has a name, its own history, monikers, and signs of identification (logos and team graphics). Clans do occasionally cluster into trans-national organizations, assuming a shared identity across national boundaries and adopting a loose federalist structure. Generally, however, clans comprise players in the same country, because proximity reduces network lag – a real factor in games that require quick responses.

Although most clans revolve around first-person combat games, there are hundreds of clans vying against one another in real-time strategy games such as *Age of Empires, HomeWorld*, and *StarCraft*. (*StarCraft* alone has 219 competing clans.) Because strategy games are more tactically complex than squad-based combat, clans in this genre tend to maintain more elaborate Web sites that go into some detail about the clan's history, rules, chain of command, custom maps, and treaties with other clans. (Some clans even create password-protected areas for their allies to access strategic and diplomatic communication – the smoky back rooms of strategy gaming.)

The clan network may seem anarchic: It is fiercely competitive and has no centralized authority. But beneath the gruesome aesthetics and inter-mural bravado, it is a highly cooperative system that runs far more efficiently than any "official" organization of similar scale, because clans, and the players that comprise them, have a clear set of shared goals. Regardless of who wins or loses, they are mutually dependent on the shared spaces where gaming occurs, whether those spaces are maintained by gamers for gamers, like ClanBase, or owned and operated by game publishers, like Sony, Electronic Arts, or Blizzard Entertainment, the developer of hit games including *StarCraft*, *Warcraft*, and *Diablo II*.

In online worlds such as *Everquest*, *Asheron's Call* or *Dark Age of Camelot*, the environment itself demands group formation. With dangerous monsters roaming around, a solo player doesn't last long in the wild; parties of four to six form in the interest of sheer survival, and ripen into war buddies as battles are fought and won. In addition, larger groups of players agglomerate into guilds ranging from a few dozen to upwards of a hundred affiliated characters. Like clans in the combat and strategy genres, these groups are tribal. They have their own rites of passage and leadership structures. They form alliances with or declare wars on other guilds.

Above and beyond their well-honed skills, players have group identities that keep them rooted in the environment, long after they've mastered the intricacies of game play. Hacking and slashing aside, online games construct a multi-layered social context that's more structurally sophisticated than any "grown-up" online community. If solo Web-browsing is socially one-dimensional (an individual), and online discussion groups are two-dimensional (a circle), then massively multiplayer games are four-dimensional; the experience plays out on multiple scales: the individual player, small adventuring groups, tribal organizations like clans and guilds, the online world as a whole.

"In a game, you're set up within natural factions," says Bill Roper, vice president of Blizzard Entertainment's San Mateo development studio. "For example, in *World of Warcraft*, there are specific races like orcs and humans and dwarves. They all have their likes and dislikes, and that's part of the game world. When we extend that into a massively multiplayer game, you have these built-in friendships and animosities in the way the game is designed, so you start naturally allying yourself along those lines. So now it becomes: How am I affected as an individual? How am I affected with the people I like to run around with? How does that then affect the larger group, like the clan of us, or the guild of us, which may be 30, 40, 50, 100 people? And then how does it affect all of my people – all of the orcs, or all of the humans? One thing that games offer that you don't get elsewhere is that what you do has several layers of ramifications."

"Many-to-many" is a common buzzword in technology circles. But usually, it really means one-to-one-to-one: I can interact with many people, and so can other individuals. There's very little in the way of true group-to-group interaction. And yet this is one of the most compelling aspects of the online game experience – not me against you, but my team against your team, or my team playing something that your team built. The functional unit is not the individual; it's the pack. Group cohesion keeps players in the game, as in the real world: Clans, guilds, packs, teams, buddy lists, book clubs, the people you forward a joke to – that's where the leverage is.

Collaboration is part of it, but that's missing half the equation. Collaboration assumes that people interact across an inward-facing project circle. In contrast, games assume that groups face other groups. There is a lot of kinetic potential in that intersection. Perhaps it's because most software is engineered in the West, where the individual is the prime unit, but the discourse around online identity allows only for personal identity, with little or no acknowledgement of group identity. Games encompass both kinds of identity, in player culture and in the applications themselves (i.e., character names reflect their social affiliations). That creates another, very meaningful layer of context that's particularly resonant in non-Western cultures. It is not a coincidence that the capital of online games is not the United States or Europe, but Asia. As technology permeates the non-Western world, it would be useful to consider what applications might emerge from the more nuanced and complete set of social assumptions that are taken for granted in game design.

PERSISTENCE AND ACCRETION

The business value of social context is especially important for companies such as Electronic Arts, Sony, and Microsoft, which maintain persistent multiplayer worlds that support hundreds of thousands of gamers on a subscription basis. Unlike most games, whose playing fields exist only while participants are actively engaged, multiplayer online worlds such as *Everquest*, *Ultima Online* or *Asheron's Call* persist, whether or not any particular player is logged on at any given time. The virtual environment is not something that vanishes when you stop playing: There are forces (some internal, some resulting from other players' actions) continuously at work. This persistence gives the game depth, and is psychologically magnetic: The player is compelled to return habitually (even compulsively) to the environment, lest some new opportunity or crisis arise in his absence.

Compared to transient multiplayer environments (i.e. combat and strategy games), the experience is qualitatively different. The world is dynamic, and therefore less predictable. More importantly, the game extends over the course of days, weeks or months.

The persistence of the environment allows players to develop their characters' identities within these worlds, which all hew to the conventions of role-playing games (RPG's). In an RPG, a player's progress is represented not by geographical movement (as in console adventure games such as *Mario Bros.* or *Tomb Raider*, where the object is to get from point A to point B, defeating enemies along the way), but by the development of his character, who earns experience points by overcoming in-game challenges. At certain milestone point-tallies, the character is promoted to a new experience level, gaining strength, skill, and access to new weapons and tactics – but also attracting more powerful enemies. The better the player becomes, the more challenging his opponents become. Thus, the player scales a well-constructed learning curve over several months as he builds his level-

1 character into a highly skilled, fully equipped level-50 powerhouse. Not surprisingly, players are highly invested in the characters they have built up.

As in Slashdot's "karma" system or eBay's reputation ratings, "leveling up" is a big motivating factor for players: It's the game's way of validating their cumulative accomplishments with something quantifiable, if not tangible. It's not enough for players to amass knowledge and skill; they have to see it manifest in the physics of the game. The accretion of value in persistent worlds changes the psychology of leisure: You haven't "spent" 1000 hours playing a game; you've "built up your character." You've made progress! Accretion transforms idle time into something that feels industrious. It turns spending into earning. You see the same psychological dynamics with frequent flyer miles – and the same sort of behavior. Travelers go to great lengths – sometimes thousands of unnecessary miles - to build their characters up from Blue to Silver or Gold Elite, in order to get double mileage and dedicated check-in and the mystical power of free confirmed upgrades.

Accretion elicits emotional investment. In the physical world, travelers cherish their stamp-filled passports, as well as their frequent flyer miles. Scouts have badges. Skateboarders and rock climbers proudly point to their scars; ballet dancers save their scuffed and worn toe shoes. But outside of games, there are very few online experiences that leave you with any sense of lasting value. You are only spending attention, not investing it. One reason the Web seems so rootless and superficial is the lack of accretion; there are very few mechanisms that render people's investment of time, attention, and emotional energy into persistent artifacts of quantifiable value.

Where those artifacts exist, they not only represent experiential value but also are often parlayed into real-world financial value, as players monetize their time. The player accounts of high-level online game characters, which may be cultivated over years, sell for hundreds of dollars on eBay, itself a massively multiplayer game (like any market), and a role-playing game at that. Persistent world characters are, after all, statistical profiles of a player's cumulative experience in the world. When it comes to interaction design, eBay and MMPs are siblings separated at birth – which is why they mesh so well.

Of course, not all of this MMP-to-real-world arbitrage is strictly legal under the games' end-user licensing agreements, and most people create characters to use rather than to sell. But the thing that divides persistent worlds from other online games is this: They're a service, not a product, and if the administrators crack down too hard, or make themselves look like jerks, players will leave and take their $12.95 a month with them. There's a recognition that if you allow people to build value in a persistent world, this sort of asset trading will occur on some level – *Ultima Online* has even started offering "buffed" characters as a premium service; for $29.95, you can buy an advanced level wizard straight from the company store. Other services include the ability to change your character's name for $29.99 (an implicit acknowledgement that characters are transferred to people who want to graft pre-established personas onto second-hand avatars). As long as players' activities aren't compromising the experience for others, MMP companies eventually let them wag the dog.

CONCLUSION

From the perspective of someone trying to model and/or simulate networked organizations, these phenomena point to a few salient and generalizable insights:

1) The importance of groups, not just individuals, as a fundamental unit of analysis.

Many computer-based models of organizations, to include agent-based models and simulations, use the individual as a fundamental unit of analysis (looking at the to/from lines in e-mail headers, or the actions of individual nodes in a supply chain or manufacturing process, for instance). While this level of granularity is critical, there are also phenomena that differentiate groups as distinct entities with distinct identities – as more than a cluster of individuals. Groups, and their behavior AS groups in response to other groups, bears further analysis. This is an important model design issue in the crowd simulation described in Chapter 17 (Loftin et al., 2005).

2) Determining where value can accrete in a distributed environment.

Distributed organizations, like centralized ones, strive to create value. But in a distributed context, where communication may be more asynchronous and mediated, the question is, where does that value creation occur, and where does value accrete – what does it stick to, in the ether of cyberspace? How much of the social energy is channeled into building things that have value outside the organization? Does the value of those products accrete outside the organization, and if so, how is that constructive activity captured and recycled by the organization? Conversely, how much of the organization's social energy is "lost to heat," which is to say, fed into processes that stoke the group's internal politics but create little or nothing of persistent value (flame-prone company discussion boards being a prime example). Within the bounds of computational social science, how does one differentiate hand waving from constructive discourse? Server traffic logs and Outlook folders don't necessarily get you there.

3) Modeling the membrane

There is a serious and difficult question about "where do you stop" in the modeling or simulation of a networked organization. Beyond partners and suppliers, between whom the lines increasingly blur, there are competitors (who may or may not be cooperating with the organization on other fronts). Increasingly and perhaps more importantly, there is an progressively more informed and active marketplace of customers who have their own group dynamics and who often become an inextricable component of the product or service offering (e.g. Amazon's reader-reviewers). In an increasingly interdependent world where technology, security concerns, and globalization escalate the complexity of outward-facing relationships, the organization's membrane may well be the tail that wags the dog.

REFERENCES

Massively multiplayer games are a field whose literature is online rather than in books. The following references provides links to the most important repositories of writing and discussion in this domain:

Bartle, R. (19XX). "Hearts, Clubs, Diamonds, Spades: Players Who Suit MUDS" http://www.mud.co.uk/richard/hcds.htm

Brown, J.S., & Duguid, P. (2000). The Social Life of Information. Boston: Harvard Business School Press.

Castranova, E. (19XX). "Virtual Worlds: A First-Hand Account of Market and Society on the Cyberian Frontier," http://papers.ssrn.com/sol3/delivery.cfm/SSRN_ID294828_code020114590.pdf?abstractid=294828

Gamasutra (www.gamasutra.com) is the online counterpart to Game Developer Magazine. It contains a large archive of technical and design articles published by professionals in the game industry. Free registration is required.

Koster, R. is the creative director of Sony Online Entertainment (Everquest, Star Wars Galaxies). He maintains an archive of papers about massively multiplayer game design at http://www.legendmud.org/raph/.

Loftin, R. B., Petty, M.D. Gaskins, R.C. & McKenzie, F.D. (2005). Modeling Crowd Behavior For Military Simulation Applications. In W.B. Rouse & K.R. Boff, Eds., Organizational Simulation: From Modeling and Simulation to Games and Entertainment. New York: Wiley.

Morningstar, C., & Farmer, R. (19XX). "The Lessons of Lucasfilm's Habitat," http://www.fudco.com/chip/lessons.html

The MUD-DEV MAILING LIST is a high-level discussion conducted by designers working in the online game industry. Users can subscribe at www.kanga.nu/lists/listinfo/mud-dev

AUTHOR INDEX

SUBJECT INDEX